Strong Coupling Gauge Theories in LHC Era

Strong Coupling Gauge Theories in LHC Era

Proceedings of the Workshop in Honor of
Toshihide Maskawa's 70th Birthday and
35th Anniversary of Dynamical Symmetry Breaking in SCGT

Nagoya University, Japan 8 – 11 December 2009

editors

H Fukaya • M Harada
M Tanabashi • K Yamawaki

Nagoya University, Japan

NEW JERSEY • LONDON • SINGAPORE • BEIJING • SHANGHAI • HONG KONG • TAIPEI • CHENNAI

Published by

World Scientific Publishing Co. Pte. Ltd.
5 Toh Tuck Link, Singapore 596224
USA office: 27 Warren Street, Suite 401-402, Hackensack, NJ 07601
UK office: 57 Shelton Street, Covent Garden, London WC2H 9HE

British Library Cataloguing-in-Publication Data
A catalogue record for this book is available from the British Library.

STRONG COUPLING GAUGE THEORIES IN LHC ERA
Proceedings of the Workshop in Honor of Toshihide Maskawa's 70th Birthday and 35th Anniversary of Dynamical Symmetry Breaking in SCGT

ISBN-13 978-981-4329-51-4
ISBN-10 981-4329-51-7

Printed in Singapore by World Scientific Printers.

PREFACE

Spontaneous symmetry breaking has been a central paradigm of particle physics. It was actually born as dynamical symmetry breaking in analogy with the BCS theory of superconductor in the original work by Y. Nambu about half a century ago. The dynamical symmetry breaking was realized in QCD, a prototype of strong coupling gauge theory (SCGT). It was founded by T. Maskawa and H. Nakajima back in 1974 that in order for the gauge theories to develop spontaneous chiral symmetry breaking the gauge coupling must be strong, larger than a certain non-zero critical coupling, in the ladder Schwinger-Dyson equation with the coupling being nonrunning.

This was the beginning of the SCGT with conformal/scale-invariant dynamics. In 1986 it was founded by K. Yamawaki, M. Bando and K. Matumoto that this conformal/scale-invariant dynamics has a large anomalous dimension of unity at criticality, so as to resolve the FCNC problem in the technicolor (TC) models. Subsequently, the first Nagoya SCGT workshop was held in 1988, largely focusing on such a dynamical symmetry breaking with conformality/criticality and/or large anomalous dimension. Along this line new types of dynamical symmetry breaking such as the top quark condensate were proposed soon after the workshop. Since then the Nagoya SCGT workshop series were further held in 1990, 1996, 2002, 2006 and this time, the sixth in 2009.

Actually, over last 30 years theorists have been desperately seeking physics beyond the standard model. Now it is the era when the LHC will make a decisive test of various theoretical proposals. Certainly the new SCGT is one of such attractive possibilities.

Apart from the LHC experiments, there has been remarkable progress of lattice gauge theories in recent years, so that the conformal dynamics so far difficult to be analyzed in reliable ways could be solved by computers to give definite predictions at the level to be compared with the LHC experiments.

We then organized the sixth Nagoya SCGT workshop "Strong Coupling Gauge Theories in LHC Era (SCGT 09) in December 8-11, 2009. The SCGT 09 was the first SCGT workshop after Y. Nambu, M. Kobayashi and T. Maskawa shared the 2008 Nobel Prize in physics. It should be mentioned that Nambu is the initiator of the dynamical symmetry breaking, and actually was a repeated participant of the Nagoya SCGT workshops, while Maskawa and Kobayashi are both alumni of Nagoya University (both undergraduate and graduate). Moreover, Maskawa was to be the Director of the new institute, Kobayashi-Maskawa Institute (KMI), Nagoya University, starting April 2010. The SCGT 09 was then held in honor of Maskawa's 70th birthday (February 7, 2010) and 35th anniversary of the Maskawa-Nakajima paper initiating the dynamical symmetry breaking in conformal/scale-invariant gauge dynamics of SCGT.

During four days of sessions, we had 92 participants (28 from abroad), with 44 talks in the main program and 15 talks for the poster sessions. The topics were broad, not just conformal gauge dynamics but various aspects of strong coupling gauge theories including QCD in varieties of approaches covering lattice studies, holography, effective theories, and so on. Also several experimental talks were presented. We do believe that these presentations included in the Proceedings are very useful for the future developments in particle physics in the revolutionary era of LHC.

The workshop was financially supported by the Nagoya University Global COE Program "Quest for Fundamental Principles in the Universe: from Particles to the Solar System and the Cosmos" from the Ministry of Education, Culture, Sports, Science and Technology, Japan, and also by Daiko Foundation and by Nagoya University Foundation. We would like to express sincere thanks to these organizations for generous supports.

We would also like to thank all the members of the International Advisory Committee, the Organizing Committee, the Local Organizing Committee, and the SCGT 09 Secretariat and the GCOE support stuff office, and the graduate students at Nagoya University for their devoted contributions to make the workshop successful.

Editors
Hidenori Fukaya
Masayasu Harada
Masaharu Tanabashi
Koichi Yamawaki

OPENING ADDRESS

Good morning, ladies and gentlemen. On behalf of Nagoya University, I would like to express hearty welcome to all of you attending the 2009 Nagoya Global COE workshop on "Strong Coupling Gauge Theories in LHC Era", SCGT 09. I heard that this workshop is made in honor of Professor Toshihide Maskawa's 70th birthday and also for celebration of the 35th anniversary of his paper which is cornerstone of the subject of this workshop, namely the symmetry breaking in the strong coupling gauge theories. I understand that the subject is to reveal the Origin of Mass, the most urgent problem in particle physics today and the main research target of the largest accelerator experiments, LHC, just getting in operation.

I would like to comment that we, at Nagoya University, are proud of our alumni, Professor Maskawa and Professor Kobayashi, who shared 2008 Nobel Prize in physics for the symmetry breaking of CP symmetry.

We are also proud of Professor Maskawa's work on another type of symmetry breaking which has grown up into the present active field, the subject of this workshop, where Nagoya group has made major contributions. This is the sixth of the Nagoya SCGT workshops following the meetings in 1988, 1990, 1996, 2002 and 2006, ranging over two decades. The Nagoya SCGT workshops appear to be prestigious in the sense that every meeting had attendance of at least one of the Nobel laureates; Professor Maskawa (1988, 2009), Professor Kobayashi (1990), Professor Nambu (1988, 1990, 1996, 2002) and Professor 't Hooft (2006).

Here I should mention that this building, here you are, was constructed to celebrate Professor Noyori's 2001 Nobel Prize in chemistry. We are also proud of our alumnus Professor Osamu Shimomura who won the 2008 Nobel Prize in chemistry. It is my pleasure to announce to you that, next April, Professor Maskawa will come back to Nagoya University as Director of a newly created institute, Kobayashi-Maskawa Institute for the Origin of Particles and the Universe. The new building of the institute will be ready in 2011 spring. We are hoping that particle physics at Nagoya will be boosted even more.

I hope that the discussions at this workshop will contribute to the next Nobel Prize(s) in the near future. Thank you for your attention and please enjoy year stay at Nagoya and Nagoya University.

Michinari Hamaguchi, D. Med
President of Nagoya University

WORKSHOP ORGANIZATION

International Advisory Committee:

T.W. Appelquist (Yale)	W.A. Bardeen (FNAL)
S.J. Brodsky (SLAC)	R.S. Chivukula (MSU)
C.T. Hill (FNAL)	M. Kobayashi (JSPS)
J.B. Kogut (DOE/Maryland)	T. Maskawa (KSU/Nagoya)
V.A. Miransky (W. Ontario)	Y. Nambu (Chicago)
M. Rho (Saclay)	E.H. Simmons (MSU)
R. Sundrum (Johns Hopkins)	G. 't Hooft (Utrecht)
V.I. Zakharov (MPI/ITEP)	

Organizing Committee:

Chairperson	K. Yamawaki (Nagoya)
Co-chairperson	M. Harada (Nagoya)
	M. Tanabashi (Nagoya)

K-I. Aoki (Kanazawa)	H. Fukaya (Nagoya)
M. Hayakawa (Nagoya)	Y. Hosotani (Osaka)
K. Kanaya (Tsukuba)	Y. Kikukawa (Tokyo-Komaba)
R. Kitano (Tohoku)	K.-I. Kondo (Chiba)
T. Kugo (Kyoto)	T. Onogi (Osaka)
T. Sakai (Nagoya)	K. Tobe (Nagoya)

Sponsored by Nagoya University Global COE Program "Quest for Fundamental Principles in the Universe: from Particles to the Solar System and the Cosmos."

Financial support also from Daiko Foundation and Nagoya University Foundation.

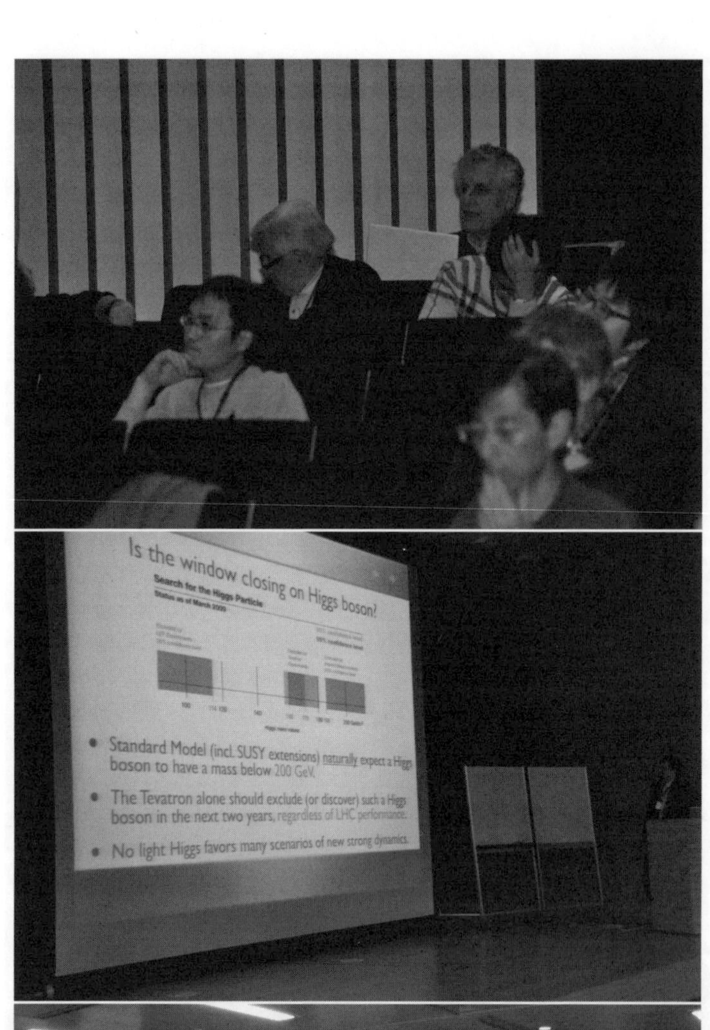
Is the window closing on Higgs boson?
Search for the Higgs Particle
Standard Model (incl. SUSY extensions) naturally expect a Higgs boson to have a mass below 200 GeV.
The Tevatron alone should exclude (or discover) such a Higgs boson in the next two years, regardless of LHC performance.
No light Higgs favors many scenarios of new strong dynamics.

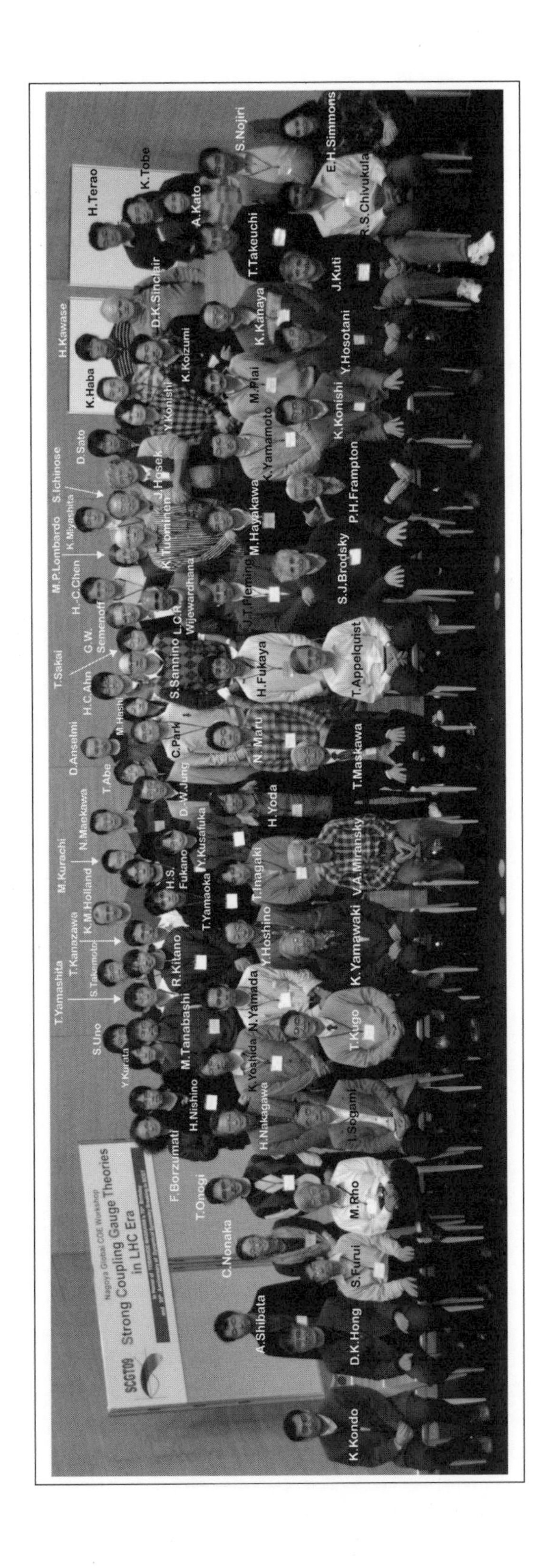
SCGT09
Nagoya Global COE Workshop
Strong Coupling Gauge Theories in LHC Era
K.Kondo
D.K.Hong
A.Shibata
S.Furui
C.Nonaka
M.Rho
T.Onogi
F.Borzumati
I.Sogami
H.Nakagawa
H.Nishino
Y.Kurata
T.Kugo
K.Yoshida
M.Tanabashi
S.Uno
T.Yamashita
N.Yamada
S.Takemoto
T.Kanazawa
R.Kitano
Y.Hoshino
K.Yamawaki
T.Yamaoka
K.M.Holland
M.Kurachi
H.S. Fukano
T.Inagaki
V.A.Miransky
Y.Kusafuka
N.Maekawa
H.Yoda
D.-W.Jung
T.Abe
D.Anselmi
C.Park
N. Maru
T.Maskawa
M.Hashi
H.C.Ahn
T.Sakai
G.W. Semenoff
S.Sannino
H.Fukaya
T.Appelquist
L.C.R. Wijewardhana
H.-C.Chen
J.T.Pleming
M.P.Lombardo
K.Miyashita
S.Ichinose
K.Tuominen
S.J.Brodsky
M.Hayakawa
P.H.Frampton
J.Hosek
D.Sato
K.Yamamoto
Y.Konishi
K.Konishi
M.Piai
K.Haba
H.Kawase
K.Koizumi
Y.Hosotani
K.Kanaya
D.K.Sinclair
J.Kuti
T.Takeuchi
H.Terao
A.Kato
R.S.Chivukula
K.Tobe
S.Nojiri
E.H.Simmons

CONTENTS

AdS/QCD, Light-Front Holography, and the Nonperturbative Running Coupling

Stanley J. Brodsky

SLAC National Accelerator Laboratory, Stanford University
Stanford, California 94309, USA
E-mail: sjbth@slac.stanford.edu

Guy de Téramond

Universidad de Costa Rica
San José, Costa Rica
E-mail: gdt@asterix.crnet.cr

Alexandre Deur

Thomas Jefferson National Accelerator Facility
Newport News, VA 23606, USA
E-mail: deurpam@jlab.org

The combination of Anti-de Sitter space (AdS) methods with light-front (LF) holography provides a remarkably accurate first approximation for the spectra and wavefunctions of meson and baryon light-quark bound states. The resulting bound-state Hamiltonian equation of motion in QCD leads to relativistic light-front wave equations in terms of an invariant impact variable ζ which measures the separation of the quark and gluonic constituents within the hadron at equal light-front time. These equations of motion in physical space-time are equivalent to the equations of motion which describe the propagation of spin-J modes in anti–de Sitter (AdS) space. The eigenvalues give the hadronic spectrum, and the eigenmodes represent the probability distributions of the hadronic constituents at a given scale. A positive-sign confining dilaton background modifying AdS space gives a very good account of meson and baryon spectroscopy and form factors. The light-front holographic mapping of this model also leads to a non-perturbative effective coupling $\alpha_s^{AdS}(Q^2)$ which agrees with the effective charge defined by the Bjorken sum rule and lattice simulations. It displays a transition from perturbative to nonperturbative conformal regimes at a momentum scale ~ 1 GeV. The resulting β-function appears to capture the essential characteristics of the full β-function of QCD, thus giving further support to the application of the gauge/gravity duality to the confining dynamics of strongly coupled QCD.

Keywords: AdS/CFT correspondence, AdS/QCD, light-front QCD, light-front quantization, nonperturbative QCD coupling.

1. Introduction

The AdS/CFT correspondence[1] between a gravity or string theory on a higher dimensional Anti–de Sitter (AdS) space-time with conformal gauge field theories

(CFT) in physical space-time has brought a new set of tools for studying the dynamics of strongly coupled quantum field theories, and it has led to new analytical insights into the confining dynamics of QCD. The AdS/CFT duality provides a gravity description in a $(d+1)$-dimensional AdS space-time in terms of a flat d-dimensional conformally-invariant quantum field theory defined at the AdS asymptotic boundary.[2] Thus, in principle, one can compute physical observables in a strongly coupled gauge theory in terms of a classical gravity theory. Since the quantum field theory dual to AdS_5 space in the original correspondence[1] is conformal, the strong coupling of the dual gauge theory is constant, and its β-function is zero. Thus one must consider a deformed AdS space in order to simulate color confinement and have a running coupling $\alpha_s^{AdS}(Q^2)$ for the gauge theory side of the correspondence. As we shall review here, a positive-sign confining dilaton background modifying AdS space gives a very good account of meson and baryon spectroscopy and their elastic form factors. The light-front holographic mapping of this model also leads to a non-perturbative effective coupling $\alpha_s^{AdS}(Q^2)$ which agrees with the effective charge defined by the Bjorken sum rule and lattice simulations.[3]

In the standard applications of AdS/CFT methods, one begins with Maldacena's duality between the conformal supersymmetric $SO(4,2)$ gauge theory and a semiclassical supergravity string theory defined in a 10 dimension $\mathrm{AdS}_5 \times S^5$ space-time.[1] The essential mathematical tool underlying Maldacena's observation is the fact that the effects of scale transformations in a conformal theory can be mapped to the z dependence of amplitudes in AdS_5 space. QCD is not conformal but there is in fact much empirical evidence from lattice, Dyson Schwinger theory and effective charges that the QCD β function vanishes in the infrared.[4] The QCD infrared fixed point arises since the propagators of the confined quarks and gluons in the loop integrals contributing to the β function have a maximal wavelength.[5] The decoupling of quantum loops in the infrared is analogous to QED where vacuum polarization corrections to the photon propagator decouple at $Q^2 \to 0$.

We thus begin with a conformal approximation to QCD to model an effective dual gravity description in AdS space. One uses the five-dimensional AdS_5 geometrical representation of the conformal group to represent scale transformations within the conformal window. Confinement can be introduced with a sharp cut-off in the infrared region of AdS space, as in the "hard-wall" model,[6] or, more successfully, using a dilaton background in the fifth dimension to produce a smooth cutoff at large distances as in the "soft-wall" model.[7] The soft-wall AdS/CFT model with a positive-sign dilaton-modified AdS space leads to the potential $U(z) = \kappa^4 z^2 + 2\kappa^2(L+S-1)$,[8] in the fifth dimension coordinate z. We assume a dilaton profile $\exp(+\kappa^2 z^2)$,[8–11] with opposite sign to that of Ref. 7. The resulting spectrum reproduces linear Regge trajectories, where $\mathcal{M}^2(S, L, n)$ is proportional to the internal spin, orbital angular momentum L and the principal quantum number n.

The modified metric induced by the dilaton can be interpreted in AdS space as a gravitational potential for an object of mass m in the fifth dimension: $V(z) = mc^2\sqrt{g_{00}} = mc^2 R\, e^{\pm\kappa^2 z^2/2}/z$. In the case of the negative solution, the potential decreases monotonically, and thus an object in AdS will fall to infinitely large values of z. For the positive solution, the potential is non-monotonic and has an absolute minimum at $z_0 = 1/\kappa$. Furthermore, for large values of z the gravitational potential increases exponentially, confining any object to distances $\langle z \rangle \sim 1/\kappa$.[8] We thus will choose the positive sign dilaton solution. This additional warp factor leads to a well-defined scale-dependent effective coupling. Introducing a positive-sign dilaton background is also relevant for describing chiral symmetry breaking,[10] since the expectation value of the scalar field associated with the quark mass and condensate does not blow-up in the far infrared region of AdS, in contrast with the original model.[7]

Glazek and Schaden[12] have shown that a harmonic oscillator confining potential naturally arises as an effective potential between heavy quark states when one stochastically eliminates higher gluonic Fock states. Also, Hoyer[13] has argued that the Coulomb and a linear potentials are uniquely allowed in the Dirac equation at the classical level. The linear potential becomes a harmonic oscillator potential in the corresponding Klein-Gordon equation.

Light-front (LF) quantization is the ideal framework for describing the structure of hadrons in terms of their quark and gluon degrees of freedom. The light-front wavefunctions (LFWFs) of bound states in QCD are relativistic generalizations of the Schrödinger wavefunctions, but they are determined at fixed light-front time $\tau = x^+ = x^0 + x^3$, the time marked by the front of a light wave,[14] rather than at fixed ordinary time t. They play the same role in hadron physics that Schrödinger wavefunctions play in atomic physics.[15] The simple structure of the LF vacuum provides an unambiguous definition of the partonic content of a hadron in QCD.

Light-front holography[8,16–20] connects the equations of motion in AdS space and the Hamiltonian formulation of QCD in physical space-time quantized on the light front at fixed LF time. This correspondence provides a direct connection between the hadronic amplitudes $\Phi(z)$ in AdS space with LF wavefunctions $\phi(\zeta)$ describing the quark and gluon constituent structure of hadrons in physical space-time. In the case of a meson, $\zeta = \sqrt{x(1-x)\mathbf{b}_\perp^2}$ is a Lorentz invariant coordinate which measures the distance between the quark and antiquark; it is analogous to the radial coordinate r in the Schrödinger equation. In effect ζ represents the off-light-front energy shell or invariant mass dependence of the bound state; it allows the separation of the dynamics of quark and gluon binding from the kinematics of constituent spin and internal orbital angular momentum.[19] Light-front holography thus provides a connection between the description of hadronic modes in AdS space and the Hamiltonian formulation of QCD in physical space-time quantized on the light-front at fixed LF time τ.

The mapping between the LF invariant variable ζ and the fifth-dimension AdS coordinate z was originally obtained by matching the expression for electromagnetic current matrix elements in AdS space[21] with the corresponding expression for the current matrix element, using LF theory in physical space time.[16] It has also been shown that one obtains the identical holographic mapping using the matrix elements of the energy-momentum tensor,[18,22] thus verifying the consistency of the holographic mapping from AdS to physical observables defined on the light front.

The resulting equation for the mesonic $q\bar{q}$ bound states at fixed light-front time has the form of a single-variable relativistic Lorentz invariant Schrödinger equation

$$\left(-\frac{d^2}{d\zeta^2} - \frac{1-4L^2}{4\zeta^2} + U(\zeta)\right)\phi(\zeta) = \mathcal{M}^2\phi(\zeta), \tag{1}$$

where the confining potential is $U(\zeta) = \kappa^4\zeta^2 + 2\kappa^2(L+S-1)$ in the soft-wall model. Its eigenvalues determine the hadronic spectra and its eigenfunctions are related to the light-front wavefunctions of hadrons for general spin and orbital angular momentum. This LF wave equation serves as a semiclassical first approximation to QCD, and it is equivalent to the equations of motion which describe the propagation of spin-J modes in AdS space. The resulting light-front wavefunctions provide a fundamental description of the structure and internal dynamics of hadronic states in terms of their constituent quark and gluons. There is only one parameter, the mass scale $\kappa \sim 1/2$ GeV, which enters the confinement potential. In the case of mesons $S = 0, 1$ is the combined spin of the q and $\bar{q}$, L is their relative orbital angular momentum as determined by the hadronic light-front wavefunctions.

The concept of a running coupling $\alpha_s(Q^2)$ in QCD is usually restricted to the perturbative domain. However, as in QED, it is useful to define the coupling as an analytic function valid over the full space-like and time-like domains. The study of the non-Abelian QCD coupling at small momentum transfer is a complex problem because of gluonic self-coupling and color confinement. We will show that the light-front holographic mapping of classical gravity in AdS space, modified by a positive-sign dilaton background, leads to a non-perturbative effective coupling $\alpha_s^{AdS}(Q^2)$ which is in agreement with hadron physics data extracted from different observables, as well as with the predictions of models with built-in confinement and lattice simulations.

2. The Hadron Spectrum and Form Factors in Light-Front AdS/QCD

The meson spectrum predicted by Eq. (1) has a string-theory Regge form $\mathcal{M}^2 = 4\kappa^2(n + L + S/2)$; *i.e.*, the square of the eigenmasses are linear in both L and n, where n counts the number of nodes of the wavefunction in the radial variable ζ. This is illustrated for the pseudoscalar and vector meson spectra in Fig. 1, where the data are from Ref. 23. The pion ($S = 0, n = 0, L = 0$) is massless for zero quark mass, consistent with the chiral invariance of massless QCD. Thus one can compute the hadron spectrum by simply adding $4\kappa^2$ for a unit change in the radial quantum number, $4\kappa^2$ for a change in one unit in the orbital quantum number L and $2\kappa^2$ for a change of one unit of spin S. Remarkably, the same rule holds for three-quark baryons as we shall show below.

In the light-front formalism, one sets boundary conditions at fixed τ and then evolves the system using the light-front (LF) Hamiltonian $P^- = P^0 - P^3 = id/d\tau$.

The invariant Hamiltonian $H_{LF} = P^+P^- - P_\perp^2$ then has eigenvalues $\mathcal{M}^2$ where $\mathcal{M}$ is the physical mass. Its eigenfunctions are the light-front eigenstates whose Fock state projections define the frame-independent light-front wavefunctions. The eigensolutions of Eq. (1) provide the light-front wavefunctions of the valence Fock state of the hadrons $\psi(x, \mathbf{b}_\perp)$ as illustrated for the pion in Fig. 2 for the soft (a) and hard wall (b) models. The resulting distribution amplitude has a broad form $\phi_\pi(x) \sim \sqrt{x(1-x)}$ which is compatible with moments determined from lattice gauge theory. One can then immediately compute observables such as hadronic form factors (overlaps of LFWFs), structure functions (squares of LFWFs), as well as the generalized parton distributions and distribution amplitudes which underly hard exclusive reactions. For example, hadronic form factors can be predicted from the overlap of LFWFs in the Drell-Yan West formula. The prediction for the space-like pion form factor is shown in Fig. 2(c). The pion form factor and the vector meson poles residing in the dressed current in the soft wall model require choosing a value of κ smaller by a factor of $1/\sqrt{2}$ than the canonical value of κ which determines the mass scale of the hadronic spectra. This shift is apparently due to the fact that the transverse current in $e^+e^- \to q\bar{q}$ creates a quark pair with $L^z = \pm 1$ instead of the $L^z = 0$ $q\bar{q}$ composition of the vector mesons in the spectrum.

Individual hadrons in AdS/QCD are identified by matching the power behavior of the hadronic amplitude at the AdS boundary at small z to the twist τ of its interpolating operator at short distances $x^2 \to 0$, as required by the AdS/CFT dictionary. The twist also equals the dimension of fields appearing in chiral super-multiplets;[24] thus the twist of a hadron equals the number of constituents plus the relative orbital angular momentum. One then can apply light-front holography to relate the amplitude eigensolutions in the fifth dimension coordinate z to the LF wavefunctions in the physical space-time variable ζ.

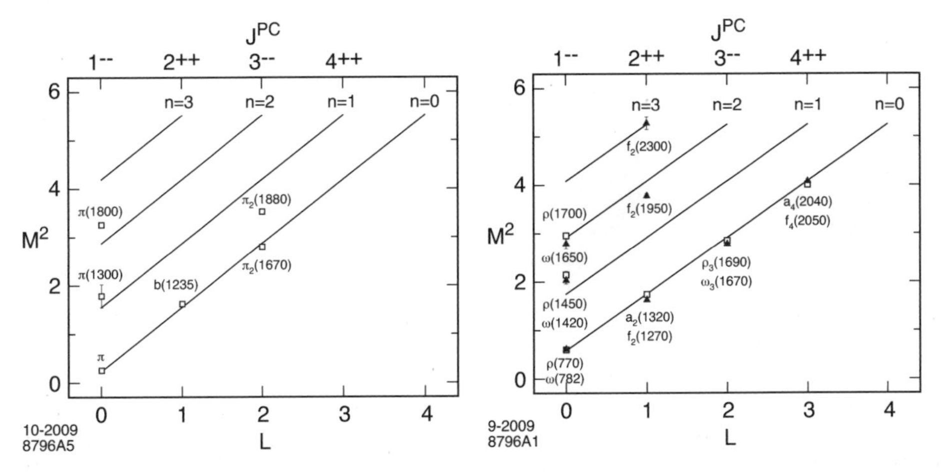

Fig. 1. Parent and daughter Regge trajectories for (a) the π-meson family with $\kappa = 0.6$ GeV; and (b) the $I=1$ ρ-meson and $I=0$ ω-meson families with $\kappa = 0.54$ GeV.

Equation (1) was derived by taking the LF bound-state Hamiltonian equation of motion as the starting point.[19] The term L^2/ζ^2 in the LF equation of motion (1) is derived from the reduction of the LF kinetic energy when one transforms to two-dimensional cylindrical coordinates (ζ, φ), in analogy to the $\ell(\ell+1)/r^2$ Casimir term in Schrödinger theory. One thus establishes the interpretation of L in the AdS equations of motion. The interaction terms build confinement corresponding to the dilaton modification of AdS space.[19] The duality between these two methods provides a direct connection between the description of hadronic modes in AdS space and the Hamiltonian formulation of QCD in physical space-time quantized on the light-front at fixed LF time τ.

The identification of orbital angular momentum of the constituents is a key element in the description of the internal structure of hadrons using holographic principles. In our approach quark and gluon degrees of freedom are explicitly introduced in the gauge/gravity correspondence,[25] in contrast with the usual AdS/QCD framework[26,27] where axial and vector currents become the primary entities as in effective chiral theory. Unlike the top-down string theory approach, one is not limited to hadrons of maximum spin $J \leq 2$, and one can study baryons with finite color $N_C = 3$. Higher spin modes follow from shifting dimensions in the AdS wave equations. In the soft-wall model the usual Regge behavior is found $\mathcal{M}^2 \sim n + L$, predicting the same multiplicity of states for mesons and baryons as observed experimentally.[28] It is possible to extend the model to hadrons with heavy quark constituents by introducing nonzero quark masses and short-range Coulomb corrections. For other recent calculations of the hadronic spectrum based on AdS/QCD, see Refs. 29–40. Other recent computations of the pion form factor are given in Refs. 41, 42.

For baryons, the light-front wave equation is a linear equation determined by the LF transformation properties of spin 1/2 states. A linear confining potential

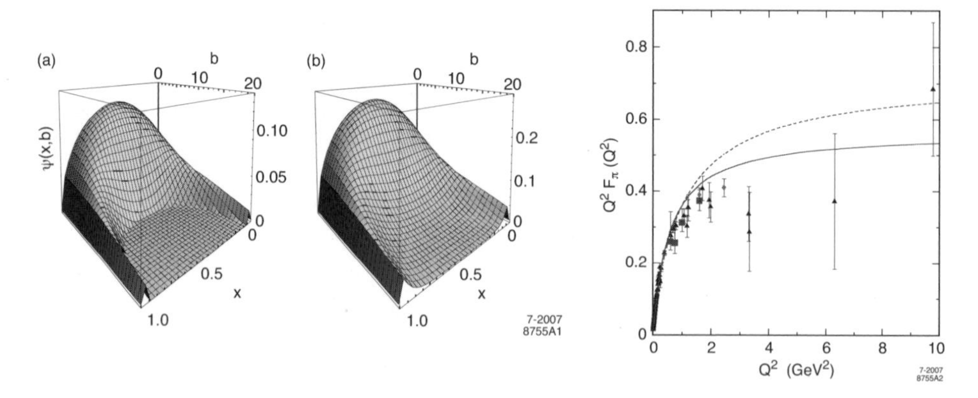

Fig. 2. Pion LF wavefunction $\psi_\pi(x, \mathbf{b}_\perp)$ for the AdS/QCD (a) hard-wall ($\Lambda_{QCD} = 0.32$ GeV) and (b) soft-wall ($\kappa = 0.375$ GeV) models. (c) Space-like scaling behavior for $Q^2 F_\pi(Q^2)$. The continuous line is the prediction of the soft-wall model for $\kappa = 0.375$ GeV. The dashed line is the prediction of the hard-wall model for $\Lambda_{\rm QCD} = 0.22$ GeV. The triangles are the data compilation of Baldini.

$U(\zeta) \sim \kappa^2\zeta$ in the LF Dirac equation leads to linear Regge trajectories.[43] For fermionic modes the light-front matrix Hamiltonian eigenvalue equation $D_{LF}|\psi\rangle = \mathcal{M}|\psi\rangle$, $H_{LF} = D_{LF}^2$, in a 2×2 spinor component representation is equivalent to the system of coupled linear equations

$$
\begin{aligned}
-\frac{d}{d\zeta}\psi_- - \frac{\nu+\frac{1}{2}}{\zeta}\psi_- - \kappa^2\zeta\psi_- &= \mathcal{M}\psi_+, \\
\frac{d}{d\zeta}\psi_+ - \frac{\nu+\frac{1}{2}}{\zeta}\psi_+ - \kappa^2\zeta\psi_+ &= \mathcal{M}\psi_-,
\end{aligned} \tag{2}
$$

with eigenfunctions

$$
\begin{aligned}
\psi_+(\zeta) &\sim z^{\frac{1}{2}+\nu}e^{-\kappa^2\zeta^2/2}L_n^{\nu}(\kappa^2\zeta^2), \\
\psi_-(\zeta) &\sim z^{\frac{3}{2}+\nu}e^{-\kappa^2\zeta^2/2}L_n^{\nu+1}(\kappa^2\zeta^2),
\end{aligned} \tag{3}
$$

and eigenvalues $\mathcal{M}^2 = 4\kappa^2(n+\nu+1)$.

The baryon interpolating operator $\mathcal{O}_{3+L} = \psi D_{\{\ell_1} \dots D_{\ell_q}\psi D_{\ell_{q+1}} \dots D_{\ell_m\}}\psi$, $L = \sum_{i=1}^{m} \ell_i$, is a twist 3, dimension $9/2 + L$ with scaling behavior given by its twist-dimension $3 + L$. We thus require $\nu = L + 1$ to match the short distance scaling behavior. Higher spin modes are obtained by shifting dimensions for the fields. Thus, as in the meson sector, the increase in the mass squared for higher baryonic state is $\Delta n = 4\kappa^2$, $\Delta L = 4\kappa^2$ and $\Delta S = 2\kappa^2$, relative to the lowest ground state, the proton. Since our starting point to find the bound state equation of motion for baryons is the light-front, we fix the overall energy scale identical for mesons and baryons by imposing chiral symmetry to the pion[20] in the LF Hamiltonian equations. By contrast, if we start with a five-dimensional action for a scalar field in presence of a positive sign dilaton, the pion is automatically massless.

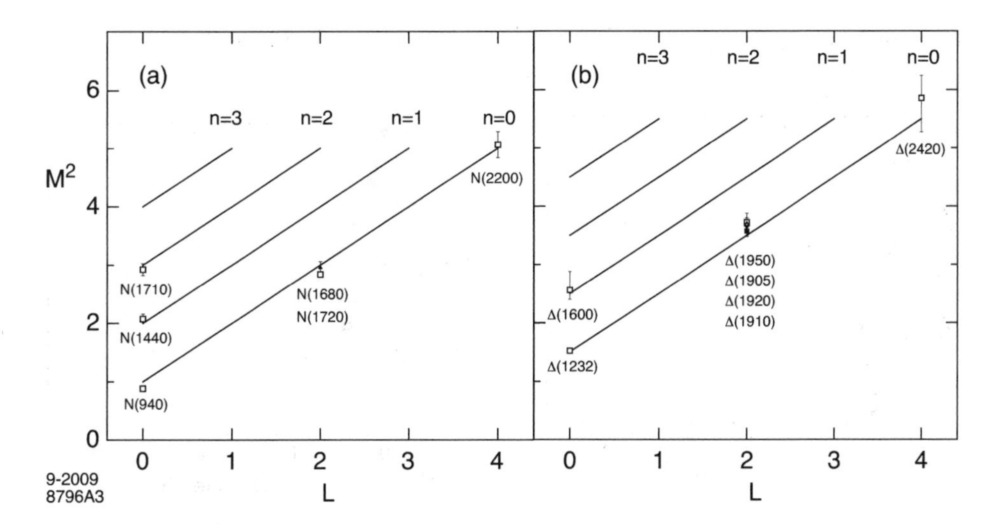

Fig. 3. **56** Regge trajectories for the N and Δ baryon families for $\kappa = 0.5$ GeV.

The predictions for the **56**-plet of light baryons under the $SU(6)$ flavor group are shown in Fig. 3. As for the predictions for mesons in Fig. 1, only confirmed PDG[23] states are shown. The Roper state $N(1440)$ and the $N(1710)$ are well accounted for in this model as the first and second radial states. Likewise the $\Delta(1660)$ corresponds to the first radial state of the Δ family. The model is successful in explaining the important parity degeneracy observed in the light baryon spectrum, such as the $L=2$, $N(1680)-N(1720)$ degenerate pair and the $L=2$, $\Delta(1905), \Delta(1910), \Delta(1920), \Delta(1950)$ states which are degenerate within error bars. Parity degeneracy of baryons is also a property of the hard wall model, but radial states are not well described in this model.[44]

As an example of the scaling behavior of a twist $\tau=3$ hadron, we compute the spin non-flip nucleon form factor in the soft wall model.[43] The proton and neutron Dirac form factors are given by

$$F_1^p(Q^2) = \int d\zeta\, J(Q,\zeta)\, |\psi_+(\zeta)|^2, \tag{4}$$

$$F_1^n(Q^2) = -\frac{1}{3}\int d\zeta\, J(Q,\zeta)\left[|\psi_+(\zeta)|^2 - |\psi_-(\zeta)|^2\right], \tag{5}$$

where $F_1^p(0)=1,\ F_1^n(0)=0$. The non-normalizable mode $J(Q,z)$ is the solution of the AdS wave equation for the external electromagnetic current in presence of a dilaton background $\exp(\pm\kappa^2 z^2)$.[17,45] Plus and minus components of the twist 3 nucleon LFWF are

$$\psi_+(\zeta)=\sqrt{2}\kappa^2\, \zeta^{3/2} e^{-\kappa^2\zeta^2/2}, \quad \Psi_-(\zeta)=\kappa^3\, \zeta^{5/2} e^{-\kappa^2\zeta^2/2}. \tag{6}$$

The results for $Q^4 F_1^p(Q^2)$ and $Q^4 F_1^n(Q^2)$ and are shown in Fig. 4.[46]

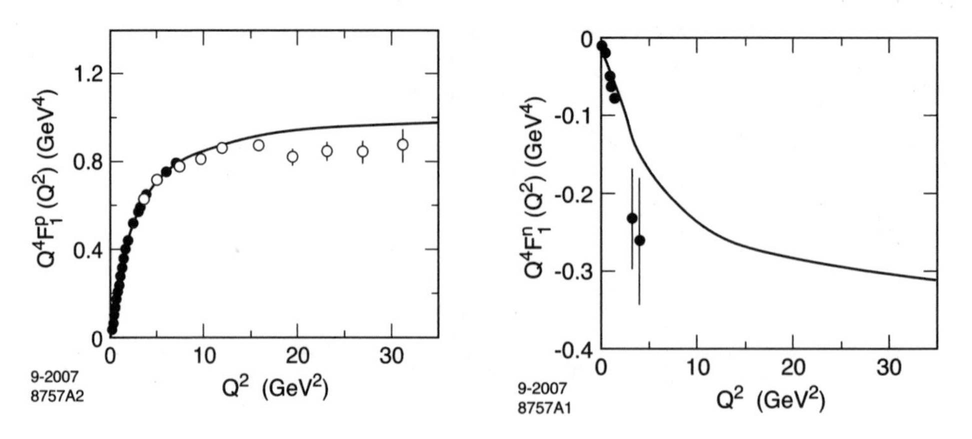

Fig. 4. Predictions for $Q^4 F_1^p(Q^2)$ and $Q^4 F_1^n(Q^2)$ in the soft wall model for $\kappa = 0.424$ GeV.

3. Nonperturbative Running Coupling from Light-Front Holography

The definition of the running coupling in perturbative quantum field theory is scheme-dependent. As discussed by Grunberg,[47] an effective coupling or charge can be defined directly from physical observables. Effective charges defined from different observables can be related to each other in the leading-twist domain using commensurate scale relations (CSR).[48] The potential between infinitely heavy quarks can be defined analytically in momentum transfer space as the product of the running coupling times the Born gluon propagator: $V(q) = -4\pi C_F \alpha_V(q)/q^2$. This effective charge defines a renormalization scheme – the α_V scheme of Appelquist, Dine, and Muzinich.[49] In fact, the holographic coupling $\alpha_s^{AdS}(Q^2)$ can be considered to be the nonperturbative extension of the α_V effective charge defined in Ref. 49. We can also make use of the g_1 scheme, where the strong coupling $\alpha_{g_1}(Q^2)$ is determined from the Bjorken sum rule.[50] The coupling $\alpha_{g_1}(Q^2)$ has the advantage that it is the best-measured effective charge, and it can be used to extrapolate the definition of the effective coupling to large distances.[51] Since α_{g_1} has been measured at intermediate energies, it is particularly useful for studying the transition from small to large distances.

We will show[3] how the LF holographic mapping of effective classical gravity in AdS space, modified by a positive-sign dilaton background, can be used to identify an analytically simple color-confining non-perturbative effective coupling $\alpha_s^{AdS}(Q^2)$ as a function of the space-like momentum transfer $Q^2 = -q^2$. This coupling incorporates confinement and agrees well with effective charge observables and lattice simulations. It also exhibits an infrared fixed point at small Q^2 and asymptotic freedom at large Q^2. However, the fall-off of $\alpha_s^{AdS}(Q^2)$ at large Q^2 is exponential: $\alpha_s^{AdS}(Q^2) \sim e^{-Q^2/\kappa^2}$, rather than the perturbative QCD (pQCD) logarithmic fall-off. We also show in Ref. 3 that a phenomenological extended coupling can be defined which implements the pQCD behavior.

As will be discussed below, the β-function derived from light-front holography becomes significantly negative in the non-perturbative regime $Q^2 \sim \kappa^2$, where it reaches a minimum, signaling the transition region from the infrared (IR) conformal region, characterized by hadronic degrees of freedom, to a pQCD conformal ultraviolet (UV) regime where the relevant degrees of freedom are the quark and gluon constituents. The β-function is always negative: it vanishes at large Q^2 consistent with asymptotic freedom, and it vanishes at small Q^2 consistent with an infrared fixed point.[5,52]

Let us consider a five-dimensional gauge field G propagating in AdS_5 space in presence of a dilaton background $\varphi(z)$ which introduces the energy scale κ in the five-dimensional action. At quadratic order in the field strength the action is

$$S = -\frac{1}{4}\int d^5x\,\sqrt{g}\,e^{\varphi(z)}\,\frac{1}{g_5^2}\,G^2, \tag{7}$$

where the metric determinant of AdS$_5$ is $\sqrt{g} = (R/z)^5$, $\varphi = \kappa^2 z^2$ and the square of the coupling g_5 has dimensions of length. We can identify the prefactor

$$g_5^{-2}(z) = e^{\varphi(z)} g_5^{-2}, \tag{8}$$

in the AdS action (7) as the effective coupling of the theory at the length scale z. The coupling $g_5(z)$ then incorporates the non-conformal dynamics of confinement. The five-dimensional coupling $g_5(z)$ is mapped, modulo a constant, into the Yang-Mills (YM) coupling g_{YM} of the confining theory in physical space-time using light-front holography. One identifies z with the invariant impact separation variable ζ which appears in the LF Hamiltonian: $g_5(z) \to g_{YM}(\zeta)$. Thus

$$\alpha_s^{AdS}(\zeta) = g_{YM}^2(\zeta)/4\pi \propto e^{-\kappa^2\zeta^2}. \tag{9}$$

In contrast with the 3-dimensional radial coordinates of the non-relativistic Schrödinger theory, the natural light-front variables are the two-dimensional cylindrical coordinates (ζ, ϕ) and the light-cone fraction x. The physical coupling measured at the scale Q is the two-dimensional Fourier transform of the LF transverse coupling $\alpha_s^{AdS}(\zeta)$ (9). Integration over the azimuthal angle ϕ gives the Bessel transform

$$\alpha_s^{AdS}(Q^2) \sim \int_0^\infty \zeta d\zeta \, J_0(\zeta Q) \, \alpha_s^{AdS}(\zeta), \tag{10}$$

in the $q^+ = 0$ light-front frame where $Q^2 = -q^2 = -\mathbf{q}_\perp^2 > 0$ is the square of the space-like four-momentum transferred to the hadronic bound state. Using this ansatz we then have from Eq. (10)

$$\alpha_s^{AdS}(Q^2) = \alpha_s^{AdS}(0) \, e^{-Q^2/4\kappa^2}. \tag{11}$$

In contrast, the negative dilaton solution $\varphi = -\kappa^2 z^2$ leads to an integral which diverges at large ζ. We identify $\alpha_s^{AdS}(Q^2)$ with the physical QCD running coupling in its nonperturbative domain.

The flow equation (8) from the scale dependent measure for the gauge fields can be understood as a consequence of field-strength renormalization. In physical QCD we can rescale the non-Abelian gluon field $A^\mu \to \lambda A^\mu$ and field strength $G^{\mu\nu} \to \lambda G^{\mu\nu}$ in the QCD Lagrangian density $\mathcal{L}_{\rm QCD}$ by a compensating rescaling of the coupling strength $g \to \lambda^{-1} g$. The renormalization of the coupling $g_{phys} = Z_3^{1/2} g_0$, where g_0 is the bare coupling in the Lagrangian in the UV-regulated theory, is thus equivalent to the renormalization of the vector potential and field strength: $A^\mu_{ren} = Z_3^{-1/2} A^\mu_0$, $G^{\mu\nu}_{ren} = Z_3^{-1/2} G^{\mu\nu}_0$ with a rescaled Lagrangian density $\mathcal{L}^{ren}_{\rm QCD} = Z_3^{-1} \mathcal{L}^0_{\rm QCD} = (g_{phys}/g_0)^{-2} \mathcal{L}_0$. In lattice gauge theory, the lattice spacing a serves as the UV regulator, and the renormalized QCD coupling is determined from the normalization of the gluon field strength as it appears in the gluon propagator. The inverse of the lattice size L sets the mass scale of the resulting running coupling. As is the case in lattice gauge theory, color confinement in AdS/QCD reflects nonperturbative dynamics at large distances. The QCD couplings defined from lattice gauge theory and the soft wall holographic model are thus similar in concept, and both schemes are expected to have similar properties in the nonperturbative domain, up to a rescaling of their respective momentum scales.

4. Comparison of the Holographic Coupling with Other Effective Charges

The effective coupling $\alpha^{AdS}(Q^2)$ (solid line) is compared in Fig. 5 with experimental and lattice data. For this comparison to be meaningful, we have to impose the same normalization on the AdS coupling as the g_1 coupling. This defines α_s^{AdS} normalized to the g_1 scheme: $\alpha_{g_1}^{AdS}\left(Q^2 = 0\right) = \pi$. Details on the comparison with other effective charges are given in Ref. 53.

The couplings in Fig. 5 (a) agree well in the strong coupling regime up to $Q \sim 1$ GeV. The value $\kappa = 0.54$ GeV is determined from the vector meson Regge trajectory.[8] The lattice results shown in Fig. 5 from Ref. 54 have been scaled to match the perturbative UV domain. The effective charge α_{g_1} has been determined in Ref. 53 from several experiments. Fig. 5 also displays other couplings from different observables as well as α_{g_1} which is computed from the Bjorken sum rule[50] over a large range of momentum transfer (cyan band). At $Q^2 = 0$ one has the constraint on the slope of α_{g_1} from the Gerasimov-Drell-Hearn (GDH) sum rule[55] which is also shown in the figure. The results show no sign of a phase transition, cusp, or other non-analytical behavior, a fact which allows us to extend the functional dependence of the coupling to large distances. As discussed below, the smooth behavior of the AdS strong coupling also allows us to extrapolate its form to the perturbative domain.

The hadronic model obtained from the dilaton-modified AdS space provides a semi-classical first approximation to QCD. Color confinement is introduced by the harmonic oscillator potential, but effects from gluon creation and absorption are not included in this effective theory. The nonperturbative confining effects vanish

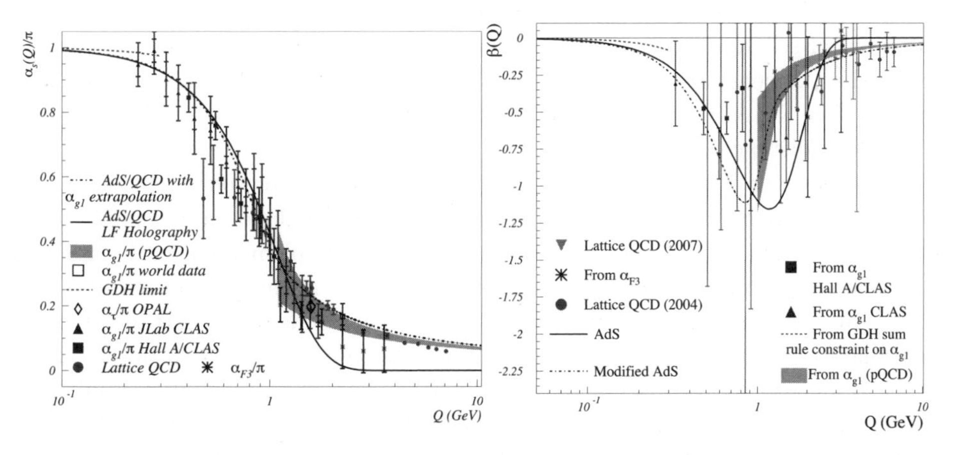

Fig. 5. (a) Effective coupling from LF holography for $\kappa = 0.54$ GeV compared with effective QCD couplings extracted from different observables and lattice results. (b) Prediction for the β function compared to lattice simulations, JLab and CCFR results for the Bjorken sum rule effective charge.

exponentially at large momentum transfer (Eq. (11)), and thus the logarithmic fall-off from pQCD quantum loops will dominate in this regime.

The running coupling α_s^{AdS} given by Eq. (11) is obtained from a color-confining potential. Since the strong coupling is an analytical function of the momentum transfer at all scales, we can extend the range of applicability of α_s^{AdS} by matching to a perturbative coupling at the transition scale, $Q \sim 1$ GeV, where pQCD contributions become important. In order to have a fully analytical model, we write

$$\alpha_{Modified,g_1}^{AdS}(Q^2) = \alpha_{g_1}^{AdS}(Q^2) g_+(Q^2) + \alpha_{g_1}^{fit}(Q^2) g_-(Q^2), \tag{12}$$

where $g_\pm(Q^2) = 1/(1 + e^{\pm(Q^2 - Q_0^2)/\tau^2})$ are smeared step functions which match the two regimes. The parameter τ represents the width of the transition region. Here $\alpha_{g_1}^{AdS}$ is given by Eq. (11) with the normalization $\alpha_{g_1}^{AdS}(0) = \pi$ – the plain black line in Fig. 5 – and $\alpha_{g_1}^{fit}$ in Eq. (12) is the analytical fit to the measured coupling α_{g_1}.[53] The couplings are chosen to have the same normalization at $Q^2 = 0$. The smoothly extrapolated result (dot-dashed line) for α_s is also shown on Fig. 5. We use the parameters $Q_0^2 = 0.8$ GeV^2 and $\tau^2 = 0.3$ GeV^2.

The β-function for the nonperturbative effective coupling obtained from the LF holographic mapping in a positive dilaton modified AdS background is

$$\beta^{AdS}(Q^2) = \frac{d}{d \log Q^2} \alpha^{AdS}(Q^2) = \frac{\pi Q^2}{4\kappa^2} e^{-Q^2/(4\kappa^2)}. \tag{13}$$

The solid line in Fig. 5 (b) corresponds to the light-front holographic result Eq. (13). Near $Q_0 \simeq 2\kappa \simeq 1$ GeV, we can interpret the results as a transition from the nonperturbative IR domain to the quark and gluon degrees of freedom in the perturbative UV regime. The transition momentum scale Q_0 is compatible with the momentum transfer for the onset of scaling behavior in exclusive reactions where quark counting rules are observed.[56] For example, in deuteron photo-disintegration the onset of scaling corresponds to momentum transfer of 1.0 GeV to the nucleon involved.[57] Dimensional counting is built into the AdS/QCD soft and hard wall models since the AdS amplitudes $\Phi(z)$ are governed by their twist scaling behavior z^τ at short distances, $z \to 0$.[6]

Also shown on Fig. 5 (b) are the β-functions obtained from phenomenology and lattice calculations. For clarity, we present only the LF holographic predictions, the lattice results from,[54] and the experimental data supplemented by the relevant sum rules. The width of the aqua band is computed from the uncertainty of α_{g_1} in the perturbative regime. The dot-dashed curve corresponds to the extrapolated approximation given by Eq. (12). Only the point-to-point uncorrelated uncertainties of the JLab data are used to estimate the uncertainties, since a systematic shift cancels in the derivative. Nevertheless, the uncertainties are still large. The β-function extracted from LF holography, as well as the forms obtained from the works of Cornwall,[52] Bloch, Fisher *et al.*,[58] Burkert and Ioffe[59] and Furui and Nakajima,[54] are seen to have a similar shape and magnitude.

Judging from these results, we infer that the actual β-function of QCD will extrapolate between the non-perturbative results for $Q < 1$ GeV and the pQCD results for $Q > 1$ GeV. We also observe that the general conditions

$$\beta(Q \to 0) = \beta(Q \to \infty) = 0, \tag{14}$$

$$\beta(Q) < 0, \text{ for } Q > 0, \tag{15}$$

$$\frac{d\beta}{dQ}\Big|_{Q=Q_0} = 0, \tag{16}$$

$$\frac{d\beta}{dQ} < 0, \text{ for } Q < Q_0, \ \frac{d\beta}{dQ} > 0, \text{ for } Q > Q_0 \tag{17}$$

are satisfied by our model β-function obtained from LF holography.

Eq. (14) expresses the fact that QCD approaches a conformal theory in both the far ultraviolet and deep infrared regions. In the semiclassical approximation to QCD without particle creation or absorption, the β-function is zero and the approximate theory is scale invariant in the limit of massless quarks.[60] When quantum corrections are included, the conformal behavior is preserved at very large Q because of asymptotic freedom and near $Q \to 0$ because the theory develops a fixed point. An infrared fixed point is in fact a natural consequence of color confinement:[52] since the propagators of the colored fields have a maximum wavelength, all loop integrals in the computation of the gluon self-energy decouple at $Q^2 \to 0$.[5] Condition (15) for Q^2 large, expresses the basic anti-screening behavior of QCD where the strong coupling vanishes. The β-function in QCD is essentially negative, thus the coupling increases monotonically from the UV to the IR where it reaches its maximum value: it has a finite value for a theory with a mass gap. Equation (16) defines the transition region at Q_0 where the beta function has a minimum. Since there is only one hadronic-partonic transition, the minimum is an absolute minimum; thus the additional conditions expressed in Eq. (17) follow immediately from Eqs. (14–16). The conditions given by Eqs. (14–17) describe the essential behavior of the full β-function for an effective QCD coupling whose scheme/definition is similar to that of the V-scheme.

5. Conclusions

As we have shown, the combination of Anti-de Sitter space (AdS) methods with light-front (LF) holography provides a remarkably accurate first approximation for the spectra and wavefunctions of meson and baryon light-quark bound states. We obtain a connection between a semiclassical first approximation to QCD, quantized on the light-front, with hadronic modes propagating on a fixed AdS background. The resulting bound-state Hamiltonian equation of motion in QCD leads to relativistic light-front wave equations in the invariant impact variable ζ which measures the separation of the quark and gluonic constituents within the hadron at equal light-front time. This corresponds to the effective single-variable relativistic Schrödinger-like equation in the AdS fifth dimension coordinate z, Eq. (1). The eigenvalues give the hadronic spectrum, and the eigenmodes represent the probability distributions of the hadronic constituents at a given scale. As we have shown, the light-front holographic mapping of effective classical gravity in AdS space, modified by a positive-sign dilaton background, provides a very good description of the spectrum and form factors of light mesons and baryons.

There are many phenomenological applications where detailed knowledge of the QCD coupling and the renormalized gluon propagator at relatively soft momentum transfer are essential. This includes the rescattering (final-state and initial-state interactions) which create the leading-twist Sivers single-spin correlations in semi-inclusive deep inelastic scattering,[61,62] the Boer-Mulders functions which lead to anomalous $\cos 2\phi$ contributions to the lepton pair angular distribution in the unpolarized Drell-Yan reaction,[63] and the Sommerfeld-Sakharov-Schwinger correction to heavy quark production at threshold.[64] The confining AdS/QCD coupling from light-front holography can lead to a quantitative understanding of this factorization-breaking physics.[65]

We have also shown that the light-front holographic mapping of effective classical gravity in AdS space, modified by the same positive-sign dilaton background predicts the form of a non-perturbative effective coupling $\alpha_s^{AdS}(Q)$ and its β-function. The AdS/QCD running coupling is in very good agreement with the effective coupling α_{g_1} extracted from the Bjorken sum rule. Surprisingly, the Furui and Nakajima lattice results[54] also agree better overall with the g_1 scheme rather than the V scheme. Our analysis indicates that light-front holography captures the essential dynamics of confinement. The holographic β-function displays a transition from nonperturbative to perturbative regimes at a momentum scale $Q \sim 1$ GeV. It appears to captures the essential characteristics of the full β-function of QCD, thus giving further support to the application of the gauge/gravity duality to the confining dynamics of strongly coupled QCD.

Acknowledgments

Presented by SJB at SCGT09, 2009 International Workshop on Strong Coupling Gauge Theories in the LHC Era, Nagoya, December 8-11, 2009. We thank Volker Burkert, John Cornwall, Sadataka Furui, Philipp Hägler, Wolfgang Korsch, G. Peter Lepage, Takemichi Okui, Joannis Papavassiliou and Anatoly Radyushkin for helpful comments. This research was supported by the Department of Energy contracts DE–AC02–76SF00515 and DE-AC05-84ER40150. SLAC-PUB-13998. JLAB-PHY-10-1130.

References

1. J. M. Maldacena, Adv. Theor. Math. Phys. **2**, 231 (1998) [Int. J. Theor. Phys. **38**, 1113 (1999)] [arXiv:hep-th/9711200].
2. S. S. Gubser, I. R. Klebanov and A. M. Polyakov, Phys. Lett. B **428**, 105 (1998) [arXiv:hep-th/9802109]; E. Witten, Adv. Theor. Math. Phys. **2**, 253 (1998) [arXiv:hep-th/9802150].
3. S. J. Brodsky, G. F. de Teramond and A Deur, arXiv:1002.3948 [hep-ph].
4. A. Deur, V. Burkert, J. P. Chen and W. Korsch, Phys. Lett. B **665**, 349 (2008) [arXiv:0803.4119 [hep-ph]].

5. S. J. Brodsky and R. Shrock, Phys. Lett. B **666**, 95 (2008) [arXiv:0806.1535 [hep-th]].
6. J. Polchinski and M. J. Strassler, Phys. Rev. Lett. **88**, 031601 (2002) [arXiv:hep-th/0109174].
7. A. Karch, E. Katz, D. T. Son and M. A. Stephanov, Phys. Rev. D **74**, 015005 (2006) [arXiv:hep-ph/0602229].
8. G. F. de Teramond and S. J. Brodsky, arXiv:0909.3900 [hep-ph].
9. O. Andreev and V. I. Zakharov, Phys. Rev. D **74**, 025023 (2006) [arXiv:hep-ph/0604204].
10. F. Zuo, arXiv:0909.4240 [hep-ph].
11. S. S. Afonin, arXiv:1001.3105 [hep-ph].
12. S. D. Glazek and M. Schaden, Phys. Lett. B **198**, 42 (1987).
13. P. Hoyer, arXiv:0909.3045 [hep-ph].
14. P. A. M. Dirac, Rev. Mod. Phys. **21**, 392 (1949).
15. S. J. Brodsky, H. C. Pauli and S. S. Pinsky, Phys. Rept. **301**, 299 (1998) [arXiv:hep-ph/9705477].
16. S. J. Brodsky and G. F. de Teramond, Phys. Rev. Lett. **96**, 201601 (2006) [arXiv:hep-ph/0602252];
17. S. J. Brodsky and G. F. de Teramond, Phys. Rev. D **77**, 056007 (2008) [arXiv:0707.3859 [hep-ph]].
18. S. J. Brodsky and G. F. de Teramond, Phys. Rev. D **78**, 025032 (2008) [arXiv:0804.0452 [hep-ph]].
19. G. F. de Teramond and S. J. Brodsky, Phys. Rev. Lett. **102**, 081601 (2009) [arXiv:0809.4899 [hep-ph]].
20. G. F. de Teramond and S. J. Brodsky, arXiv:1001.5193 [hep-ph].
21. J. Polchinski and M. J. Strassler, JHEP **0305**, 012 (2003) [arXiv:hep-th/0209211].
22. Z. Abidin and C. E. Carlson, Phys. Rev. D **77**, 095007 (2008) [arXiv:0801.3839 [hep-ph]].
23. C. Amsler *et al.* (Particle Data Group), Phys. Lett. B **667**, 1, (2008).
24. N. J. Craig and D. Green, JHEP **0909**, 113 (2009) [arXiv:0905.4088 [hep-ph]].
25. S. J. Brodsky and G. F. de Teramond, Phys. Lett. B **582**, 211 (2004) [arXiv:hep-th/0310227].
26. J. Erlich, E. Katz, D. T. Son and M. A. Stephanov, Phys. Rev. Lett. **95**, 261602 (2005) [arXiv:hep-ph/0501128].
27. L. Da Rold and A. Pomarol, Nucl. Phys. B **721**, 79 (2005) [arXiv:hep-ph/0501218].
28. E. Klempt and A. Zaitsev, Phys. Rept. **454**, 1 (2007) [arXiv:0708.4016 [hep-ph]].
29. H. Boschi-Filho, N. R. F. Braga and H. L. Carrion, Phys. Rev. D **73**, 047901 (2006) [arXiv:hep-th/0507063].
30. N. Evans and A. Tedder, Phys. Lett. B **642**, 546 (2006) [arXiv:hep-ph/0609112].
31. D. K. Hong, T. Inami and H. U. Yee, Phys. Lett. B **646**, 165 (2007) [arXiv:hep-ph/0609270].
32. P. Colangelo, F. De Fazio, F. Jugeau and S. Nicotri, Phys. Lett. B **652**, 73 (2007) [arXiv:hep-ph/0703316].
33. H. Forkel, Phys. Rev. D **78**, 025001 (2008) [arXiv:0711.1179 [hep-ph]].
34. A. Vega and I. Schmidt, Phys. Rev. D **78** (2008) 017703 [arXiv:0806.2267 [hep-ph]].
35. K. Nawa, H. Suganuma and T. Kojo, Mod. Phys. Lett. A **23**, 2364 (2008) [arXiv:0806.3040 [hep-th]].
36. W. de Paula, T. Frederico, H. Forkel and M. Beyer, Phys. Rev. D **79**, 075019 (2009) [arXiv:0806.3830 [hep-ph]].
37. P. Colangelo, F. De Fazio, F. Giannuzzi, F. Jugeau and S. Nicotri, Phys. Rev. D **78** (2008) 055009 [arXiv:0807.1054 [hep-ph]].

38. H. Forkel and E. Klempt, Phys. Lett. B **679**, 77 (2009) [arXiv:0810.2959 [hep-ph]].
39. H. C. Ahn, D. K. Hong, C. Park and S. Siwach, Phys. Rev. D **80**, 054001 (2009) [arXiv:0904.3731 [hep-ph]].
40. Y. Q. Sui, Y. L. Wu, Z. F. Xie and Y. B. Yang, arXiv:0909.3887 [hep-ph].
41. H. J. Kwee and R. F. Lebed, JHEP **0801**, 027 (2008) [arXiv:0708.4054 [hep-ph]].
42. H. R. Grigoryan and A. V. Radyushkin, Phys. Rev. D **76** (2007) 115007 [arXiv:0709.0500 [hep-ph]]; Phys. Rev. D **78** (2008) 115008 [arXiv:0808.1243 [hep-ph]].
43. S. J. Brodsky and G. F. de Teramond, arXiv:0802.0514 [hep-ph].
44. G. F. de Teramond and S. J. Brodsky, Phys. Rev. Lett. **94**, 201601 (2005) [arXiv:hep-th/0501022].
45. H. R. Grigoryan and A. V. Radyushkin, Phys. Rev. D **76** (2007) 095007 [arXiv:0706.1543 [hep-ph]].
46. The data compilation is from M. Diehl, Nucl. Phys. Proc. Suppl. **161**, 49 (2006) [arXiv:hep-ph/0510221].
47. G. Grunberg, Phys. Lett. B **95**, 70 (1980); Phys. Rev. D **29**, 2315 (1984); Phys. Rev. D **40**, 680 (1989).
48. S. J. Brodsky and H. J. Lu, Phys. Rev. D **51**, 3652 (1995) [arXiv:hep-ph/9405218]; S. J. Brodsky, G. T. Gabadadze, A. L. Kataev and H. J. Lu, Phys. Lett. B **372**, 133 (1996) [arXiv:hep-ph/9512367].
49. T. Appelquist, M. Dine and I. J. Muzinich, Phys. Lett. B **69**, 231 (1977).
50. J. D. Bjorken, Phys. Rev. **148**, 1467 (1966).
51. A. Deur, arXiv:0907.3385 [nucl-ex].
52. J. M. Cornwall, Phys. Rev. D **26**, 1453 (1982).
53. A. Deur, V. Burkert, J. P. Chen and W. Korsch, Phys. Lett. B **650**, 244 (2007) [arXiv:hep-ph/0509113].
54. S. Furui and H. Nakajima, Phys. Rev. D **70**, 094504 (2004); S. Furui, arXiv:0908.2768 [hep-lat].
55. S. D. Drell and A. C. Hearn, Phys. Rev. Lett. **16**, 908 (1966); S. B. Gerasimov, Sov. J. Nucl. Phys. **2** (1966) 430 [Yad. Fiz. **2** (1966) 598].
56. S. J. Brodsky and G. R. Farrar, Phys. Rev. Lett. **31**, 1153 (1973).
57. H. Gao and L. Zhu, AIP Conf. Proc. **747**, 179 (2005) [arXiv:nucl-ex/0411014].
58. J. C. R. Bloch, Phys. Rev. D **66**, 034032 (2002) [arXiv:hep-ph/0202073]; P. Maris and P. C. Tandy, Phys. Rev. C **60**, 055214 (1999) [arXiv:nucl-th/9905056]; C. S. Fischer and R. Alkofer, Phys. Lett. B **536**, 177 (2002) [arXiv:hep-ph/0202202]; C. S. Fischer, R. Alkofer and H. Reinhardt, Phys. Rev. D **65**, 094008 (2002) [arXiv:hep-ph/0202195]; R. Alkofer, C. S. Fischer and L. von Smekal, Acta Phys. Slov. **52**, 191 (2002) [arXiv:hep-ph/0205125]; M. S. Bhagwat, M. A. Pichowsky, C. D. Roberts and P. C. Tandy, Phys. Rev. C **68**, 015203 (2003) [arXiv:nucl-th/0304003].
59. V. D. Burkert and B. L. Ioffe, Phys. Lett. B **296**, 223 (1992); J. Exp. Theor. Phys. **78**, 619 (1994) [Zh. Eksp. Teor. Fiz. **105**, 1153 (1994)].
60. G. Parisi, Phys. Lett. B **39**, 643 (1972).
61. S. J. Brodsky, D. S. Hwang and I. Schmidt, Phys. Lett. B **530**, 99 (2002) [arXiv:hep-ph/0201296].
62. J. C. Collins, Phys. Lett. B **536**, 43 (2002) [arXiv:hep-ph/0204004].
63. D. Boer, S. J. Brodsky and D. S. Hwang, Phys. Rev. D **67**, 054003 (2003) [arXiv:hep-ph/0211110].
64. S. J. Brodsky, A. H. Hoang, J. H. Kuhn and T. Teubner, Phys. Lett. B **359**, 355 (1995) [arXiv:hep-ph/9508274].
65. J. Collins and J. W. Qiu, Phys. Rev. D **75**, 114014 (2007) [arXiv:0705.2141 [hep-ph]].

New Results on Non-Abelian Vortices — Further Insights into Monopole, Vortex and Confinement

K. Konishi

Department of Physics, "E. Fermi", University of Pisa, and INFN, Sez. di Pisa
Largo Pontecorvo, 3, 56127, Pisa, Italy
E-mail: konishi@df.unipi.it
http://www.df.unipi.it/~konishi/

We discuss some of the latest results concerning the non-Abelian vortices. The first concerns the construction of non-Abelian BPS vortices based on general gauge groups of the form $G = G' \times U(1)$. In particular detailed results about the vortex moduli space have been obtained for $G' = SO(N)$ or $USp(2N)$. The second result is about the "fractional vortices", i.e., vortices of the minimum winding but having substructures in the tension (or flux) density in the transverse plane. Thirdly, we discuss briefly the monopole-vortex complex.

Keywords: Monopoles, vortex, non-Abelian gauge theories, duality.

1. Introduction

The last few years have witnessed a remarkable progress in our understanding of the non-Abelian vortices and their relation to monopoles, both of which are thirty-year old problems in theoretical physics, and which can bear important implications to some deep issues such as quark confinement. The plan of this talk is: (i) a very brief review of non-Abelian monopoles; (ii) a brief review of '03-'07 results on non-Abelian vortices; (iii) a new result on non-Abelian vortices based on general gauge groups; (iv) the fractional vortices; and (v) a brief discussion on the monopole-vortex complex and non-Abelian duality.

It has become customary to think of quark confinement as a dual superconductor, in which (chromo-) electric fields are confined in a medium in which a magnetic charge is condensed. The original suggestion by 't Hooft and Mandelstam is essentially Abelian: the effective low-energy degrees of freedom are magnetic monopoles arising from the Yang-Mills gauge fields. It is however possible that the dual superconductor relevant to quark confinement is of a non-Abelian kind, in which case we must better understand the quantum mechanical properties of these degrees of freedom. We would like to understand how the 't Hooft-Polyakov monopoles[1] (arising from a gauge symmetry breaking, $G \to H$) and Abrikosov-Nielsen-Olesen vortices[2] (of a broken gauge theory $H \to \mathbb{1}$) are generalized in situations where the relevant gauge group H is non-Abelian.

The key developments which allowed us a qualitatively better understanding of these solitons are the following. First, the Seiberg-Witten solutions of $\mathcal{N} = 2$ supersymmetric gauge theories[3,4] revealed the quantum-mechanical behavior of the magnetic monopoles in an unprecedented fashion. In the presence of matter fields (quarks and squarks) these theories have, typically, vacua with non-Abelian dual gauge symmetry in the infrared.[5,6] Thus in these systems non-Abelian monopoles do exist and play a central role in confinement and dynamical symmetry breaking. Second, the discovery of non-Abelian vortex solutions,[7,8] i.e., soliton vortices with continuous, non-Abelian moduli, has triggered an intense research activity on the classical and quantum properties of these solitons, leading to a rich variety of new interesting results.[9–11]

2. Monopoles

When the gauge symmetry is spontaneously broken

$$G \overset{\langle \phi_1 \rangle \neq 0}{\longrightarrow} H \tag{1}$$

where H is some non-Abelian subgroup of G, the system possesses a set of regular magnetic monopole solutions in the semi-classical approximation. They are natural generalizations of the Abelian 't Hooft-Polyakov monopoles,[1] found originally in the $G = SO(3)$ theory broken to $H = U(1)$ by a Higgs mechanism. The gauge field looks asymptotically as

$$F_{ij} = \epsilon_{ijk} B_k = \epsilon_{ijk} \frac{r_k}{r^3} (\beta \cdot \mathbf{H}), \tag{2}$$

in an appropriate gauge, where $\mathbf{H}$ are the diagonal generators of H in the Cartan subalgebra. A straightforward generalization of the Dirac's quantization condition leads to[12]

$$2\,\beta \cdot \alpha \in \mathbf{Z} \tag{3}$$

where α are the root vectors of H. In the simplest such case, $G = SU(3)$, $H = SU(2) \times U(1)/\mathbb{Z}_2 \sim U(2)$, a straightforward idea that the degenerate monopole solutions to be a doublet of the unbroken $SU(2)$ leads however to the well-known difficulties.[14,15]

On the other hand, the quantization condition Eq. (3) implies that the monopoles should transform, if any, under the dual of $U(2)$: the individual solutions are labelled by β which live in the weight vector space of $\tilde{H}$, generated by the dual roots,

$$\alpha^* = \frac{\alpha}{\alpha \cdot \alpha}. \tag{4}$$

As transformation groups of fields H and $\tilde{H}$ are relatively non-local, the sought-for transformations of monopoles must look as non-local field transformations from the point of view of the original theory.[16]

But the most significant fact is that fully quantum mechanical monopoles appears in the low-energy dual description of a wide class of $\mathcal{N}=2$ supersymmetric QCD.[5,6] There must be ways to understand the physics of non-Abelian monopoles starting from a more familiar, semiclassical soliton picture.

3. Vortices

Attempts to understand the semi-classical origin of the non-Abelian monopoles appearing in the so-called r-vacua of the $\mathcal{N}=2$ supersymmetric $SU(N)$ gauge theory, has eventually led to the discovery of the non-Abelian *vortices*.[7,8] They are natural generalizations of the Abrikosov-Nielsen-Olesen (ANO) vortex. Unlike the ANO vortex, however, the non-Abelian vortices carry continuous zeromodes, i.e., they have a nontrivial moduli.

The simplest model in which these vortices appear is an $SU(N)\times U(1)$ gauge theory with $N_f=N$ flavors of squarks in the fundamental representation. The secret of the non-Abelian vortices lies in the so-called color-flavor locked phase, in which the squark fields (written as $N\times N$ color-flavor mixed matrix) takes the VEV of the form,

$$\langle q(x)\rangle = v\,\mathbb{1}_{N\times N}\,. \tag{5}$$

The $SU(N)$ gauge symmetry is completely broken, but the color-flavor mixed diagonal symmetry remains unbroken.

The vortex configuration in this vacuum involves scalar fields of the form,

$$q(x)=v\begin{pmatrix} e^{i\phi}f(\rho) & 0 & 0 & 0\\ 0 & g(\rho) & 0 & 0\\ & & \ddots & \\ 0 & 0 & \ddots & 0\\ 0 & 0 & 0 & g(\rho)\end{pmatrix} \tag{6}$$

where ρ,ϕ,z (the static vortex does not depend on z) are the cylindrical coordinates. The gauge fields take appropriate form, in order to ensure that the kinetic term tends to zero asymptotically, $\mathcal{D}q(x)\to 0$. In Eq. (6) the first flavor of the squark winds, but the full solution $A_i, q(x)$ can be rotated in the color flavor $SU(N)$ transformations,

$$A_i, \to U(A_i+i\partial_i)U^\dagger, \qquad q(x)\to Uq(x)U^\dagger\,,$$

leaving the tension invariant.

In other words, individual vortices break the exact $SU(N)_{C+F}$ symmetry of the system, developing therefore non-Abelian orientational zeromodes. Its nature is seen from the fact that the vortex Eq. (6) breaks the global symmetry as

$$SU(N)\to SU(N-1)\times U(1)/\mathbb{Z}_{N-1}; \tag{7}$$

the vortex moduli is given by

$$CP^{N-1}\sim \frac{SU(N)}{SU(N-1)\times U(1)/\mathbb{Z}_{N-1}}\,. \tag{8}$$

They are Nambu-Goldstone modes, which however can propagate only inside the vortex: far from it they are massive.

The quantum properties of the non-Abelian orientational modes (the effective CP^{N-1} sigma model), the study of non-Abelian vortices of higher winding numbers, the generalization to the cases of larger number of flavors and the study of the resulting, much richer vortex moduli spaces, the question of vortex stability in the presence of small non-BPS corrections, extension to more general class of gauge theories, etc. have been the subjects of an intense research activity recently.

4. Non-Abelian vortices with general gauge groups

One of the new results by us[17] is the construction of non-Abelian vortex solutions based on a general gauge group $G' \times U(1)$, where $G' = SU(N), SO(N), USp(2N)$, etc. As in models based on $SU(N)$ gauge groups studied extensively in the last few years, we work with simple models which have the structure of the bosonic sector of $N = 2$ supersymmetric models. The model contains a FI (Fayet-Iliopoulos)-like term in the $U(1)$ sector, allowing the system to develop stable vortices. A crucial aspect is that we work in a fully Higgsed vacuum, but with an unbroken color-flavor diagonal symmetry. We take as our model system

$$\mathcal{L} = \mathrm{Tr}_c\Big[-\frac{1}{2e^2} F_{\mu\nu}F^{\mu\nu} - \frac{1}{2g^2}\hat{F}_{\mu\nu}\hat{F}^{\mu\nu} + \mathcal{D}_\mu \mathcal{H} \, (\mathcal{D}^\mu \mathcal{H})^\dagger \\ -\frac{e^2}{4}\left|X^0 t^0 - 2\xi t^0\right|^2 - \frac{g^2}{4}\left|X^a t^a\right|^2 \Big] ,$$

with the field strength, gauge fields and covariant derivative denoted as

$$F_{\mu\nu} = F^0_{\mu\nu} t^0 , \quad F^0_{\mu\nu} = \partial_\mu A^0_\nu - \partial_\nu A^0_\mu , \quad \hat{F}_{\mu\nu} = \partial_\mu A_\nu - \partial_\nu A_\mu + i\,[A_\mu, A_\nu] , \\ A_\mu = A^a_\mu t^a , \quad \mathcal{D}_\mu = \partial_\mu + i A^0_\mu t^0 + i A^a_\mu t^a ,$$

A^0_μ is the gauge field associated with $U(1)$ and A^a_μ are the gauge fields of G'. The matter scalar fields are written as an $N \times N_\mathrm{F}$ complex color (vertical) – flavor (horizontal) mixed matrix H. It can be expanded as $X = HH^\dagger = X^0 t^0 + X^a t^a + X^\alpha t^\alpha$, $X^0 = 2\,\mathrm{Tr}_c\left(HH^\dagger t^0\right)$, $X^a = 2\,\mathrm{Tr}_c\left(HH^\dagger t^a\right)$. t^0 and t^a stand for the $U(1)$ and G' generators, respectively, and finally, $t^\alpha \in \mathfrak{g}'_\perp$, where $\mathfrak{g}'_\perp$ is the orthogonal complement of the Lie algebra $\mathfrak{g}'$ in $\mathfrak{su}(N)$. The traces with subscript c are over the color indices. e and g are the $U(1)$ and G' coupling constants, respectively, while ξ is a real constant.

We choose the maximally "color-flavor-locked" vacuum of the system,

$$\langle H \rangle = \frac{v}{\sqrt{N}} \mathbf{1}_N , \qquad \xi = \frac{v^2}{\sqrt{2N}} . \tag{9}$$

We have taken $N_\mathrm{F} = N$ which is the minimal number of flavors allowing for such a vacuum. Note that, unlike the $U(N)$ model studied extensively in the last several years, the vacuum is not unique in these cases (i.e., with a general gauge group), even with such a minimum choice of the flavor multiplicity. This difference may be traced to the fact that groups such as $SO(N) \times U(1)$ and $USp(N) \times U(1)$ form strictly smaller subgroups of $U(N)$.

The existence of a continuous vacuum degeneracy implies the emergence of vortices of semi-local type; this aspect will be crucial in the discussion of the fractional vortices in the second part of this talk. However, for now, we stick to the particular vacuum Eq. (9) and consider vortices and their moduli in this theory. The standard Bogomol'nyi completion reads

$$T = \int d^2x \, \mathrm{Tr}_c \Big[\frac{1}{e^2} \left| F_{12} - \frac{e^2}{2} \left(X^0 t^0 - 2\xi t^0 \right) \right|^2 + \frac{1}{g^2} \left| \hat{F}_{12} - \frac{g^2}{2} X^a t^a \right|^2 \\ + 4 \left| \bar{\mathcal{D}} H \right|^2 - 2\xi F_{12} t^0 \Big] \geq -\xi \int d^2x \, F_{12}^0 \,, \tag{10}$$

where $\bar{\mathcal{D}} \equiv \frac{\mathcal{D}_1 + i\mathcal{D}_2}{2}$, $z = x^1 + ix^2$. In the BPS limit one has

$$T = 2\sqrt{2N}\pi\xi\nu = 2\pi v^2 \nu \,, \qquad \nu = -\frac{1}{2\pi\sqrt{2N}} \int d^2x \, F_{12}^0 \,, \tag{11}$$

where ν is the $U(1)$ winding number of the vortex. This leads immediately to the vortex BPS equations

$$\bar{\mathcal{D}} H = \bar{\partial} H + i\bar{A} H = 0 \,, \tag{12}$$

$$F_{12}^0 = e^2 \left[\mathrm{Tr}_c \left(H H^\dagger t^0 \right) - \xi \right] , \qquad F_{12}^a = g^2 \, \mathrm{Tr}_c \left(H H^\dagger t^a \right) . \tag{13}$$

The matter BPS equation (12) can be solved by the Ansatz

$$H = S^{-1}(z, \bar{z}) H_0(z) \,, \qquad \bar{A} = -i S^{-1}(z, \bar{z}) \bar{\partial} S(z, \bar{z}) \,, \tag{14}$$

where S belongs to the complexification of the gauge group, $S \in \mathbb{C}^* \times G'^{\mathbb{C}}$. $H_0(z)$, holomorphic in z, is the *moduli matrix*, which contains all moduli parameters of the vortices.

A gauge invariant object can be constructed from S as $\Omega = S S^\dagger$. This can be conveniently split into the $U(1)$ part and the G' part, so that $S = s\,S'$ and analogously $\Omega = \omega\,\Omega'$, $\omega = |s|^2$, $\Omega' = S' S'^\dagger$. The tension (11) can be rewritten as

$$T = 2\pi v^2 \nu = 2v^2 \int d^2x \, \partial\bar{\partial} \log\omega \,, \qquad \nu = \frac{1}{\pi} \int d^2x \, \partial\bar{\partial} \log\omega \,, \tag{15}$$

and ν determines the asymptotic behavior of the Abelian field as

$$\omega = s s^\dagger \sim |z|^{2\nu} \,, \qquad \text{for } |z| \to \infty \,.$$

The minimal vortex solutions can then be written down by making use of the holomorphic invariants for the gauge group G' made of H, which we denote as $I^i_{G'}(H)$. If the $U(1)$ charge of the i-th invariant is n_i, $I^i_{G'}(H)$ satisfies

$$I^i_{G'}(H) = I^i_{G'}\left(s^{-1} S'^{-1} H_0 \right) = s^{-n_i} I^i_{G'}(H_0(z)) \,,$$

while the boundary condition is $I^i_{G'}(H)\Big|_{|z|\to\infty} = I^i_{\rm vev}\, e^{i\nu n_i \theta}$, where $\nu\, n_i$ is the number of the *zeros* of $I^i_{G'}$. This leads then to the following asymptotic behavior

$$I^i_{G'}(H_0) = s^{n_i} I^i_{G'}(H) \stackrel{|z|\to\infty}{\longrightarrow} I^i_{\rm vev} z^{\nu n_i} \,.$$

It shows that $I^i_{G'}(H_0(z))$, being holomorphic in z, are actually polynomials. Therefore $\nu\, n_i$ must be positive integers for all i:

$$\nu\, n_i \in \mathbb{Z}_+ \qquad \longrightarrow \qquad \nu = \frac{k}{n_0}\,, \qquad k \in \mathbb{Z}_+\,,$$

with $n_0 \equiv \gcd\left\{n_i\, | I^i_{\rm vev} \neq 0\right\}$. The $U(1)$ gauge transformation $e^{2\pi i/n_0}$ leaves $I^i_{G'}(H)$ invariant and thus the true gauge group is

$$G = \left[U(1) \times G'\right] / \mathbb{Z}_{n_0}\,,$$

where $\mathbb{Z}_{n_0}$ is the center of the group G'. The minimal winding in $U(1)$ found here, $\frac{1}{n_0}$, corresponds to the minimal element of $\pi_1(G) = \mathbb{Z}$: it represents a minimal loop in our group manifold G. As a result we find the following non-trivial constraints for H_0

$$I^i_{G'}(H_0) = I^i_{\rm vev}\, z^{\frac{kn_i}{n_0}} + O\left(z^{\frac{kn_i}{n_0}-1}\right)\,.$$

4.1. *GNO quantization for non-Abelian vortices*

Certain special solutions of a given theory can be found readily, as follows. It turns out that each such solution is characterized by a *weight vector of the dual group*, and is parametrized by a set of integers ν_a ($a = 1, \cdots, {\rm rank}(G')$)

$$H_0(z) = z^{\nu \mathbf{1}_N + \nu_a \mathcal{H}_a} \in U(1)^{\mathbb{C}} \times G'^{\mathbb{C}}\,, \tag{16}$$

where $\nu = k/n_0$ is the $U(1)$ winding number and $\mathcal{H}_a$ are the generators of the Cartan subalgebra of $\mathfrak{g}'$. H_0 must be holomorphic in z and *single-valued*, which gives the constraints for a set of integers ν_a

$$(\nu \mathbf{1}_N + \nu_a \mathcal{H}_a)_{ll} \in \mathbb{Z}_{\geq 0} \quad \forall l\,.$$

Suppose that we now consider scalar fields in an r-representation of G'. The constraint is equivalent to

$$\nu + \nu_a \mu_a^{(i)} \in \mathbb{Z}_{\geq 0} \quad \forall i\,, \tag{17}$$

where $\vec{\mu}^{(i)} = \mu_a^{(i)}$ ($i = 1, 2, \cdots, \dim(r)$) are the weight vectors for the r-representation of G'. Subtracting pairs of adjacent weight vectors, one arrives at the quantization condition

$$\vec{\nu} \cdot \vec{\alpha} \in \mathbb{Z}\,, \tag{18}$$

for every *root vector* α of G'.

Table 1. Some pairs of dual groups.

G'	$\tilde{G}'$
$SU(N)$	$SU(N)/\mathbb{Z}_N$
$U(N)$	$U(N)$
$SO(2M)$	$SO(2M)$
$USp(2M)$	$SO(2M+1)$
$SO(2M+1)$	$USp(2M)$

Now Eq. (18) is formally identical to the well-known GNO *monopole* quantization condition,[12] as well as to the naïve vortex flux quantization rule.[13] There is however a crucial difference here, from these earlier results. Because of an exact flavor (color-flavor diagonal $G_{\rm C+F}$) symmetry, broken by individual vortex solutions, our vortices possess continuous orientational moduli. These zero modes are normalizable, unlike those encountered in the earlier attempts to define quantum "non-Abelian monopoles".

These non-Abelian modes of our vortices—they are a kind of Nambu-Goldstone modes—can fluctuate and propagate along the vortex length. In systems with a hierarchical symmetry breaking,

$$G_0 \to G = G' \times U(1) \to \mathbf{1},$$

where our $G = G' \times U(1)$ model might emerge as the low-energy approximation, these orientational zero modes get absorbed by massive monopoles at the vortex extremities. This process endows the monopoles with fully quantum-mechanical non-Abelian (GNO-dual) charges, as has been suggested by the author and others in several occasions,[16] but we shall not dwell on this subject further here.

The solution of the quantization condition (18) is that

$$\vec{\tilde{\mu}} \equiv \vec{\nu}/2 \ ,$$

is any of the *weight vectors* of the dual group of G'. The dual group, denoted as $\tilde{G}'$, is defined by the dual root vectors[12] $\vec{\alpha}^* = \vec{\alpha}/(\vec{\alpha}\cdot\vec{\alpha})$. Examples of dual pairs of groups G', $\tilde{G}'$, are shown in Table 1. Note that (17) is actually stronger than (18), the l.h.s. must be a nonnegative integer. This positive quantization condition allows for a few weight vectors only. For concreteness, let us consider scalar fields in the fundamental representation, and choose a basis where the Cartan generators of $G' = SO(2M), SO(2M+1), USp(2M)$ are given by

$$\mathcal{H}_a = \mathrm{diag}\Big(\underbrace{0,\cdots,0}_{a-1},\frac{1}{2},\underbrace{0,\cdots,0}_{M-1},-\frac{1}{2},0,\cdots,0\Big) \ , \qquad (19)$$

with $a = 1,\cdots,M$. In this basis, special solutions H_0 have the form for $G' = SO(2M)$ and $USp(2M)$

$$H_0^{(\tilde{\mu}_1,\cdots,\tilde{\mu}_M)} = \mathrm{diag}\left(z^{k_1^+},\cdots,z^{k_M^+},z^{k_1^-},\cdots,z^{k_M^-}\right) \ , \qquad (20)$$

while for $SO(2M+1)$

$$H_0^{(\tilde{\mu}_1,\cdots,\tilde{\mu}_M)} = \mathrm{diag}\left(z^{k_1^+},\cdots,z^{k_M^+},z^{k_1^-},\cdots,z^{k_M^-},z^k\right) , \tag{21}$$

where $k_a^{\pm} = \nu \pm \tilde{\mu}_a$.

For example, in the cases of $G' = SO(4), USp(4)$ with a $\nu = 1/2$ vortex, there are four special solutions with $\vec{\tilde{\mu}} = (\frac{1}{2},\frac{1}{2}),(\frac{1}{2},-\frac{1}{2}),(-\frac{1}{2},\frac{1}{2}),(-\frac{1}{2},-\frac{1}{2})$

$$H_0^{(\frac{1}{2},\frac{1}{2})} = \mathrm{diag}(z,z,1,1) = z^{\frac{1}{2}\mathbf{1}_4+1\cdot\mathcal{H}_1+1\cdot\mathcal{H}_2}, \tag{22}$$

$$H_0^{(\frac{1}{2},-\frac{1}{2})} = \mathrm{diag}(z,1,1,z) = z^{\frac{1}{2}\mathbf{1}_4+1\cdot\mathcal{H}_1-1\cdot\mathcal{H}_2}, \tag{23}$$

$$H_0^{(-\frac{1}{2},\frac{1}{2})} = \mathrm{diag}(1,z,z,1) = z^{\frac{1}{2}\mathbf{1}_4-1\cdot\mathcal{H}_1+1\cdot\mathcal{H}_2}, \tag{24}$$

$$H_0^{(-\frac{1}{2},-\frac{1}{2})} = \mathrm{diag}(1,1,z,z) = z^{\frac{1}{2}\mathbf{1}_4-1\cdot\mathcal{H}_1-1\cdot\mathcal{H}_2}. \tag{25}$$

These four vectors are the same as the weight vectors of two Weyl spinor representations $\mathbf{2}\oplus\mathbf{2}'$ of $\tilde{G}' = SO(4)$ for $G' = SO(4)$, and the same as those of the Dirac spinor representation $\mathbf{4}$ of $\tilde{G}' = Spin(5)$ for $G' = USp(4)$.

The weight vectors corresponding to the $k=1$ vortex in various gauge groups are shown in Fig. 1. In all cases the results found are consistent with the GNO duality.

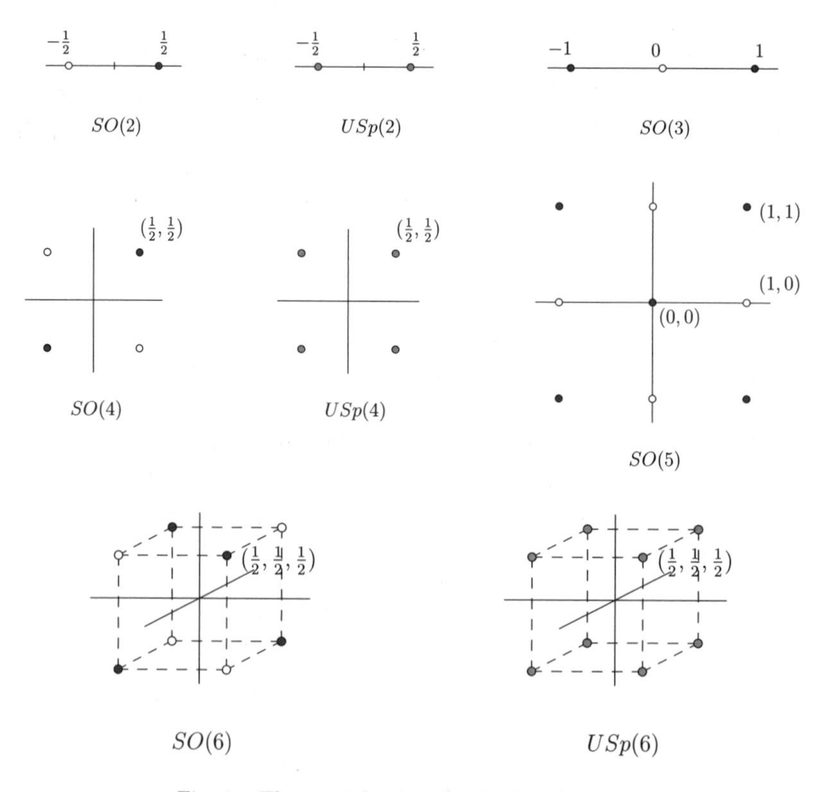

Fig. 1. The special points for the $k=1$ vortex.

5. Fractional vortices and lumps

Another exciting recent result concerns the fractional vortex and lumps.[18] We have pointed out above that in a general class of gauge theories the vacuum is not unique, even if the Fayet-Iliopoulos term is present and even if the number of the flavors is the minimum possible for a "color-flavor" locked vacuum to exist. In other words, there is a nontrivial vacuum degeneracy, or the *vacuum moduli* $\mathcal{M}$. In the first part of the talk, we were mainly interested in the *vortex moduli* $\mathcal{V}$, in a particular, maximally color-flavor locked vacuum. Here we are going to consider all possible vortices—the vortex moduli $\mathcal{V}$—on all possible points of the vacuum moduli $\mathcal{M}$ at the same time.

There are in fact two crucial ingredients for the fractional vortex: the *vacuum degeneracy* and the *BPS saturated* nature of the vortices. The first point was emphasized just above: the situation is schematically illustrated in Fig. 2. Even if we restrict ourselves to the minimally winding vortex solutions only, the vortices represent non-trivial fiber bundles over the vacuum moduli $\mathcal{M}$.

The BPS-saturated nature of the vortices, on the other hand, implies that the vortex equations become first-order equations. The matter equations of motion are solved by the moduli-matrix Ansatz. The other equations–the gauge field equations–reduce, in the strong coupling limit or, anyway, sufficiently far from the vortex center, to the vacuum equations for the scalar fields. In other words, the vortex solutions approximate the sigma model lumps.

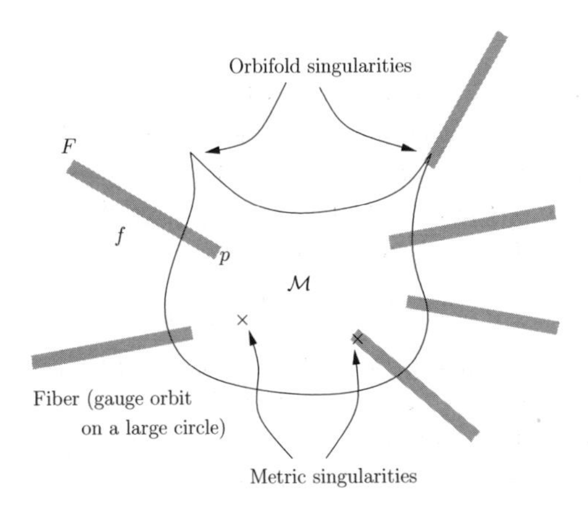

Fig. 2. Vacuum moduli $\mathcal{M}$, fiber F over it, and possible singularities.

5.1. *Structures of the vacuum moduli*

A vortex solution is defined on each point of $\mathcal{M}$, in the sense that the scalar configuration along a sufficiently large circle (S^1) surrounding it traces a non-trivial closed orbit in the fiber F (hence a point in $\mathcal{M}$). The existence of a vortex solution requires that

$$\pi_1(F, f) \neq \mathbf{1} , \tag{26}$$

where f is a point in M, the minima of the potential. The field configuration on a disk D^2 encircled by S^1 traces $\mathcal{M}$, apart from points at finite radius where it goes off M (hence from $\mathcal{M}$). Thus it represents an element of $\pi_2(\mathcal{M}, p)$, where p is the gauge orbit containing f, or $p = \pi(f)$: π is the projection of the fiber onto a point of the base $\mathcal{M}$. The exact sequence of homotopy groups for the fiber bundle reads

$$\cdots \to \pi_2(M, f) \to \pi_2(\mathcal{M}, p) \to \pi_1(F, f) \to \pi_1(M, f) \to \pi_1(\mathcal{M}, p) \to \cdots$$

where $\pi_2(M/F, f) \sim \pi_2(\mathcal{M}, p)$. See Fig. 3.

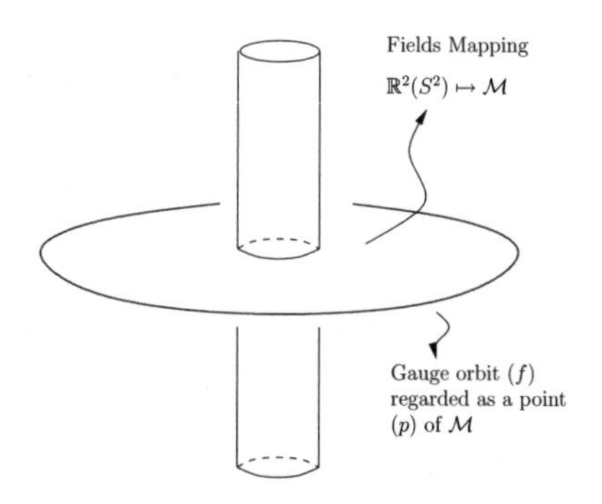

Fig. 3.

Given the points f, p and the space $\mathcal{M}$, the vortex solution is still not unique. Any exact symmetry of the system broken by an individual vortex solution gives rise to vortex zero modes (moduli), $\mathcal{V}$. Our main interest here however is the vortex moduli which arises from the non-trivial vacuum moduli $\mathcal{M}$ itself. Due to the BPS nature of our vortices, the gauge field equation

$$F_{12}^I = g_I^2 \left(q^\dagger T^I q - \xi^I \right) ,$$

reduces, in the strong-coupling limit (or in any case, sufficiently far from the vortex center), to the vacuum equation defining M. This means that a vortex configuration can be approximately seen as a non-linear σ-model (NLσM) lump with target space $\mathcal{M}$, as was already anticipated. Various distinct maps $S^2 \mapsto \mathcal{M}$ of the same homotopy class correspond to physically inequivalent solutions; each of these corresponds to a vortex with the equal tension

$$T_{\min} = -\xi^I \int d^2x \, F_{12}^I > 0 ,$$

by their BPS nature. They represent non-trivial *vortex moduli.*

The semi-local vortices of the so-called extended-Abelian Higgs (EAH) model arise precisely this way. In an Abelian Higgs model with N flavors of (scalar) electrons, $M = S^{2N-1}$, $F = S^1$, $\mathcal{M} = S^{2N-1}/S = \mathbb{C}P^{N-1}$, and the exact homotopy sequence tells us that $\pi_2(\mathbb{C}P^{N-1})$ and $\pi_1(S^1)$ are isomorphic: each (i.e. minimum) vortex solution corresponds to a minimal σ-model lump solution.

In most cases discussed in our paper,[18] however, the base space $\mathcal{M}$ will be various kinds of *singular manifolds*: a manifold with singularities, unlike in the EAH model. The nature of the singularities depends on the system and on the particular point(s) of $\mathcal{M}$. Our BPS degenerate vortices represent (generalized) fiber bundles with the singular manifolds $\mathcal{M}$ as the base space.

5.2. *Two mechanisms for fractional vortex–lump*

There are two distinct mechanisms leading to the appearance of a fractional vortex. The first is related to the presence of orbifold singularities in $\mathcal{M}$. For example, let us consider a $\mathbb{Z}_2$ point p_0 such as the one appearing in a simple $U(1)$ model with two scalars, one of which has charge 2. At this singularity, both $\pi_2(\mathcal{M}, p)$ and $\pi_1(F, f)$ make a discontinuous change. The minimum element of $\pi_1(F_0, f_0)$ is half of that of $\pi_1(F, f)$ defined off the singularity, and similarly for $\pi_2(\mathcal{M}, p_0)$ with respect to $\pi_2(\mathcal{M}, p)$, $p \neq p_0$. Even though the exact homotopy sequence continues to hold on and off the orbifold point, the vortex defined near such a point will look like a doubly-wound vortex, with two centers (if the vortex moduli parameters are chosen appropriately). Analogous multi-peak vortex solution occurs near a $\mathbb{Z}_N$ orbifold point of $\mathcal{M}$.

Another cause for the appearance of fractional peaks is simple and very general: a deformed sigma model geometry. This phenomenon can be best seen by considering our system in the strong coupling limit. Even if the base point p is a perfectly generic, regular point of $\mathcal{M}$, not close to any singularity, the field configurations in the transverse plane (S^2) trace the whole vacuum moduli space $\mathcal{M}$. The energy distribution reflects the nontrivial structure of $\mathcal{M}$ as the volume of the target space is mapped into the transverse plane, $\mathbb{C}$

$$E = 2\int_{\mathbb{C}} \frac{\partial^2 K}{\partial\phi^I \partial\phi^{\dagger J}} \partial\phi^I \bar{\partial}\phi^{\dagger \bar{J}} = 2\int_{\mathbb{C}} \bar{\partial}\partial K \ .$$

The field configuration may hit for instance one of the singularities (conic or not), or simply the regions of large scalar curvature. Such phenomena thus occur very generally if the underlying sigma model has a deformed geometry [a]. At such points the energy density will show a peak, not necessarily at the vortex center. Even at finite coupling, the vortex tension density will exhibit a similar substructure.

5.3. *Some models*

A simple model showing the fractional vortex is an extended Abelian Higgs model, with two scalar fields A and B with charges 2 and 1, respectively. Depending on the point of $\mathcal{M}$ (which is $\mathbb{C}P^1$) the minimum vortex shows doubly-peaked substructure

[a] Basically the same phenomenon was found also by Collie and Tong.[19]

clearly, see Fig. 4. The fractional vortex structure in this model nicely illustrates the first mechanism discussed above: the point $B = 0$ is a Z_2 orbifold point, since there the only nonvanishing field, A, having charge 2, must wind only half of the $U(1)$ to be single-valued.

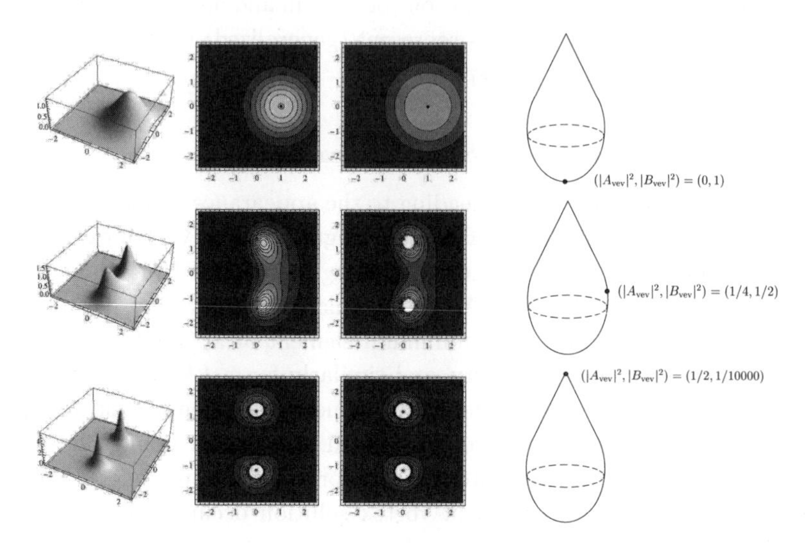

Fig. 4. The energy (the left-most and the 2nd left panels) and the magnetic flux (the 2nd right panels) density, together with the boundary values (A, B) (the right-most panel) for the minimal vortex.

Another interesting model is a $U_1(1) \times U_2(1)$ gauge theory with three flavors of scalar electrons $H = (A, B, C)$ with charges $Q_1 = (2, 1, 1)$ for $U(1)_1$ and $Q_2 = (0, 1, -1)$ for $U(1)_2$. Even though the model has the same $\mathbb{C}P^1$ as the vacuum moduli $\mathcal{M}$ as the first model, the vortex properties are quite different. This model turns out to provide a good example of fractional vortex of the second type (deformed sigma-model geometry).

Fractional vortex occurs also in non-Abelian gauge theories, such as the one with gauge group $G = SO(N) \times U(1)$. An illustrative example of fractional vortex in an $SO(6) \times U(1)$ model is shown in Fig. 6.

6. Why non-Abelian vortices imply non-Abelian monopoles

A more recent work of our research group concerns the monopole-vortex complex solitons occurring in systems with hierarchical gauge symmetry breaking,

$$G \stackrel{\langle\phi_1\rangle\neq 0}{\longrightarrow} H \stackrel{\langle\phi_2\rangle\neq 0}{\longrightarrow} \mathbb{1}\,, \qquad |\langle\phi_1\rangle| \gg |\langle\phi_2\rangle|\,. \tag{27}$$

The homotopy-group sequence

$$\cdots \to \pi_2(G) \to \pi_2(G/H) \to \pi_1(H) \to \pi_1(G) \to \cdots\,. \tag{28}$$

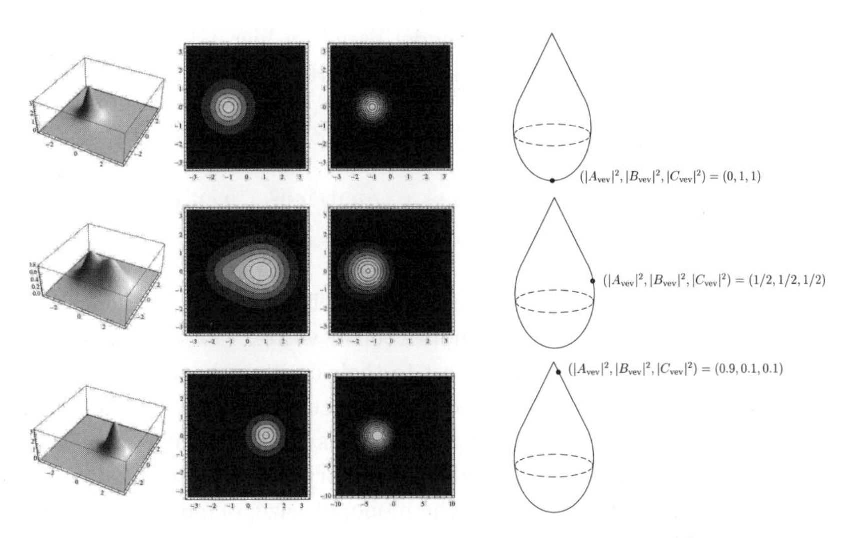

Fig. 5. The energy density (left-most) and the magnetic flux density $F_{12}^{(1)}$ (2nd from the left), $F_{12}^{(1)}$ (2nd from the right) and the boundary condition (right-most).

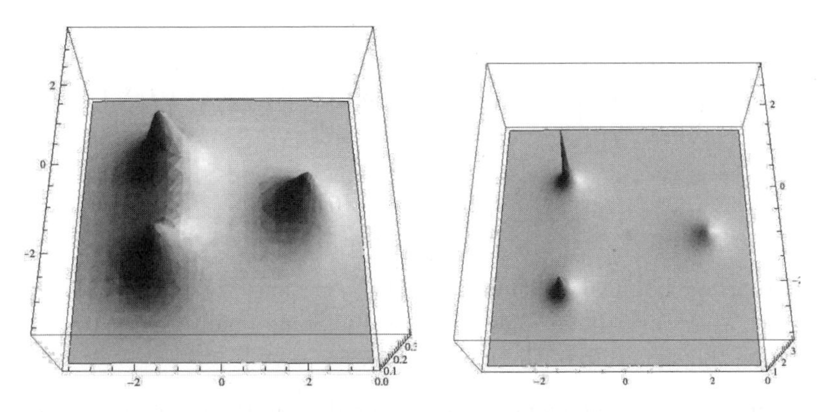

Fig. 6. The energy density of three fractional vortices (lumps) in the $U(1) \times SO(6)$ model in the strong coupling approximation. The positions are $z_1 = -\sqrt{2} + i\sqrt{2}$, $z_2 = -\sqrt{2} - i\sqrt{2}$, $z_3 = 2$. The two figures correspond to two different choices of certain size moduli parameters.

tells us that the properties of the regular monopoles arising from the breaking $G \to H$ are related to the vortices of the low-energy system. In particular, the fact that $\pi_2(G) = \mathbb{1}$ for any group G, implies that

$$\pi_2(G/H) \sim \pi_1(H)/\pi_1(G) \,. \tag{29}$$

For instance, in the case of the symmetry breaking, $SU(N+1) \to SU(N) \times U(1)/\mathbb{Z}_N$ the first set of checks (on the Abelian and non-Abelian magnetic flux matching) have been done[20] soon after the discovery of the non-Abelian vortex in the $U(N)$ theory. We wish to study more carefully the monopole-vortex configurations, taking into account a small non-BPS correction term.

For instance one might study the model based on hierarchically broken gauge symmetry, $SU(3) \to SU(2) \times U(1) \to \mathbb{1}$, with the Hamiltonian,

$$H = \int d^3x \Big[\frac{1}{4g^2}(F^3_{ij})^2 + \frac{1}{4g_0^2}(F^0_{ij})^2 + \frac{1}{g^2}|\mathcal{D}_i\phi^a|^2 + \frac{1}{g_0^2}|\mathcal{D}_i\phi^0|^2 + |\mathcal{D}_i q|^2 \\ + g_0^2|\mu\phi^8 + \sqrt{2}Q^\dagger t^8 Q|^2 + g^2|\mu\phi^3 + \sqrt{2}Q^\dagger t^3 Q|^2 + 2\,Q^\dagger\lambda^\dagger\lambda Q \Big] , \tag{30}$$

describing the system after the first breaking. Such a low-energy theory is of the type studied in our original work on non-Abelian vortex,[8] except for small terms involving $\phi^3(x)$ and $\phi^8(x)$ (which were set to their constant in that paper). In fact, the model is the same as the one studied by Auzzi et. al. recently.[21] The system has unbroken, exact color-flavor diagonal $SU(2)_{C+F}$ symmetry. Neglecting the fields which get mass of the order of the higher symmetry breaking scale, and the fields which go to zero exponentially outside the monopole size, one makes an Ansatz (in the monopole and vortex singular gauge):

$$A_\phi = t_3 A^3_\phi(\rho, z) + t_8 A^8_\phi(\rho, z); \qquad A^3_\phi = -\frac{1}{\rho} f_3(\rho, z), \quad A^8_\phi = -\sqrt{3}\frac{1}{\rho} f_8(\rho, z),$$

$$\phi(\mathbf{r}) = \begin{pmatrix} v & 0 & 0 \\ 0 & v & 0 \\ 0 & 0 & -2v \end{pmatrix} + \lambda(\rho, z), \qquad \lambda(\rho, z) = t_3\lambda_3(\rho, z) + t_8\lambda_8(\rho, z)\,.$$

$$q(x) = \begin{pmatrix} w_1(\rho, z) & 0 \\ 0 & w_2(\rho, z) \end{pmatrix} ,$$

with appropriate boundary conditions. The equations for the profile functions $f_3, f_8, w_1 w_2, \lambda_3, \lambda_8$ may be studied numerically. Some qualitative features can be read off from the structure of these equations.

(i) The Dirac string of the monopole is hidden deep in the vortex core; the zero of the squark field at the vortex core makes the singularity harmless.

(ii) The whole monopole-vortex complex breaks $SU(2)_{C+F}$: the orientational zeromodes develops which live on $SU(2)/U(1) \sim CP^1$.

(iii) The degeneracy between the monopole of the broken "U spin" and the monopole of the broken "V-spin", which are naïvely related by the unbroken $SU(2)$ of the high-mass-scale breaking $SU(3) \to SU(2) \times U(1)$, is *explicitly* broken in the vacuum with small squark VEV.

(iv) Nevertheless, there is a new, exact continuous degeneracy among the monopole-vortex complex configurations, related by the color-flavor symmetry (CP^1 moduli). It is possible that such an exact non-Abelian symmetry possessed by the monopole is at the origin of the non-Abelian dual gauge symmetry which emerges at low energies of the softly broken $\mathcal{N} = 2$ supersymmetric QCD.[6]

Acknowledgments

The main new results reported here are the fruit of a collaboration with M. Eto, T. Fujimori, S. B. Gudnason, T. Nagashima, M. Nitta, K. Ohashi, and W. Vinci. The last part on unpublished work on the monopole-vortex is based on a collaboration with S. B. Gudnason, D. Dorigoni, A. Michelini and M. Cipriani. I thank them all.

References

1. G. 't Hooft, Nucl. Phys. B **79**, 817 (1974), A.M. Polyakov, JETP Lett. **20**, 194 (1974).
2. A. Abrikosov, Sov. Phys. JETP **32**, 1442 (1957); H. Nielsen, P. Olesen, Nucl. Phys. B **61**, 45 (1973).
3. N. Seiberg, E. Witten, Nucl. Phys. B **426**, 19 (1994); Erratum *ibid.* B **430**, 485 (1994).
4. N. Seiberg, E. Witten, Nucl. Phys. B **431**, 484 (1994).
5. P. C. Argyres, M. R. Plesser, N. Seiberg, Nucl. Phys. B **471**, 159 (1996); P.C. Argyres, M.R. Plesser, A.D. Shapere, Nucl. Phys. B **483**, 172 (1997); K. Hori, H. Ooguri, Y. Oz, Adv. Theor. Math. Phys. **1**, 1 (1998).
6. G. Carlino, K. Konishi, H. Murayama, JHEP **0002**, 004 (2000); Nucl. Phys. B **590**, 37 (2000).
7. A. Hanany and D. Tong, JHEP **0307**, 037 (2003) [arXiv:hep-th/0306150].
8. R. Auzzi, S. Bolognesi, J. Evslin, K. Konishi and A. Yung, Nucl. Phys. B **673**, 187 (2003) [arXiv:hep-th/0307287].
9. D. Tong, "TASI lectures on solitons," arXiv:hep-th/0509216.
10. M. Shifman and A. Yung, Rev. Mod. Phys. **79**, 1139 (2007), [arXiv:hep-th/0703267].
11. M. Eto, Y. Isozumi, M. Nitta, K. Ohashi and N. Sakai, J. Phys. A **39**, R315 (2006), [arXiv:hep-th/0602170].
12. P. Goddard, J. Nuyts and D. Olive, Nucl. Phys. B **125** (1977) 1; F. A. Bais, Phys. Rev. D **18** (1978) 1206; B. J. Schroers and F. A. Bais, Nucl. Phys. B **512** (1998) 250, hep-th/9708004; E. J. Weinberg, Nucl. Phys. B **167** (1980) 500.
13. K. Konishi and L. Spanu, Int. J. Mod. Phys. A18 (2003) 249, arXiv: hep-th/0106175.
14. A. Abouelsaood, Nucl. Phys. B **226**, 309 (1983); P. Nelson, A. Manohar, Phys. Rev. Lett. **50**, 943 (1983); A. Balachandran, G. Marmo, M. Mukunda, J. Nilsson, E. Sudarshan, F. Zaccaria, Phys. Rev. Lett. **50**, 1553 (1983); P. Nelson, S. Coleman, Nucl. Phys. B **227**, 1 (1984)
15. N. Dorey, C. Fraser, T.J. Hollowood, M.A.C. Kneipp, "NonAbelian duality in N=4 supersymmetric gauge theories," [arXiv: hep-th/9512116]; Phys.Lett. B **383**, 422 (1996)
16. M. Eto, L. Ferretti, K. Konishi, G. Marmorini, M. Nitta, K. Ohashi, W. Vinci, N. Yokoi, Nucl.Phys. **B780**, 161-187 (2007), [arXiv: hep-th/0611313].
17. M. Eto, T. Fujimori, S. B. Gudnason, K. Konishi, T. Nagashima, M. Nitta, K. Ohashi, W. Vinci, JHEP 0906:004 (2009), arXiv:0903.4471 [hep-th]; M. Eto, T. Fujimori, S. B. Gudnason, K. Konishi, M. Nitta, K. Ohashi, W. Vinci, Phys. Lett: B669: 98-101 (2008), arXiv:0802.1020 [hep-th].
18. M. Eto, T. Fujimori, S. B. Gudnason, K. Konishi, T. Nagashima, M. Nitta, K. Ohashi, W. Vinci, Phys. Rev. D80:045018,2009, arXiv:0905.3540 [hep-th].
19. B. Collie, D. Tong, JHEP 0908:006,2009, e-Print: arXiv:0905.2267 [hep-th].
20. R. Auzzi, S. Bolognesi, J. Evslin and K. Konishi, Nucl. Phys. B **686**, 119 (2004), [arXiv:hep-th/0312233].
21. R. Auzzi, M. Eto, W. Vinci, JHEP 0711:090,2007, arXiv:0709.1910 [hep-th].

Study on Exotic Hadrons at *B*-Factories

Toru Iijima
Department of Physics, Nagoya University, Furo-cho, Chikusa-ku
Nagoya 464-8602, Japan

The high luminosity e^+e^- collision at the *B*-factory experiments (Belle/BaBar) have revealed rich spectra of hadron resonances in the charmonium region. Many of the newly found states do not fit to the unfilled level of the conventional $c\bar{c}$ spectrum. Some of them are thought to be exotic states, having sub-structure more complex than the quark anti-quark mesons. In fact, Belle has found some states with non-zero electric charge, that require a minimum quark content of $c\bar{c}u\bar{d}$. In this paper, we review the present status of studies on such exotic hadrons at the *B*-factory experiments.

1 Introduction

Thanks to the great success of the *B*-factory experiments at KEK (KEKB/Belle) and SLAC (PEP II/BaBar), our knowledge on *CP* violation phenomena has been greatly improved in the past years. The success resulted in the Nobel Prize of Physics in 2008 awarded to Professors Kobayashi and Maskawa. Another success of the *B*-factory experiments, which is an unexpected bonus, is the discoveries of many new hadronic states, as shown in Figure 1. Of particular interest are a number of charmonium-like meson states, collectively known as " $X\ Y\ Z$ " mesons. Their measured mass does not fit to the mass spectrum predicted by QCD-motivated conventional quark model calculations. Although they lie above the open charm threshold at 3.73 GeV, they decay preferably to charmonium and one or two pions rather than two charm mesons ($D\bar{D}$). Because of their peculiar natures, some of them are thought to be "exotics", and have been extensively discussed in literatures [1].

Quantum Chromodynamics (QCD), the fundamental theory for strong interaction, does not exclude the possible existence of hadrons with a substructure that is more complex than the ordinal three-quark baryons and quark-antiquark mesons. In fact, such quark configurations were considered already in the original Gell-Mann's paper for the quark model [2]. The possible exotic state includes;

- ***Tetraquark***: diquark and anti-diquark state, usually denoted as $[Qq][\bar{Q}\bar{q}']$, where Q is the heavy quark.
- ***Molecule***: bound state of two mesons, $[Q\bar{q}][\bar{Q}q']$.
- ***Hybrid***: bound state of a quark-antiquark pair and a number of gluons.

Although the interpretation is uncertain at this stage, the increasing number of *XYZ* meson states reported by *B*-factories as well as CLEO and CDF experiments, acquire a lot of interests in the community, together with the pentaquark state Θ^+ ($uudd\bar{s}$) reported first by the LEPS experiment at the Spring-8 [3]. Studies of such exotic hadrons have now evolved to make an interdisciplinary research area between particle and nuclear physics.

In this paper, we review the present status of studies on such exotics hadrons at the *B*-factory experiments. Results are shown mainly using the Belle data.

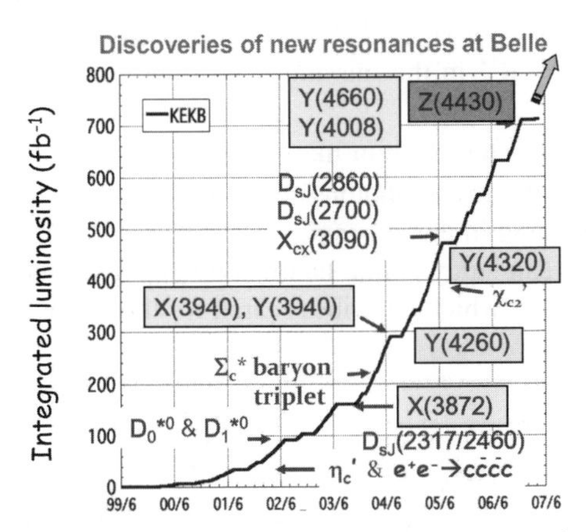

Figure 1: History for discoveries of new hadron resonances at the Belle experiment.

2 Study on exotic hadrons at *B*-Factories

Such many discoveries of exotic hadrons have been brought by the high luminosity accelerators and the excellent detectors, which are capable to reconstruct all of the interesting reactions exclusively in clean environment of e^+e^- collisions. In the case of the KEK *B*-factory, KEKB is an asymmetric energy collider of 8GeV electron and 3.5GeV positron beams, running on the Υ(4S) resonance (center-of-mass energy is 10.58GeV). The peak luminosity has reached $2.1 \times 10^{34}\text{cm}^{-2}\text{s}^{-1}$, the world highest luminosity and two times more than the design value. The integrated luminosity by December 2009 is about 950fb^{-1}, among which 710fb^{-1} is taken on the Υ(4S). This data set corresponds to production of about 800M $B\overline{B}$ pairs and 960M charm and anti-charm meson pairs. The Belle detector surrounds the interaction point with a large solid angle (about 90% of 4π), and records the production of $q\overline{q}(q = u, d, s, c, b)$ with minimum bias trigger condition. The detector has also excellent particle identification system, which allows one to distinguish the final state particles, *e*. μ. π. *K*, *p* and γ.

There are several ways to produce charmonium states at the *B*-factories (Figure 2);

- ***B decays***: the dominant *B* meson decay occurs through $b \to cW$ transition, followed by $W \to \overline{c}s$. Charmonium is produced when c and $\overline{c}$ quarks are arranged close enough in the phase space. The s quark is combined with the spectator $\overline{u}$ or $\overline{d}$ quark to make the associated $\overline{K}$ meson. The possible J^{PC} quantum number of the produced $c\overline{c}$ system is 0^{-+}, 1^{--}or 1^{++}. The well-known *X*(3872) and *Z*(4430) were found in this process.
- ***ISR process***: in e^+e^- collision, the initial state e^+ and e^- occasionally radiate a high energy γ ray, and the e^+ and e^- subsequently annihilate with a correspondingly

reduced energy. This produces a charmonium state with J^{PC}=1^{--}. The charmonium-like states, Y(4260), Y(4360) and Y(4660) were observed by this method.

- ***Double charm production***: the Belle collaboration discovered that when a J/ψ is produced in the process, $e^+e^- \to J/\psi$ +anything, the accompanying system contains another $c\overline{c}$ system with high probability. This process offers a way to search for states with ^{PC} = 0^{+-} , 0^{++} in the missing mass, therefore, independent of the decay mode.
- ***Two-photon collision***: charmonium-like state, in $J^{PC} = 0^{-+}$, 0^{++}, 2^{++} can be produced also via collisions of virtual photons, each produced by e^+ and e^- beams. The Z(3940) state was found in this process.

More aggressively, one can study exotic hadrons by changing the accelerator energy, as discussed later in Section 4, which describes search for the bottom counterpart of *XYZ*.

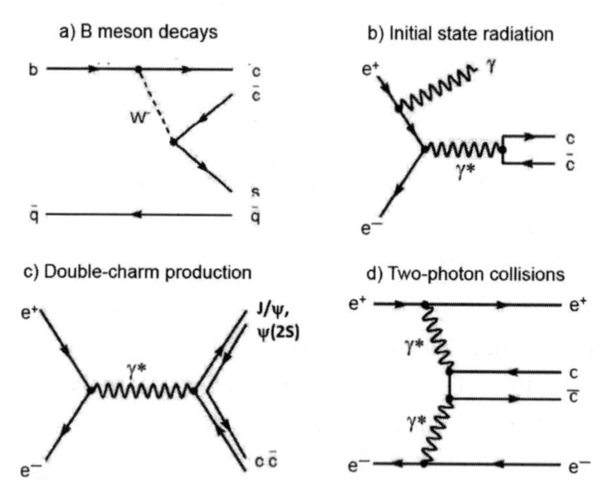

Figure 2: Processes to produce charmonium-like states at *B*-factories; a) *B* decays, b) ISR process, c) double charm production and d) two-photon collision.

3 Status of *XYZ*

Table 1 summarizes the candidate *XYZ* mesons found by now. Among the 15 claimed states, *X*(3872) was first discovered by Belle, and has been studied most extensively. There are three candidates of charged states, *Z*(4430), Z_1(4051), Z_2(4248). There are also candidates for strangeness and bottom counterparts, Y_s(2175) and Y_b(10890). In the following, we present status of X(3872) and *Z*(4430), Z_1(4051), Z_2(4248).

3.1. *X(3872)*

The *X*(3872) was first observed in 2003 by Belle as a narrow peak near 3872 MeV in the $\pi^+\pi^- J/\psi$ invariant mass distribution in $B^- \to K^- \pi^+\pi^- J/\psi$ decays [4]. It was subsequently observed also by CDF [5], D0 [6] and BaBar[7]. The world average values for its mass and width are M = 3872$\pm$0.8 MeV and Γ= 3.0 +2.1/-1.7 MeV [8]. Both Belle [9] and BaBar [10] have reported evidence for the radiative transition $X(3872) \to \gamma J/\psi$, which

State	M(MeV)	Γ(MeV)	J^{PC}	Decay	Production
$Y_s(2175)$	2175 ± 8	58 ± 26	1^{--}	$\phi f_0(980)$	ISR
$X(3872)$	3871.4 ± 0.6	<2.3	1^{++}	$\pi^+\pi^- J/\psi, \gamma J/\psi$	B decay
$X(3875)$	3875.5 ± 1.5	$3.0^{+2.1}_{-1.7}$		$D^0\bar{D}^0\pi^0$	B decay
$Z(3940)$	3929 ± 5	29 ± 10	2^{++}	$D\bar{D}$	Two-photon
$X(3940)$	3942 ± 9	37 ± 17	J^{P+}	$D\bar{D}^*$	Double-charm
$Y(3940)$	3943 ± 17	87 ± 34	J^{P+}	$\omega J/\psi$	B decay
$Y(4008)$	4008^{+82}_{-49}	226^{+97}_{-80}	1^{-+}	$\pi^+\pi^- J/\psi$	ISR
$Z(4051)$	4051^{+24}_{-43}	82^{+51}_{-28}	?	$\pi^+\chi_{c1}$	B decay
$X(4160)$	4156 ± 29	139^{+113}_{-65}	J^{P+}	$D^*\bar{D}^*$	Double-charm
$Z(4248)$	4248^{+185}_{-45}	177^{+320}_{-72}	?	$\pi^+\chi_{c1}$	B decay
$Y(4260)$	4264 ± 12	83 ± 22	1^{--}	$\pi^+\pi^- J/\psi$	ISR
$Y(4350)$	4361 ± 13	74 ± 18	1^{--}	$\pi^+\pi^-\psi'$	ISR
$Z(4430)$	4433 ± 5	45^{+35}_{-18}	?	$\pi^+\psi'$	B decay
$Y(4660)$	4664 ± 12	48 ± 15	1^{--}	$\pi^+\pi^-\psi'$	ISR
$Y_b(10890)$	10889.6 ± 2.3	$54.7^{+8.9}_{-7.6}$	1^{--}	$\pi^+\pi^-\Upsilon(nS)$	$e^+e^-\to Y_b$

Table 1: Summary of the $X\,Y\,Z$ states which have been claimed by now.

indicates that the $X(3872)$ has $C = +$. This implies that the dipion in the $X(3872)\to\pi^+\pi^-$ J/ψ decay has $C = -$, suggesting it originates from a ρ. In fact, the dipion invariant mass distribution measured by CDF is consistent with originating from $\rho\to\pi^+\pi^-$ [11]. The decay of a charmonium state to $\rho\ J/\psi$ would violate isospin, and the evidence for this process provides a strong argument against a charmonium explanation for this state. Angular correlations among the final state particles from $X(3872)\to\pi^+\pi^-\ J/\psi$ rule out all J^{PC} assignments for $X(3872)$ other than $J^{PC} = 1^{++}$ and 2^{-+} [12].

The possibility to interpret $X(3872)$ as a tetraquark state was proposed by Maiani *et al.* [13]. This scenario predicts nearly degenerate states including charged states which have not observed yet. It also predicts slight mass difference between two states produced by charged and neutral B decays. The neutral B decay, $B^0\to X(3872)\ K^0$ has been observed by Belle, and the mass difference is found to be ;

$$\delta M_X = M(X \text{ from } B^+) - M(X \text{ from } B^0) = 0.18 \pm 0.89 \pm 0.26 \text{ MeV},$$

which is consistent with zero, as shown in Figure 3 [14]. This measurement disfavors the tetraquark interpretation.

The molecule explanation seems to be the most likely interpretation of the $X(3872)$, due to its proximity to the D^0D^{*0} threshold at 3871.8 ± 0.3 MeV. The decay

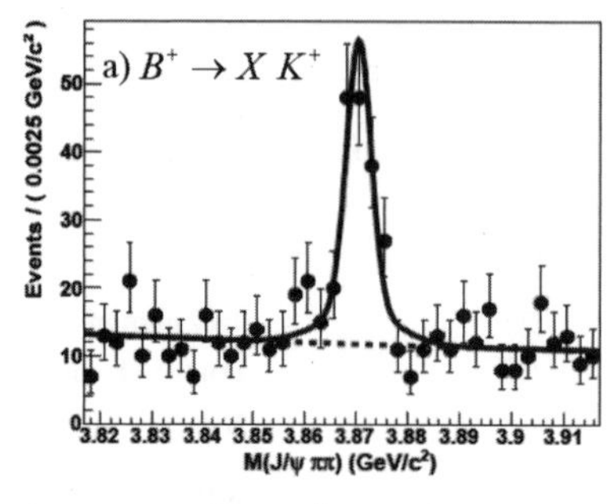

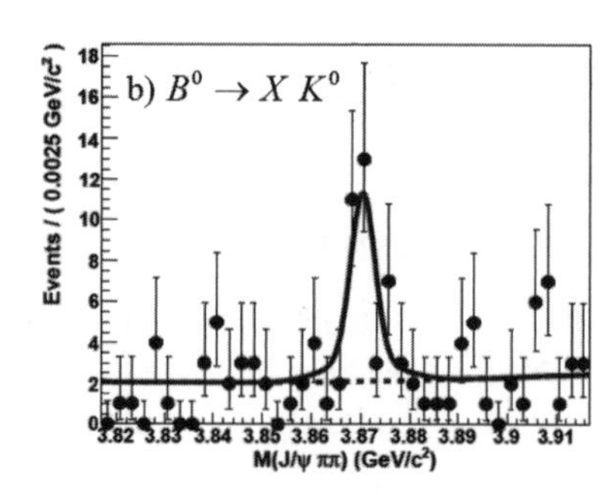

Figure 3: Measured $J/\psi\pi^+\pi^-$ invariant mass distribution from charged (left) and neutral (right) B decays.

$X(3872) \to D^0\bar{D}^0\pi^0$ has been seen by Belle [15] and the decay $X(3872) \to D^0\bar{D}^0\gamma$ by BaBar [16]. These decay implies that the X(3872) decays predominantly via $D^0\bar{D}^{*0}$.

3.2. $Z^+(4430)$, $Z_1^+(4051)$, $Z_2^+(4248)$

The Belle collaboration has claimed three charmonium-like states with an electric charge, which cannot be conventional $c\bar{c}$ states. They require the minimum configuration of $c\bar{c}u\bar{d}$. They represent clear evidence for some sort of multiquark state, either a molecule or tetraquark.

The Z^+(4430) was first observed in 2008 in the $\pi^+\psi'$ invariant mass distribution from the $B \to K\,\pi^+\psi'$decay [17] in the 6.57M $B\bar{B}$ sample. More recently, Belle performed a complete Dalitz analysis, which confirms the Z$^+$(4430) with $M = 4443^{+15+17}_{-12-13}$ MeV and $\Gamma = 109^{+86+57}_{-43-52}$ MeV [18]. The Belle observation was not confirmed by the BaBar analysis using 4.55M $B\bar{B}$ sample [19], although two $\pi^+\psi'$ invariant mass distribution was consistent within statistical errors.

The Belle collaboration has also observed two resonance structures in the $\pi^+\chi_{c1}$ mass distribution from the the $B \to K\pi^+\chi_{c1}$ decay with $M_1 = 4051 \pm 14^{+20}_{-41}$ MeV, $\Gamma_1 = 82^{+21+47}_{-17-22}$ MeV, and $M_2 = 4248^{+44+180}_{-29-35}$ MeV, $\Gamma_2 = 177^{+54+316}_{-39-61}$ MeV [20].

4 *XYZ*-like states in the *s*- and *b*-quark sectors

Having observed so many interesting charmonium-like states, it would be natural to ask whether or not there are counterpart states in the *s*- and *b*-quark sectors. Recent results show that such candidates exist.

In a BaBar study of the ISR process $e^+e^- \to \gamma_{ISR}\,\pi^+\pi^-\phi$, where the $\pi^+\pi^-$ comes from $f_0(980) \to \pi^+\pi^-$, a bump is seen at 2175MeV in the $\pi^+\pi^-\phi$ mass distribution [21]. This bump has been confirmed in an ISR measurement by Belle [22] and seen in $J/\psi \to \eta f_0\phi$ decays by BES [23]. Although a conventional $s\bar{s}$ assignment cannot be ruled out, this state has properties similar to what one would expect for a s-quark counterpart of the Y(4260).

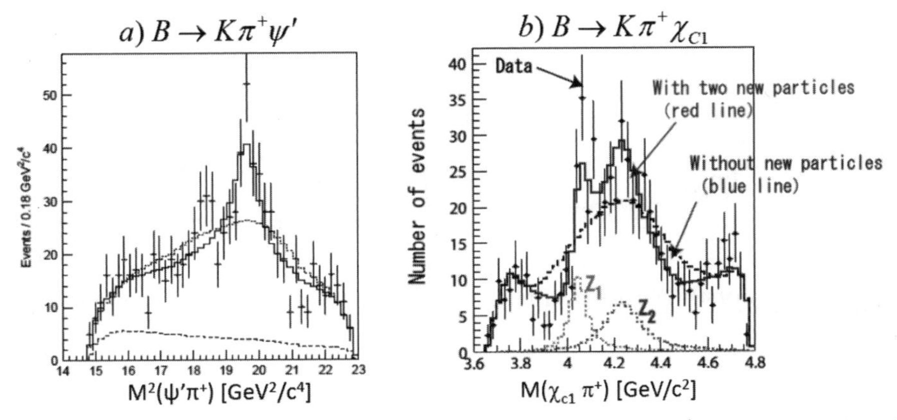

Figure 4: a) The $\pi^+\psi'$ invariant mass distribution from the $B \to K\pi^+\psi'$ decay showing the Z(4430) peak (left). b) The $\pi^+\chi_{c1}$ mass distribution from the the $B \to K\pi^+\chi_{c1}$ decay showing the $Z_1(4051)$ and $Z_2(4248)$ peaks.

The Belle collaboration has reported measurements of the energy dependence of the $e^+e^- \to \pi^+\pi^- Y(\text{nS})$ (n=1,2 and 3) cross section around E_{CM} = 10.9 GeV, and found peaks in all three channels at 10.899 GeV [24]. The mass and width values of the peak are quite distinct from those of the nearby Y(5S) bottomonium ($b\bar{b}$) state, and the cross section values are more than two-orders-of-magnitude above expectations for a conventional ($b\bar{b}$) state, for a conventional $b\bar{b}$ system. One interpretation for this peak is that it is a b-quark sector equivalent of the 1^{--} Y states seen in the c-quark sector [25].

5 Summary

The B factory experiments at KEK (KEKB/Belle) and SLAC (PEP II/BaBar) have observed many new hadrons, which cannot be explained by conventional $q\bar{q}$ meson pictures. The Belle collaboration has observed states with non-zero electric charge, which require the minimum of contents of four quarks $c\bar{c}u\bar{d}$ .There is some evidence for similar structures in the s- and b-quark sectors. More studies on both experimental and theoretical sides are necessary to interpret these states. Both B factories have large data samples at hand, and will continue providing interesting results in coming years. In future, super B-factories will provide opportunities to search for much more states, and study detailed properties (spin, parity, decay modes etc.) of the observed states.

Acknowledgments

This work is partially supported by a Grant-in-Aid for Scientific Research on Priority Areas "Elucidation of New Hadrons with a Variety of Flavors" from the ministry of Education, Culture, Sports, Science and Technology of Japan.

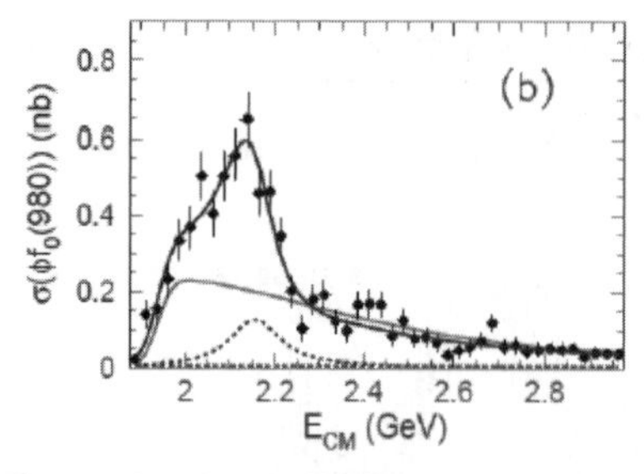

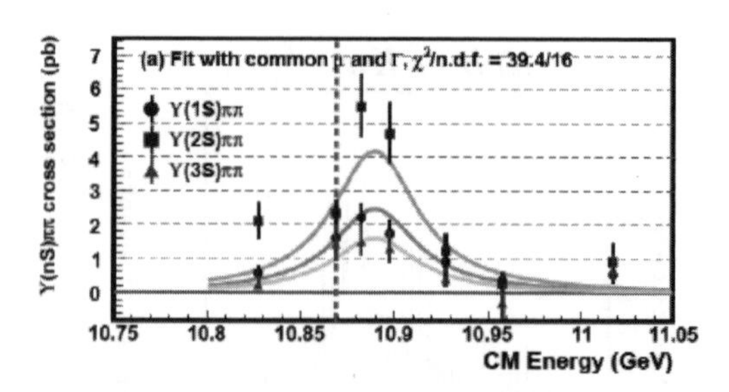

Figure 5: The $f_0(980)$ ϕ mass distribution for $e^+e^- \to \gamma_{ISR}\ \pi^+\pi^-\phi$ (left), and the energy dependence of $e^+e^- \to \pi^+\pi^-\ Y(\text{nS})$ (n=1,2 and 3) cross section around E_{CM} = 10.9 GeV.

References

1. Recent review can be found in, for example, Stephan Godfrey and Stephan L. Olsen, Ann.Rev.Nucl.Part.Sci. **58**, 51 (2008) , available also as arXiv:0801.3867.
2. M. Gell-Mann, Phys. Lett. **8**, 214 (1964).
3. T. Nakano *et al.* (LEPS Collaboration), Phys. Rev. Lett. **91**, 012002 (2003); T. Nakano *et al.* (LEPS Collaboration), Phys. Rev. C **79**, 025210 (2009).
4. S. K. Choi *et al.* (Belle Collaboration), Phys. Rev. Lett. **91**, 262001 (2003).
5. D. E. Acosta *et al.* (CDF II Collaboration), Phys. Rev. Lett. **93**, 072001 (2004).
6. V. M. Abazov *et al.* (D0 Collaboration), Phys. Rev. Lett. **93**, 162002 (2004).
7. B. Aubert *et al.* (BaBar Collaboration), Phys. Rev. D **71**, 071103 (2005).
8. C. Amsler *et al.* (Particle Data Group), Phys. Lett. B **667**, 1 (2008).
9. K. Abe *et al.* (Belle Collaboration), arXiv: hep-ex/0505037.
10. B. Aubert *et al.* (BaBar Collaboration), Phys. Rev. D **74**, 071101 (2006).
11. A. Abulencia *et al.* (CDF Collaboration), Phys. Rev. Lett. **96**, 231801 (2006).
12. A. Abulencia *et al.* (CDF Collaboration), Phys. Rev. Lett. **98**, 132002 (2007).
13. L. Maiani, F. Piccinini, A. D. Polosa and V. Riquer, Phys. Rev. D **71**, 014028 (2005)
14. I. Adachi *et al.* (Belle Collaboration), arXiv: hep-ex/0809.1224.
15. G. Gokhroo *et al.* (Belle Collaboration), Phys. Rev. Lett. **97**, 162002 (2006).
16. B. Aubert *et al.* (BaBar Collaboration), Phys. Rev. D **77**, 011102 (2008).
17. S. K. Choi *et al.* (Belle Collaboration), Phys. Rev. Lett. **100**, 142001 (2008).
18. R. Mizuk *et al.* (Belle Collaboration), Phys. Rev. D **80**, 031104 (2009).
19. B. Aubert *et al.* (BaBar Collaboration), Phys. Rev. D **79**, 112001 (2009).
20. R. Mizuk *et al.* (Belle Collaboration), Phys. Rev. D **78**, 072004 (2008).
21. B. Aubert *et al.* (BaBar Collaboration), Phys. Rev. D **74**, 091193 (2006).
22. I. Adachi *et al.* (Belle Collaboration), arXiv: hep-ex/0808.0008.
23. M. Ablikim *et al.* (BES Collaboration), Phys. Rev. Lett. **100**, 102003 (2008).
24. I. Adachi *et al.* (Belle Collaboration), arXiv: hep-ex/0808.2445.
25. W. S. Hou, Phys. Rev. D **74**, 017504 (2006).

Cold Compressed Baryonic Matter with Hidden Local Symmetry and Holography

Mannque Rho

Institut de Physique Théorique, CEA Saclay, 91191 Gif-sur-Yvette Cédex, France
and
Department of Physics, Hanyang University, 133-791 Seoul, Korea
E-mail: mannque.rho@cea.fr
http://ipht.cea.fr, //hadron.hanyang.ac.kr

I describe a novel phase structure of cold dense baryonic matter predicted in a hidden local symmetry approach anchored on gauge theory and in a holographic dual approach based on the Sakai-Sugimoto model of string theory. This new phase is populated with baryons with half-instanton quantum number in the gravity sector which is dual to half-skyrmion in gauge sector in which chiral symmetry is restored while light-quark hadrons are in the color-confined phase. It is suggested that such a phase that aries at a density above that of normal nuclear matter and below or at the chiral restoration point can have a drastic influence on the properties of hadrons at high density, in particular on short-distance interactions between nucleons, e.g., multi-body forces at short distance and hadrons – in particular kaons – propagating in a dense medium. Potentially important consequences on the structure of compact stars will be predicted.

1. The Issue

Baryonic matter near the nuclear matter density $n_0 \approx 0.16\ \mathrm{fm}^{-3}$ is very well understood, thanks to many years of nuclear experimental and theoretical efforts, but there is a dearth of experimental information beyond n_0. Unlike in high temperature where lattice QCD calculations come in tandem with experiments performed at relativistic heavy-ion colliders, there are no reliable theoretical tools available for dense matter for which lattice gauge calculation suffers from the notorious sign problem. Thus beyond n_0, one knows very little of what's going on. At asymptotically high density, perturbative QCD with weak coupling allows a clear-cut prediction, but the density required there is so high that it is hardly likely to be relevant to the physics of dense matter in terrestrial laboratories or in compact stars. A large number of model calculations are nonetheless available in the literature with a plethora of predictions for nucleon stars, quark stars etc., but the problem with these predictions is that constrained or aided neither by experiments nor by theory, it is difficult to assess the reliability of the calculations with wildly varying results at densities going beyond that of the nuclear matter.

The prospect for the future, however, is pretty good, in particular experimentally. Indeed the forthcoming accelerators devoted to the physics of dense matter, such as FAIR/GSI, will help build realistic phenomenological models just like the ones in standard nuclear physics where the wealth of accurate data played a pivotal role in arriving at realistic nuclear structure models with no guidance from the fundamental theory of strong interactions QCD. It is now established, thanks to the recent development in QCD-based effective field theories of nuclei and nuclear matter, that what the nuclear physicists have been doing was on the right track all along. While awaiting the advent of experimental inputs for going toward dense matter, the challenge for the theorists is: Make predictions, relying solely on theoretical schemes that are as faithful as possible to QCD proper, for the poorly understood regime between the normal nuclear matter density n_0 and the chiral transition density n_χ – at which the restoration of chiral symmetry is expected to take place. In this talk, I would like to describe one such (theoretical) effort being carried out in the context of the World Class University (WCU) Program at Hanyang University in Seoul, Korea, that is anchored on the exploitation of the notion of hidden local symmetry in strong interactions and its holographic generalization following from gauge/string duality. What I will describe is just the beginning. As such, what's predicted at present cannot be immediately confronted with nature. But it will be easily subjected to confirmation or falsification when the data become available.

This talk is not strictly in the main line of this conference where the principal focus is on strong coupling phenomena in the LHC era. My talk concerns rather physics under extreme conditions at a much lower energy scale which will be probed at other laboratories. The basic concept involved, however, such as hidden local symmetry and gauge/string duality, turns out to be closely related to the main theme of this conference.

2. Hidden Local Symmetry (HLS)

2.1. *HLS Lagrangian*

Hidden local symmetry (HLS) in the strong interactions[a] is a flavor gauge symmetry, the effective degrees of which consist of Goldstone bosons (or more realistically pseudo-Goldstone bosons in Nature), i.e., the pions ($\pi = \frac{1}{2}\vec{\tau} \cdot \vec{\pi}$) and the vector fields V_μ valued in $U(2)$ (i.e., $V_\mu = \frac{1}{2}\vec{\tau} \cdot \vec{\rho}_\mu + \frac{1}{2}\omega_\mu$), which elevates the energy scale of the chiral symmetric interactions from the current algebra scale with soft pions to the vector-meson scale $m_V \sim 800$ MeV.

There are two ways to view how hidden gauge symmetry in the strong interactions arises. One is to view the local symmetry as an "emergent" symmetry that exploits the redundancy of local gauge symmetry and the other is to view it as descending (via Kaluza-Klein reduction) from higher dimensions.

In Nature, in addition to the pions and the vector mesons V_μ, there are other light mesonic excitations such as scalar mesons and axial-vector mesons. As for the baryons – that are indispensable for dense matter, they will be generated as

[a]Here I will be focusing mainly on the two-flavor (u,d) case in the chiral limit, i.e., with zero quark mass, but will introduce the strangeness (s) quantum number as needed.

solitons from the HLS Lagrangian, instead of positing them as is done in standard treatments. This approach is most natural from the point of view of large N_c QCD.

Now limited to the minimum number of degrees of freedom, the gauge-invariant Lagrangian takes the simple form[b]

$$\mathcal{L} = \frac{F_\pi^2}{2} \mathrm{Tr}\,\{|D_\mu \xi_L|^2 + |D_\mu \xi_R|^2 + \frac{\gamma}{2}|D_\mu U|^2\} \; - \frac{1}{2}\,\mathrm{Tr}\,[V_{\mu\nu}V^{\mu\nu}] + \cdots \tag{1}$$

where $U = e^{2i\pi/F_\pi} = \xi_L^\dagger \xi_R$, $\xi_L = e^{-i\frac{\pi}{F_\pi}} e^{i\frac{\sigma}{F_\sigma}}$, $\xi_R = e^{i\frac{\pi}{F_\pi}} e^{i\frac{\sigma}{F_\sigma}}$, $D_\mu \xi$ is the covariant derivative, and $\gamma = 1/a - 1$ with $a \equiv F_\sigma^2/F_\pi^2$. In this form, this is an elegant, though highly nonlinear, Lagrangian. However being an effective Lagrangian defined with a cutoff, higher derivative terms at loop orders indicated by the ellipsis in (1) must figure at the quantum level.

In matter-free space, namely, at zero temperature ($T = 0$) and zero density ($n = 0$), the HLS Lagrangian is gauge-equivalent to the standard and well-tested chiral Lagrangian with the pions alone, and as such leads to the same chiral perturbative procedure for pion interactions at very low energy as the standard (pion-only) chiral Lagrangian.[1] At tree order, the hidden local symmetry is of no special power or advantage; gauge fixed to unitary gauge, it will simply reproduce the current algebra results of the standard chiral Lagrangian. The local symmetry, however, picks up its power at higher loop orders since it then renders – in principle – feasible a systematic power counting including the vector mesons, a task that is highly cumbersome if not impossible in a gauge-fixed theory.[2] What is perhaps not properly appreciated among nuclear theorists investigating hadronic matter under extreme conditions (at high T and/or high n) is that hidden local symmetry can incorporate *with ease* certain properties of hadronic matter that involve highly correlated many-body interactions that are difficult to access by models that do not possess flavor local symmetry. The purpose of this talk is to discuss what transpires when HLS in its simplest form and in certain approximations is applied to cold dense matter beyond the nuclear matter density. I will present certain intriguing predictions that differ from standard phenomenological many-body models found in the literature.

It has been shown at one-loop order in matter-free space that the HLS theory matched to QCD (in terms of correlators) flows to a unique fixed point called "vector manifestation" when temperature or density approaches the critical point where chiral restoration takes place[1c]. In the presence of dense matter, the parameters of the Lagrangian governed by renormalization group equations flow as density increases.[d] When matched to the QCD vector and axial vector correlators via the

[b]In what follows, the parametric pion decay constant will be denoted F_π while the physical pion decay constant will be written as f_π.

[c]It is easy to see that this statement should hold to higher loop orders although there is no mathematically rigorous proof. In practice, higher orders are not very useful at present in view of the large number of unknown parameters that arise at higher orders. This means that the results I will show cannot be taken to be quantitatively accurate, even if qualitatively sound.

[d]Unless specifically mentioned otherwise, similar arguments apply to temperature. In contrast to temperature, to take into account the density, fermion degrees of freedom, here baryons, need be introduced. This problem will be addressed below.

in-medium OPE, that is, with the quark condensate $\langle\bar{q}q\rangle$ and the gluon condensate $\langle G^2\rangle$ endowed with background density dependence, the parameters of the HLS Lagrangian (1) tend, as density approaches the chiral transition density n_χ, to

$$g^\star \propto \langle\bar{q}q\rangle^\star \to 0, \ a^\star \equiv F_\sigma^\star/F_\pi^\star \to 1 \tag{2}$$

where the asterisk stands for density dependence. In contrast, there turns out to be nothing special with the flow of the decay constants $F^\star_{\sigma,\pi}$ for varying densities although the physical decay constant $f_\pi^\star$, which receives radiative corrections, should go to zero as $\langle\bar{q}q\rangle^\star$ goes to zero. The consequence of the VM point (2) is that the parametric mass $m_V^\star$ vanishes

$$m_V^\star \approx a^\star F_\pi^\star g^\star \propto g^\star \to 0. \tag{3}$$

Equations (2) and (3) are the essential ingredients of the hidden local symmetry theory with the VM – denoted "HLS/VM" for short – that I will resort to in what follows.

Now the important practical question is: Where and how are these fixed point values manifested? In other words, how can one "see" the presence of these fixed points that are the potential signals for the chiral symmetry structure of the hadronic vacuum? This represents one key question nowadays asked by nuclear physicists.

To answer this question, one should recognize the caveats in applying this Lagrangian to Nature. First of all, the Lagrangian itself is a highly truncated one, with the minimum number of relevant degrees of freedom, so cannot be expected to work accurately for certain processes that require other degrees of freedom. Secondly, to confront experimental observables, dense loop corrections must be calculated in conformity with chiral perturbation theory, what is most important, with the parameters of the Lagrangian *running with density* and satisfying thermodynamic and symmetry constraints. This means that the signals of HLS/VM cannot in general be singled out in a simple manner. They are likely to be compounded with density dependence brought in by dense loop corrections. As in the case of high-temperature processes in relativistic heavy-ion collisions, e.g., dilepton production, a careful weeding-out of "trivial effects" of many-body nature will be required before one can see the desired effects. This issue is discussed in detail in .[3]

2.2. *Gauge-fixed Lagrangian*

I will take the Lagrangian (1) in unitary gauge. It has the form

$$\begin{aligned}\mathcal{L}_{hls} &= \frac{f_\pi^2}{4}\mathrm{Tr}(\partial_\mu U^\dagger \partial^\mu U) - \frac{f_\pi^2}{4} a \mathrm{Tr}[\ell_\mu + r_\mu + i(g/2)(\vec{\tau}\cdot\vec{\rho}_\mu + \omega_\mu)]^2 \\ &\quad -\tfrac{1}{4}\vec{\rho}_{\mu\nu}\cdot\vec{\rho}^{\mu\nu} - \tfrac{1}{4}\omega_{\mu\nu}\omega^{\mu\nu} + \cdots,\end{aligned} \tag{4}$$

with $U = \exp(i\vec{\tau}\cdot\vec{\pi}/f_\pi) \equiv \xi^2$, $\ell_\mu = \xi^\dagger\partial_\mu\xi$, and $r_\mu = \xi\partial_\mu\xi^\dagger$, $\vec{\rho}_{\mu\nu} = \partial_\mu\vec{\rho}_\nu - \partial_\nu\vec{\rho}_\mu + g\vec{\rho}_\mu \times \vec{\rho}_\nu$ and $\omega_{\mu\nu} = \partial_\mu\omega_\nu - \partial_\nu\omega_\mu$. Because of the presence of the isoscalar vector meson ω, there is an additional Lagrangian which is parity-odd

$$\mathcal{L}_{hWZ} = +\tfrac{3}{2} g \omega_\mu B^\mu \tag{5}$$

with the baryon current

$$B^\mu = \frac{1}{24\pi^2} \varepsilon^{\mu\nu\alpha\beta} \mathrm{Tr}(U^\dagger \partial_\nu U U^\dagger \partial_\alpha U U^\dagger \partial_\beta U). \tag{6}$$

This is a part of the gauged Wess-Zumino term that survives for $N_f < 3$ while the topological 5D Wess-Zumino term is absent in the two-flavor case I am discussing. In what follows, it will be called "hWZ term" – "h" standing for homogenous. In general the hWZ Lagrangian has four (homogeneous) terms with arbitrary coefficients but if one imposes vector dominance in the isoscalar channel and consider the ρ meson to be so massive that one can substitute $\rho_\mu^a = i\mathrm{Tr}\,[\tau^a(l_\mu + r_\mu)]$, then it reduces to one term of the form (5).[5e]

2.3. *Describing dense matter*

The Lagrangian $\mathcal{L} = \mathcal{L}_{hls} + \mathcal{L}_{hWZ}$ contains no *explicit* baryon degrees of freedom. However there is a topological baryon current B_μ in (5), so one can think of the ω field as a chemical potential conjugate to the baryon charge. I will parallel this structure below to a holographic dual form in 5D where a similar structure arises.

One might attempt to describe dense baryonic matter with (4) and (5) in the mean field. However this approach simply does not work in the absence of explicit baryon degrees of freedom. One natural way is to generate skyrmions as solitons from the Lagrangian. In principle, there can be soliton solutions with baryon number going from 1 to infinity, the latter corresponding to nuclear matter. Indeed, there have been attempts to build finite nuclei with mass number up to, say, ~ 27 which would correspond to a skyrmion with baryon charge $B = 27$.[6] The resulting theory with skyrmions coupled to fluctuating vector and pion fields, properly quantized, could be expressed in terms of local baryon fields coupled to the mesons satisfying the same symmetry structure as the starting theory. This is in the same spirit as the baryon chiral Lagrangian where local baryon fields are coupled to pions in a chirally invariant away, which is the basis of chiral perturbation theory for pion-nucleon interactions. Given such a baryon chiral Lagrangian, then one can do a mean field calculation for many-body systems that would correspond to what is called Walecka mean field theory in nuclear physics. This strategy, backed by experimental data, can be found to be successful up to the density for which data are available. But how to apply it to a matter beyond the nuclear matter density is

[e]There are two caveats in this. One is that vector dominance is most likely violated with a approaching 1 in hot and dense matter.[4] The second is that the "large ρ mass limit" is at odds with the VM where the vector meson mass goes to zero. These two caveats can be avoided if one works with all the hWZ terms which make the calculation considerably more involved. The merit of the form (5) is that it is consistent with the Chern-Simons term in 5D that I will consider in the holographic approach discussed below.

not known since there have been no experiments that probed the given high density regime. This is because at higher densities, higher dimension operators must enter, which correspond, in the localized form, to rendering the constants of the mean field Lagrangian density-dependent and the density dependence is not known by theory alone.

What is in principle more powerful and closer to QCD proper is to treat many-body systems in terms of the solitonic solution of the Lagrangian. The soliton of winding number A is the solution for an A-body baryonic system. The system with $A = \infty$ will then be a nuclear matter. Nobody has been able to perform an analytic or continuum calculation of such multi-body systems but there have been a variety of calculations putting the skyrmions on a crystal lattice. At high density, one expects the system to be in a crystal and so this is a natural approach to the problem although at lower density nuclear matter is known to be in a liquid than in a crystal. In fact, it has been established that nearly independently of detailed structure of the Lagrangian, there is a phase change from a matter of skyrmions at low density to a matter consisting of half-skyrmions at higher density,[8] and this phase change takes place almost independently of whether the Lagrangian contains local gauge fields or not. The resulting half-skyrmion phase possesses certain scale invariance. What could be different is the critical density at which this takes place, which depends on the parameters of the Lagrangian as well as the degrees of freedom involved.

The crystal structure of dense skyrmion matter with a variety of different ansatze for the skyrmoion configuration is reviewed recently by Park and Vento.[7] For latter purpose, I will restrict myself to the approach[9] based on the Atiyah-Manton ansatz,[10] since it is closely connected to the instanton picture that arises in 5D Yang-Mills Lagrangian of holographic dual QCD.

The Atiyah-Manton ansatz for the chiral field U is given by the holonomy in (Euclidean) time direction

$$U(\vec{x}) = CS\left\{P\exp\left[-\int_{-\infty}^{\infty} A_4(\vec{x},t)dr\right]\right\}C^\dagger \tag{7}$$

where A_4 is the (Euclidean) time component of an instanton field of charge B – which is A for the A-baryon system, P is the path ordering, S is a constant matrix to make U approach 1 at infinity and C denotes an overall $SU(2)$ rotation. The homotopy assures that the static soliton configuration carries the same baryon number as the total charge of the instanton. Note that here YM gauge field A_μ is a fictitious field, not present in the theory (1). In holographic QCD described with a 5D YM field, A_4 is the gauge field present in the 5th direction in the theory.

In,[9] the ansatz (7) is employed in putting A instantons in an FCC crystal to compute the ground state energies as a function of the crystal lattice size L[f]. By shrinking the size L, the multi-skyrmion system is *by fiat* compressed. In Nature,

[f]In FCC, the baryon number density is given by $n = 1/2L^3$.

gravity does the job of squeezing. In Fig. 1 is shown the local baryon number density profile before and after the splitting. These results are obtained with the Skyrme Lagrangian with massive pions but the generic features are the same with the HLS Lagrangian (1). What is important with this phase change is that the half-skyrmion phase has $\langle\bar{q}q\rangle \propto \mathrm{Tr}U = 0$ but $f_\pi \neq 0$ whereas the skyrmion phase has $\langle\bar{q}q\rangle \neq 0$ and $f_\pi \neq 0$, which is symptomatic of chiral symmetry in the Nambu-Goldstone mode. What is novel here is that the chiral symmetry is "restored" whereas there is propagating pion with non-vanishing decay constant. This means that quarks are still confined in that phase although chiral symmetry is "restored." This feature is quite robust. This result implies that in the half-skyrmion phase, the nucleon mass going roughly like $f_\pi^\star$ will not drop appreciably as does the vector meson mass. This will have an important consequence on nuclear physics at high density.

I should mention that a similar half-skyrmion phase structure is found in condensed matter physics (in 3D).[11] There the half-skyrmions get deconfined, that is, they are unbound. In the strong-interaction case at hand, one cannot say whether the half-skyrmions are bound or not [g].

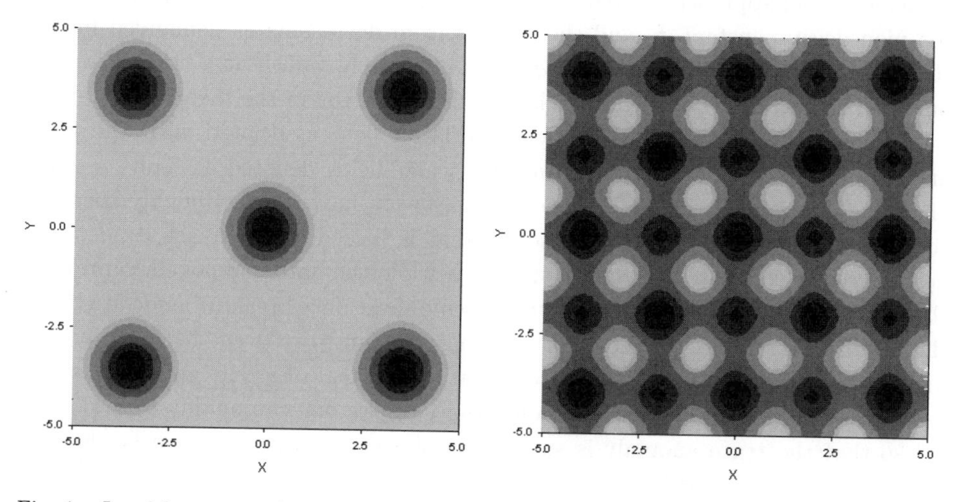

Fig. 1. Local baryon number densities at $L = 3.5$ and $L = 2.0$. For $L = 2.0$ the system is a half-skyrmion (or nearly half-skyrmion with mass) in a CC crystal configuration.

2.4. *Meson fluctuations and dilatons*

Given the background with skyrmions and half-skyrmions, the interesting question is how mesons behave in dense medium. This question has been addressed with the Lagrangian given by Eqs. (4) and (5). What is found there is illuminating: It

[g]In holographic QCD, it appears that the half-instantons are bound although maximally separated by Coulomb repulsion.[12]

indicates that without additional degree(s) of freedom, the VM property of HLS fails to manifest with the unitary gauge Lagrangian in medium.[13] It is found that a dilaton field is rquired.[14]

What happens to meson properties in dense medium is crucially controlled by the hWZ term (5). In mean field in medium, it contributes

$$\mathcal{L}_{hWZ} = +\tfrac{3}{2} g\omega_0(x)n(x) \tag{8}$$

where $n(x)$ is the nuclear matter distribution. Now the ω meson gives rise to a Coulomb potential, so the hWZ term leads to a repulsive interaction. What is most important is that this repulsive interaction turns out to dominate over other terms, and the repulsion diverges as density increases. In order to prevent the energy per baryon coming from the hWZ term from diverging, m_ω^* has to increase sufficiently fast. And since $m_\omega^* \sim f_\pi^* g$ in this model, for a fixed g, f_π^* must therefore *increase.* This is at variance with Nature: QCD predicts that it should *decrease* and go to zero – in the chiral limit – at the chiral transition. The problem here is that the VM property – that $g \to 0$ as density approaches the critical – is missing in this gauge-fixed mean field approach. Stated concisely, the presence of the ω field prevents both the pion decay constant f_π and the ω mass from decreasing as demanded by the HLS/VM. The quark condensate $\langle\bar{q}q\rangle$ does ultimately vanish at a critical density but with an increasing f_π. The ω field thus brings havoc in the theory.

This defect can be remedied by the dilaton field associated with the trace anomaly of QCD[14h]. In,[14] two-component scalar fields denoted as "soft" χ_S and "hard" χ_H satisfying the QCD trace anomaly were introduced. Roughly speaking the soft component represents the scalar that is locked to the quark condensate, hence directly connected to chiral symmetry, while the hard component represents the glue-ball degrees of freedom tied to the scale-invariance breaking associated with the asymptotic running of the color coupling constant g_c. Given that χ_H involves high-energy excitations at a scale much greater than that of χ_S, it can be integrated out for our purpose. Thus retaining only the χ_S field, one can modify Eqs. (1) and (5) so that the trace anomaly is suitably implemented. How this can be done in consistency with the QCD trace anomaly is described in.[16]

---

[h]There is no elegant and simple way known of introducing scalar fields in the HLS theory. What I describe here may not be the most efficient way for HLS *per se* but it is one simple and elegant way for an effective field theory. The role of the dilaton field χ_S in what is referred to as Brown-Rho scaling[15] has been incorrectly understood by numerous authors who have argued against it on theoretical grounds.
Although it is not in the main line of discussions of this talk, it is worth mentioning here that the dilaton introduced therewith resolves one embarrassing feature of the Skyrme model for the nucleon, that is that, with the parameters of the Lagrangian given by the meson sector, the soliton mass comes out much too high, say, ~ 1.5 GeV. When the dilaton is put with a relatively low mass, say, $\lesssim 1$ Gev, then the soliton mass drops to near the physical nucleon mass. The scalar degree of freedom carries a strong attraction that cannot be ignored in the single baryon as well as baryonic matter structure.

With the dilaton included consistently with the scale anomaly, the effective HLS Lagrangian in unitary gauge is

$$\mathcal{L} = \mathcal{L}_{hls\chi} + +\mathcal{L}_{hWZ} \tag{9}$$

where

$$\mathcal{L}_{HLS\chi} = \frac{f_\pi^2}{4}(\chi/f_\chi)^2 \mathrm{Tr}(\partial_\mu U^\dagger \partial^\mu U) - \frac{f_\pi^2}{4} a(\chi/f_\chi)^2 \mathrm{Tr}[\ell_\mu + r_\mu + i(g/2)(\vec{\tau}\cdot\vec{\rho}_\mu + \omega_\mu)]^2$$
$$-\tfrac{1}{4}\vec{\rho}_{\mu\nu}\cdot\vec{\rho}^{\mu\nu} - \tfrac{1}{4}\omega_{\mu\nu}\omega^{\mu\nu} + \tfrac{1}{2}\partial_\mu\chi\partial^\mu\chi + V(\chi) \tag{10}$$
$$\mathcal{L}_{hWZ} = \tfrac{3}{2} g(\chi/f_\chi)^n \omega_\mu B^\mu \tag{11}$$

with $\chi \equiv \chi_S$ where f_χ is the χ decay constant and $V(\chi) = B\chi^4 \ln\frac{\chi}{f_\chi e^{1/4}}$ is the potential that gives the energy-momentum tensor whose trace gives the soft component of the trace anomaly. The exponent n which cannot be fixed by first principles can be arbitrary with $n > 1$ without violating scale symmetry.[16] In the analysis of,[14] $n = 3$ was taken for simplicity, but $n = 2$ gives an equally satisfying result at least qualitatively.

The lesson from this model is that the dilaton field plays a crucial role in making HLS theory compatible with nature. It makes both the ω meson mass and the pion decay constant decrease as density increases in consistency with HLS/VM. An additional important consequence of the dilaton is that it reveals a novel phase with half-skyrmions as the relevant degrees of freedom where the quark condensate vanishes, $\langle\bar{q}q\rangle = 0$, while $f_\pi \neq 0$. This phase appears at a density $n_{1/2}$ above the nuclear matter density n_0 but below the chiral transition density n_χ. This phase is characterized by the restored chiral symmetry but with quarks *confined.* At what density this phase appears depends on the parameters of the Lagrangian that defines the background but *not* on the effective mass of the dilaton.

3. Holographic Dense Matter

Before going to describing the predictions of the new structure uncovered in HLS/VM, let me describe the approach taken from the gravity dual of dense matter in a holographic QCD model. There is a huge literature on the approach to dense and hot matter from the AdS/CFT angle which I cannot possibly review here. Let me focus uniquely on a particular holographic QCD model that is closely related to what is described above in HLS/VM, i.e., the Sakai-Sugimoto model.[17] This is the only model that I know of that has the chiral symmetry of QCD in the chiral limit. It describes well – in the large N_c and large 't Hooft constant λ limit where the duality should be reliable and in the probe approximation with $N_f/n_c \ll 1$ – both meson properties[17] and baryon, specially nucleon, properties[18] which are known to be well described in the quenched approximation in lattice calculations. What is particularly relevant to this discussion is that the model gives a fairly accurate description of nucleon static properties and provides the first "derivation" of the vector dominance of the nucleon form factor as emphasized in.[19] The latter resolves a long-standing puzzle why Sakurai's vector dominance is strongly violated in the nucleon form factors while it works very well in the meson form factors[i].

3.1. *Sakai-Sugimoto model*

The Sakai-Sugimoto action[17] valid in the large N_c and λ limit and in the probe (or quenched) approximation can be reduced to the form

$$S = S_{YM} + S_{CS} \tag{12}$$

where

$$S_{YM} = -\int dx^4 dw \, \frac{1}{2e^2(w)} \, \mathrm{tr} F_{mn} F^{mn} + \cdots , \tag{13}$$

with $(a, b) = 0, 1, 2, 3, w$ is the 5D YM action coming from DBI action where the contraction is with respect to the flat metric $dx_\mu dx^\mu + dw^2$. The position-dependent electric coupling $e(w)$ is given in terms of N_c, λ and the Kaluza-Klein mass M_{KK} (see Hong et al in[18]) and

$$S_{CS} = \frac{N_c}{24\pi^2} \int_{4+1} \omega_5(A) \tag{14}$$

with $d\omega_5(A) = \mathrm{tr} F^3$ is the Chern-Simons action that encodes anomalies. There are two important features in this hQCD model. One is that an action of the same form arises as an "emergent" gauge action from low-energy chiral Lagrangians with the 5th dimension "deconstructed".[2] The other is that when viewed in 4D by Kaluza-Klein reduction, it contains an infinite tower of both vector and axial-vector mesons. Now with the low-energy strong interaction dynamics given in the form (12), baryons arise as instantons from S_{YM} whose size shrinks to zero as $\sim \lambda^{-1/2}$ as $\lambda \to \infty$.[18] The shrinking to zero size of the instanton is prevented by the Chern-Simons term which takes the form in the presence of the instanton

$$S_{CS} = \frac{N_c}{8\pi^2} \int \hat{A}_0 \mathrm{tr}(F \wedge F) \tag{15}$$

where $\hat{A}$ is the $U(1)$ gauge field (analogous to the abelian gauge field, ω, in HLS). The Chern-Simons term provides a Coulomb repulsion that stabilizes the instanton just as the ω field does to the skyrmion in HLS.

When applied to dense matter, the instanton density distribution $\mathrm{tr}(F \wedge F)$ in (15) is nothing but the baryon number density distribution and hence $\hat{A}_0$ can be interpreted as a variable conjugate to density, i.e., the baryon chemical potential μ_B.

One can have a simple but instructive idea of what this theory contains for dense matter by noting the close parallel between (4) in 4D and (13) in 5D and between

[i] I should point to the remarkable observation by Hashimoto *et al.* in[18] that the form factor of the nucleon vector-dominated by the infinite tower of vector mesons can be re-expressed by a dipole form as found empirically!

(5) in 4D and (15) in 5D. The instanton in (13) stabilized by the $\hat{A}_0$ field in the CS term (15) is an analog of the skyrmion in (4) stabilized by the ω field in (5). This suggests that if the action (12) is treated in mean field for the point-like instantons as baryons balanced by the Coulomb force given by the CS term, there will be the same defect found with the HLS theory without the dilaton degree of freedom that *simulates* the HLS/VM property. One expects that the pion decay constant in this holographic theory will *increase* instead of decrease with increasing density. Indeed this was found in a recent analysis of the action (12) in cold dense matter.[20]

3.2. *Half-instanton/dyonic phase*

There is at present no quantitative work in this holographic QCD model of the sort described above for the crystal structure in HLS/VM. But one can reveal a phase that is a gravity dual to the half-skyrmion phase.

Given that the action (12) can support instanton solutions for many-baryon systems with the instanton number A, the pertinent question one can ask vis-à-vis with dense matter is what happens to the instanton at high density. This question was recently answered with the discovery that there can arise a half-instanton matter that is in a dyonic salt form in which chiral symmetry is restored while the baryons are color-singlet hadrons, very much like in the half-skyrmion matter. In fact, the two half-instantons fractionized from one instanton are found to be bound in a bcc configuration, separated by Coulomb repulsion. There is no quantitative description of the dyonic salt configuration at the moment but one can make a plausible qualitative argument using geometry, which, I believe, is robust.[12]

The basic idea is that the flavor instanton in the action (12) consists of a superposition of constituents that are BPS dyons in the leading $N_c\lambda$ order. In close analogy to colorons[21,22] with the colored instanton splitting into constituent dyons as a mechanism for color deconfinement where the Polyakov line plays the central role, here it is also a nontrivial holonomy that works to split the flavor instanton into BPS dyons of opposite charges $e = g = \pm 1/2$. In,[12] geometric reasoning is used for the splitting phenomenon and a dynamical mechanism with topological repulsion balancing the Coulomb forces mediated by the charges of the constituents is invoked for the half-and-half non-BPS structure. It is estimated that the splitting occurs at a density comparable to what was observed in the skyrmion-half-skyrmion transition[7j]. Also the binding energy between the two dyons of opposite charges comes out roughly about 180 MeV, similar to the binding of half-skyrmions in the gauge model.[7]

The arguments leading to the instanton-dyon phase transition is admittedly heuristic and needs to be supported, in the absence of analytical method, by nu-

[j]In the skyrmion case, the transition density $n_{1/2}$ was found to be sensitively dependent on the parameters of the Lagrangian. For the parameters fixed in the meson sector, it comes out to be somewhere between ~ 1.3 times and ~ 3 times the nuclear matter density. This range is roughly what is found in this hQCD model.

merical work. But there is no reason to think it is not qualitatively correct. It is easy to understand that with the action (12), were it not for the warping in the conformal direction w in the charge $e(w)$, the constituents of the instanton would be BPS dyons, and hence there would be no reason why the instanton should split into two equal $1/2$ instantons. It is the interaction encoded in the curved geometry that turns the BPS dyons with the instanton charge partitioned in v and $1-v$ with v arbitrary into the half-and-half configuration. In this connection, what is highly relevant is the observation made by Hashimoto et al that at short distances, the curvature effect can be ignored and many-body forces between instantons get suppressed.[23] In the standard nuclear physics picture, short-range repulsion in three-body forces (ignoring four-body forces) is understood to be responsible for pushing phase transitions (e.g., kaon condensation, chiral restoration etc.) in nuclear matter to much higher densities. If holographic duality were to hold in this phenomenon, this feature of many-body forces in the gravity description would imply in the gauge sector that the many-body repulsion problematic in standard nuclear theory is simply absent at high density. This would have an extremely important implication on the EOS of compact stars. In HLS, many-body forces are also suppressed by the vector manifestation of the vanishing hidden gauge coupling, thus making a valuable prediction that is inaccessible to QCD proper. The two predictions are qualitatively the same although they seem to involve different mechanisms, RG flow matched to QCD for the gauge sector and the infinite tower encoded in the instanton structure in the gravity sector. It would be interesting to "see" the connection if there is any.

4. Signals of Half-Skyrmion/Dyonic Phase

The search for *direct* signals of the HLS/VM has been mainly focused on dilepton production in relativistic heavy ion collisions and in EM interactions. The former is for high temperature matter and the latter precision measurement in finite nuclei at nuclear matter density and at zero temperature. The expectation was that via vector dominance, dileptons will be profusely produced from the ρ vector meson in medium with its mass shifted downwards due to the HLS/VM. The searches so far made, however, came out more or less negative, that is, the mass shift associated with the HLS/VM has not been observed. It should, however, be pointed out that this by no means implies that the HLS/VM prediction is absent in nature. In fact, one of the most remarkable – and unexpected – predictions of HLS/VM, which is found in none of the treatments by the theorists in the field, is that vector dominance is maximally violated as $a \to 1$ and $g \to 0$, and the dileptons, ironically, tend to largely decouple from the ρ vector meson, so the observed dileptons carry practically no information on chiral symmetry properties. In fact, what's observed in the ρ spectral function are predominantly mundane nuclear effects due to many-body interactions and do not exhibit clear-cut signals for the chiral structure of the vacuum change that one wants to isolate. The same suppression mechanism takes place both in T and in density. This could indicate that singling out of a "smoking-gun" signal for the manifestation of chiral symmetry in dense (and hot) medium at the forthcoming laboratory FAIR/GSI will not be an easy task.

Let me turn to a case which appears to be more promising, i.e., the property of kaons in cold dense matter. Particularly interesting is the behavior of negatively charged kaons (i.e., anti-kaons) fluctuating in the medium described as a dense solitonic background. This concerns two very important issues in physics of dense hadronic matter. One is the possibility that K^- can trigger strongly correlated mechanisms to compress hadronic matter in finite nuclei to high density as proposed by Akaishi et al[24k]. The other issue is the role of kaon condensation in dense compact star matter which has ramifications on the minimum mass of black holes in the Universe and cosmological natural selection.[26] It would be of the greatest theoretical interest to address this problem by means of the instanton matter given in hQCD, which has the potential to also account for shorter-distance degrees of freedom via an infinite tower of vector and axial vector mesons. However, no numerical works in this framework are presently available. I shall therefore take the simple Skyrme model with the vector field V_μ integrated out, implemented with three important ingredients, viz, the Wess-Zumino term, the soft dilaton field χ that accounts for the scale symmetry tied to spontaneously broken chiral symmetry as explained above and the kaon mass term. Though perhaps oversimplified, I expect it to capture the generic qualitative feature of the novel phenomenon.

The objective is to see how kaons behave on top of the background provided by the skyrmion matter as density is raised. I will assume that the back-reaction of the kaon fluctuation on the background is ignorable. The basic idea is to treat the kaon as "heavy," not light as in chiral perturbation theory, and wrap it with the skyrmion in the way that Callan and Klebanov did for hyperons.[27] The appropriate ansatz is [l]

$$U(\vec{x},t) = \sqrt{U_K(\vec{x},t)}U_0(\vec{x})\sqrt{U_K(\vec{x},t)}, \tag{16}$$

where U_K and U_0 are, respectively, the fluctuating kaon (doublet) field and the $SU(2)$ soliton field.

The results of this model recently worked out in[29] are summarized in Fig. 2. The mass of the dilaton that figures importantly in this approach is presently unknown. What's taken here are two values, one corresponding to the lowest observed scalar mass $\sim$ 600 MeV and the other a value that is used in mean-field theory calculations of nuclear matter using an effective Lagrangian that implements Landau Fermi liquid theory incorporating HLS/VM. Noteworthy in these results is that up to

[k]I must mention that there is a considerable controversy on this matter, pros and cons to the Akaishi-Yamazaki idea. For the most recent summary of both theoretical and experimental status, see.[25] Ultimately experiments will be the jury, and the verdict will have to await the results of those experiments that are being – and will be – performed.

[l]This ansatz is more convenient for treating kaon fluctuations than the one taken by Callan and Klebanov where the kaon field is sandwiched by the square-root of the soliton field. The two ansatze are found to give an equally good description of the hyperons.[28]

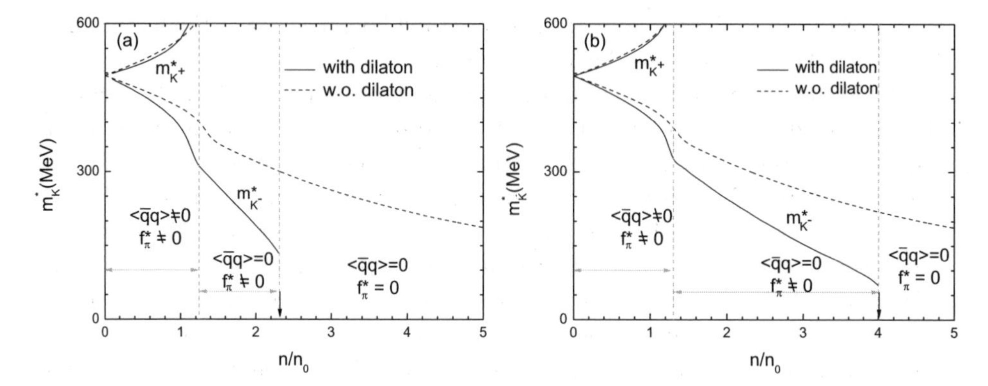

Fig. 2. $m^*_{K^\pm}$ vs. n/n_0 in dense skyrmion matter that contains three phases as indicated, with the middle being the half-skyrmion phase. The parameters are fixed at $f_\pi = 93$ MeV, $\sqrt{2}ef_\pi = m_\rho = 780$ MeV and dilaton mass $m_\chi = 600$ MeV (left panel) and 720 MeV (right panel).

the threshold of the half-skyrmion phase, the antikaon mass behaves more or less according to chiral perturbation theory;[25] however once the half-skyrmion phase sets in, the kaon mass drops more steeply, going to zero at about the same density as the chiral transition density $n_\chi \sim (2.3-4)n_0$. This strong attraction that results from the combination of the topological WZ term *and* the dilaton field is most probably inaccessible in standard chiral perturbation theory, and hence missing from the chiral perturbation treatments in the literature. This additional attraction could help lead toward the Akaishi-Yamazaki scenario of deeply bound kaonic nuclei. Implementing the mechanism found here in finite nuclei is not an easy task that needs to be worked out.

Another physically interesting question is what does the half-skyrmion matter do to compact stars? What we have done above is to subject the kaon to the skyrmion background given in the large N_c limit. In the large N_c limit, there is no distinction between symmetric matter and asymmetric matter. Compact stars have neutron excess which typically engenders repulsion at densities above n_0, and hence the asymmetry effect needs to be taken into account. This effect will arise when the system is collective-quantized, which gives rise to the leading $1/N_c$ correction to the energy of the bound system. Most of this correction could be translated into a correction to the effective mass of the kaon, so the rapid dropping of the kaon will have an important impact on the physics of compact stars.

Acknowledgments

I am grateful for the invitation by Koichi Yamawaki to give this talk. I would also like to acknowledge delightful collaborations with Hyun Kyu Lee, Byung-Yoon Park, Sang-Jin Sin and Ismail Zahed. This work was partly supported by the WCU project of Korean Ministry of Education, Science and Technology (R33-2008-000-10087-0).

References

1. M. Harada and K. Yamawaki, *Phys. Rept.* **381**, 1 (2003).
2. H. Arkani-Hamed, H. Georgi and M.D. Schwartz, *Ann. Phys.* **305**, 96 (2003).
3. G. E. Brown et al, *Prog. Theor. Phys.* **121**, 1209 (2009); M. Rho, "Dileptons get nearly 'blind' to mass-scaling effects in hot and/or dense matter," arXiv:0912.3116 [nucl-th].
4. M. Harada and C. Sasaki, *Nucl. Phys.* A **736**, 300 (2004).
5. U.-G. Meissner, *Phys. Rept.* **161**, 213 (1988).
6. R.A. Battye, N.S. Manton and P.M. Sutcliffe, "Skyrmions and nuclei," in *The Multifaceted Skyrmion* (World Scientific 2010), ed. G.E. Brown and M. Rho.
7. B.-Y. Park and V. Vento, "Skyrmion approach to finite density and temperature," in *The Multifaceted Skyrmion* (World Scientific 2010), ed. G.E. Brown and M. Rho.
8. A. S. Goldhaber and N. S. Manton, *Phys. Lett.* B **198**, 231 (1987); A. D. Jackson and J. J. M. Verbaarschot, *Nucl. Phys.* A **484**, 419 (1988); M. Kugler and S. Shtrikman, *Phys. Rev.* D **40**, 3421 (1989).
9. B. Y. Park, D. P. Min, M. Rho and V. Vento, *Nucl. Phys.* A **707**, 381 (2002).
10. M. F. Atiyah and N. S. Manton, *Phys. Lett.* B **222**, 438 (1989).
11. See, e.g., K. Moon, "Skyrmions and merons in bailayer quantum Hall systems" and T. Senthil et al, "Deconfined quantum critical points" in *The Multifaceted Skyrmion* (World Scientific 2010), ed. G.E. Brown and M. Rho.
12. M. Rho, S. J. Sin and I. Zahed, "Dense QCD: A holographic dyonic salt," arXiv:0910.3774 [hep-th].
13. B. Y. Park, M. Rho and V. Vento, *Nucl. Phys.* A **736**, 129 (2004).
14. B. Y. Park, M. Rho and V. Vento, *Nucl. Phys.* A **807**, 28 (2008).
15. G. E. Brown and M. Rho, *Phys. Rev. Lett.* **66**, 2720 (1991).
16. H. K. Lee and M. Rho, *Nucl. Phys.* A **829**, 76 (2009).
17. T. Sakai and S. Sugimoto, *Prog. Theor. Phys.* **113**, 843 (2005); *Prog. Theor. Phys.* **114**, 1083 (2005).
18. D. K. Hong, M. Rho, H. U. Yee and P. Yi, *Phys. Rev.* D **76**, 061901 (2007). H. Hata, T. Sakai, S. Sugimoto and S. Yamato, *Prog. Theor. Phys.* **117**, 1157 (2007); K. Hashimoto, T. Sakai and S. Sugimoto, *Prog. Theor. Phys.* **120**, 1093 (2008); K. Y. Kim and I. Zahed, *JHEP* **0809**, 007 (2008).
19. D. K. Hong, M. Rho, H. U. Yee and P. Yi, *Phys. Rev.* D **77**, 014030 (2008).
20. K. Y. Kim, S. J. Sin and I. Zahed, *JHEP* **0801**, 002 (2008).
21. T. C. Kraan and P. van Baal, *Nucl. Phys.* B **533**, 627 (1998); *Phys. Lett.* B **435**, 389 (1998).
22. D. Diakonov, "Topology and confinement," arXiv:0906.2456 [hep-ph].
23. K. Hashimoto, N. Iizuka and T. Nakatsukasa, "N-body nuclear forces at short distances in holographic QCD," arXiv:0911.1035 [hep-th].
24. Y. Akaishi, A. Dote and T. Yamazaki, *Phys. Lett.* B **613**, 140 (2005).
25. A. Gal, "Overview of strangeness nuclear physics," arXiv:0904.4009 [nucl-th]; W. Weise, "Antikaon interactions with nucleons and nuclei," arXiv:1001.1300 [nucl-th].
26. G. E. Brown, C. H. Lee and M. Rho, *Phys. Rev. Lett.* **101**, 091101 (2008).
27. C.G. Callan and I. Klebanov, *Nucl. Phys.* **B262**, 365 (1985).
28. E. M. Nyman and D. O. Riska, *Rept. Prog. Phys.* **53**, 1137 (1990).
29. B. Y. Park, J. I. Kim and M. Rho, "Kaons in dense half-skyrmion matter," arXiv:0912.3213 [hep-ph].

Aspects of Baryons in Holographic QCD

T. Sakai

Department of Physics, Nagoya University
Nagoya 464-8602, Japan
E-mail: tsakai@eken.phys.nagoya-u.ac.jp

We discuss some recent results on a holographic description of QCD on the basis of a D4-D8-$\overline{\text{D8}}$-brane configuration in type IIA superstring theory. A special attention is paid to studies of the baryon dynamics in that model. Using the fact that baryons in large N_c QCD should be regarded as a soliton, it is seen that this model reproduces the experimental data quite nicely.

Keywords: Gauge/string duality, holographic QCD.

1. Introduction

It has been now realized that the gauge/string correspondence provides us with a powerful machinery for analyzing the strong coupling dynamics of gauge theory. Initiated by the pioneering work by Maldacena (for a review, see Ref. 1), there have been lots of attempts to apply this idea to quantum chromodynamics(QCD). In Ref. 2, we proposed a model on the basis of a D4-D8-$\overline{\text{D8}}$-brane configuration in type IIA superstring theory. This model defines a string dual of large N_c QCD with N_f massless flavor quarks. We reach the conclusion that the low energy dynamics of the hadron physics is described by a five-dimensional Yang-Mills-Chern-Simons(YM-CS) gauge theory with the gauge group $U(N_f)$:

$$S = S_{\text{YM}} + S_{\text{CS}} , \tag{1}$$

$$S_{\text{YM}} = -\kappa \int d^4x dz \ \text{tr} \left[\frac{1}{2} h(z) \mathcal{F}_{\mu\nu}^2 + k(z) \mathcal{F}_{\mu z}^2 \right] , \tag{2}$$

$$S_{\text{CS}} = \frac{N_c}{24\pi^2} \int_{M^4 \times \mathbb{R}} \omega_5^{U(N_f)}(\mathcal{A}) = \frac{N_c}{24\pi^2} \int_{M^4 \times \mathbb{R}} \text{tr} \left(\mathcal{A}\mathcal{F}^2 - \frac{i}{2}\mathcal{A}^3\mathcal{F} - \frac{1}{10}\mathcal{A}^5 \right) , \tag{3}$$

with

$$\kappa = \frac{\lambda N_c}{216\pi^3} , \quad \lambda = g_{\text{YM}}^2 N_c , \quad h(z) = (1+z^2)^{-1/3} , \quad k(z) = 1 + z^2$$

From Kaluza-Klein(KK) reduction of the five-dimensional gauge potential along the z direction, this model yields not only the massless pion associated with spontaneous chiral symmetry breaking, but also an infinite tower of massive vector mesons such as $\rho, a_1, \cdots$. It is found that this model reproduces many quantitative aspects of the hadron physics by tuning two "free parameters" λ and M_{KK}. The latter is the only dimensionful parameter of our model, and we mostly work in units of $M_{\text{KK}} = 1$. Moreover, it is worth emphasizing that the YM part of the action (1) is equal to a five-dimensional gauge theory that are constructed from hidden local symmetry (for a review, see Ref. 3) with an infinite number of hidden local gauge group incorporated, each of which corresponds to a massive vector meson.[4]

Contrary to the holographic description of the mesons as KK modes, baryons in our model can be described as a soliton.[5] This idea has been well-known in large N_c QCD,[6] and in fact, was first applied to the Skyrmion for making a quantitative analysis of baryons.[7] Compared to the studies that have been done so far, Ref. 5 has an advantage in that our model allows us to construct a soliton solution that takes into account the effects from an infinite number of massive mesons as well as the massless pion. In fact, we can find out a systematic framework for describing negative parity baryons.

The main purpose of this contribution is to present recent developments in the holographic description of baryons that are obtained in Refs. 5, 8, 9. In the next section, we give a brief review of a holographic description of large N_c QCD on the basis of an intersecting D-brane configuration in type IIA superstring theory. A focus is given on how to obtain the currents associated with the chiral flavor symmetry $U(N_f)_L \times U(N_f)_R$. In section 3, we solve a soliton solution corresponding to a baryon. It is argued that a YM instanton of a fixed size plays a central role. Using these results, we work out the baryon form factors and then compare these results with experiments. We then sketch a computation of the short distance behavior of nuclear force using our model. We conclude this paper with summary in section 4.

2. D-branes and holographic QCD

We start with the following brane configuration in type IIA superstring[2]

	0	1	2	3	(4)	5	6	7	8	9
N_c D4	∘	∘	∘	∘	∘					
N_f D8-$\overline{\text{D8}}$	∘	∘	∘	∘		∘	∘	∘	∘	∘

(4)

Here x_4 is compactified as $x_4 \sim x_4 + 2\pi/M_{\text{KK}}$.

N_c D4-branes are wrapped around S^1 with the antiperiodic boundary condition imposed on the world volume fermions. These define four-dimensional $SU(N_c)$ pure YM theory with the UV cut-off given by M_{KK}, with the bare coupling constant at M_{KK} equal to $\lambda = g_{\text{YM}}^2 N_c$. The weak coupling description of the YM theory is valid for $\lambda \ll 1$. N_f D8-$\overline{\text{D8}}$-branes are responsible for the chiral flavor symmetry $U(N_f)_L \times U(N_f)_R$: it is found that open strings that extend from D4-branes to D8- and $\overline{\text{D8}}$-branes lead to the chiral fermions q_L and q_R, respectively. We thus obtain QCD with massless flavor quarks.

When $\lambda \gg 1$, the weak coupling description of QCD is not valid any more. For the purpose of understanding the strong coupling phenomena, the gauge/string duality is useful. The point here is that D-branes allow two different description of

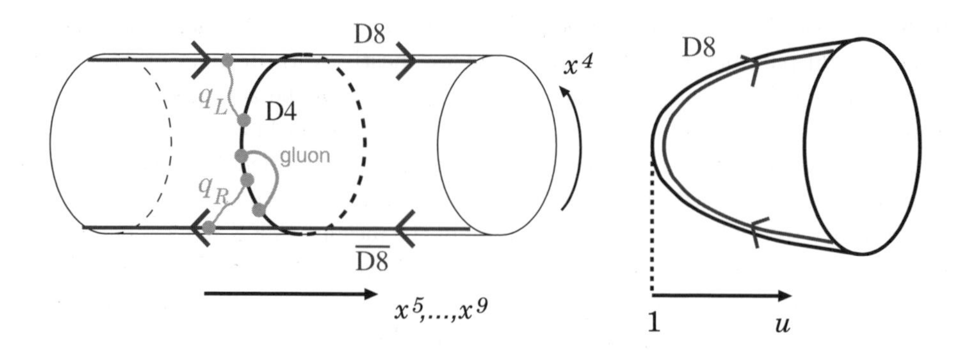

Fig. 1. D4-D8-$\overline{\text{D8}}$-brane configuration

a common physics. One is the open string picture, where D-branes can be identified with a hypersurface on which open strings can end. The other is to regard D-branes as a source of gravity so that they create a curved background as a solution of type IIA supergravity(SUGRA). The SUGRA solution of the D4-branes is obtained by Witten.[10] This is reliable when $N_c \to \infty$ with $\lambda \gg 1$, which ensures that the SUGRA approximation of the D4-brane solution is reliable. Now add to this background the D8-$\overline{\text{D8}}$-brane pairs. It is in general highly difficult to find a SUGRA solution of intersecting D-branes. To evade this problem, we work in the probe approximation $N_f \ll N_c$, for which the added D8-$\overline{\text{D8}}$ pairs can be treated as a probe with their backreaction into the D4-background neglected. From the gauge theory point of view, this is identical to neglecting the quark loop effects. As shown in Ref. 2, the probe D8-$\overline{\text{D8}}$ pairs must be interpolated with each other smoothly because of the curved D4 background. We emphasize that this is a manifestation of spontaneous chiral symmetry breaking $U(N_f)_L \times U(N_f)_R \to U(N_f)_V$. It can be seen that the embedding of the smoothly connected D8-branes into the Witten's D4 geometry is specified by $y = 0$, such that the two-dimensional cigar-like geometry appearing at Fig. 1 is mapped to a two-dimensional plane spanned by y, z. Then, the world volume coordinate of the D8-branes is given by x^μ, z and S^4. By taking into account only the KK zero mode along the S^4, we end up with the five-dimensional YM-CS theory given in (1).

Let us first discuss how the meson degrees of freedom arise from the five-dimensional gauge potential. To this end, we expand it in terms of a complete set of the wavefunctions of z:

$$\mathcal{A}_\mu(x,z) = \sum_{n\geq 1} B_\mu^{(n)}(x)\,\psi_n(z)\;,$$

$$\mathcal{A}_z(x,z) = \sum_{n\geq 1} \varphi^{(n)}(x)\phi_n(z) + \varphi^{(0)}(x)\phi_0(z)\;, \tag{5}$$

where

$$-k^{1/3}\partial_z(k\,\partial_z\psi_n)=\lambda_n\psi_n\ ,\quad \kappa\int dz k^{-1/3}\,\psi_n\psi_m=\delta_{nm}\ ,\quad \phi_n\propto\partial_z\psi_n\ . \tag{6}$$

Then, the four-dimensional gauge field $B_\mu^{(n)}(x)$ can be identified with the vector mesons, and $\varphi^{(0)}(x)$ as the massless pion. The nonzero mode of $\varphi^{(n)}(x)$ are eaten up by $B_\mu^{(n)}(x)$ to gain a nonvanishing mass. It is important that we can assign P and C quantum numbers of the mesons uniquely, because these symmetries originate from a discrete symmetry of the D4-D8-$\overline{\text{D8}}$-brane system. For details, see Ref. 11. The mass spectrum of the mesons is summarized as

	$\varphi^{(0)}$	$v_\mu^n=B_\mu^{(2n-1)}$	$a_\mu^n=B_\mu^{(2n)}$
J^{PC}	0^{-+}	1^{--}	1^{++}
mass^2	0	$\lambda_{2n-1}M_{\text{KK}}^2$	$\lambda_{2n}M_{\text{KK}}^2$

Next, we discuss spontaneous breaking of chiral flavor symmetry. Note that the field profile in (5) is left unchanged under the constant gauge transformation acting at $z=\pm\infty$:

$$g_\pm=g(z=\pm\infty)\ , \tag{7}$$

because the wavefunctions ψ and φ are normalizable so that they dump rapidly enough as z grows. We interpret $g_\pm$ as the elements of chiral flavor symmetry group $U(N_f)_{L,R}$. This flavor symmetry can be gauged by introducing the corresponding gauge potential $\mathcal{A}_{L,R}^\mu$, which should be identified with the gauge potential $\mathcal{A}(x,z)$ at $z=\pm\infty$:

$$\mathcal{A}^\mu(x,z\to\pm\infty)=\mathcal{A}_{L,R}^\mu(x)\ . \tag{8}$$

This can be incorporated by using the nonnormalizable zero mode $\psi_0(z)$. Using the standard AdS/CFT dictionary, we find that the chiral flavor current can be read off of the linear coupling to $\mathcal{A}_{L,R}^\mu(x)$as[8]

$$\mathcal{J}_{L\mu}=-\kappa\left(k(z)\,\mathcal{F}_{\mu z}^{\text{cl}}\right)\Big|_{z=+\infty}\ ,\qquad \mathcal{J}_{R\mu}=+\kappa\left(k(z)\,\mathcal{F}_{\mu z}^{\text{cl}}\right)\Big|_{z=-\infty}\ . \tag{9}$$

Here the superscript cl refers to the fact that the field strength $\mathcal{F}$ is evaluated on-shell.

For more details of a study of the meson physics, see also Ref. 12.

3. Baryons in holographic QCD

As argued before, baryons in large N_c QCD may be regarded as a soliton. In fact, Adkins, Nappi and Witten found that the semiclassical quantization of the Skyrmion leads to quantitative results in baryon properties that are consistent with experiments.[7] Here the baryon number is identified with the topological number associated with the mapping $S^3 \to SU(2)$. Now the question is what happens when we incorporate the effects of heavy mesons into the baryon soliton. There have been some attempts toward it so far (see e.g. Ref. 13). In Ref. 5, we derived a baryon solution of the five-dimensional YM-CS theory (1). The point is that as in the case of the Skyrmion, the baryon number is equal to a topological number. To see this, recall that the baryon number is defined by the baryon number current as

$$J_B^\mu = \frac{1}{N_c}\left(\widehat{J}_L^\mu + \widehat{J}_R^\mu\right) = -\kappa \int dz\, k(z) \widehat{F}_{\rm cl}^{\mu z} \, . \tag{10}$$

Here the quantities without hat denote the $U(1)$ part of $U(N_f)$ in (1), and the factor $1/N_c$ is due to the fact that a single baryon composed of N_c quarks carries a unit baryon number. For simplicity, we focus on the case $N_f = 2$. Using the equations of motion of (1), the baryon number can be rewritten as

$$Q_B = \int d^3x\, J_B^0 = \frac{1}{64\pi^2}\, \epsilon_{MNPQ} \int d^3x dz\, {\rm tr}(F_{MN} F_{PQ}) \, . \tag{11}$$

This is exactly the instanton number. It is also interesting that the instanton number can be shown to equal the winding number of the Skyrmion.[5]

In the context of holographic QCD, the baryon solutions can be obtained by finding a topological soliton with a nontrivial instanton number. However, since the theory (1) is defined on a curved background with the CS five-form added, it is difficult to find an analytic expression of the solitons. Instead, we work in the large λ limit to make the problem tractable. That is, define the new variables

$$\begin{aligned}
&\widetilde{x}^M = \lambda^{+1/2} x^M \, , \quad \widetilde{x}^0 = x^0 \, , \\
&\widetilde{\mathcal{A}}_0(t, \widetilde{x}) = \mathcal{A}_0(t, x) \, , \quad \widetilde{\mathcal{A}}_M(t, \widetilde{x}) = \lambda^{-1/2} \mathcal{A}_M(t, x) \, , \\
&\widetilde{\mathcal{F}}_{MN}(t, \widetilde{x}) = \lambda^{-1} \mathcal{F}_{MN}(t, x) \, , \quad \widetilde{\mathcal{F}}_{0M}(t, \widetilde{x}) = \lambda^{-1/2} \mathcal{F}_{0M}(t, x) \, ,
\end{aligned} \tag{12}$$

and regard the quantities with tildes as being $\mathcal{O}(\lambda^0)$. Hereafter, we omit the tilde for simplicity. We then find that for $\lambda \gg 1$, the equations of motion for the $SU(2)$ part reduce to

$$D_M F_{0M} + \frac{1}{64\pi^2 a} \epsilon_{MNPQ} \widehat{F}_{MN} F_{PQ} + \mathcal{O}(\lambda^{-1}) = 0 \, , \tag{13}$$

$$D_N F_{MN} + \mathcal{O}(\lambda^{-1}) = 0 \, . \tag{14}$$

Also, the equations of motion for the $U(1)$ part become

$$\partial_M \widehat{F}_{0M} + \frac{1}{64\pi^2 a} \epsilon_{MNPQ} \left\{ {\rm tr}(F_{MN} F_{PQ}) + \frac{1}{2} \widehat{F}_{MN} \widehat{F}_{PQ} \right\} + \mathcal{O}(\lambda^{-1}) = 0 \, , \tag{15}$$

$$\partial_N \widehat{F}_{MN} + \mathcal{O}(\lambda^{-1}) = 0 \, . \tag{16}$$

A baryon solution with a unit baryon number reads for the $SU(2)$ sector

$$A_0 = 0 \, , \quad A_M = A_M^{\rm BPST}(x^M) \, , \tag{17}$$

where $M, N = 1, 2, 3, z$, and $A_M^{\rm BPST}(x^M)$ refers to the BPST one-instanton solution in a flat $\mathbb{R}^4$. For the $U(1)$ part,

$$\widehat{A}_0 = \frac{27\pi}{\xi^2}\left[1 - \frac{\rho^4}{(\rho^2+\xi^2)^2}\right] , \quad \widehat{A}_M = 0 , \tag{18}$$

where $\xi = \sqrt{(\vec{x}-\vec{X})^2 + (z-Z)^2}$, with $\vec{X}, Z$ being the center of the instanton, and ρ the instanton size. The parameters ρ and Z are not genuine moduli but they must be fixed by minimizing the energy of the soliton solution, which is given by

$$M(\rho, Z) = M_0\left[1 + \lambda^{-1}\left(\frac{\rho^2}{6} + \frac{3^6\pi}{5}\frac{1}{\rho^2} + \frac{Z^2}{3}\right) + \mathcal{O}(\lambda^{-2})\right] , \tag{19}$$

with $M_0 = 8\pi^2\kappa$. In particular, this implies that the baryon solution is given by a YM instanton with a fixed size of $\mathcal{O}(\lambda^{1/2})$ in the original coordinate. This gives a natural explanation of the observation by Atiyah and Manton in Ref. 14. They discuss that the Skyrmion can be well approximated with a YM instanton that is path-ordered integrated along an artificial fifth direction. In Ref. 14, the physical meaning of the extra direction was unclear, while in our model it is nothing but the holographic direction in the gauge/string duality.

Now that we find the soliton solution, we next formulate a Lagrangian that characterizes the dynamics of the collective coordinates. For this purpose, we promote the instanton moduli to time-dependent variables. The kinetic term is given by a one-dimensional nonlinear σ model whose target space is equal to the the moduli space of the $SU(2)$ one-instanton solution. On top of this, we have the potential term of ρ and Z, which is given by the soliton mass given above. In summary, the Lagrangian reads

$$L = L_X + L_Z + L_y + \mathcal{O}(\lambda^{-1}) , \tag{20}$$

$$\begin{aligned} L_X &= -M_0 + \frac{m_X}{2}\dot{\vec{X}}^2 , \\ L_Z &= \frac{m_Z}{2}\dot{Z}^2 - \frac{m_Z\omega_Z^2}{2}Z^2 , \\ L_y &= \frac{m_y}{2}\dot{y}_I^2 - \frac{m_y\omega_\rho^2}{2}\rho^2 - \frac{Q}{\rho^2} = \frac{m_y}{2}\left(\dot{\rho}^2 + \rho^2\dot{a}_I^2\right) - \frac{m_y\omega_\rho^2}{2}\rho^2 - \frac{Q}{\rho^2} , \end{aligned} \tag{21}$$

where

$$m_X = m_Z = m_y/2 = 8\pi^2\kappa\lambda^{-1}, \quad \omega_Z^2 = \frac{2}{3} , \quad \omega_\rho^2 = \frac{1}{6} , \quad Q = \frac{N_c^2}{5m_X} . \tag{22}$$

Here $y_I = \rho a_I$, $(I = 1, 2, 3, 4)$ with $\sum_{I=1}^4 a_I a_I = 1$. a_I are an element of $SU(2)$ that specifies the global gauge transformation of the BPST instanton. Quantization of this system is straightforward with the energy eigenstate given by

$$|l, n_\rho, n_z\rangle . \tag{23}$$

This belongs to $(\boldsymbol{l+1}, \boldsymbol{l+1})$ of $SU(2)_I \times SU(2)_J = SO(4)$ with $l = 1, 3, 5, \cdots$ and $n_\rho, n_z = 0, 1, 2, \cdots$. Here $SU(2)_{I,J}$ is the isospin and spin group, respectively. n_ρ and n_z denote the quantum excitation number associated with ρ and Z, respectively. The energy eigenvalue is

$$\begin{aligned} M_{l,n_\rho,n_z} =& M_0 + \sqrt{\frac{(l+1)^2}{6} + \frac{2}{15}N_c^2} + \frac{2(n_\rho + n_z) + 2}{\sqrt{6}} \\ \simeq& M_0 + \sqrt{\frac{2}{15}}N_c + \frac{1}{4}\sqrt{\frac{5}{6}}\frac{(l+1)^2}{N_c} + \frac{2(n_\rho + n_z) + 2}{\sqrt{6}} \,. \end{aligned} \quad (24)$$

A comment is in order. In comparison with the analysis in the Skyrme model, we notice two new quantum numbers n_ρ and n_z. This enables us to describe higher excited baryons including negative parity baryons. In fact, it is found that the eigenstate $|l, n_\rho, n_z\rangle$ carries the parity $P = (-1)^{n_z}$. Note that the collective motion due to Z arises only after incorporating an infinite number of heavy mesons into the effective Lagrangian of the pion.

Now we propose the following dictionary that relates our prediction with experimental data:

(n_ρ, n_z)	$(0,0)$	$(1,0)$ $(0,1)$	$(1,1)$ $(2,0)/(0,2)$	$(2,1)/(0,3)$ $(1,2)/(3,0)$
$N\,(l=1)$	940^+	1440^+ 1535^-	1655^- 1710^+, ?	2090^-_*, ? 2100^+_*, ?
$\Delta\,(l=3)$	1232^+	1600^+ 1700^-	1940^-_* 1920^+, ?	?, ? ?, ?

Here the experimental values are taken from PDG.[15] $*$ indicates a poor evidence of the existence of a meson. Using this rule, we compare mass differences of the two results:

$$m_\Delta - m_N = \begin{cases} M_{l=3,n_\rho=n_z=0} - M_{l=1,n_\rho=n_z=0} \simeq 569\text{MeV (our model)} \\ 1232 - 940 = 292\text{MeV (experiments)} \end{cases}$$

$$m_{N(1440)} - m_N = \begin{cases} M_{l=1,n_\rho=0,n_z=1} - M_{l=1,n_\rho=n_z=0} \simeq 774\text{MeV (our model)} \\ 1440 - 940 = 500\text{MeV (experiments)} \end{cases}$$

Here we have used $M_{\text{KK}} \simeq 949$MeV, which is due to fitting our prediction of the ρ mass with experiments. One of the interesting features of our result is that both $N'(1440)$ and $N^*(1530)$ are predicted to have the same mass. This is hard to explain in the phenomenological quark model: it predicts $m_{N'(1440)} - m_N \simeq 2(m_{N^*(1530)} - m_N)$. A resolution of this puzzle is made by a recent lattice simulation of QCD.[16] Our prediction is consistent with this result. Moreover, since our mass formula depends on n_ρ and n_z only through $n_\rho + n_z$, the baryon states with a common value of $n_\rho + n_z$ are all predicted to be degenerate.

3.1. *Baryon form factors*

In this subsection, we compute the baryon form factors using the formalism given above. See Ref. 8 for details. For closely related works, see also Refs. 17–21.

Consider the matrix element of the vector and axial-vector current $\mathcal{J}^{\mu C}_{V,A}(x)$ for a spin-1/2 baryon state B:

$$\begin{aligned}&\langle \vec{p}', B, s'|\mathcal{J}^{\mu C}_{V}(0)|\vec{p}, B, s\rangle\\ &= i(2\pi)^{-3}\,\overline{u}(\vec{p}', s')\frac{\tau^C}{2}\left[\gamma^\mu F_1^{(C)}(k^2) - \frac{1}{2m_B}\,\sigma^{\mu\nu}k_\nu\,F_2^{(C)}(k^2)\right]u(\vec{p}, s)\end{aligned}$$

$$\begin{aligned}&\langle \vec{p}', B, s'|\mathcal{J}^{\mu C}_{A}(0)|\vec{p}, B, s\rangle\\ &= (2\pi)^{-3}\,\overline{u}(\vec{p}', s')\frac{\tau^C}{2}\left[i\gamma_5\gamma^\mu\,g_A^{(C)}(k^2) + \frac{1}{2m_B}\,k^\mu\gamma_5\,g_P^{(C)}(k^2)\right]u(\vec{p}, s)\end{aligned}$$

with $k = p - p'$. Here, the baryon state $|B\rangle$ is given by the mass eigenstates found before. The current $\mathcal{J}_{V,A}$ can be obtained from the formula (9). Hence the left hand side of the two equations is easy to compute in our model. Actually, we find

$$\begin{aligned}&\widehat{F}_1(\vec{k}^2) = N_c\sum_{n\geq 1}\frac{g_{v^n}\langle\psi_{2n-1}(Z)\rangle}{\vec{k}^2+\lambda_{2n-1}}\,, \quad \widehat{F}_2(\vec{k}^2) = N_c\left(\frac{g_{I=0}}{2}-1\right)\sum_{n\geq 1}\frac{g_{v^n}\langle\psi_{2n-1}(Z)\rangle}{\vec{k}^2+\lambda_{2n-1}}\,,\\ &F_1(\vec{k}^2) = \sum_{n\geq 1}\frac{g_{v^n}\langle\psi_{2n-1}(Z)\rangle}{\vec{k}^2+\lambda_{2n-1}}\,, \quad F_2(\vec{k}^2) = \frac{g_{I=1}}{2}\sum_{n\geq 1}\frac{g_{v^n}\langle\psi_{2n-1}(Z)\rangle}{\vec{k}^2+\lambda_{2n-1}}\,, \end{aligned}\tag{26}$$

$$\begin{aligned}&\widehat{g}_A(\vec{k}^2) = \frac{N_c}{4M_0}\sum_{n\geq 1}\frac{g_{a_n}\langle\partial_Z\psi_{2n}(Z)\rangle}{\vec{k}^2+\lambda_{2n}}\,, \quad \widehat{g}_P(k^2) = \frac{2m_B}{k^2}\,\widehat{g}_A(k^2)\,,\\ &g_A(\vec{k}^2) = \frac{8\pi^2\kappa}{3}\langle\rho^2\rangle\sum_{n\geq 1}\frac{g_{a_n}\langle\partial_Z\psi_{2n}(Z)\rangle}{\vec{k}^2+\lambda_{2n}}\,, \quad g_P(k^2) = \frac{2m_B}{k^2}\,g_A(k^2)\,,\end{aligned}\tag{27}$$

with

$$g_{I=0} = \frac{m_B}{M_0}\,, \quad g_{I=1} = \frac{32\pi^2\kappa m_B}{3}\langle\rho^2\rangle\,,$$

$\langle\rho^2\rangle = \langle n_\rho|\rho^2|n_\rho\rangle$, $\langle\psi_{2n-1}(Z)\rangle = \langle n_z|\psi_{2n-1}(Z)|n_z\rangle$, $\langle\partial_Z\psi_{2n}(Z)\rangle = \langle n_z|\partial_Z\psi_{2n}(Z)|n_z\rangle$.

It is interesting to note that these results exhibit the vector meson dominance. In other words, they can be reproduced by the effective action of the form

$$\mathcal{L}^{v}_{\rm int} = \sum_{n\geq 1}\left(\widehat{g}_{v^n BB}\,\widehat{v}^n_\mu\,\overline{B}i\gamma^\mu\frac{\tau^0}{2}B + g_{v^n BB}\,v^{na}_\mu\,\overline{B}i\gamma^\mu\frac{\tau^a}{2}B\right)$$

$$+\frac{1}{4m_B}\sum_{n\geq 1}\left(\widehat{h}_{v^n BB}\big(\partial_\mu\widehat{v}^n_\nu - \partial_\nu\widehat{v}^n_\mu\big)\overline{B}\sigma^{\mu\nu}\frac{\tau^0}{2}B + h_{v^n BB}\big(\partial_\mu v^{na}_\nu - \partial_\nu v^{na}_\mu\big)\overline{B}\sigma^{\mu\nu}\frac{\tau^a}{2}B\right),$$

$$\mathcal{L}^{a}_{\rm int} = \sum_{n\geq 1}\left(\widehat{g}_{a^n BB}\,\widehat{a}^n_\mu\,\overline{B}i\gamma_5\gamma^\mu\frac{\tau^0}{2}B + g_{a^n BB}\,a^{na}_\mu\,\overline{B}i\gamma_5\gamma^\mu\frac{\tau^a}{2}B\right)$$

$$+\,2i\left(\widehat{g}_{\pi BB}\,\widehat{\pi}\,\overline{B}\gamma_5\frac{\tau^0}{2}B + g_{\pi BB}\,\pi^a\,\overline{B}\gamma_5\frac{\tau^a}{2}B\right)\,.$$

The whole tower of the cubic coupling constants can be fixed uniquely in holographic QCD.

By noting that the electromagnetic current equals

$$J^\mu_{\rm em} = J^{a=3,\mu}_V + \frac{1}{N_c}\widehat{J}^\mu\,, \tag{28}$$

the electromagnetic form factors of the proton and neutron of Sachs type is found to take the form

$$G^{\rm p}_E(\vec{k}^2) = \frac{2}{g_{\rm p}}G^{\rm p}_M(\vec{k}^2) = \frac{2}{g_{\rm n}}G^{\rm n}_M(\vec{k}^2) = \sum_{n\geq 1}\frac{g_{v^n}\langle\psi_{2n-1}(Z)\rangle}{\vec{k}^2+\lambda_{2n-1}}\,,\quad G^{\rm n}_E(\vec{k}^2) = 0\,, \tag{29}$$

with the g-factors given by

$$g_{\rm p,n} = \frac{1}{2}\left(g_{I=0}\pm g_{I=1}\right)\,. \tag{30}$$

We now compare these results with experiments. First, it is known experimentally that the electric form factor of the proton is well approximated with the dipole behavior

$$G^{\rm p}_E(\vec{k}^2)\Big|_{\rm exp} \simeq \left(1+\frac{\vec{k}^2}{\Lambda^2}\right)^{-2}\,, \tag{31}$$

with $\Lambda^2 = 0.758\,{\rm GeV}^2$ from experiments. We can verify that $G^{\rm p}_E(\vec{k}^2)$ obtained above matches this dipole profile well. Next, the result $G^{\rm n}_E(\vec{k}^2) = 0$ is also in good agreement with experiments. Note that the vanishing form factor is due to cancellation between the isotriplet and isosinglet form factors. This is reminiscent of the old work by Nambu, which predicted the ω meson, the lightest isosinglet vector meson, from the requirement of $G^{\rm n}_E(\vec{k}^2) = 0$. In our model, the cancellation of the isotriplet and isosinglet form factors occurs for an infinite number of isotriplet and isosinglet vector mesons.

In summary, we list our prediction of the physical quantities of baryons

	our model	Skyrmion	experiment
$\langle r^2\rangle^{1/2}_{I=0}$	0.742 fm	0.59 fm	0.806 fm
$\langle r^2\rangle^{1/2}_{\mathrm{M},\,I=0}$	0.742 fm	0.92 fm	0.814 fm
$\langle r^2\rangle_{\mathrm{E,p}}$	$(0.742\ \mathrm{fm})^2$	∞	$(0.875\ \mathrm{fm})^2$
$\langle r^2\rangle_{\mathrm{E,n}}$	0	$-\infty$	$-0.116\ \mathrm{fm}^2$
$\langle r^2\rangle_{\mathrm{M,p}}$	$(0.742\ \mathrm{fm})^2$	∞	$(0.855\ \mathrm{fm})^2$
$\langle r^2\rangle_{\mathrm{M,n}}$	$(0.742\ \mathrm{fm})^2$	∞	$(0.873\ \mathrm{fm})^2$
$\langle r^2\rangle^{1/2}_{A}$	0.537 fm	–	0.674 fm
μ_p	2.18	1.87	2.79
μ_n	−1.34	−1.31	−1.91
$\left\lvert\frac{\mu_p}{\mu_n}\right\rvert$	1.63	1.43	1.46
g_A	0.734	0.61	1.27
$g_{\pi NN}$	7.46	8.9	13.2
$g_{\rho NN}$	5.80	–	$4.2 \sim 6.5$

Here we used as an input $M_{\mathrm{KK}} = 949$ MeV , $\kappa = 0.00745$, which follows from a fitting with the experimental value of the ρ mass and the pion decay constant. In most cases, our predictions are improved compared with those from the Skyrme model analyses.

3.2. *Nuclear force from string theory*

We end this section with a sketch of Ref. 9. The purpose of this paper is to compute the short distance potential of two nuclei. A direct evaluation of it by means of QCD is one of the outstanding issues in the hadron physics. That was first done in Ref. 22 using a lattice simulation. Here we see that holographic QCD is powerful enough to enable us to obtain an exact expression of it. For related works, see also Refs. 23, 24.

The idea is to construct a two-baryon solution and then compute the energy of this system. As in a single baryon case, the exact solution of the two-baryon system is hard to find. We thus work in the $\lambda \gg 1$ case, where the solution can be approximated with a two-instanton solution in a flat $\mathbb{R}^4$. Collective motion is described by a Lagrangian of the 16 variables, which are time-dependent moduli of the two-instanton solution. The kinetic term is given by a one-dimensional nonlinear ω-model on the two-instanton moduli space. The potential takes the form:

$$U = 2M_0 + U_{SU(2)} + U_{U(1)}\ , \tag{32}$$

$$U_{SU(2)} = \frac{N_c}{216\pi^3}\int d^3x dz\ \mathrm{tr}\left(-\frac{z^2}{6}F_{ij}^2 + z^2 F_{iz}^2\right) = \frac{N_c}{6\cdot 216\pi^3}\int d^3x dz\ \mathrm{tr}(z^2 F_{MN}^2)\ , \tag{33}$$

$$U_{U(1)} = \frac{N_c}{512\pi^3}\int d^3x dz\ \widehat{F}_{0M}^2\ . \tag{34}$$

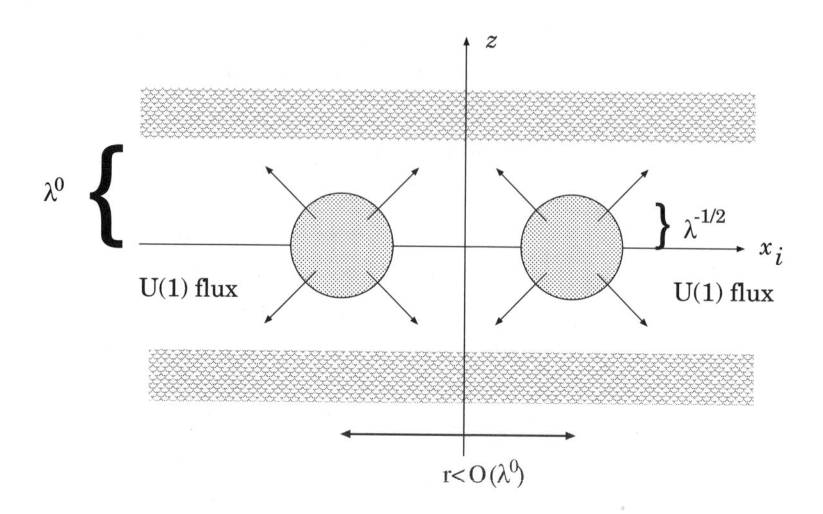

Fig. 2. Sketch of two-baryon system in the original coordinate

Although $U_{SU(2)}$ can be computed analytically, $U_{U(1)}$ cannot. Assuming R, the relative distance of the baryons in a flat $\mathbb{R}^4$, to be $O(\lambda^{-1/2}) < R < O(\lambda^0)$, the total Hamiltonian reduces to

$$H = H(1)_{\rm ins} + H(2)_{\rm ins} + \frac{1}{R^2} U^{(1)} + \mathcal{O}(R^{-3}) \, .$$

Here $H(i)_{\rm ins}$, $(i = 1, 2)$ is the Hamiltonian of a single baryon we have worked with so far. $U^{(1)}$ measures the potential energy of two baryons. In order to evaluate it, consider the matrix element

$$V = \Big\langle (I_1', J_1'); (I_2', J_2') \Big| \frac{1}{R^2} U^{(1)} \Big| (I_1, J_1); (I_2, J_2) \Big\rangle \, .$$

Here I_i and J_i denotes the third component of the isospin and spin of the ith baryon. $\big|(I_1, J_1); (I_2, J_2)\big\rangle$ and $\big|(I_1', J_1'); (I_2', J_2')\big\rangle$ are an eigenfunction of $H(1)_{\rm ins} + H(2)_{\rm ins}$, being given by the tensor product of two wavefunctions. It can be shown that

$$\begin{aligned} V &= V_{\rm C}(|\vec{r}|) + S_{12} V_{\rm T}(|\vec{r}|) \, , \\ V_{\rm C}(|\vec{r}|) &= \pi \left(\frac{27}{2} + \frac{32}{5} (\vec{I}_1 \cdot \vec{I}_2)(\vec{J}_1 \cdot \vec{J}_2) \right) \frac{N_c}{\lambda} \frac{1}{|\vec{r}|^2} \, , \\ V_{\rm T}(|\vec{r}|) &= \frac{8\pi}{5} (\vec{I}_1 \cdot \vec{I}_2) \frac{N_c}{\lambda} \frac{1}{|\vec{r}|^2} \, , \end{aligned} \tag{35}$$

where S_{12} is the tensor operator defined by

$$S_{12} = 3(\vec{\sigma}_1 \cdot \vec{r})(\vec{\sigma}_2 \cdot \vec{r})/\vec{r}^2 - \vec{\sigma}_1 \cdot \vec{\sigma}_2 \, . \tag{36}$$

$\vec{r}$ is the relative position of the two baryons in $\mathbb{R}^3$. We interpret this as a short distance behavior of nuclear force between two nuclei. One of the most interesting features of this result is that it behaves as $1/|\vec{r}|^2$. This is identical to the typical behavior of Coulomb potential in four spatial dimensions. In fact, the $U(1)$ gauge potential in (1) accounts for this Coulomb behavior.

4. Summary

In this paper, we gave an overview of recent developments in the holographic description of QCD on the basis of the D4-D8-$\overline{\text{D8}}$-brane configuration. A particular emphasis is put on baryon properties. By a semiclassical analysis of the baryon solution, we worked out the physical quantities of baryons. It was shown that this model reproduces many of the experimental results quantitatively. Furthermore, it is worth stressing that our model unites many insightful ideas that have been studies so far independently, such as hidden local symmetry, baryons as instantons, the role of the isosinglet vector mesons in the short distance repulsive core of nuclear force, etc. These successes may be regarded as a strong evidence that the gauge/string duality is a powerful tool to analyze the strong coupling dynamics of nonabelian gauge theories. A thorough investigation in this line would be worthwhile.

Acknowledgments

We would like to thank S. Sugimoto, H. Hata, S. Yamato and K. Hashimoto for fruitful collaborations.

References

1. O. Aharony, S. S. Gubser, J. M. Maldacena, H. Ooguri and Y. Oz, "Large N field theories, string theory and gravity," Phys. Rept. **323** (2000), 183, hep-th/9905111.
2. T. Sakai and S. Sugimoto, "Low energy hadron physics in holographic QCD," Prog. Theor. Phys. **113** (2005), 843, hep-th/0412141.
3. M. Bando, T. Kugo and K. Yamawaki, "Nonlinear Realization and Hidden Local Symmetries," Phys. Rept. **164**, 217 (1988).
4. D. T. Son and M. A. Stephanov, "QCD and dimensional deconstruction," Phys. Rev. D **69**, 065020 (2004) [arXiv:hep-ph/0304182].
5. H. Hata, T. Sakai, S. Sugimoto and S. Yamato, "Baryons from instantons in holographic QCD," Prog. Theor. Phys. **117**, 1157 (2007) [arXiv:hep-th/0701280].
6. E. Witten, "Current Algebra, Baryons, And Quark Confinement," Nucl. Phys. B **223**, 433 (1983).
7. G. S. Adkins, C. R. Nappi and E. Witten, "Static Properties Of Nucleons In The Skyrme Model," Nucl. Phys. B **228**, 552 (1983).
8. K. Hashimoto, T. Sakai and S. Sugimoto, "Holographic Baryons : Static Properties and Form Factors from Gauge/String Duality," Prog. Theor. Phys. **120**, 1093 (2008) [arXiv:0806.3122 [hep-th]].
9. K. Hashimoto, T. Sakai and S. Sugimoto, "Nuclear Force from String Theory," Prog. Theor. Phys. **122**, 427 (2009) [arXiv:0901.4449 [hep-th]].
10. E. Witten, "Anti-de Sitter space, thermal phase transition, and confinement in gauge theories," Adv. Theor. Math. Phys. **2** (1998), 505; hep-th/9803131.
11. T. Imoto, T. Sakai and S. Sugimoto, "Mesons as Open Strings in a Holographic Dual of QCD," arXiv:1005.0655 [hep-th].

12. T. Sakai and S. Sugimoto, "More on a holographic dual of QCD," Prog. Theor. Phys. **114** (2005), 1083, hep-th/0507073.
13. U. G. Meissner, "Low-Energy Hadron Physics From Effective Chiral Lagrangians With Vector Mesons," Phys. Rept. **161**, 213 (1988).
14. M. F. Atiyah and N. S. Manton, "SKYRMIONS FROM INSTANTONS," Phys. Lett. B **222** (1989), 438.
15. W. M. Yao *et al.* [Particle Data Group], "Review of particle physics," J. of Phys. G **33** (2006), 1.
16. K. Sasaki, S. Sasaki and T. Hatsuda, "Spectral analysis of excited nucleons in lattice QCD with maximum entropy method," Phys. Lett. B **623**, 208 (2005) [arXiv:hep-lat/0504020].
17. D. K. Hong, M. Rho, H. U. Yee and P. Yi, "Chiral dynamics of baryons from string theory," Phys. Rev. D **76**, 061901 (2007) [arXiv:hep-th/0701276].
18. D. K. Hong, M. Rho, H. U. Yee and P. Yi, "Dynamics of Baryons from String Theory and Vector Dominance," JHEP **0709**, 063 (2007) [arXiv:0705.2632 [hep-th]].
19. D. K. Hong, M. Rho, H. U. Yee and P. Yi, "Nucleon Form Factors and Hidden Symmetry in Holographic QCD," Phys. Rev. D **77**, 014030 (2008) [arXiv:0710.4615 [hep-ph]].
20. J. Park and P. Yi, "A Holographic QCD and Excited Baryons from String Theory," JHEP **0806**, 011 (2008) [arXiv:0804.2926 [hep-th]].
21. H. Hata, M. Murata and S. Yamato, "Chiral currents and static properties of nucleons in holographic QCD," Phys. Rev. D **78**, 086006 (2008) [arXiv:0803.0180 [hep-th]].
22. N. Ishii, S. Aoki and T. Hatsuda, "The nuclear force from lattice QCD," Phys. Rev. Lett. **99** (2007), 022001, nucl-th/0611096; "Nuclear Force from Monte Carlo Simulations of Lattice Quantum Chromodynamics," arXiv:0805.2462.
   H. Nemura, N. Ishii, S. Aoki and T. Hatsuda, "Hyperon-nucleon force from lattice QCD," arXiv:0806.1094.
   S. Aoki, J. Balog, T. Hatsuda, N. Ishii, K. Murano, H. Nemura and P. Weisz, "Energy dependence of nucleon-nucleon potentials," arXiv:0812.0673.
23. K. Y. Kim and I. Zahed, "Nucleon-Nucleon Potential from Holography," JHEP **0903**, 131 (2009) [arXiv:0901.0012 [hep-th]].
24. Y. Kim, S. Lee and P. Yi, "Holographic Deuteron and Nucleon-Nucleon Potential," JHEP **0904**, 086 (2009) [arXiv:0902.4048 [hep-th]].

Nuclear Force from String Theory

K. Hashimoto

Nishina Center, RIKEN
2-1 Hirosawa, Wako, Saitama 351-0198, Japan
E-mail: koji@riken.jp

We compute nuclear forces at short distance, by applying gauge/gravity correspondence to large N_c QCD. This talk is based on work in collaboration with T. Sakai and S. Sugimoto.

Keywords: Holographic QCD, nuclear force, AdS/CFT correspondence.

1. From String Theory to Nuclear Physics

The application of AdS/CFT correspondence to QCD has been quite successful, at least for delivering qualitative understanding of strong coupling physics expected in QCD. However, it is important to notice that most of the researches of this application have been on hadron physics, not on nuclear physics. In fact, there is a big gap between particle physics community and nuclear physics community, and this gap exists because of the strong coupling of QCD, we can say. Once we can solve QCD at strong coupling, then it should reveal various fundamental inputs in nuclear physics. The AdS/CFT correspondence, although it can be applied only for large N_c limit of QCD-like theories, may bring some hope for this quite important direction of merger of two physics communities.

Light nuclei are studied by a few body quantum mechanics under an assumption that nucleons are point particles, with given nuclear forces. Once QCD is solved at low energy, *i.e.* at strong coupling, to give the force among baryons, then we can use it for the basis of the nuclear physics. Since holographic QCD, the application of the AdS/CFT correspondence to QCD, can solve a certain limit of QCD-like theories, it is very natural to try to understand the nuclear forces from this duality point of view. In this talk, I describe such computations of the nuclear force in holographic QCD (based on the work[1,2] which have been done in collaboration with T. Sakai and S. Sugimoto).[a]

On the other hand, heavy nuclei are very difficult to treat, as a complicated quantum many body problem. One approach which I propose along the line of the AdS/CFT correspondence is to use this large number for the number of the

[a]See also related works.[3,4]

nucleons as a large N limit of holography. Using the AdS/CFT applied to heavy nuclei, we can get a gravitational dual description of the nuclei naturally. This I call "holographic nuclei",[5,6] and the details will be delivered elsewhere.

It has been known for many decades that nucleons always experience a repulsive force at short distances. This is known as a "repulsive core" of the nucleon. It is one of the keystones if any approximation of strongly coupled QCD may exhibit this repulsive core or not. In this talk, I show that the holographic QCD can reproduce beautifully the repulsive core of the nucleon.

2. Derivation of Nuclear Force at Short Range

The idea of how to derive the nuclear forces in holographic QCD is quite simple. We rely on the following two key ideas.

- By the AdS/CFT, strong coupling QCD is rephrased by a higher dimensional $U(N_f)$ non-Abelian gauge theory, where N_f is the number of flavors in the large N_c QCD.
- Mesons are the gauge fields of this theory, while baryons are solitons of it.

There are various holographic QCD models, but here we use the so-called D4-D8 model (Sakai-Sugimoto model).[7,8] Then the gauge theory is effectively a 4+1-dimensional Yang-Mills-Chern-Simons $U(N_f)$ theory in a curved spacetime. Solitons are Yang-Mills instantons in this theory (instantons are localized in 4-dimensional space, while we have 5 dimensional spacetime as a starting point, so in effect the instanton is a particle-like excitation, identified as a baryon). Fortunately, we know generic 2-instanton solution in flat space, so we can use the generic solution to explore the nuclear forces. Quantization of a single instanton soliton in this curved geometry was done by Hata et al.,[9] and this is used to specify the baryon states.

Here we show only the results of our computation, for simplicity.

$$V = V_{\rm C}(r) + (3\hat{\mathbf{r}}\cdot\sigma_1\,\hat{\mathbf{r}}\cdot\sigma_2 - \sigma_1\cdot\sigma_2)\,V_{\rm T}(r). \tag{1}$$

The central and the tensor forces are written as

$$V_{\rm C}(r) = \pi\left(\frac{27}{2} + \frac{8}{5}\sigma_1\cdot\sigma_2\,\vec{I}_1\cdot\vec{I}_2\right)\frac{1}{g_{\rm QCD}^2 M_{\rm KK}}\frac{1}{r^2}, \tag{2}$$

$$V_{\rm T}(r) = \frac{8\pi}{5}\vec{I}_1\cdot\vec{I}_2\frac{1}{g_{\rm QCD}^2 M_{\rm KK}}\frac{1}{r^2}. \tag{3}$$

As one can see in the central force, whatever the spin $(\sigma_i/2)$ or the isospin $\vec{I}_i$ is, it is repulsive. So we found a repulsive potential for any nucleons.

Note that our nuclear force has only two parameters, $g_{\rm QCD}$ and $M_{\rm KK}$. This is a peculiar feature of this holographic QCD model. These two numbers are usually used to fit the ρ meson mass and the pion decay constant, then all the other observables are predictions of the model, and this nuclear force is one of the predictions.

3. Conclusion and Discussions

We computed[2] nuclear forces at short distances, by solving a large N_c QCD-like theory using the AdS/CFT correspondence. This would serve as a first analytic result reproducing the repulsive core of nucleons, via a solution of the strongly coupled QCD-like theory.

In the other work[1] we computed static properties of baryons, including meson-baryon couplings. This in fact gives a long-range nuclear force via the meson exchange picture.

Along the direction of this research, it would be quite natural to extend the computation to a system with a larger number of nucleons. Indeed, the 3-body nuclear force is one of key issues in nuclear physics: it is known that it exists from few-body calculations of energy levels of light nuclei, while the intrinsic nature of the force is to be revealed. I have tackled this problem with the AdS/CFT correspondence with my collaborators and found that the 3-body forces are suppressed.[10] It would be very interesting to apply the AdS/CFT correspondence to various mysteries in nuclear physics.

References

1. K. Hashimoto, T. Sakai and S. Sugimoto, Prog. Theor. Phys. **120** (2008) 1093 [arXiv:0806.3122 [hep-th]].
2. K. Hashimoto, T. Sakai and S. Sugimoto, Prog. Theor. Phys. **122** (2009) 427 [arXiv:0901.4449 [hep-th]].
3. D. K. Hong, M. Rho, H. U. Yee and P. Yi, JHEP **0709** (2007) 063 [arXiv:0705.2632 [hep-th]].
4. Y. Kim, S. Lee and P. Yi, JHEP **0904** (2009) 086 [arXiv:0902.4048 [hep-th]].
5. K. Hashimoto, Prog. Theor. Phys. **121** (2009) 241 [arXiv:0809.3141 [hep-th]].
6. K. Hashimoto, JHEP **0912** (2009) 065 [arXiv:0910.2303 [hep-th]].
7. T. Sakai and S. Sugimoto, Prog. Theor. Phys. **113** (2005) 843 [arXiv:hep-th/0412141].
8. T. Sakai and S. Sugimoto, Prog. Theor. Phys. **114** (2005) 1083 [arXiv:hep-th/0507073].
9. H. Hata, T. Sakai, S. Sugimoto and S. Yamato, Prog. Theor. Phys. **117** (2007) 1157 [arXiv:hep-th/0701280].
10. K. Hashimoto, N. Iizuka and T. Nakatsukasa, arXiv:0911.1035 [hep-th].

Integrating Out Holographic QCD Back to Hidden Local Symmetry*

Masayasu Harada[a,†], Shinya Matsuzaki[b,§] and Koichi Yamawaki[a,‡]

[a] *Department of Physics, Nagoya University*
Nagoya, 464-8602, Japan
[†] *E-mail: harada@hken.phys.nagoya-u.ac.jp*
[‡] *E-mail: yamawaki@eken.phys.nagoya-u.ac.jp*

[b] *Department of Physics, Pusan National University*
Busan 609-735, Korea
[§] *E-mail: synya@pusan.ac.kr*

We develop a previously proposed gauge-invariant method to integrate out infinite towers of vector and axialvector mesons arising as Kaluza-Klein (KK) modes in a class of holographic models of QCD (HQCD). We demonstrate that HQCD can be reduced to the chiral perturbation theory (ChPT) with the hidden local symmetry (HLS) (so-called HLS-ChPT) having only the lowest KK mode identified as the HLS gauge boson, and the Nambu-Goldstone bosons. The $\mathcal{O}(p^4)$ terms in the HLS-ChPT are completely determined by integrating out infinite towers of vector/axialvector mesons in HQCD: Effects of higher KK modes are fully included in the coefficients. As an example, we apply our method to the Sakai-Sugimoto model.

1. Introduction

Holography, based on gauge/gravity duality,[1] has been of late fashion to reveal a part of features in strongly coupled gauge theories involving the application to QCD (so-called holographic QCD (HQCD)). There are two types of holographic approaches: One is called "top-down" approach starting with a stringy setting; the other is called "bottom-up" approach beginning with a five-dimensional gauge theory defined on an AdS (anti-de Sitter space) background. It is a key point to notice that in whichever approach one eventually employs a five-dimensional gauge model with a characteristic induced-metric and some boundary conditions on a brane configuration.

In the low-energy region, any model of HQCD is reduced to a certain effective hadron model in four-dimensions. Such effective models include vector and axialvector mesons as infinite towers of Kaluza-Klein (KK) modes together with the Nambu-Goldstone bosons (NGBs) associated with the spontaneous chiral symmetry breaking. Green functions in QCD are evaluated straightforwardly from the effective model following the holographic dictionary. Full sets of infinite towers

---

*Talk given by S. M.

of exchanges of KK modes (vector and axialvector mesons) contribute to Green functions involving current correlators and form factors and could mimic ultraviolet behaviors in QCD, although such a hadronic description would not be reliable above a certain high-energy scale. This implies that appropriate/gauge-invariant holographic results require calculations including *full* set of KK towers, which, however, would not be practical because of forms written in terms of an infinite sum.

It was pointed out[2] that the infinite tower of KK modes is interpreted as a set of gauge bosons of the hidden local symmetries (HLSs).[3,4] This implies an interesting possibility that, in the low-energy region, any holographic models can be reduced to the simplest HLS model, provided that the infinite tower of KK modes is integrated out keeping only the lowest one identified with the ρ meson and its flavor partners. Effects from the higher KK modes would then be *fully* incorporated into higher derivative terms ($\mathcal{O}(p^4)$ terms) in the HLS effective field theory as an extension of the conventional chiral perturbation theory (ChPT),[5] so-called the HLS-ChPT[4,6] which is manifestly gauge-invariant formulation and makes it possible to calculate any Green functions order by order in derivative expansion. Once holographic models are expressed in terms of the HLS-ChPT, one can even calculate meson-loop corrections of subleading order in $1/N_c$ expansion. This would give a new insight into the HQCD which as it stands is valid only in the large N_c limit. Indeed, in the previous work,[7] we proposed a consistent method of integrating out the infinite towers of vector and axialvector mesons in the Sakai-Sugimoto (SS) model[8,9] into the HLS-ChPT, with the $\mathcal{O}(p^4)$ terms explicitly given by the integrated higher modes effects. Then we were able to do the first calculations of $1/N_c$ corrections to the SS model.

In this talk, we report the results of our work[12] developing in detail the integrating-out method proposed in Ref. 7. First of all, we work in a class of holographic models to introduce our integrating-out method. Next, as an example, we apply our procedure to the Sakai-Sugimoto (SS) model[8,9] to give the HLS model with a full set of $\mathcal{O}(p^4)$ terms determined. We then calculate the pion electromagnetic form factor to demonstrate how powerful our formulation is even before including the $1/N_c$ effects through loop. As we will see explicitly, the momentum-dependence of the form factor is evaluated including *full* set of contributions from KK modes without performing infinite sums. Our method can straightforwardly be applied to other types of holographic models such as those given in Refs. 10 and 11. More details are presented in Ref. 12.

2. A gauge-invariant way to integrate out HQCD

In this section, starting with a class of HQCD models including the SS model,[8,9] we introduce a way to obtain a low-energy effective model in four-dimension described only by the lightest vector meson identified as ρ meson based on the HLS together with the NGBs. Suppose that the fifth direction, spanned by the coordinate z, extends from minus infinity to plus infinity ($-\infty < z < \infty$) [a]. We employ a five-dimensional gauge theory which has a vectorial $U(N)$ gauge symmetry defined on a certain background associated with the gauge/gravity duality. As far as gauge-invariant sector such as the Dirac-Born-Infeld part of the SS model[8,9] is concerned, the five-dimensional action in large N_c limit can be written as [b]

$$S_5 = N_c \int d^4x dz \left(-\frac{1}{2} K_1(z) \mathrm{tr}[F_{\mu\nu}F^{\mu\nu}] + K_2(z) M_{\mathrm{KK}}^2 \mathrm{tr}[F_{\mu z}F^{\mu z}] \right), \quad (1)$$

where $K_{1,2}(z)$ denote a set of metric-functions of z constrained by the gauge/gravity duality. M_{KK} is a typical mass scale of KK modes of the gauge field A_M with $M = (\mu, z)$. The boundary condition of A_M is chosen as $A_M(x^\mu, z = \pm\infty) = 0$. A transformation which does not change this boundary condition satisfies $\partial_M g(x^\mu, z)|_{z=\pm\infty} = 0$, where $g(x^\mu, z)$ is the transformation matrix of the gauge symmetry. This implies an emergence of global chiral $U(N)_L \times U(N)_R$ symmetry in four-dimension characterized by the transformation matrices $g_{R,L} = g(z = \pm\infty)$.

Following Refs. 7, 8, and 9, we work in $A_z = 0$ gauge. There still exists a four-dimensional gauge symmetry under which $A_\mu(x^\mu, z)$ transforms as $A_\mu \to h \cdot A_\mu \cdot h^\dagger - i\partial_\mu h \cdot h^\dagger$ with $h = h(x^\mu)$. This gauge symmetry is identified[7–9] with the HLS.[3,4] In the $A_z \equiv 0$ gauge, the NGB fields $\pi(x^\mu)$ disappear from the chiral field $U = e^{2i\pi/F_\pi}$ since $U \to 1$. They are instead included[7] in $A_\mu(x^\mu, z)$ at the boundary as $A_\mu|_{z=\pm\infty} = \alpha_\mu^{R,L} = i\xi_{R,L}\partial_\mu \xi_{R,L}^\dagger$, where $\xi_{L,R}$ form the chiral field U as $U = \xi_L^\dagger \cdot \xi_R$. Since $\xi_{L,R} \to h \cdot \xi_{L,R} \cdot g_{L,R}^\dagger$ [3,4], $\alpha_\mu^{R,L}$ transform as $\alpha_\mu^{R,L} \to h \cdot \alpha_\mu^{R,L} \cdot h^\dagger - i\partial_\mu h \cdot h^\dagger$.

We introduce infinite towers of massive KK modes for vector ($V_\mu^{(n)}(x^\mu)$) and axialvector ($A_\mu^{(n)}(x^\mu)$) meson fields, where we treat $V_\mu^{(n)}$ as the HLS gauge fields transforming under the HLS as $V_\mu^{(n)} \to h \cdot V_\mu^{(n)} \cdot h^\dagger - i\partial_\mu h \cdot h^\dagger$, while $A_\mu^{(n)}$ as matter fields transforming as $A_\mu^{(n)} \to h \cdot A_\mu^{(n)} \cdot h^\dagger$. The five-dimensional gauge field $A_\mu(x^\mu, z)$ is now expanded as [c]

$$A_\mu(x^\mu, z) = \alpha_\mu^R(x^\mu)\phi^R(z) + \alpha_\mu^L(x^\mu)\phi^L(z) + \sum_{n=1}^{\infty} \left(A_\mu^{(n)}(x^\mu)\psi_{2n}(z) - V_\mu^{(n)}(x^\mu)\psi_{2n-1}(z) \right). \quad (2)$$

The functions $\{\psi_{2n-1}(z)\}$ and $\{\psi_{2n}(z)\}$ are the eigenfunctions satisfying the eigenvalue equation obtained from the action (1): $-K_1^{-1}(z)\partial_z(K_2(z)\partial_z \psi_n(z)) = \lambda_n \psi_n(z)$, where λ_n denotes the nth eigenvalue. On the other hand, the gauge-invariance requires the functions $\phi^{R,L}(z)$ to be different from the eigenfunctions: From the transformation properties for $A_\mu(x^\mu, z)$, $\alpha_\mu^{R,L}$, $A_\mu^{(n)}$, and $V_\mu^{(n)}$ we see that the functions

---

[a] In an application to another type of HQCD,[10] the z coordinate is defined on a finite interval, which is different from the z coordinate used here. They are related by an appropriate coordinate transformation as done in Refs. 8, 9.

[b] Models of HQCD having the left- and right-bulk fields such as F_L, F_R[10,11] can be described by the same action as in Eq.(1) with a suitable z-coordinate transformation prescribed.

[c] In Eq.(2) we put a relative minus sign in front of the HLS gauge fields $V_\mu^{(n)}$ for a convention.

$\phi^{R,L}(z)$, $\{\psi_{2n-1}(z)\}$, and $\{\psi_{2n}(z)\}$ are constrained as

$$\phi^R(z) + \phi^L(z) - \sum_{n=1}^{\infty} \psi_{2n-1}(z) = 1\,. \tag{3}$$

Using this, we may rewrite Eq.(2) to obtain $A_\mu(x^\mu, z) = \alpha_{\mu||}(x^\mu) + \alpha_{\mu\perp}(x^\mu)(\phi^R(z) - \phi^L(z)) + \sum_{n=1}^{\infty} A_\mu^{(n)}(x^\mu)\psi_{2n}(z) + \sum_{n=1}^{\infty} \left(\alpha_{\mu||}(x^\mu) - V_\mu^{(n)}(x^\mu)\right) \psi_{2n-1}(z)$, where $\alpha_{\mu||,\perp} = \frac{\alpha_\mu^R \pm \alpha_\mu^L}{2}$ transform under the HLS as $\alpha_{\mu||} \to h \cdot \alpha_{\mu||} \cdot h^\dagger - i\partial_\mu h \cdot h^\dagger$ and $\alpha_{\mu\perp} \to h \cdot \alpha_{\mu\perp} \cdot h^\dagger$, respectively. Note that $\alpha_{\mu\perp}$ includes the NGB fields as $\alpha_{\mu\perp} = \frac{1}{F_\pi}\partial_\mu \pi + \cdots$. The corresponding wave function $(\phi^R - \phi^L)$ should therefore be the eigenfunction for the zero mode, ψ_0: $\phi^R(z) - \phi^L(z) = \psi_0(z)$. From this and Eq.(3) we see that the wave functions ϕ^R and ϕ^L are not the eigenfunctions but are given as $\phi^{R,L}(z) = \frac{1}{2}\left[1 + \sum_{n=1}^{\infty} \psi_{2n-1}(z) \pm \psi_0(z)\right]$.

By substituting Eq.(2) with Eq.(3) into the action (1), the five-dimensional theory is now described by the NGB fields along with the infinite towers of the vector and axialvector meson fields in four dimensions.

We first naively try to truncate towers of the vector and axialvector meson fields simply eliminating $V_\mu^{(n)}$ and $A_\mu^{(n)}$ for $n > N$. Then we find

$$S_5^{\text{truncation}} \ni \int dz d^4x\, K_2(z) \sum_{n=N+1}^{\infty} \lambda_{2n-1}\psi_{2n-1}^2(z) \text{tr}[\alpha_{\mu||}(x^\mu)]^2\,, \tag{4}$$

which explicitly breaks the chiral symmetry as well as the HLS, because $\alpha_{\mu||} \to h \cdot \alpha_{\mu||} \cdot h^\dagger - i\partial_\mu h \cdot h^\dagger$. Naive truncation of tower of vector meson fields thus forces us to encounter the explicit violation of the chiral symmetry.

Now we shall propose a method to truncate towers of vector and axialvector meson fields in a gauge-invariant manner. Consider a low-energy effective theory below the mass of $n = N+1$ level. Such an effective theory can be obtained by integrating out mesons with $n \geq N+1$ via the equations of motion. Neglecting terms including the derivatives acting on the heavy fields $V_\mu^{(k)}$ and $A_\mu^{(k)}$ with $k > N$, the equations of motion for them take the following forms: $V_\mu^{(k)} = \alpha_{\mu||}$, $A_\mu^{(k)} = 0$ $(k = N+1, N+2, \cdots, \infty)$. Putting these solutions into the action, we have, instead of Eq.(4),

$$S_5^{\text{integrate out}} \ni \int dz d^4x\, K_2(z) \sum_{n=1}^{N} \lambda_{2n-1}\psi_{2n-1}^2(z) \text{tr}[\alpha_{\mu||}(x^\mu) - V_\mu^{(n)}(x^\mu)]^2\,, \tag{5}$$

which is certainly gauge-invariant. Note also that the gauge-invariance now requires not the constraint in Eq.(3) but $\phi^R(z) + \phi^L(z) - \sum_{n=1}^{N} \psi_{2n-1}(z) = 1$.

Let us now consider a low-energy effective model obtained by integrating out all the higher vector and axialvector meson fields except the lowest vector meson field $V_\mu^{(1)} \equiv V_\mu$. Following the gauge-invariant way proposed above, the expansion of $A_\mu(x^\mu, z)$ is expressed as

$$A_\mu(x^\mu, z) = \alpha_{\mu\perp}(x^\mu)\psi_0(z) + (\hat{\alpha}_{\mu||}(x^\mu) + V_\mu(x^\mu)) + \hat{\alpha}_{\mu||}(x^\mu)\psi_1(z)\,, \tag{6}$$

where $\hat{\alpha}_{\mu||} = -V_\mu + \alpha_{\mu||}$. One can further introduce the external gauge fields by gauging the global chiral $U(N)_L \times U(N)_R$ symmetry. (For details, see Ref. 12.) Then we obtain the low-energy effective model including only the lightest HLS field V_μ and the NGB fields π described by the HLS-ChPT with $\mathcal{O}(p^4)$ terms:[4,6] The $\mathcal{O}(p^4)$ terms include the effects from infinite towers of higher KK modes and are completely determined as explicitly shown in Ref. 12; sum rules such as those introduced in Ref. 9 are also fully built in the HLS-ChPT Lagrangian. Our formulation is thus more practical and useful.

Finally, we once again emphasize that our methodology presented here is applicable to any models of HQCD.

3. Application to Sakai-Sugimoto Model

In this section, we apply our methodology to the Sakai-Sugimoto (SS) model[8,9] based on $D8/\bar{D}8/D4$ brane configuration. The five-dimensional gauge-invariant portion (so-called the Dirac-Born-Infeld (DBI) part) of the low-energy effective action in the SS model is given by[8,9]

$$S_{\rm SS}^{\rm DBI} = N_c G \int d^4x dz \left(-\frac{1}{2} K^{-1/3}(z) {\rm tr}[F_{\mu\nu}F^{\mu\nu}] + K(z) M_{\rm KK}^2 {\rm tr}[F_{\mu z}F^{\mu z}] \right), \quad (7)$$

where $K(z) = 1 + z^2$ is the induced metric of the five-dimensional space-time; the overall coupling G is the rescaled 't Hooft coupling expressed as $G = N_c g_{\rm YM}^2/(108\pi^3)$ with $g_{\rm YM}$ being the gauge coupling of the $U(N_c)$ gauge symmetry on the N_c D4-branes;[8,9] the mass scale M_{KK} is related to the scale of the compactification of the N_c D4-branes onto the S^1. Comparing Eq.(7) with Eq.(1), we read off $K_1(z) = GK^{-1/3}(z)$, $K_2(z) = GK(z)$, so that we find the equation of motion, $-K^{1/3}(z)\partial_z\left(K(z)\partial_z\psi_n\right) = \lambda_n\psi_n$ with the eigenvalues λ_n and the eigenfunctions ψ_n of the KK modes of the five-dimensional gauge field $A_\mu(x^\mu, z)$.

Application to the Chern-Simons term is straightforward that is explicitly demonstrated in Ref. 12.

As emphasized in the end of the previous section, without introducing any sum rules, we are able to calculate amplitudes straightforwardly from the effective Lagrangian which includes contributions from *full* set of higher KK modes. To see it more explicitly, as an example, we shall study the pion electromagnetic (EM) form factor $F_V^{\pi^\pm}$ at tree-level of the present model. $F_V^{\pi^\pm}$ is readily constructed from the Lagrangian written in terms of the HLS-ChPT presented in Ref. 12: $F_V^{\pi^\pm}(Q^2)|_{\rm HLS} = g_{\gamma\pi\pi}(Q^2) + \frac{g_\rho(Q^2)g_{\rho\pi\pi}(Q^2)}{m_\rho^2+Q^2}$, where $Q^2 = -p^2$ denotes a momentum-squared in space-like region and[12] $g_{\gamma\pi\pi}(Q^2) = \left(1 - \frac{a}{2}\right) + \frac{ag^2z_6}{4}\frac{Q^2}{m_\rho^2}$, $g_\rho(Q^2) = \frac{m_\rho^2}{g}\left(1 + g^2z_3\frac{Q^2}{m_\rho^2}\right)$, $g_{\rho\pi\pi}(Q^2) = \frac{1}{2}ag\left(1 + \frac{g^2z_4}{2}\frac{Q^2}{m_\rho^2}\right)$. The applicable momentum range should be restricted to $0 \le Q^2 \ll \{m_{\rho'}^2, m_{\rho''}^2, \cdots\}$ since we have integrated out higher KK modes keeping only the ρ meson. Note that our form factor $F_V^{\pi^\pm}(Q^2)|_{\rm HLS}$ automatically ensures the EM gauge-invariance, $F_V^{\pi^\pm}(0)|_{\rm HLS} = 1$:

One can easily show[12] that if towers of vector and axialvector mesons had naively been truncated as in Eq.(4) one would have $F_V^{\pi^\pm}(0)|_{\rm truncation} \neq 1$, leading to a violation of the EM gauge-invariance. (It turns out[12] that higher KK modes actually play the crucial role to maintain the EM gauge-invariance.) We further rewrite $F_V^{\pi^\pm}(Q^2)|_{\rm HLS}$ as [d]

$$F_V^{\pi^\pm}(Q^2)|_{\rm HLS} = \left(1 - \frac{1}{2}\tilde{a}\right) + \tilde{z}\frac{Q^2}{m_\rho^2} + \frac{g_\rho g_{\rho\pi\pi}}{m_\rho^2 + Q^2}\,, \tag{8}$$

where $\tilde{a} = a\left(1 - \frac{g^2 z_4}{2} - g^2 z_3 + \frac{(g^2 z_3)(g^2 z_4)}{2}\right)$ and $\tilde{z} = \frac{1}{4}a\left(g^2 z_6 + (g^2 z_3)(g^2 z_4)\right)$[12] which are expressed in terms of the five-dimensional theory as [e][12] $\tilde{a} = \frac{2 g_\rho g_{\rho\pi\pi}}{m_\rho^2} = \frac{\pi}{4}\lambda_1 \frac{\langle\psi_1\rangle\langle\psi_1(1-\psi_0^2)\rangle}{\langle\psi_1^2\rangle}$, $\tilde{z} = \frac{\pi}{8}\lambda_1\left(\frac{\langle\psi_1\rangle\langle\psi_1(1-\psi_0^2)\rangle}{\langle\psi_1^2\rangle} - \langle 1 - \psi_0^2\rangle\right)$. These $\tilde{a}$ and $\tilde{z}$ are calculated independently of any inputs to be $\tilde{a} \simeq 2.62$ and $\tilde{z} \simeq 0.08$, where we have used $\lambda_1 \simeq 0.669$.

Using the expression (8) and the values of $\tilde{a}$ and $\tilde{z}$, we evaluate the momentum-dependence of $F_V^{\pi^\pm}$ which was actually not possible in the original SS model[9] because of the form written in terms of the infinite summation. In Fig. 1 we show the predicted curve of $F_V^{\pi^\pm}$ with respect to Q^2 together with the experimental data from Ref. 13. The χ^2-fit results in good agreement with the data (χ^2/d.o.f $= 147/53 \simeq 2.77$). Comparison with the result derived from the lowest vector meson dominance (LVMD) hypothesis with $\tilde{a} = 2$ and $\tilde{z} = 0$ is shown by a dashed curve in the left panel of Fig. 1. The right panel of Fig. 1 shows a comparison with the result obtained by fitting the parameters $(\tilde{a}, \tilde{z})$ to the experimental data. It is interesting to note that the best-fit values of $\tilde{a}$ and $\tilde{z}$ are quite close to those in the predicted curve. This fact reflects that the predicted curve fits well with the experimental data.

4. Summary

In this talk, we developed a methodology to integrate out arbitrary parts of infinite towers of vector and axialvector mesons arising as KK modes in a class of HQCD models. It was shown that our method is gauge-invariant in contrast to a naive truncation [See Eq.(4)]. It was demonstrated that any models of HQCD in the low-energy region can be described by the HLS-ChPT. We applied our method to the SS model and evaluated the momentum-dependence of the pion EM form factor as an example, which demonstrated power of our formulation. The predicted form factor was shown to be fitted well with the experimental data in the low-energy (space-like momentum) region. This was difficult in the original SS model due to the forms of the form factors written in terms of infinite sum of vector meson exchanges. More on phenomenological applications of our formulation to the SS model is presented in Ref. 12.

---

[d] Here $g_\rho \equiv g_\rho(Q^2 = -m_\rho^2)$ and $g_{\rho\pi\pi} \equiv g_{\rho\pi\pi}(Q^2 = -m_\rho^2)$.
[e] $\langle A\rangle \equiv \int dz K^{-1/3}(z) A(z)$.

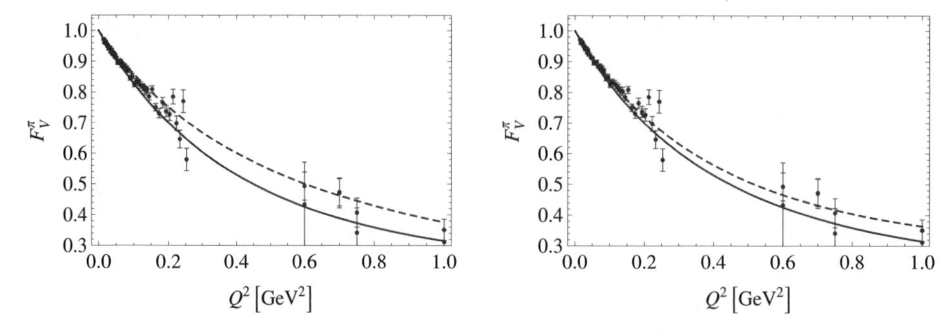

Fig. 1. The predicted curve of the pion EM form factor $F_V^{\pi^\pm}$ with respect to space-like momentum-squared Q^2 (denoted by solid line) fitted with the experimental data from Ref. 13 with χ^2/d.o.f $= 147/53 \simeq 2.77$. In the left panel, the dashed curve corresponds to the form factor in the LVMD hypothesis of the HLS model with $\tilde{a} = 2$ and $\tilde{z} = 0$ taken. (χ^2/d.o.f $= 226/53 \simeq 4.26$). The dashing curve in the right panel corresponds to the form factor fitted with the experimental data, yielding the best fit values of $\tilde{a}$ and $\tilde{z}$, $\tilde{a}|_{\rm best} = 2.44$, $\tilde{z}|_{\rm best} = 0.08$ (χ^2/d.o.f $= 81/51 \simeq 1.56$).

Acknowledgments

This work was supported in part by the JSPS Grant-in-Aid for Scientific Research; (B) 18340059 (K.Y.), (C) 20540262 (M.H.), Innovative Areas #2104 "Quest on New Hadrons with Variety of Flavors" (M.H.), the Global COE Program "Quest for Fundamental Principles in the Universe" (K.Y. and M.H.), and the Daiko Foundation (K.Y.). S.M. is supported by the Korea Research Foundation Grant funded by the Korean Government (KRF-2008-341-C00008).

References

1. J. M. Maldacena, Adv. Theor. Math. Phys. **2**, 231 (1998); Int. J. Theor. Phys. **38**, 1113 (1999); S. S. Gubser, I. R. Klebanov and A. M. Polyakov, Phys. Lett. B **428**, 105 (1998); E. Witten, Adv. Theor. Math. Phys. **2**, 253 (1998); O. Aharony, S. S. Gubser, J. M. Maldacena, H. Ooguri and Y. Oz, Phys. Rept. **323**, 183 (2000).
2. C. T. Hill, S. Pokorski and J. Wang, Phys. Rev. D **64**, 105005 (2001) [arXiv:hep-th/0104035].
3. M. Bando, T. Kugo, S. Uehara, K. Yamawaki and T. Yanagida, Phys. Rev. Lett. **54**, 1215 (1985); M. Bando, T. Kugo and K. Yamawaki, Nucl. Phys. B **259**, 493 (1985); M. Bando, T. Fujiwara, and K. Yamawaki, Prog. Theor. Phys. **79** (1988) 1140; M. Bando, T. Kugo and K. Yamawaki, Phys. Rept. **164**, 217 (1988).
4. M. Harada and K. Yamawaki, Phys. Rept. **381**, 1 (2003).
5. J. Gasser and H. Leutwyler, Annals Phys. **158**, 142 (1984); Nucl. Phys. B **250**, 465 (1985).
6. M. Tanabashi, arXiv:hep-ph/9306237; Phys. Lett. B **316**, 534 (1993).
7. M. Harada, S. Matsuzaki and K. Yamawaki, Phys. Rev. D **74**, 076004 (2006) [arXiv:hep-ph/0603248].
8. T. Sakai and S. Sugimoto, Prog. Theor. Phys. **113**, 843 (2005) [arXiv:hep-th/0412141].
9. T. Sakai and S. Sugimoto, Prog. Theor. Phys. **114**, 1083 (2006) [arXiv:hep-th/0507073].
10. L. Da Rold and A. Pomarol, Nucl. Phys. B **721**, 79 (2005) [arXiv:hep-ph/0501218].

11. J. Erlich, E. Katz, D. T. Son and M. A. Stephanov, Phys. Rev. Lett. **95**, 261602 (2005) [arXiv:hep-ph/0501128].
12. M. Harada, S. Matsuzaki, and K. Yamawaki, in preparation.
13. S. R. Amendolia *et al.* [NA7 Collaboration], Nucl. Phys. B **277**, 168 (1986); J. Volmer *et al.* [The Jefferson Lab F(pi) Collaboration], Phys. Rev. Lett. **86**, 1713 (2001); V. Tadevosyan *et al.* [Jefferson Lab F(pi) Collaboration], Phys. Rev. C **75**, 055205 (2007) [arXiv:nucl-ex/0607007]; T. Horn *et al.* [Jefferson Lab F(pi)-2 Collaboration], Phys. Rev. Lett. **97**, 192001 (2006) [arXiv:nucl-ex/0607005].

Holographic Heavy Quarks and the Giant Polyakov Loop

Gianluca Grignani
Dipartimento di Fisica, Università di Perugia and I.N.F.N. Sezione di Perugia
Via Pascoli, 06123 Perugia, Italy

Joanna Karczmarek and Gordon W. Semenoff
Department of Physics and Astronomy, University of British Columbia
6224 Agricultural Road, Vancouver, British Columbia, Canada V6T 1Z1

We consider the Polyakov loop in finite temperature planar $\mathcal{N} = 4$ supersymmetric Yang-Mills theory defined on a spatial S^3 and in representations where the number of boxes in the Young Tableau k scales so that $\frac{k}{N}$ remains finite in the large N limit. We review the argument that, in the deconfined phase of the gauge theory, and for symmetric representations with row Young tableau, there is a quantum phase transition in the expectation value of the Polyakov loop operator which occurs as the size of the representation is increased.

Keywords: Holography, confinement, character.

1. Introduction

AdS/CFT holography identifies the deconfinement phase transition of planar $\mathcal{N} = 4$ Yang-Mills theory with the collapse of hot Anti-de Sitter space to an anti-de Sitter-Schwarzchild black hole.[1] The planar limit of Yang-Mills theory is achieved by taking the large N limit while holding the 't Hooft coupling $\lambda = g_{YM}^2 N$ fixed. The finite temperature gauge theory is defined on $S^3 \times S^1$ where space is the 3-sphere and S^1 is periodic Euclidean time. Since $\mathcal{N} = 4$ is an adjoint gauge theory, it has a center symmetry whose spontaneous breaking is associated with de-confinement.[6,7] As we shall review in the following, a deconfining phase transition can be identified by analyzing the gauge theory in the perturbative regime.[2,3]

In this Article, we shall review some of our recent work on the properties of the deconfined phase of the gauge theory.[4] Before that, in the next Section, we shall discuss the motivation.

2. Giant Wilson loops

In the duality between large N gauge field theory and string theory, the expectation value of the Wilson loop normally corresponds to a disc amplitude for an open fundamental string. This has been made precise for the Maldacena-Wilson loop[10] of the $\mathcal{N}=4$ theory,

$$W_{\rm M}[C] = \left\langle {\rm Tr}\mathcal{P}e^{\oint_C d\tau\left(iA_\mu(x)\dot{x}^\mu(\tau)+\phi^I(x)\theta^I|\dot{x}(\tau)|\right)}\right\rangle \tag{1}$$

where $x^\mu(\tau)$ parametrizes a closed curve C. $\phi^I(x)$, $I=1,...,6$ are the scalar quark fields of $\mathcal{N}=4$ super Yang-Mills theory and θ^I is a unit 6-vector. When λ is large, this loop is computed by finding the minimum area of a fundamental string worldsheet which has boundary located on the contour C placed at the boundary of AdS_5 and at a point θ^I on S^5. If the contour C in (1) links periodic Euclidean time in the finite temperature Yang-Mills theory, $W_{\rm M}[C]$ carries center charge and it can be nonzero only when the center symmetry is spontaneously broken. Then, the boundary of the disc must links periodic Euclidean time in AdS background. Whether such a disc exists depends on whether the time circle is contractible. It is not contractible in the hot AdS background, and it is contractable on the black hole background. This was pointed out by Witten as further evidence for the identification of the black hole transition with deconfinement.[1]

It is known that, in the zero temperature Yang-Mills theory defined on a spatial R^3, an interesting phenomenon occurs for loops in large representations. When the number of boxes k in the Young tableau is large so that $\frac{k}{N}$ is finite in the large N limit, the dual fundamental string worldsheet is replaced by a probe D-brane with k units of world-volume electric flux,[11] studying $\frac{1}{2}$-BPS loops, where some results are known for all values of the coupling constant.[12] For the anti-symmetric representation, the dual is a D5-brane whose world volume is a direct product of $AdS_2 \subset AdS_5$ and $S^4 \subset S^5$. For a symmetric representation, it is a D3-brane with world volume $AdS_2 \times S^2 \subset AdS_5$.

The interesting question of whether these D-branes exist in the finite temperature geometry, particularly where they would be dual to a gauge theory loop linking periodic Euclidean time was studied by Hartnoll and Kumar[5] who searched for solutions of the appropriate Born-Infeld actions on the AdS black hole background. For the D5-brane wrapped on $S^4 \subset S^5$ which corresponds to a totally antisymmetric representation on the gauge theory side, there seem to be solutions for any $\frac{k}{N}$ with the usual cutoff at $k=N$ dictated by the maximum size of an antisymmetric representation in gauge theory side and a maximum radius for embedding S^4 in S^5 supergravity. However, in the case of the D3-brane, which should correspond to a totally symmetric representation, Hartnoll and Kumar could not find any solutions at all. This fundamental difference between the two cases is what motivated our work on the gauge theory which is summarized in Ref.[4] and which we are reviewing here.

3. Effective field theory

The confinement problem in $\mathcal{N}=4$ super Yang-Mills theory on spatial S^3 can be studies at weak coupling,[2,3] Due to the curvature of S^3, the vector, scalar and spinor fields are gapped. The temporal component of the gauge field A_0 has a zero mode on the S^3. Then, in the regime where the temperature is much less than the gap, $T = \frac{1}{\beta} << 1$ (we choose the S^3 to have unit radius), an effective field theory for the zero-mode of A_0 can be found by integrating out all of the other degrees of freedom. This effective field theory can be used to study the realization of the center symmetry. It turns out to be a unitary matrix model where the order parameter for center symmetry breaking is the expectation value of the trace of the matrix,

$$\langle \mathrm{Tr} U \rangle = \frac{\int [dU] e^{-S_{\mathrm{eff}}[U]} \mathrm{Tr} U}{\int [dU] e^{-S_{\mathrm{eff}}[U]}} \tag{2}$$

where U is a unitary matrix, $[dU]$ is the Haar measure. To one-loop order, the effective action is

$$S_{\mathrm{eff}}[U] = -\sum_{n=1}^{\infty} \left[z_B(x^n) + (-1)^{n+1} z_F(x^n) \right] \frac{|\mathrm{Tr} U^n|^2}{n} \tag{3}$$

where

$$x = e^{-\frac{\beta}{R}} \ , \ z_B(x) = \frac{6x + 12x^2 - x^3}{(1-x)^3} \ , \ z_F(x) = \frac{16 x^{\frac{3}{2}}}{(1-x)^3} \, . \tag{4}$$

The effective action inherits symmetries from its parent theory: gauge invariance:

$$S_{\mathrm{eff}}[U] = S_{\mathrm{eff}}[VUV^{-1}] \tag{5}$$

and center symmetry: $S_{\mathrm{eff}}[U] = S_{\mathrm{eff}}[cU]$. The expectation value (2) transforms under the center transformation and would average to zero when the center symmetry is not broken.

Gauge invariance (5) allows one to diagonalize the unitary matrices to form a model of the eigenvalues. It can then be solved by a saddle-point method in the large N limit. It is found that it has a phase transition. The expectation value vanishes in the low temperature phase and it is nonzero in the high temperature phase. This is interpreted as a spontaneous breaking of center symmetry which occurs with the deconfinement transition. Note that, even though it is normally thought of as a strong coupling phenomenon, it is seen in this theory at weak coupling. For the effective action (3), the phase transition occurs at $T_C \simeq 0.38$ which is marginal to the regime $T << 1$. We will assume that it is within the range of validity of the effective field theory technique. In the following we will explore the deconfined phase. We will assume that we are at temperatures just above the critical one and we will assume that the effective matrix model gives an accurate description of the physics there.

4. Higher representations

The unitary matrix model can be used to calculate the center symmetry order parameter in any irreducible representation R of the $SU(N)$ gauge group,

$$\langle \mathrm{Tr}_R U(x) \rangle = \frac{\int [dU] e^{-S_{\mathrm{eff}}[U]} \mathrm{Tr}_R \, U}{\int [dU] e^{-S_{\mathrm{eff}}[U]}} \, . \tag{6}$$

This quantity can have interesting behavior which depends on the size and nature of the representation. Two types of representation are easy to analyze: a completely symmetric representation $\mathcal{S}_k$ whose Young tableau is a single row with k boxes and completely antisymmetric representation $\mathcal{A}_k$ whose Young tableau is a single column with k boxes.

Note that we have not normalized the operator $\mathrm{Tr}_R U$ by dividing by a factor of the dimension of the representation. Our reason for not doing so is to be able to compare our results directly with holographic duality where the appropriate operator is the un-normalized one.

The center charge of a representation is equal to the number of boxes in the Young Tableau corresponding to that representation, modulo N. Thus, both representations $\mathcal{S}_k$ and $\mathcal{A}_k$ have center charge k mod N. The expectation value (6) is therefore expected to vanish in the confined phase when this charge is non-zero. On the other hand, the expectation value can be non-zero in the deconfined phase.

If the matrix were diagonal, $U = \mathrm{diag}[e^{i\phi_1}, ..., e^{i\phi_N}]$, the permutation symmetry can be used to order the eigenvalues in a completely symmetric or completely antisymmetric representation so that they occur in order of non-decreasing index:

$$\mathrm{Tr}_{\mathcal{S}_k} U = \sum_{a_1 \leq a_2 \leq ... \leq a_k} e^{i\phi_{a_1}} e^{i\phi_{a_2}} ... e^{i\phi_{a_k}} \ , \ \ \mathrm{Tr}_{\mathcal{A}_k} U = \sum_{a_1 < a_2 < ... < a_k} e^{i\phi_{a_1}} e^{i\phi_{a_2}} ... e^{i\phi_{a_k}} \tag{7}$$

It is convenient to obtain these expressions from generating functions[9]

$$\mathrm{Tr}_{\mathcal{S}_k} U = \oint \frac{dt}{2\pi i t^{k+1}} \prod_{a=1}^{N} \frac{1}{1 - te^{i\phi_a}} \ , \ \ \mathrm{Tr}_{\mathcal{A}_k} U = \oint \frac{dt}{2\pi i t^{k+1}} \prod_{a=1}^{N} \left(1 + te^{i\phi_a}\right) \tag{8}$$

where the contour integral over t projects onto the term in a Taylor expansion of the integrand which contains k eigenvalues. The contour encircles the origin. It can be moved away from the origin if it does not cross singularities of the integrand. In the case of the anti-symmetric representation (8) when N is finite, the integrand is a polynomial and the contour can be moved anywhere. For the symmetric representation (8) it should remain within the unit circle. The covariant expressions for the free energies are gotten by taking logs of (8)

$$\beta \Gamma_{\mathcal{S}_k/\mathcal{A}_k} = -\frac{1}{N} \ln \frac{1}{2\pi i} \oint dt \frac{1}{t^{k+1}} \langle \exp\left[\mp \mathrm{Tr} \ln(1 \mp tU)\right] \rangle \ , \tag{9}$$

(9) have center charge k modulo N.

In the large N limit, the quantities in (9) can be computed using two saddle point approximations. The first integrates over unitary matrices in (6). At large N, the eigenvalues of U become classical variables and their distribution is found by minimizing S_{eff} plus a Jacobian from the unitary integral measure. As long as $k \ll N^2$, the loop operators in (9) do not modify the eigenvalue distribution in the leading order at large N. It is given by a density $\rho(\phi)$. In the large N limit the expectation values in Eq. (9) are computed using the eigenvalue density,[a]

$$\beta\Gamma_{\mathcal{S}_k/\mathcal{A}_k} = -\frac{1}{N}\ln\frac{1}{2\pi i}\oint dt\frac{1}{t}\exp\left(\mp N\int_{-\pi}^{\pi}d\phi\rho(\phi)\ln(1\mp te^{i\phi}) - k\ln t\right). \quad (10)$$

The second use of a saddle-point approximation is to evaluate the integral over t in (10). Let $\hat{t}$ satisfy the saddle-point equation

$$R_{\mathcal{S}_k}(\hat{t}) \equiv \int_{-\pi}^{\pi}d\phi\rho(\phi)\frac{\hat{t}e^{i\phi}}{1-\hat{t}e^{i\phi}} = \frac{k}{N}\ , \quad R_{\mathcal{A}_k}(\hat{t}) \equiv \int_{-\pi}^{\pi}d\phi\rho(\phi)\frac{\hat{t}e^{i\phi}}{1+\hat{t}e^{i\phi}} = \frac{k}{N}\,. \quad (11)$$

The functions $R_{\mathcal{S}_k/\mathcal{A}_k}(t)$ in (11) are related to the resolvent of the matrix model and are holomorphic functions of t with cut singularities on the unit circle determined by the support of $\rho(\phi)$. Once the solution $\hat{t}$ is determined, the free energy is given by

$$\beta\Gamma_{\mathcal{S}_k/\mathcal{A}_k} = \pm\int_{-\pi}^{\pi}d\phi\rho(\phi)\ln(1\mp\hat{t}e^{i\phi}) + \frac{k}{N}\ln\hat{t}\,. \quad (12)$$

Before we proceed further, let us consider a simple example, the confined phase. Center symmetry is an invariance under a simultaneous translation of all eigenvalues $\phi_a \to \phi_a + 2\pi/N$. In the center-symmetric confined phase, the distribution is translation invariant, eigenvalues are uniformly distributed on the unit circle and $\rho_{\text{conf}} = \frac{1}{2\pi}$. We can integrate over ϕ in the saddle-point equations (11),

$$R_{\mathcal{S}_k}(\hat{t}) = \begin{cases} 0 & |\hat{t}|<1 \\ -1 & |\hat{t}|>1 \end{cases} = \frac{k}{N}\ , \quad R_{\mathcal{A}_k}(\hat{t}) = \begin{cases} 0 & |\hat{t}|<1 \\ 1 & |\hat{t}|>1 \end{cases} = \frac{k}{N}\,. \quad (13)$$

In the case of the symmetric representation, the saddle-point equation (13) has a solution only when $\frac{k}{N} = 0$. We interpret the absence of a solution when $\frac{k}{N} \neq 0$ as meaning that the expectation value vanishes. Certainly, if there is no saddle-point of a periodic function of a variable ϕ, the integration is not dominated by any particular value of ϕ and ϕ must be integrated over its entire range. This would average the expectation value of any operator with non-zero center charge to zero. It is in the other case, when there is a saddle point, where the large N limit forces one to evaluate the integrand at the saddle point and the expectation value is generically non-zero. Similarly, for the anti-symmetric representation, the saddle-point equation (13) has a solution only when either $\frac{k}{N} = 0$ or $\frac{k}{N} = 1$, the two cases where the antisymmetric representation is center neutral. This is also interpreted as confinement, the expectation value vanishes in all other cases. Note that it has an expected $k \to N-k$ duality, though it comes from interchanging two saddle points, one with $|\hat{t}| < 1$ and one with $|\hat{t}| > 1$. Neither of these saddle-points alone exhibit this duality.

[a]We will argue that using the leading order N^0 density $\rho(\phi)$ in (10) is sufficient to obtain $\beta\Gamma_{\mathcal{S}_k/\mathcal{A}_k}$ to leading order N^0 accuracy.

5. Giant loop phase transition

In the following, we will focus on the symmetric representation $\mathcal{S}_k$. Consider, for example, the eigenvalue distribution $\rho(\phi) = \frac{1}{2\pi}(1 + 2p\cos\phi)$ which solves the large N limit of the matrix model (2) precisely at the critical point. It also describes the strong-coupling phase of large N 2-dimensional lattice Yang-Mills theory.[8] The parameter $p = \frac{1}{N}\langle \text{Tr } U\rangle = \int d\phi\rho(\phi)e^{i\phi}$ is the fundamental representation loop. Positivity of the density requires $0 \le p \le \frac{1}{2}$. This distribution depends on ϕ and therefore is deconfined.

There is one solution of $R_{\mathcal{S}_k}(\hat{t}) = \frac{k}{N}$ in the region $|\hat{t}| < 1$ at $\hat{t} = \frac{k}{N}/p$. (If $\frac{k}{N}$ and p are such that $|\hat{t}| > 1$, both $R_{\mathcal{S}_k}$ and $\Gamma_{\mathcal{S}_k}$ should be extended there by analytic continuation.) The free energy is

$$\Gamma_{\mathcal{S}_k} = \frac{k}{N}\ln\left[\frac{k/N}{ep}\right] , \tag{14}$$

where $e = 2.718\ldots$. $\Gamma_{\mathcal{S}_k}$ has the interesting feature that, as $\frac{k}{N}$ is increased, it changes sign from negative to positive. This results in a phase transition which occurs when $\frac{k}{N} = \left(\frac{k}{N}\right)_{\text{crit}} = ep$. When $\frac{k}{N} < \left(\frac{k}{N}\right)_{\text{crit}}$, $\Gamma_{\mathcal{S}_k}$ is negative and the loop expectation value, $e^{-N\Gamma}$, is exponentially large. When $\frac{k}{N} > \left(\frac{k}{N}\right)_{\text{crit}}$, $\Gamma_{\mathcal{S}_k}$ is positive and the loop vanishes for $N \to \infty$. This phase transition implies that, even in the deconfined phase, where the fundamental representation loop is non-zero, sufficiently large symmetric representations are still confined.

To see this behavior in another example, consider the semi-circle distribution which, for $|\phi| < 2\arcsin\sqrt{2-2p}$, is

$$\rho(\phi) = \frac{\cos\frac{\phi}{2}}{\pi(2-2p)}\sqrt{2 - 2p - \sin^2\frac{\phi}{2}} \tag{15}$$

and which vanishes in the gap $2\arcsin\sqrt{2-2p} \le |\phi| \le \pi$. We still use the fundamental loop, p, as a parameter and now $\frac{1}{2} \le p \le 1$. This is the distribution in the weak coupling phase of 2-dimensional lattice Yang-Mills theory.[8] It is also an approximation to the deconfined distribution for weakly coupled $\mathcal{N} = 4$ Yang-Mills theory.[3] For sufficiently weak coupling, it is accurate near the phase transition where $p = \frac{1}{2}$. The saddle point computation can be done explicitly near $t = 0$ and analytically continued. The free energy is

$$\begin{aligned}\Gamma_{\mathcal{S}_k} &= (2\theta\cosh\theta - \sinh\theta)\,\frac{\sinh\theta + \sqrt{\sinh^2\theta + 2 - 2p}}{2-2p} \\ &\quad - \frac{1}{2} - \ln\left[\frac{\sinh\theta + \sqrt{\sinh^2\theta + 2 - 2p}}{2-2p}\right] ,\end{aligned} \tag{16}$$

where θ is defined by $\hat{t} = e^{2\theta}$ and is determined by the saddle-point equation

$$\frac{k}{N} + \frac{1}{2} = \cosh\theta\left[\frac{\sinh\theta + \sqrt{\sinh^2\theta + 2 - 2p}}{2-2p}\right] , \tag{17}$$

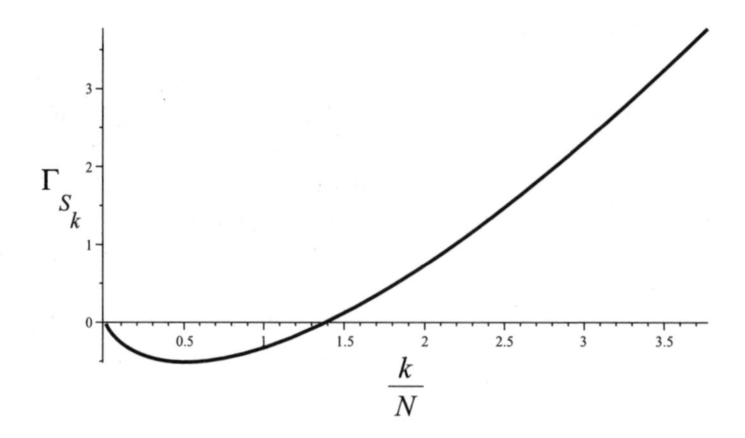

Fig. 1. The free energy $\Gamma_{\mathcal{S}_k}(\theta)$ as a function of $\frac{k}{N}$ in the semi-circle distribution with $p = 0.51$.

which can be solved for $\sinh(\theta)$. The free energy is plotted in Fig. 1. With $p = 0.51$, the free energy becomes positive at $\theta \simeq 0.50$ which corresponds to $\frac{k}{N}_{\text{crit.}} \simeq 1.4$. In fact, we can see that the phase transition occurs over the whole allowed range of p and is surprisingly insensitive to the value of p.

Now, we turn to the antisymmetric representation. For a large class of distributions gapped around $\phi = \pi$ and with $\frac{dR_{\mathcal{A}_k}(\hat{t})}{d\hat{t}} > 0$, which includes the semi-circle distribution (15), it can be shown that that $\Gamma_{\mathcal{A}_k}$ is always negative and the phase transition that we are discussing does not occur. We refer the reader to Ref. [4] for the details.

Our conclusion is that, in the weakly coupled Yang-Mills theory, the symmetric and antisymmetric representations have markedly different behavior. The symmetric one has a phase transition to a phase where the loop expectation value vanishes, even in the deconfined phase, whereas the antisymmetric one does not. This is qualitatively similar to what is seen in the string theory dual[5] where the D5-brane dual to the antisymmetric representation exists for all allowed values of $\frac{k}{N}$ whereas the D3-brane dual to the antisymmetric representation is absent. The apparent absence of a critical value of $\frac{k}{N}$ at strong coupling implies tha $\left[\frac{k}{N}\right]_{\text{crit.}}(T, \lambda)$ becomes coupling constant dependent and goes to zero faster than $\frac{4}{\sqrt{\lambda}}$ as $\lambda \to \infty$.

References

1. E. Witten, Adv. Theor. Math. Phys. **2**, 505 (1998) [arXiv:hep-th/9803131].
2. B. Sundborg, Nucl. Phys. B **573**, 349 (2000) [arXiv:hep-th/9908001].
3. O. Aharony, J. Marsano, S. Minwalla, K. Papadodimas and M. Van Raamsdonk, Adv. Theor. Math. Phys. **8**, 603 (2004) [arXiv:hep-th/0310285].
4. G. Grignani, J. L. Karczmarek and G. W. Semenoff, arXiv:0904.3750 [hep-th].
5. S. A. Hartnoll and S. Prem Kumar, Phys. Rev. D **74**, 026001 (2006) [arXiv:hep-th/0603190].
6. A. M. Polyakov, Phys. Lett. B **72**, 477 (1978).

7. L. Susskind, Phys. Rev. D **20**, 2610 (1979).
8. D. J. Gross and E. Witten, Phys. Rev. D **21** (1980) 446.
9. S. A. Hartnoll and S. P. Kumar, JHEP **0608**, 026 (2006) [arXiv:hep-th/0605027].
10. J. M. Maldacena, Phys. Rev. Lett. **80**, 4859 (1998) [arXiv:hep-th/9803002].
11. N. Drukker and B. Fiol, JHEP **0502**, 010 (2005) [arXiv:hep-th/0501109]. S. Yamaguchi, Int. J. Mod. Phys. A **22**, 1353 (2007) [arXiv:hep-th/0601089]; S. Yamaguchi, JHEP **0605**, 037 (2006) [arXiv:hep-th/0603208]. J. Gomis and F. Passerini, JHEP **0608**, 074 (2006) [arXiv:hep-th/0604007]; J. Gomis and F. Passerini, JHEP **0701**, 097 (2007) [arXiv:hep-th/0612022]; K. Okuyama and G. W. Semenoff, JHEP **0606**, 057 (2006) [arXiv:hep-th/0604209];
12. J. K. Erickson, G. W. Semenoff and K. Zarembo, Nucl. Phys. B **582**, 155 (2000) [arXiv:hep-th/0003055]; V. Pestun, arXiv:0712.2824 [hep-th].

Effect of Vector–Axial-Vector Mixing to Dilepton Spectrum in Hot and/or Dense Matter*

Masayasu Harada

Department of Physics, Nagoya University, Nagoya, 464-8602, Japan
E-mail: harada@hken.phys.nagoya-u.ac.jp

Chihiro Sasaki

Frankfurt Institute for Advanced Studies, J.W. Goethe University, D-60438 Frankfurt, Germany
Physik-Department, Technische Universität München, D-85747 Garching, Germany
E-mail: sasaki@fias.uni-frankfurt.de

In this write-up we summarize main results of our recent analyses on the mixing between transverse ρ and a_1 mesons in hot and/or dense matter. We show that the axial-vector meson contributes significantly to the vector spectral function in hot matter through the mixing. In dense baryonic matter, we include a mixing through a set of $\omega\rho a_1$-type interactions. We show that a clear enhancement of the vector spectral function appears below $\sqrt{s} = m_\rho$ for small three-momenta of the ρ meson, and thus the vector spectrum exhibits broadening.

1. Introduction

In-medium modifications of hadrons have been extensively explored in the context of chiral dynamics of QCD.[3,4] Due to an interaction with pions in the heat bath, the vector and axial-vector current correlators are mixed. At low temperatures or densities a low-energy theorem based on chiral symmetry describes this V-A mixing.[5] The effects to the thermal vector spectral function have been studied through the theorem,[6] or using chiral reduction formulas based on a virial expansion.[7]

In Ref. 1, it was shown that the effects of the V-A mixing, and how the axial-vector mesons affect the spectral function near the chiral phase transition, within an effective field theory. The analysis was carried out assuming several possible patterns of chiral symmetry restoration: dropping or non-dropping ρ meson mass along with changing a_1 meson mass, both considered to be options from a phenomenological point of view.

In Ref. 2, we studied a novel effect of the V-A mixing through a set of $\omega\rho a_1$-type interactions at finite baryon density, which was introduced by a Chern-Simons term in a holographic QCD model.[8] We focused on the V-A mixing at tree level and its consequence on the in-medium spectral functions which are the main input to the

*This talk is based on the work done in Refs. 1, 2.

experimental observables. We showed that the mixing produces a clear enhancement of the vector spectral function below $\sqrt{s} = m_\rho$, and that the vector spectral function is broadened due to the mixing. We also discussed its relevance to dilepton measurements.

In this write-up, we summarize main results of these papers especially focusing on the effect to the vector spectral function.

2. Effects of Vector–Axial-vector Mixing in Hot Matter

We start with showing the vector spectral function at $T/T_c = 0.6$, without any dropping masses, in Fig. 1 (left). Two cases are compared; one includes the V-A mixing and the other does not. Both the spectral functions has a peak at M_ρ. The effects of V-A mixing can be seen as a shoulder at $\sqrt{s} = M_{a_1} - m_\pi$ and a bump above $\sqrt{s} = M_{a_1} + m_\pi$.

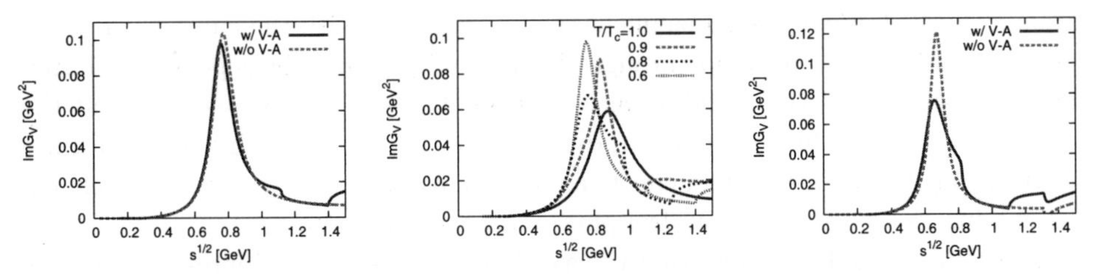

Fig. 1. Left figure shows the vector spectral function at temperature $T/T_c = 0.6$. The solid (red) curve includes the effect of V-A mixing, while the dashed (green) curve does not. The middle figure shows the spectral function (option (A)) for $m_\pi = 140\,\text{MeV}$ at several temperatures $T/T_c = 0.6$-1.0. The right figure shows the vector spectral function (option (B)) for $m_\pi = 140$ MeV in type (I) at temperature $T/T_c = 0.8$.

In Ref. 1, two possible cases of chiral symmetry restoration are studied:

(A) Dropping a_1 meson mass but non-dropping ρ mass;
(B) Dropping ρ and a_1 meson masses.

In both cases, the "flash temperature"[9] T_f is introduced for controling how the mesons experience partial restoration of chiral symmetry. Then, the masses of vector and/or axial-vector mesons are assumed to have temperature dependences only above T_f. In option (A) at the chiral limit, the dropping a_1 meson mass for $T > T_f$ is taken as

$$\frac{M_{a_1}^2(T) - M_\rho^2}{M_{a_1}^2(T=0) - M_\rho^2} = \frac{T_c^2 - T^2}{T_c^2 - T_f^2}\,, \tag{1}$$

where T_c is the critical temperature of the chiral symmetry restoration. The vector spectral function for this option (A), with the effect of pion mass included, is shown at several temperatures in Fig. 1 (middle). Below T_c one observes the previously mentioned threshold effects moving downward with increasing temperature. It is remarkable that at T_c the spectrum shows almost no traces of a_1-ρ-π threshold effects: The ρ to a_1 mass ratio becomes almost 1 at $T = T_c$ even though the effect of M_π is included. Furthermore, one can show that a_1-ρ-π coupling constant becomes very tiny, $g_{a_1\rho\pi} \sim 0.06\, m_\pi$. This indicates that at T_c *the a_1 meson mass nearly equals the ρ meson mass and the a_1-ρ-π coupling almost vanishes even in the presence of explicit chiral symmetry breaking.*

In option (B), two types of temperature dependences were used in Ref. 1:

$$\text{(I)}: \frac{M_\rho^2(T)}{M_\rho^2(T=0)} = \frac{T_c^2 - T^2}{T_c^2 - T_f^2}, \quad \frac{M_{a_1}^2(T) - M_\rho^2(T)}{M_{a_1}^2(T=0) - M_\rho^2(T=0)} = \left(\frac{T_c^2 - T^2}{T_c^2 - T_f^2}\right)^2, \quad (2)$$

$$\text{(II)}: \frac{M_\rho^2(T)}{M_\rho^2(T=0)} = \left(\frac{T_c^2 - T^2}{T_c^2 - T_f^2}\right)^2, \quad \frac{M_{a_1}^2(T) - M_\rho^2(T)}{M_{a_1}^2(T=0) - M_\rho^2(T=0)} = \frac{T_c^2 - T^2}{T_c^2 - T_f^2}. \quad (3)$$

Figure 1 (right) shows the vector spectrum using the type (I) parameterization at $T = 0.8\, T_c$. The feature that the a_1 meson suppresses the vector spectral function through the V-A mixing remains unchanged. Compared with the curve for $T/T_c = 0.8$ in Fig. 1 (middle), a bump through the V-A mixing and the ρ peak are shifted downward since both the ρ and a_1 masses drop. The self-energy has a cusp at the threshold $2\, M_\rho$ and this appears as a dip at $\sqrt{s} \sim 1.3$ GeV. The influence of finite m_π turns out to be in threshold effects as before. We find that the nearly vanishing V-A mixing as seen for the non-dropping ρ mass, option (A).

The result given here shows that the axial-vector meson contributes significantly to the vector spectral function; the presence of the a_1 reduces the vector spectrum around M_ρ and enhances it around M_{a_1}. A major change with both dropping ρ and a_1 masses is a systematic downward shift of the vector spectrum. We observe that the a_1-ρ-π coupling almost vanishes at the critical temperature T_c and thus the V-A mixing becomes very tiny.

3. Effects of Vector–Axial-vector Mixing in Dense Matter

In this section, we summarize main points shown in Ref. 2.

At finite baryon density a system preserves parity but violates charge conjugation invariance. Chiral Lagrangians thus in general build in the term

$$\mathcal{L}_{\rho a_1} = 2C\, \epsilon^{0\nu\lambda\sigma} \mathrm{tr}\left[\partial_\nu V_\lambda \cdot A_\sigma + \partial_\nu A_\lambda \cdot V_\sigma\right], \quad (4)$$

for the vector V^μ and axial-vector A^μ mesons with the total anti-symmetric tensor $\epsilon^{0123} = 1$ and a parameter C. This mixing results in the dispersion relation[8]

$$p_0^2 - \bar{p}^2 = \frac{1}{2}\left[m_\rho^2 + m_{a_1}^2 \pm \sqrt{(m_{a_1}^2 - m_\rho^2)^2 + 16C^2\bar{p}^2}\right], \quad (5)$$

which describes the propagation of a mixture of the transverse ρ and a_1 mesons with non-vanishing three-momentum $|\vec{p}| = \bar{p}$. The longitudinal polarizations, on the other hand, follow the standard dispersion relation, $p_0^2 - \bar{p}^2 = m_{\rho,a_1}^2$. When the mixing vanishes as $\bar{p} \to 0$, Eq. (5) with lower sign provides $p_0 = m_\rho$ and it with upper sign does $p_0 = m_{a_1}$. In the following, we call the mode following the dispersion relation with the lower sign in Eq. (5) "the ρ meson", and that with the upper sign "the a_1 meson".

The mixing strength C in Eq. (4) can be estimated assuming the ω-dominance in the following way: The gauged Wess-Zumino-Witten terms in an effective chiral Lagrangian include the ω-ρ-a_1 term[10] which leads to the following mixing term

$$\mathcal{L}_{\omega\rho a_1} = g_{\omega\rho a_1} \langle\omega_0\rangle \epsilon^{0\nu\lambda\sigma} \mathrm{tr}\left[\partial_\nu V_\lambda \cdot A_\sigma + \partial_\nu A_\lambda \cdot V_\sigma\right] , \tag{6}$$

where the ω field is replaced with its expectation value given by $\langle\omega_0\rangle = g_{\omega NN} \cdot n_B/m_\omega^2$. One finds with empirical numbers $C = g_{\omega\rho a_1}\langle\omega_0\rangle \simeq 0.1\,\mathrm{GeV}$ at normal nuclear matter density. As we will show below, this is too small to have an importance in the correlation functions. In a holographic QCD approach, on the other hand, the effects from an infinite tower of the ω-type vector mesons are summed up to give $C \simeq 1\,\mathrm{GeV} \cdot (n_B/n_0)$ with normal nuclear matter density $n_0 = 0.16\ \mathrm{fm}^{-3}$.[8] In the following we assume an actual value of C in QCD in the range $0.1 < C < 1\,\mathrm{GeV}$. Some importance of the higher Kaluza-Klein (KK) modes *even in vacuum* in the context of holographic QCD can be seen in the pion electromagnetic form factor at the photon on-shell: This is saturated by the lowest four vector mesons in a top-down holographic QCD model.[11,12] In hot and dense environment those higher members get modified and the masses might be somewhat decreasing evidenced in an in-medium holographic model.[13] This might provide a strong V-A mixing $C > 0.1\,\mathrm{GeV}$ in three-color QCD and the dilepton measurements may give a good testing ground.

In Fig. 2, we show the dispersion relations (5) for the transverse modes together with those for the longitudinal modes with $C = 1$ and $0.5\,\mathrm{GeV}$. This shows that, when $C = 0.5\,\mathrm{GeV}$, there are only small changes for both ρ and a_1 mesons, while a substantial change for ρ meson when $C = 1\,\mathrm{GeV}$. For very large $\bar{p}$ the longitudinal and transverse dispersions are in parallel with a finite gap, $\pm C$.

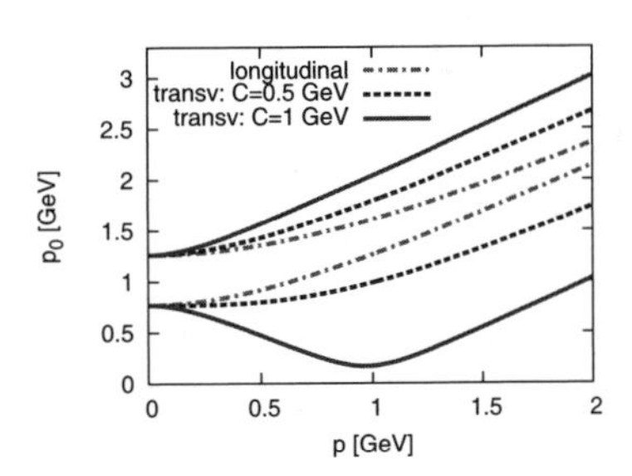

Fig. 2. The dispersion relations for the ρ (lower 3 curves) and a_1 (upper 3 curves) mesons for $C = 0.5$, and $1\,\mathrm{GeV}$.

In Fig. 3, we plot the integrated spectrum over three momentum, which is a main ingredient in dilepton production rates. Figure 3 (left) shows a clear enhancement of the spectrum below $\sqrt{s} = m_\rho$ due to the mixing. This enhancement becomes much suppressed when the ρ meson is moving with a large three-momentum as shown in Fig. 3 (right). The upper bump now emerges more remarkably and becomes a clear indication of the in-medium effect from the a_1 via the mixing.

As an application of the above in-medium spectrum, we calculate the production rate of a lepton pair emitted from dense matter through a decaying virtual photon.

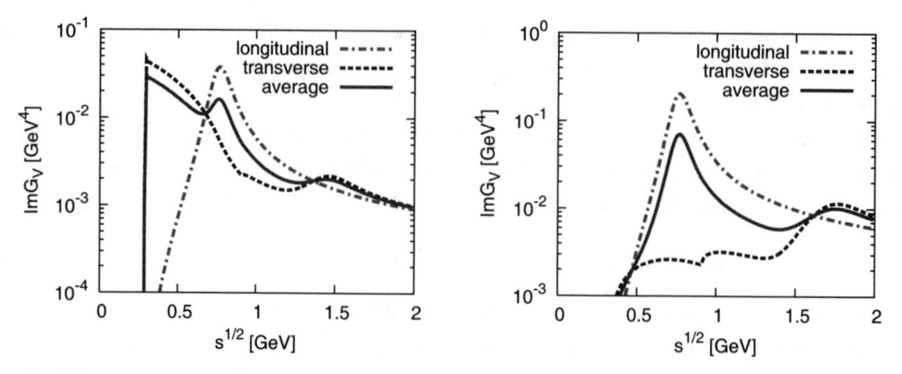

Fig. 3. The vector spectral function for $C = 1\,\mathrm{GeV}$. The curves of the left figure are calculated integrating over $0 < \bar{p} < 0.5\,\mathrm{GeV}$, and those of the right figure over $0.5 < \bar{p} < 1\,\mathrm{GeV}$. Here we use the values of masses given by $m_\pi = 0.14\,\mathrm{GeV}$, $m_\rho = 0.77\,\mathrm{GeV}$, $m_{a_1} = 1.26\,\mathrm{GeV}$, and the widths given by the imaginary part of one-loop diagrams in a chiral Lagrangian approach as[2,14] with the on-shell values of $\Gamma_\rho(s = m_\rho^2) = 0.15\,\mathrm{GeV}$ and $\Gamma_{a_1}(s = m_{a_1}^2) = 0.33\,\mathrm{GeV}$.

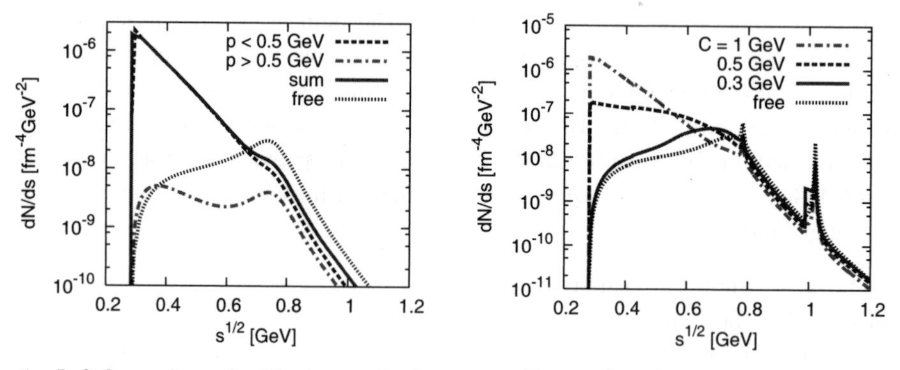

Fig. 4. Left figure shows the dilepton production rate at $T = 0.1\,\mathrm{GeV}$ for $C = 1\,\mathrm{GeV}$. Integration over $0 < \bar{p} < 0.5\,\mathrm{GeV}$ (dashed) and $0.5 < \bar{p} < 1\,\mathrm{GeV}$ (dashed-dotted) was carried out. The right figure shows the dilepton production rate at $T = 0.1\,\mathrm{GeV}$ with various mixing strength C. Integration over $0 < \bar{p} < 1\,\mathrm{GeV}$ was done. We use the constant widths with values of $\Gamma_\omega = 8.49\,\mathrm{MeV}$, $\Gamma_\phi = 4.26\,\mathrm{MeV}$, $\Gamma_{f_1(1285)} = 24.3\,\mathrm{MeV}$ and $\Gamma_{f_1(1420)} = 54.9\,\mathrm{MeV}$.

Figure 4 (left) presents the integrated rate at $T = 0.1\,\mathrm{GeV}$ for $C = 1\,\mathrm{GeV}$. One clearly observes a strong three-momentum dependence and an enhancement below $\sqrt{s} = m_\rho$ due to the Bose distribution function which result in a strong spectral broadening. The total rate is mostly governed by the spectrum with low momenta $\bar{p} < 0.5\,\mathrm{GeV}$ due to the large mixing parameter C. When density is decreased, the mixing effect gets irrelevant and consequently in-medium effect in low $\sqrt{s}$ region is reduced in compared with that at higher density. The calculation performed in hadronic many-body theory in fact shows that the ρ spectral function with a low momentum carries details of medium modifications.[15] One may have a chance to observe it in heavy-ion collisions with certain low-momentum binning at J-PARC, GSI/FAIR and RHIC low-energy running.

It is straightforward to introduce other V-A mixing between ω-$f_1(1285)$ and ϕ-$f_1(1420)$. In Fig. 4 (right) we plot the integrated rate at $T = 0.1\,\mathrm{GeV}$ with several mixing strength C which are phenomenological option. One observes that the enhancement below m_ρ is suppressed with decreasing mixing strength. This forms into a broad bump in low $\sqrt{s}$ region and its maximum moves toward m_ρ. Similarly, some contributions are seen just below m_ϕ. This effect starts at threshold $\sqrt{s} = 2m_K$. Self-consistent calculations of the spectrum in dense medium will provide a smooth change and this eventually makes the ϕ meson peak somewhat broadened.

Finally, we remark that the importance of the mixing effect studied here relies on the coupling strength C. Holographic QCD predicts an extremely strong mixing $C \sim 1\,\mathrm{GeV}$ at $n_B = n_0$ which leads to vector meson condensation at $n_B \sim 1.1\,n_0$.[8] This may be excluded by known properties of nuclear matter and therefore in reality the strength C will be smaller. We have discussed a possible range of C to be $0.1 < C < 1\,\mathrm{GeV}$ based on higher excitations and their in-medium modifications. The parameter C does carry an unknown density dependence. This will be determined in an elaborated treatment of hadronic matter along with the underlying QCD dynamics. If $C \sim 0.1\,\mathrm{GeV}$ at n_0 were preferred as the lowest-omega dominance, the mixing effect is irrelevant there. However, it becomes more important at higher densities, e.g. $C = 0.3\,\mathrm{GeV}$ at $n_B/n_0 = 3$ which leads to a distinct modification from the spectrum in free space.

4. Summary

In this write-up, we summarized main results of our recent analyses on the mixing between ρ and a_1 mesons in hot and/or dense matter.

The analysis in Ref. 1 shows that the axial-vector meson contributes significantly to the vector spectral function in hot matter through the mixing: the presence of the a_1 reduces the vector spectrum around M_ρ and enhances it around M_{a_1}. The effect of dropping mass of a_1 with or without dropping ρ mass associated with the chiral symmetry restoration is studied. It is shown that the a_1-ρ-π coupling almost vanishes at the critical temperature T_c and thus the V-A mixing becomes very tiny.

In Ref. 2, we studied a novel effect of the V-A mixing through a set of $\omega\rho a_1$-type interactions at finite baryon density. We showed that the mixing produces a clear enhancement of the vector spectral function below $\sqrt{s} = m_\rho$, and that the vector spectral function is broadened due to the mixing.

It is an interesting issue to address a change of the vector correlator with the V-A mixing at finite baryon density in section 3 toward chiral symmetry restoration. The mixing (4) is chirally symmetric and thus does not vanish at the chiral restoration in contrast to the vanishing V-A mixing near the critical temperature T_c without baryon density in section 2. A spontaneous breaking of Lorentz invariance via the omega condensation could increase the mixing strength C near chiral restoration.[16] Furthermore, if meson masses drop due to partial restoration of chiral symmetry assuming a second- or weak first-order transition in high baryon density but low temperature region, the ground state near the critical point may favor vector condensation even for a moderate mixing strength. This will be reported elsewhere.

Acknowledgment

The work of M.H. is supported in part by the JSPS Grant-in-Aid for Scientific Research (c) 20540262, Grant-in-Aid for Scientific Research on Innovative Areas (No. 2104) "Quest on New Hadrons with Variety of Flavors" from MEXT and the Global COE Program of Nagoya University "Quest for Fundamental Principles in the Universe (QFPU)" from JSPS and MEXT of Japan. The work of C.S. is supported in part by the DFG cluster of excellence "Origin and Structure of the Universe".

References

1. C. Sasaki, M. Harada and W. Weise, Prog. Theor. Phys. Suppl. **174**, 173 (2008); Phys. Rev. D **78**, 114003 (2008); Nucl. Phys. A **827**, 350C (2009).
2. M. Harada and C. Sasaki, Phys. Rev. C **80**, 054912 (2009).
3. See, e.g., V. Bernard and U. G. Meissner, Nucl. Phys. A **489**, 647 (1988); T. Hatsuda and T. Kunihiro, Phys. Rept. **247**, 221 (1994); R. D. Pisarski, hep-ph/9503330; G. E. Brown and M. Rho, Phys. Rept. **269**, 333 (1996); F. Klingl, N. Kaiser and W. Weise, Nucl. Phys. A **624**, 527 (1997); F. Wilczek, hep-ph/0003183; G. E. Brown and M. Rho, Phys. Rept. **363**, 85 (2002).
4. R. Rapp and J. Wambach, Adv. Nucl. Phys. **25**, 1 (2000), R. S. Hayano and T. Hatsuda, arXiv:0812.1702 [nucl-ex], R. Rapp, J. Wambach and H. van Hees, arXiv:0901.3289 [hep-ph].
5. M. Dey, V. L. Eletsky and B. L. Ioffe, Phys. Lett. B **252**, 620 (1990), B. Krippa, Phys. Lett. B **427**, 13 (1998).
6. E. Marco, R. Hofmann and W. Weise, Phys. Lett. B **530**, 88 (2002), M. Urban, M. Buballa and J. Wambach, Phys. Rev. Lett. **88**, 042002 (2002).
7. J. V. Steele, H. Yamagishi and I. Zahed, Phys. Lett. B **384**, 255 (1996); Phys. Rev. D **56**, 5605 (1997), K. Dusling, D. Teaney and I. Zahed, Phys. Rev. C **75**, 024908 (2007), K. Dusling and I. Zahed, Nucl. Phys. A **825**, 212 (2009).
8. S. K. Domokos and J. A. Harvey, Phys. Rev. Lett. **99**, 141602 (2007).
9. G. E. Brown, C. H. Lee and M. Rho, Nucl. Phys. A **747**, 530 (2005).
10. N. Kaiser and U. G. Meissner, Nucl. Phys. A **519**, 671 (1990).
11. T. Sakai and S. Sugimoto, Prog. Theor. Phys. **113**, 843 (2005); Prog. Theor. Phys. **114**, 1083 (2005).
12. M. Harada, S. Matsuzaki and K. Yamawaki, in preparation. See also the contribution to the same proceedings.
13. K. Peeters, J. Sonnenschein and M. Zamaklar, Phys. Rev. D **74**, 106008 (2006).
14. M. Harada and C. Sasaki, Phys. Rev. D **73**, 036001 (2006).
15. F. Riek, R. Rapp, T. S. Lee and Y. Oh, Phys. Lett. B **677**, 116 (2009).
16. K. Langfeld, H. Reinhardt and M. Rho, Nucl. Phys. A **622**, 620 (1997), K. Langfeld and M. Rho, Nucl. Phys. A **660**, 475 (1999).

Infrared Behavior of Ghost and Gluon Propagators Compatible with Color Confinement in Yang-Mills Theory with the Gribov Horizon

Kei-Ichi Kondo*

Department of Physics, University of Tokyo, Tokyo 113-0033, Japan
E-mail: kondok@faculty.chiba-u.jp

We discuss how the existence of the Gribov horizon affects the deep infrared behavior of ghost and gluon propagators of the D-dimensional $SU(N)$ Yang-Mills theory in the Landau gauge. If we use a horizon function in the Gribov-Zwanziger framework to restrict the functional integral to the first Gribov region for avoiding Gribov copies, we can show that the ghost propagator behaves like free, while the gluon propagator is non-vanishing in deep infrared region (the decoupling solution) in harmony with recent results obtained from numerical simulations on lattice, the Schwinger-Dyson equation and the functional renormalization group. This result should be compared with the Gribov prediction that such a restriction leads to the vanishing gluon propagator and enhanced ghost propagator (the scaling solution). We raise some questions on current understanding on the ghost propagator and its implications for quark confinement *à la* Wilson and color confinement *à la* Kugo-Ojima.

Keywords: Ghost dressing function, Schwinger-Dyson equation, color confinement, Kugo-Ojima, Gribov-Zwanziger, horizon function.

1. Introduction

We consider the $SU(N)$ Yang-Mills theory in D-dimensional Euclidean space. For a quantum Yang-Mills theory to be well-defined, we must give a gauge fixing procedure which is able to select just one representative from each gauge orbit in the space of Yang-Mills fields. It is known that the usual Landau gauge fixing $\partial\mathscr{A} = 0$ is not sufficient for this purpose, since there are many Yang-Mills field configurations (Gribov copies) connected by gauge transformations, which are obtained as intersection points of a gauge orbit with the gauge fixing hypersurface $\Gamma := \{\mathscr{A}; \partial\mathscr{A} = 0\}$. Long ago, Gribov[1] proposed to restrict the functional integral to the subspace Ω:

$$Z_{\rm Gribov} := \int_\Omega [d\mathscr{A}]\delta(\partial\mathscr{A})\det(K[\mathscr{A}]))\exp\{-S_{YM}[\mathscr{A}]\}, \tag{1}$$

where S_{YM} is the Yang-Mills action, $K[\mathscr{A}] := -\partial_\mu D_\mu[\mathscr{A}]$ is the Faddeev-Popov operator, and Ω is the 1st Gribov region defined by

$$\Omega := \{\mathscr{A}; \partial\mathscr{A} = 0,\ -\partial D[\mathscr{A}] > 0\} \subset \Gamma. \tag{2}$$

*On sabbatical leave of absence from Department of Physics, Chiba University, Chiba 263-8522, Japan.

Ω is a bounded and convex region including the origin $\{\mathscr{A} = 0\}$. In fact, $-\partial_\mu D_\mu[\mathscr{A} = 0] = -\partial_\mu \partial_\mu > 0$. The boundary of Ω is called the Gribov horizon:

$$\partial\Omega := \{\mathscr{A}; \partial\mathscr{A} = 0, \ -\partial D[\mathscr{A}] = 0\}. \tag{3}$$

In order to realize this restriction in a simpler manner, Zwanziger proposed a framework called the Gribov-Zwanziger theory[2–4] described by

$$Z_{\rm GZ} := \int [d\mathscr{A}]\delta(\partial^\mu \mathscr{A}_\mu) \det(K[\mathscr{A}]) \exp\left\{ -S_{YM}[\mathscr{A}] - \gamma \int d^D x h(x) \right\}, \tag{4}$$

where $h(x) = h[\mathscr{A}](x)$ is called the Zwanziger horizon function and the parameter γ called the Gribov parameter is determined by solving a gap equation, commonly called the *horizon condition*,

$$\langle h(x) \rangle_{\rm GZ} = (N^2 - 1)D. \tag{5}$$

The horizon function plays the role of restricting the integration region inside the Gribov horizon $\partial\Omega$, since the horizon function is positive definite inside the 1st Gribov region and approaches infinity near the Gribov horizon:

$$\mathscr{A} \to \partial\Omega \Longrightarrow K[\mathscr{A}] \downarrow 0 \Longrightarrow \int d^D x h[\mathscr{A}](x) \uparrow \infty \Longrightarrow \exp\{-\gamma \int d^D x h(x)\} \downarrow 0 \ (\gamma > 0). \tag{6}$$

The first proposal of the horizon function is[2]

$$h(x) = \int d^D y g f^{ABC} \mathscr{A}_\mu^B(x) (K^{-1})^{CE}(x,y) g f^{AFE} \mathscr{A}_\mu^F(y). \tag{7}$$

The second proposal is the covariant derivative form:[4]

$$h(x) = \int d^D y D_\mu[\mathscr{A}]^{AC}(x) (K^{-1})^{CE}(x,y) D_\mu[\mathscr{A}]^{AE}(y). \tag{8}$$

2. A theoretical possibility of deriving the decoupling solution

We use an exact relationship between the ghost dressing function $G(k^2)$ and the Kugo-Ojima function $u(k^2)$ with an additional function $w(k^2)$:

$$G^{-1}(k^2) = 1 + u(k^2) + w(k^2). \tag{9}$$

This identity was first derived in Ref. 7 and confirmed in Refs. 8,10. As a special case, we have a relationship between $G(0)$ and the Kugo-Ojima parameter $u(0)$:

$$G(0) = [1 + u(0) + w(0)]^{-1}. \tag{10}$$

Therefore, **the Kugo-Ojima criterion $u(0) = -1$ for color confinement is equivalent to the scaling solution $G(0) = \infty$, (Ref. 12) if and only if** $w(0) = 0$. Here u and w are defined from the modified 1-particle irreducible (m1PI) part as

$$\lambda_{\mu\nu}^{AB}(k) := \langle (g\mathscr{A}_\mu \times \mathscr{C})^A (g\mathscr{A}_\nu \times \bar{\mathscr{C}})^B \rangle_k^{m1PI} = \left[\delta_{\mu\nu} u(k^2) + \frac{k_\mu k_\nu}{k^2} w(k^2) \right] \delta^{AB}, \tag{11}$$

where $u(k^2)$ agrees with the Kugo-Ojima function usually defined by

$$\langle (D_\mu \mathscr{C})^A (g\mathscr{A}_\nu \times \bar{\mathscr{C}})^B \rangle_k := \left(\delta_{\mu\nu} - \frac{k_\mu k_\nu}{k^2} \right) \delta^{AB} u(k^2). \tag{12}$$

The m1PI part is defined from the two-point function of the composite operators

$$\begin{aligned} &\langle (g\mathscr{A}_\mu \times \mathscr{C})^A (g\mathscr{A}_\nu \times \bar{\mathscr{C}})^B \rangle_k = \lambda_{\mu\nu}^{AB}(k) + \Delta_{\mu\nu}^{AB}(k), \\ &\lambda_{\mu\nu}^{AB}(k) := \langle (g\mathscr{A}_\mu \times \mathscr{C})^A (g\mathscr{A}_\nu \times \bar{\mathscr{C}})^B \rangle_k^{\mathrm{m1PI}}, \\ &\Delta_{\mu\nu}^{AB}(k) := \langle (g\mathscr{A}_\mu \times \mathscr{C})^A \bar{\mathscr{C}}^C \rangle_k^{\mathrm{1PI}} \langle \mathscr{C}^C \bar{\mathscr{C}}^D \rangle_k \langle \mathscr{C}^D (g\mathscr{A}_\nu \times \bar{\mathscr{C}})^B \rangle_k^{\mathrm{1PI}}. \end{aligned} \tag{13}$$

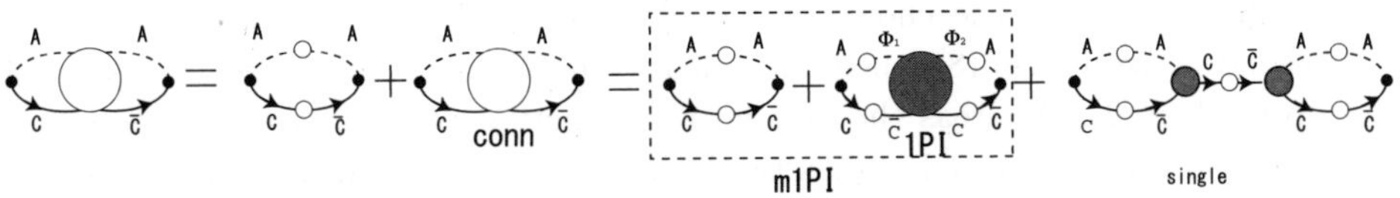

Fig. 1. Diagrammatic representation of $\langle (g\mathscr{A}_\mu \times \mathscr{C})^A (g\mathscr{A}_\nu \times \bar{\mathscr{C}})^B \rangle_k$, $\langle (g\mathscr{A}_\mu \times \mathscr{C})^A (g\mathscr{A}_\nu \times \bar{\mathscr{C}})^B \rangle_k^{\mathrm{conn}}$, $\langle (g\mathscr{A}_\mu \times \mathscr{C})^A (g\mathscr{A}_\nu \times \bar{\mathscr{C}})^B \rangle_k^{\mathrm{1PI}}$ and $\langle (g\mathscr{A}_\mu \times \mathscr{C})^A (g\mathscr{A}_\nu \times \bar{\mathscr{C}})^B \rangle_k^{\mathrm{m1PI}}$.

We propose a trick which enables one to incorporate the Gribov horizon directly into the self-consistent Schwinger-Dyson equation in the gauge-fixed Yang-Mills theory,

$$G^{-1}(k^2) = \frac{\langle h(0) \rangle}{(\mathrm{dim} G) D} + u(k^2) + w(k^2). \tag{14}$$

We give a main result using the 1st horizon function.[8,10,11] We consider $G(0)$ and $u(0)$ as functions of $w(0)$. By applying the horizon condition (5) to the horizon function (7) and the identity (10), we have

$$G(0) = 1 - (D-1)w(0)/2 + \sqrt{[1-(D-1)w(0)/2]^2 - 1 + D} > 0, \tag{15}$$

and $u(0) = -1 - w(0) + G^{-1}(0)$ reads

$$u(0) = -1 - w(0) - \frac{1}{6}\left\{ 2 - 3w(0) - \sqrt{12 + [2-3w(0)]^2} \right\}. \tag{16}$$

This implies that *the horizon condition determines the boundary value* $G(0)$ *as a function of* $w(0)$. *Consequently, we have one-parameter family of solutions parameterized by* $w(0)$. Both $G(0)$ and $u(0)$ are monotonically decreasing functions in $w(0)$; $G(0), u(0) \to +\infty$ as $w(0) \to -\infty$, while $G(0) \to 0$ and $u(0) \to -5/3$ as $w(0) \to +\infty$. Thus the scaling solution $G(0) = +\infty$ is obtained only when $w(0) = -\infty$. Otherwise $w(0) > -\infty$, the decoupling solution $0 < G(0) < \infty$ is obtained. Using a special value as an additional input $w(0) = 0$, we obtain

$$G(0) = 1 + \sqrt{D} > 0, \quad u(0) = (-D \pm \sqrt{D})/(D-1). \tag{17}$$

In particular, for $D = 4$,

$$G(0) = 3 > 0, \quad u(0) = -2/3 > -1 \quad (D=4). \tag{18}$$

This leads to the **decoupling solution**, $0 < G(0) < \infty$, (Ref. 13) which agrees with current data of numerical simulations. See Refs. 8–11 for related references.

3. Critical points

We raise some questions on current understanding on the ghost propagator and its implications for quark confinement a la Wilson and color confinement a la Kugo-Ojima.

(1) **Point.1** Both horizon functions (7) and (8) play the role of restricting the functional integral to the 1st Gribov regin in the sense of (6). However, the results depend on the choice of the horizon function. If one starts from the second horizon term (8), then one is led to the scaling solution (when $w(0) = 0$ up to Point.2). Whereas the first horizon function (7) leads to the decoupling solution. The difference between two horizon functions are just total derivatives.

(2) **Point.2** The conventional claim between the ghost propagator and the color confinement: the Kugo-Ojima criterion $u(0) = -1$ is equivalent to the scaling solution $G(0) = \infty$ follows from the relation

$$G(0) = [1 + u(0)]^{-1}. \tag{19}$$

However, this is not necessarily precise, since the exact relationship between the ghost dressing function and the Kugo-Ojima parameter must be

$$G(0) = [1 + u(0) + w(0)]^{-1}. \tag{20}$$

Therefore, **the Kugo-Ojima criterion $u(0) = -1$ is equivalent to the scaling solution $G(0) = \infty$, if and only if** $w(0) = 0$. In perturbation theory, indeed, $w(0) = 0$. However, no one knows whether $w(0) = 0$ or not exactly in the nonperturbative sense.

(3) **Point.3** The conventional multiplicative renormalization scheme is incompatible with the horizon condition. The inclusion of the horizon condition make the ghost propagator finite self-consistently.

4. Conclusion and discussion

We have discussed how the existence of the Gribov horizon modifies the deep infrared behavior of the Landau gauge SU(N) Yang-Mills theory, using the Gribov-Zwanziger framework with the horizon condition $\langle h(x) \rangle = (\dim G)D$.

We have found that there exists **one-parameter family of solutions** parameterized by a parameter $w(0)$ which was assumed to be zero implicitly. The family includes both the scaling and decoupling solutions, and specification of the parameter discriminates between them.

For the first horizon function,

$$G(0) = 1 - 3w(0)/2 + \sqrt{[1 - 3w(0)/2]^2 + 3} > 0, \tag{21}$$

$$u(0) = -1 - w(0) - \frac{1}{6}\left\{2 - 3w(0) - \sqrt{12 + [2 - 3w(0)]^2}\right\}. \tag{22}$$

In particular,

$$G(0) = 3, \quad u(0) = -2/3, \quad \text{for} \quad w(0) = 0. \tag{23}$$
$$G(0) = 2, \quad u(0) = -1, \quad \text{for} \quad w(0) = 1/2. \tag{24}$$
$$G(0) = \infty, \quad u(0) = +\infty \quad \text{for} \quad w(0) = -\infty(\text{not permitted}). \tag{25}$$

For the second horizon function:

$$G(0) = w(0)^{-1}, \tag{26}$$
$$u(0) = -1 \quad \text{irrespective of} \quad w(0). \tag{27}$$

Here the scaling solution is obtained as

$$G(0) = \infty, \quad u(0) = -1 \quad \text{for} \quad w(0) = 0, \tag{28}$$

and the decoupling solution is obtained as

$$G(0) \neq \infty, \quad u(0) = -1 \quad \text{for} \quad w(0) \neq 0. \tag{29}$$

In my opinion, one must fix the following issues:

(1) The best choice of the horizon term: The horizon terms currently available are not necessarily agree with the definition of the Gribov horizon. Find the precise horizon term to specify the Gribov region in the Gribov-Zwanziger framework?

(2) BRST symmetry and color confinement: The BRST symmetry is not maintained after being restricted to the Gribov region. (a) Find the BRST-invariant horizon term. Or, (b) find a modified (local) BRST transformation to make the horizon term invariant.
It is interesting to consider how the existence of the horizon is relevant for color confinement. In the Gribov-Zwanziger theory (restricted to the 1st Gribov region), the BRST symmetry is broken by the existence of the horizon. $\delta S_{\rm GZ} = \delta \tilde{S}_\gamma \neq 0$ Nevertheless, there exists a "BRST" like symmetry (without nilpotency [Sorella,0905.1010[hep-th]] or with nilpotency [K.-I. K., 0905.1899[hep-th]]) which leaves the Gribov-Zwanziger action invariant. Then we could apply the Kugo-Ojima idea to the Gribov-Zwanziger theory, which opens the path to searching for the modified color confinement criterion *a la* Kugo and Ojima.

(3) The value of the parameter $w(0)$: In the perturbation theory, $w(0) = 0$, which is not guaranteed in the nonperturbative level. Determine the value $w(0)$ which enables one to discriminate the scaling solution and decoupling one, once a horizon term is given. [However, a parameter $w(0)$ might reflect the ambiguity originating from the Gribov copies inside the Gribov region. Then it may be undetermined.]

(4) Gribov horizon and renormalization: The conventional multiplicative renormalization scheme is incompatible with the Gribov horizon condition. We observe that the inclusion of the horizon term cancels the ultraviolet divergence in the

Schwinger-Dyson equation for the ghost propagator and the resulting (non-perturbative) self-consistent solution becomes ultraviolet finite. In other words, the horizon condition interpolates between the infrared behavior and the ultraviolet one.

(5) UV divergence (convergence): UV renormalization might be changed if the horizon condition is taken into account. Numerical simulations on finer lattice are desirable, in addition to the larger size lattice.

(6) Scaling or decoupling: It is really meaningful to discriminate between the scaling and the decoupling from physical point of view. Both solutions can give the same physical results.

(7) Quark confinement from color confinement: It is claimed[14] that both solutions (decoupling as well as scaling) lead to quark confinement by proving the vanishing of the order parameter of quark confinement, the Polyakov loop average. Therefore, one can consider both solutions are the same from the physical point of view. The difference comes just from the gauge artifact (insufficient implementation of the non-perturbative gauge fixing procedure).

Acknowledgments

This work is financially supported by Grant-in-Aid for Scientific Research (C) 21540256 from Japan Society for the Promotion of Science (JSPS).

References

1. V.N. Gribov, Nucl. Phys. B**139**, 1–19 (1978).
2. D. Zwanziger, Nucl. Phys. B**323**, 513–544 (1989).
3. D. Zwanziger, Nucl. Phys. B**378**, 525–590 (1992).
4. D. Zwanziger, Nucl. Phys. B**399**, 477–513 (1993).
5. D. Zwanziger, Nucl.Phys.B**412**, 657-730 (1994).
6. T. Kugo and I. Ojima, Suppl. Prog. Theor. Phys. **66**, 1–130 (1979).
7. T. Kugo, hep-th/9511033.
8. K.-I. Kondo, arXiv:0904.4897 [hep-th], Phys.Lett.B. **678**, 322-330 (2009).
9. K.-I. Kondo, arXiv:0905.1899 [hep-th].
10. K.-I. Kondo, arXiv:0907.3249 [hep-th], Prog. Theor. Phys. **122**, 1455-1475 (2009).
11. K.-I. Kondo, arXiv:0909.4866 [hep-th].
12. R. Alkofer and L. von Smekal, hep-ph/0007355, Phys. Rept. **353**, 281 (2001).
13. Ph. Boucaud, J.P. Leroy, A. Le Yaouanc, J. Micheli, O. Pene and J. Rodriguez-Quintero, hep-ph/0803.2161, JHEP **06**, 099 (2008).
    Ph. Boucaud, J.P. Leroy, A. Le Yaouanc, J. Micheli, O. Pene and J. Rodriguez-Quintero, arXiv:0801.2721[hep-ph], JHEP **06**, 012 (2008).
14. J. Braun, H. Gies and J.M. Pawlowski, e-Print: arXiv:0708.2413 [hep-th]]

Chiral Symmetry Breaking on the Lattice

Hidenori Fukaya [for JLQCD and TWQCD Collaborations]

Department of Physics, Nagoya University, Nagoya 464-8602, Japan
E-mail: hfukaya@eken.phys.nagoya-u.ac.jp

This talk presents a recent lattice study by JLQCD and TWQCD collaborations on spontaneous breaking of chiral symmetry. To maintain exact chiral symmetry in the zero quark mass limit, we employ the overlap Dirac operator for the dynamical quark action. The numerical cost for the overlap fermions is high but can be reduced by fixing the topological charge of the gauge fields along the Monte Carlo updates. By studying the low-lying eigenvalues of the Dirac operator, which is always UV finite, we confirm the presence of the chiral condensate. Correcting the finite size effects calculated within the chiral perturbation theory, we determine the value of the chiral condensate at a good accuracy.

Keywords: Lattice QCD, chiral symmetry.

1. Introduction

Spontaneous breaking of chiral symmetry[1] plays a key role in low-energy limit of Quantum Chromo Dynamics (QCD). The chiral condensate is believed to be the source of the symmetry breaking and give a "*constituent*" mass to the quark fields and thus to the hadron masses $\sim \mathcal{O}(1)$ GeV. Moreover, the pion can be identified as the (pseudo) Nambu-Goldstone boson and described well by chiral perturbation theory (ChPT)[2,3] of which interaction is highly restricted by the symmetries.

Due to its universal properties, the chiral symmetry breaking has also been applied to "*beyond*" QCD, or model buildings for the electro-weak symmetry breaking. The so-called Techni-Color models[4–7] are candidates for the high energy theories which may be examined in LHC experiments.[8]

It is, however, still difficult to confirm the spontaneous chiral symmetry breaking from the 1st principle calculations of QCD nor more general strong coupling gauge theories (SCGT's). The chiral condensate is 100 % non-perturbative effects : it never obtains a finite value in the perturbative QCD expansion.

The numerical lattice QCD simulation has played a central role in non-perturbative analysis of QCD. Many kinds of masses, decay constants and matrix elements of hadrons are evaluated, which confirms, comparing with the experiments, QCD as the fundamental theory. But even with lattice QCD, it is a challenging problem to examine the chiral condensate, due to mainly three difficulties:

(1) Nielsen-Ninomiya's theorem[9] : the chiral symmetry on the lattice must be broken to avoid appearance of the unphysical fermion modes (fermion doubling).
(2) UV divergence : the chiral condensate at finite quark mass generally has power divergence which depends on the regularization scheme.
(3) Finite V correction : the spontaneous symmetry breaking is never allowed at finite volume.

In this talk, we present how these three problems are solved or circumvented in our recent works and show the numerical results for the chiral condensate.

2. Three problems and solutions

2.1. *Chiral symmetry on the lattice*

Lattice QCD is a well-defined regularization of the gauge theory, which allows us to numerically perform the functional integrals. Disctetization of the space-time leads, however, to explicit breaking of a lot of symmetries the original continuum theory has.A typical example is violation of the translational invariance.

It is well-known that the chiral symmetry is one of such symmetries incompatible with the lattice discretization. Nielsen and Ninomiya[9] proved that the chiral symmetry must be broken otherwise unphysical fermion modes, known as doublers, appear, which is inconsistent with the continuum limit.

In 1998, Neuberger solved this problem.[10] He proposed a new type of subtraction operator, called overlap Dirac operator,

$$D(m) = \left(m_0 + \frac{m}{2}\right) + \left(m_0 - \frac{m}{2}\right) \gamma_5 \mathrm{sgn}[H_W(-m_0)], \tag{1}$$

where m denotes the quark mass and $H_W \equiv \gamma_5 D_W(-m_0)$ is the Hermitian Wilson-Dirac operator with a large negative mass $-m_0$ (We take $m_0 = 1.6$ in our numerical studies.). Here and in the following the parameters are given in the lattice units. In the chiral limit $m \to 0$, the overlap-Dirac operator (1) satisfies the Ginsparg-Wilson relation:[11] $D(0)\gamma_5 + \gamma_5 D(0) = D(0)\gamma_5 D(0)/m_0$, with which the quark action has exact chiral symmetry under a modified chiral transformation.[12] Moreover, it is known that the overlap-Dirac operator has an index which corresponds to the topological charge in the continuum limit.[13]

In spite of its promising features, the numerical study employing the overlap Dirac operator was far behind. It is simply due to its high numerical cost. For the approximation of the sign function in (1), a number of operations of the Wilson-Dirac operator are needed to keep a required precision. Moreover, the fermion determinant has discontinuity on the topology boundaries, configuration where the index of the overlap Dirac operator changes. This discontinuity of the determinant prevents smooth evolution of the molecular dynamics steps and requires a special treatment,[14] which needs an additional numerical cost potentially proportional to the lattice volume squared.

To avoid the problem of the large extra numerical cost, we introduce additional Wilson fermions and twisted-mass bosonic spinors to generate a weight

$$\frac{\det[H_W(-m_0)^2]}{\det[H_W(-m_0)^2+\mu^2]}, \tag{2}$$

in the functional integrals.[15–17] Both of fermions and ghosts are unphysical as their masses are of order of the lattice cutoff but prohibits the topology changes along the simulations. In our numerical studies, we set $\mu = 0.2$. This topology fixing leads to a large reduction of the numerical cost.[17]

Since 2006, JLQCD and TWQCD collaborations have succeeded in performing Monte Carlo simulations with dynamical N_f =2 and 2+1 overlap quarks.[18] We choose 5-6 different points for the up and down quark mass m_{ud} and 2 points for the strange quark mass m_s. For the gauge part, we use the Iwasaki gauge action.[19] The lattice volumes are $V = L^3T = 16^3 \times 32$ ($N_f = 2$) and $V = L^3T = 16^3 \times 48$ ($N_f = 2+1$). We also carry out a simulation on a $V = L^3T = 24^3 \times 48$ lattice at one parameter choice, in order to check the finite volume effect. The lattice scales $a^{-1} = 1.667$ GeV ($N_f = 2$) and $a^{-1} = 1.833$ GeV ($N_f = 2+1$) are determined from the heavy quark potential, using the Sommer's scale[20] $r_0 = 0.49$ fm as an input. The lattice size is then estimated as $L \sim 1.9$ fm for $N_f = 2$, and $L \sim 1.7$ fm for $N_f = 2+1$ runs. Note that we have reduced the lightest quark mass to almost the physical value $m_{ud} \sim 3$ MeV.

2.2. *Dirac spectrum and chiral symmetry breaking*

Even with an exact chiral symmetry on the lattice, the direct calculation of the chiral condensate is difficult, due to both of ultra-violet (UV) and infra-red (IR) problems. On the UV side, there is a power divergence at finite quark mass, of the form $\sim m/a^2$ where a denotes the lattice spacing. But in the limit of $m \to 0$, an IR problem appears : the chiral condensate trivially vanishes since spontaneous breaking of symmetry is never allowed at finite volume.

One can avoid both problems by analyzing Dirac operator spectrum rather than the chiral condensate itself. There is a non-perturbative relation called Banks-Casher relation[21] which relates the zero-mode density of the Dirac operator (here denoted by ρ) to the chiral condensate,

$$\langle \bar{q}q \rangle = \pi\rho(0). \tag{3}$$

It is straightforward to extend this relation to finite value of λ:

$$\rho(\lambda) = \frac{1}{\pi}\mathrm{Re}\langle \bar{q}q \rangle \bigg|_{m_v=i\lambda}. \tag{4}$$

Note here that the valence quark mass m_v is analytically continued to a pure imaginary value $i\lambda$, while the sea quark masses remain real. The UV problem is avoided by taking the real part where the pure imaginary divergence is canceled. We expect that IR problem is also softened as the eigenvalue density (at finite λ) is non-trivial even at finite volume since the valence "*mass*", $m_v = i\lambda$, is finite.

The Banks Casher relation provides a picture of how chiral symmetry breaking occurs. In the free fermion case, the spectral density shows a cubic degeneracy:

$$\rho(\lambda) = \frac{3}{4\pi^2}\lambda^3. \tag{5}$$

Once the gauge coupling is turned on, however, this highly degenerate levels are split and the eigenvalues feel repulsive force from each other. As a result, the eigenvalues are accumulated around zero, which leads to the chiral condensate.

We can further estimate how finite volume effect is seen in the spectrum. Finite volume never allows spontaneous breaking of symmetry, which means $\rho(0)$ must be zero. Since only very long-wave length modes should reflect the difference, we expect the Dirac spectrum shows a steep drop around $1/V$, above which the bulk mode part still keeps a height around the chiral condensate $\Sigma \equiv \lim_{m\to 0}\lim_{V\to\infty}\langle\bar{q}q\rangle$.

Here we should note another advantage. The right-hand-side of (4) can be analyzed using (partially quenched) chiral perturbation theory, where one can estimate the effects from finite quark masses and fixed topological charge Q, too.

2.3. *Chiral perturbation at finite volume*

Since there is a big mass gap between the pion mass and other hadron masses, the pions are most responsible for the finite size effects and chiral perturbation theory is useful to trace the volume scaling of QCD.

In fact, a number of studies have been done on the eigenvalue distribution of the Dirac operator and a sharp drop behavior around zero, as we discussed in the previous subsection, was actually found.[22–24] In the calculation a special type of expansion, the so-called ϵ-expansion of chiral Lagrangian was employed. The ϵ-expansion exactly treats the zero-mode and precisely reproduce the sharp drop around zero in the Dirac spectral density as a consequence of a non-perturbative vacuum fluctuation.

The calculation was recently extended to the next-leading-order (NLO).[25] The zero mode integral is exactly performed while the chiral logarithm, which comes from the conventional p-expansion, is also precisely treated. This means we calculate a hybrid system of zero-mode matrix model and perturbative modes. For the zero-mode integrals we have used a technique developed in Refs.26.

In this hybrid way of calculation, the spectral density in a fixed topological sector of Q at finite V and m is calculated, which is expressed by two contributions:

$$\rho_Q(\lambda) = \Sigma_{eff}\rho_Q^{\epsilon} + \rho^p, \tag{6}$$

where the first term has the same functional form as in the leading-order,[22–24] except for Σ_{eff} having a shift from NLO contributions. The second term represents the purely NLO logarithmic curve, or "chiral logs" from the non-zero modes which is the p-expansion regime. Note that ρ^p is IR finite, and topology independent.

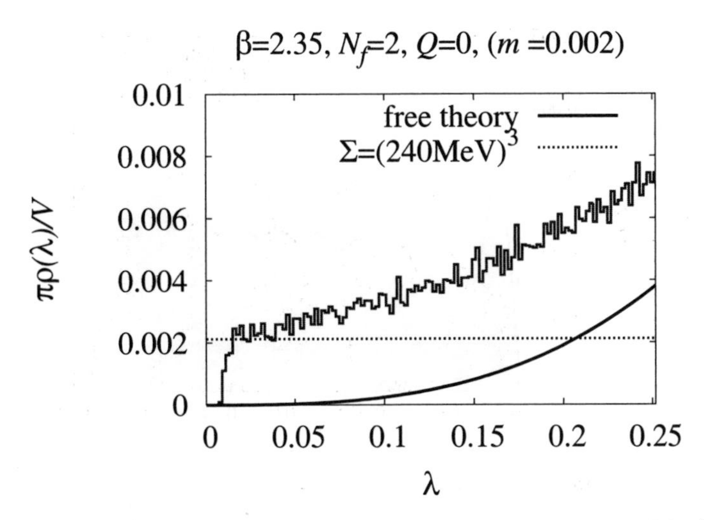

Fig. 1. Low-lying Dirac eigenvalue histogram in two-flavor lattice QCD.[27]

3. Numerical result for chiral condensate

We are now ready to calculate the chiral condensate and judge whether QCD spontaneously breaks chiral symmetry or not. The histogram shown in Fig. 1 is the lattice data for the Dirac spectrum calculated by JLQCD collaboration.[27] The result is very consistent with Banks-Casher's scenario. A significant accumulation of the low-lying modes around zero is seen in contrast to that of the free theory shown in the solid curve. Moreover, the plateau suggests a reasonable value of the chiral condensate $\Sigma^{1/3} \sim 240$ (MeV). A steep drop near zero is also seen, indicating the presence of finite volume effects.

Recently we have extended the study to the 2+1 flavor QCD and compare the results with chiral perturbation theory to NLO at finite volume. On the top panel of Fig. 2, we plot the lattice data in grey histogram and one can see that the steep drop around zero and the height of the first peak are consistent with the NLO ChPT formula. Moreover, in the bulk region, the logarithmically negative dump is also reproduced by the NLO contribution in ChPT, which cannot be explained in the LO epsilon-expansion shown in the dashed line.

In order to quantify the statistical error and quality of fitting, we use the mode-number[29] or the integrated histogram shown in the bottom panel of Fig. 2. The agreement is good upto $\lambda \sim 0.03$ which is around $m_s/2$.

We also find a good consistency with data at different masses, different volumes and different topological sectors. Details are shown in Ref.28. From the data, we determine the chiral condensate as $\Sigma = [242(4)(^{+19}_{-18})\text{MeV}]^3$ (in the $\overline{MS}$ scheme at 2 GeV) at NLO accuracy. The pion decay constant $F = 74(1)(8)$ and one of the NLO coefficient in ChPT: $L_6^r(770\text{MeV}) = -0.0001(3)(1)$ are also extracted from the sub-leading contributions.

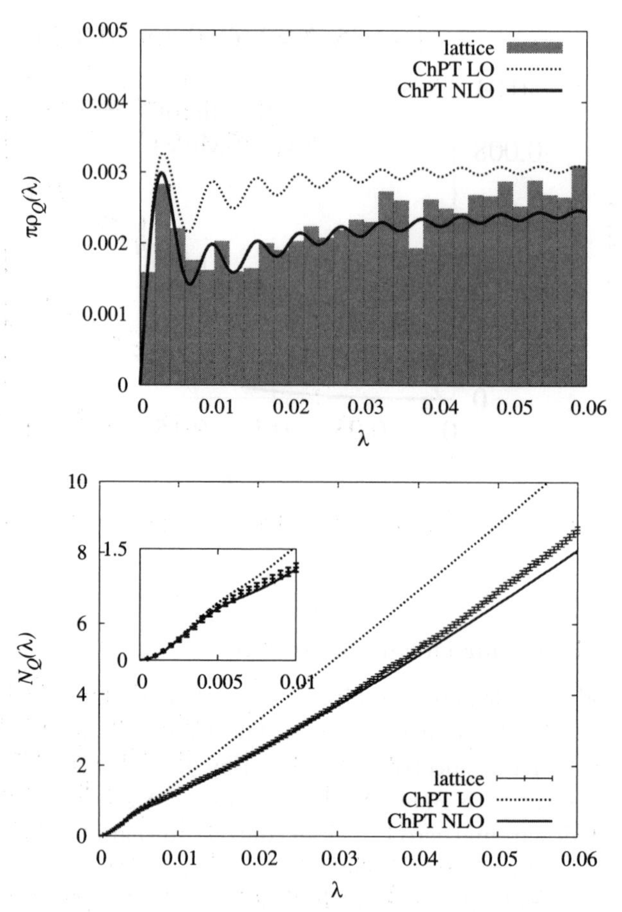

Fig. 2. Low-lying Dirac eigenvalue histogram (top) and the mode-number (cumulative histogram) (bottom) in 2+1-flavor lattice QCD.[28] ChPT result at LO (dashed) and NLO (solid) is compared.

4. Conclusion

We have performed lattice QCD simulations with overlap Dirac operator which realizes an exact chiral symmetry. To confirm the presence of chiral condensate, we have studied Dirac operator spectrum which has no UV divergence. Chiral perturbation theory is useful to convert finite volume results to the one in the infinite volume. Our lattice data are consistent with the prediction of the Banks-Casher relation and chiral perturbation including the finite volume effects. All our data strongly suggest the breaking of chiral symmetry. The low-energy constants Σ, F and L_6^r are determined from the data.

The author thanks P.H. Damgaard and members of JLQCD and TWQCD collaborations for discussions. Numerical simulations are performed on IBM System Blue Gene Solution at KEK under a support of its Large Scale Simulation Program (No. 06-13). This work was supported by the Global COE program of Nagoya Univ. Quest for Fundamental Principles in the Universe from JSPS and MEXT of Japan.

References

1. Y. Nambu and G. Jona-Lasinio, Phys. Rev. **122**, 345 (1961); Y. Nambu and G. Jona-Lasinio, Phys. Rev. **124**, 246 (1961).
2. S. Weinberg, Phys. Rev. **166**, 1568 (1968).
3. J. Gasser and H. Leutwyler, Annals Phys. **158**, 142 (1984); Nucl. Phys. B **250**, 465 (1985).
4. S. Weinberg, Phys. Rev. D **13**, 974 (1976); Phys. Rev. D **19**, 1277 (1979).
5. L. Susskind, Phys. Rev. D **20**, 2619 (1979).
6. B. Holdom, Phys. Lett. B **150**, 301 (1985).
7. K. Yamawaki, M. Bando and K. i. Matumoto, Phys. Rev. Lett. **56**, 1335 (1986).
8. S. Asai, in these proceedings.
9. H. B. Nielsen and M. Ninomiya, Nucl. Phys. B **185**, 20 (1981) [Erratum-ibid. B **195**, 541 (1982)]; Nucl. Phys. B **193**, 173 (1981).
10. H. Neuberger, Phys. Lett. B **417**, 141 (1998); Phys. Lett. B **427**, 353 (1998).
11. P. H. Ginsparg and K. G. Wilson, Phys. Rev. D **25**, 2649 (1982).
12. M. Luscher, Phys. Lett. B **428**, 342 (1998) [arXiv:hep-lat/9802011].
13. P. Hasenfratz, V. Laliena and F. Niedermayer, Phys. Lett. B **427**, 125 (1998) [arXiv:hep-lat/9801021].
14. Z. Fodor, S. D. Katz and K. K. Szabo, JHEP **0408**, 003 (2004) [arXiv:hep-lat/0311010]; T. A. DeGrand and S. Schaefer, Phys. Rev. D **71**, 034507 (2005) [arXiv:hep-lat/0412005]; T. A. DeGrand and S. Schaefer, Phys. Rev. D **72**, 054503 (2005) [arXiv:hep-lat/0506021]; N. Cundy, S. Krieg, G. Arnold, A. Frommer, T. Lippert and K. Schilling, Comput. Phys. Commun. **180**, 26 (2009) [arXiv:hep-lat/0502007].
15. T. Izubuchi and C. Dawson [RBC Collaboration], Nucl. Phys. Proc. Suppl. **106**, 748 (2002).
16. P. M. Vranas, Phys. Rev. D **74**, 034512 (2006) [arXiv:hep-lat/0606014].
17. H. Fukaya *et al.* [JLQCD Collaboration], Phys. Rev. D **74**, 094505 (2006) [arXiv:hep-lat/0607020].
18. S. Aoki *et al.* [JLQCD Collaboration], Phys. Rev. D **78**, 014508 (2008) [arXiv:0803.3197 [hep-lat]].
19. Y. Iwasaki, Nucl. Phys. B **258** (1985) 141; Y. Iwasaki and T. Yoshie, Phys. Lett. B **143**, 449 (1984).
20. R. Sommer, Nucl. Phys. B **411**, 839 (1994) [arXiv:hep-lat/9310022].
21. T. Banks and A. Casher, Nucl. Phys. B **169**, 103 (1980).
22. J. J. M. Verbaarschot and I. Zahed, Phys. Rev. Lett. **70**, 3852 (1993).
23. G. Akemann and P. H. Damgaard, Nucl. Phys. B **528**, 411 (1998).
24. P. H. Damgaard and S. M. Nishigaki, Phys. Rev. D **63**, 045012 (2001).
25. P. H. Damgaard and H. Fukaya, JHEP **0901**, 052 (2009) [arXiv:0812.2797 [hep-lat]].
26. K. Splittorff and J. J. M. Verbaarschot, Phys. Rev. Lett. **90**, 041601 (2003) [arXiv:cond-mat/0209594]; Y. V. Fyodorov and G. Akemann, JETP Lett. **77**, 438 (2003) [Pisma Zh. Eksp. Teor. Fiz. **77**, 513 (2003)] [arXiv:cond-mat/0210647]. K. Splittorff and J. J. M. Verbaarschot, Nucl. Phys. B **683**, 467 (2004) [arXiv:hep-th/0310271].
27. H. Fukaya *et al.* [JLQCD Collaboration], Phys. Rev. Lett. **98**, 172001 (2007) [arXiv:hep-lat/0702003]; H. Fukaya *et al.*, Phys. Rev. D **76**, 054503 (2007) [arXiv:0705.3322 [hep-lat]].
28. H. Fukaya *et al.* [JLQCD collaboration], arXiv:0911.5555 [hep-lat].
29. L. Giusti and M. Luscher, JHEP **0903**, 013 (2009) [arXiv:0812.3638 [hep-lat]].

Gauge-Higgs Unification: Stable Higgs Bosons as Cold Dark Matter

Yutaka Hosotani

Department of Physics, Osaka University
Toyonaka, Osaka 560-0043, Japan

In the gauge-Higgs unification the 4D Higgs field becomes a part of the extra-dimensional component of the gauge potentials. In the $SO(5) \times U(1)$ gauge-Higgs unification in the Randall-Sundrum warped spacetime the electroweak symmetry is dynamically broken through the Hosotani mechanism. The Higgs bosons become absolutely stable, and become the dark matter of the universe. The mass of the Higgs boson is determined from the WMAP data to be about 70 GeV.

Keywords: Higgs boson, dark matter, gauge-Higgs unification, Hosotani mechanism.

1. Introduction

Does the Higgs boson exist? What is it really like? What constitutes the dark matter in the universe? These are two of the most important problems in current physics. We would like to point out that these two mystery particles are really the same. In the gauge-Higgs unification scenario Higgs bosons become absolutely stable, and become the dark matter of the universe.

In the gauge-Higgs unification the 4D Higgs boson is identified with a part of the extra-dimensional component of the gauge potentials. Its couplings with others particles are controlled by the gauge principle. The 4D Higgs field corresponds to 4D fluctuations of an AB phase in the extra dimensions.[1–3] In the $SO(5) \times U(1)$ gauge-Higgs unification model in the Randall-Sundrum warped space the AB phase θ_H takes exactly the value $\frac{1}{2}\pi$ in the vacuum as a consequence of quantum dynamics. At this particular value of θ_H the 4D Higgs boson becomes absolutely stable.

The relic abundance of the Higgs bosons in the present universe is evaluated definitively with the mass of the Higgs boson as the only relevant variable parameter. Astonishingly the average mass density of the dark matter determined from the WMAP data is obtained with the Higgs mass around 70 GeV. It does not contradict with the LEP2 bound, because the ZZH coupling vanishes at $\theta_H = \frac{1}{2}\pi$. The gauge-Higgs unification scenario gives a completely new viewpoint for the Higgs boson.[4] Further it gives definitive predictions for electroweak gauge couplings of quarks and leptons, which can be tested experimentally.[5]

2. $SO(5) \times U(1)$ gauge-Higgs unification in RS

We consider an $SO(5) \times U(1)$ gauge theory in the five-dimensional Randall-Sundrum (RS) warped spacetime.[4–14] Its metric is given by

$$ds^2 = e^{-2\sigma(y)} \eta_{\mu\nu} dx^\mu dx^\nu + dy^2, \tag{1}$$

where $\eta_{\mu\nu} = \text{diag}(-1, 1, 1, 1)$, $\sigma(y) = \sigma(y + 2L)$, and $\sigma(y) = k|y|$ for $|y| \leq L$. The fundamental region in the fifth dimension is given by $0 \leq y \leq L$. The Planck brane and the TeV brane are located at $y = 0$ and $y = L$, respectively. The bulk region $0 < y < L$ is an anti-de Sitter spacetime with a cosmological constant $\Lambda = -6k^2$.

The RS spacetime has the same topology as the orbifold $M^4 \times (S^1/Z_2)$. Vector potentials $A_M(x, y)$ of the gauge group $SO(5)$ and $B_M(x, y)$ of $U(1)$ satisfy the orbifold boundary conditions

$$\begin{aligned}
&\begin{pmatrix} A_\mu \\ A_y \end{pmatrix} (x, y_j - y) = P_j \begin{pmatrix} A_\mu \\ -A_y \end{pmatrix} (x, y_j + y) P_j^{-1} , \\
&\begin{pmatrix} B_\mu \\ B_y \end{pmatrix} (x, y_j - y) = \begin{pmatrix} B_\mu \\ -B_y \end{pmatrix} (x, y_j + y) , \\
&P_j = \text{diag}(-1, -1, -1, -1, +1) , \quad (j = 0, 1),
\end{aligned} \tag{2}$$

where $y_0 = 0$ and $y_1 = L$. By the boundary conditions the $SO(5) \times U(1)$ symmetry is reduced to $SO(4) \times U(1) \simeq SU(2)_L \times SU(2)_R \times U(1)$. The symmetry $SU(2)_R \times U(1)$ is further spontaneously broken by a scalar field $\Phi(x)$ on the Planck brane to $U(1)_Y$. $\Phi(x)$ belongs to $(0, \frac{1}{2})$ representation of $SU(2)_L \times SU(2)_R$. The pattern of the symmetry reduction[11,13] is depicted in fig. 1.

With the above boundary conditions the zero modes of 4D gauge fields reside in the 4-by-4 matrix part of A_μ, whereas the zero modes of A_y in the off-diagonal part of A_y. The latter is an $SO(4)$ vector, or an $SU(2)_L$ doublet, corresponding to the Higgs doublet in the standard model. See fig. 2.

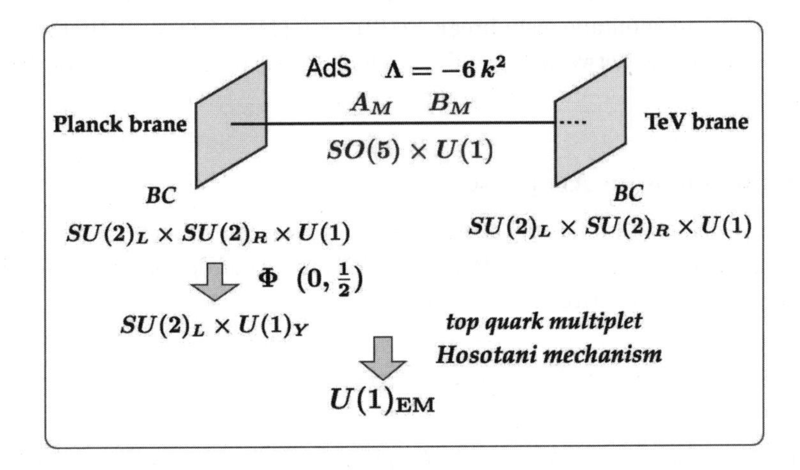

Fig. 1. The symmetry reduction in the $SO(5) \times U(1)$ gauge-Higgs unification in RS.

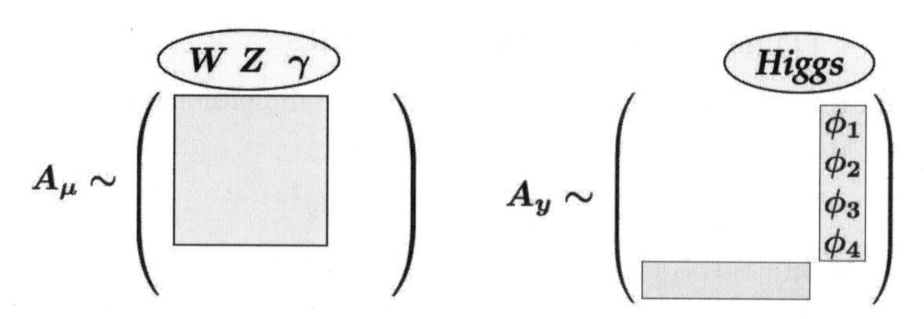

Fig. 2. W, Z, γ appear as zero modes of A_μ whereas the 4D Higgs field appears as a zero mode of A_y.

Bulk fermions for quarks and leptons are introduced as multiplets in the vectorial representation of $SO(5)$.[5,14] In the quark sector two vector multiplets are introduced for each generation. In the lepton sector it suffices to introduce one multiplet for each generation to describe massless neutrinos, whereas it is necessary to have two multiplets to describe massive neutrinos. Each vector multiplet Ψ_a is decomposed to $(\frac{1}{2},\frac{1}{2})\oplus(0,0)$ in the $SU(2)_L \times SU(2)_R$. They satisfy the orbifold boundary condition

$$\Psi_a(x, y_j - y) = P_j \Gamma^5 \Psi_a(x, y_j + y) \ , \tag{3}$$

which gives rise to chiral fermions as zero modes. The left-handed components of $(\frac{1}{2},\frac{1}{2})$ and the right-handed components of $(0,0)$ have zero modes.

In addition to the bulk fermions, right-handed brane fermions $\hat{\chi}_\alpha$ are introduced on the Planck brane. The brane fermions $\hat{\chi}_\alpha$ belong to the $(\frac{1}{2},0)$ representation of $SU(2)_L \times SU(2)_R$. These brane fermions are necessary both to have the realistic quark-lepton spectrum at low energies and to have the cancellation of 4D chiral anomalies associated to the $SO(4) \times U(1)$ gauge fields. The matter content is summarized in fig. 3. The $SO(4) \times U(1)$ gauge invariance is maintained on the Planck brane as well. The bulk fermions Ψ_a, the brane fermions $\hat{\chi}_\alpha$, and the brane scalar Φ form $SO(4) \times U(1)$ invariant interactions. When Φ develops a non-vanishing expectation value to spontaneously break $SU(2)_R \times U(1)$ to $U(1)_Y$, it simultaneously gives mass couplings between Ψ_a and $\hat{\chi}_\alpha$ which, in turn, make all exotic fermions heavy.

3. Higgs boson as an AB phase

As shown in fig. 2, the 4D Higgs field appears as a zero mode of A_y. Without loss of generality one can suppose that the A_y^{45} component develops a nonvanishing expectation value. We write

$$A_y(x, y) = \hat{\theta}_H(x) \cdot \sqrt{\frac{4k}{z_L^2 - 1}} \, h_0(y) \cdot T^{\hat{4}} + \cdots \ , \quad z_L = e^{kL} \tag{4}$$

where the zero-mode wave function is $h_0(y) = [2k/(z_L^2 - 1)]^{1/2} \, e^{2ky} \ (0 \le y \le L)$. The generator $T^{\hat{4}}$ is given by $(T^{\hat{4}}_{\rm vec})_{ab} = (i/\sqrt{2})(\delta_{a5}\delta_{b4} - \delta_{a4}\delta_{b5})$ in the vectorial

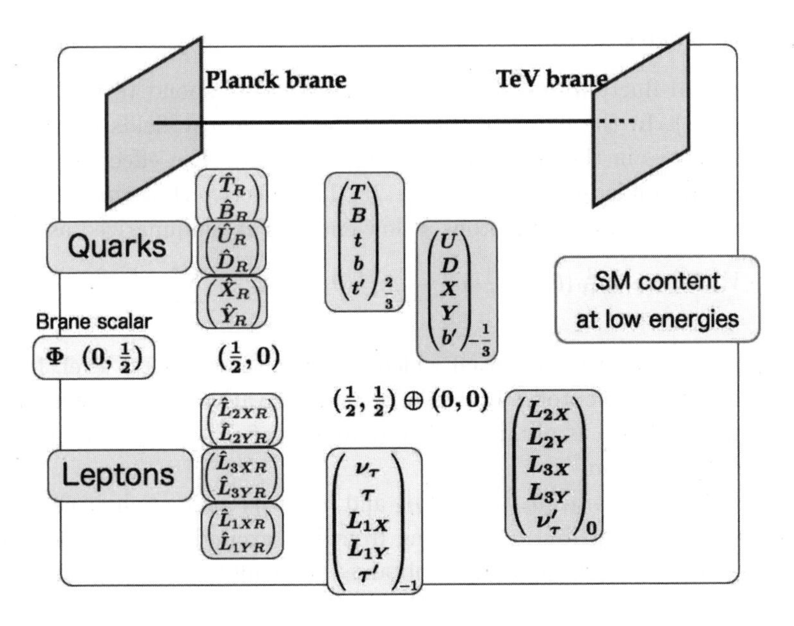

Fig. 3. Matter content in the model.

representation and $T^{\hat{4}}_{\rm sp} = (1/2\sqrt{2})I_2 \otimes \tau_1$ in the spinorial representation. $\hat{\theta}_H(x)$ is decomposed as

$$\hat{\theta}_H(x) = \theta_H + \frac{H(x)}{f_H} \quad , \quad f_H = \frac{2}{g_A}\sqrt{\frac{k}{z_L^2 - 1}} \sim \frac{2}{\sqrt{kL}}\frac{m_{\rm KK}}{\pi g} \quad . \tag{5}$$

Here the Kaluza-Klein (KK) mass scale is $m_{\rm KK} = \pi k z_L^{-1}$, and the $SO(5)$ gauge coupling g_A is related to the 4D $SU(2)_L$ weak coupling g by $g = g_A/\sqrt{L}$.

With $H(x) = 0$, in the spinorial representation,

$$P \exp\left\{ i g_A \int_0^L dy A_y \right\} = \exp\left\{ \frac{i}{2}\theta_H I_2 \otimes \tau_1 \right\} \tag{6}$$

so that the constant part θ_H of $\hat{\theta}_H(x)$ represents an Aharonov-Bohm phase in the extra-dimension. $\hat{\theta}_H$ is a phase variable. There remains the residual gauge invariance $A_M \to A'_M = \Omega A_M \Omega^{-1} - (i/g_A)\Omega\partial_M\Omega^{-1}$ which preserves the boundary conditions (2) and (3).[7,15] In general, new gauge potentials satisfy new boundary conditions given by $P'_j = \Omega(x, y_j - y)P_j\Omega(x, y_j + y)^{-1}$. The residual gauge invariance is defined with $\Omega(x, y)$ satisfying $P'_j = P_j$. Consider a large gauge transformation

$$\Omega^{\rm large}(y; \alpha) = \exp\left\{ -i\alpha \int_0^y dy \sqrt{4k/(z_L^2 - 1)}\, h_0(y) \cdot T^{\hat{4}} \right\} . \tag{7}$$

which shifts $\hat{\theta}_H(x)$ to $\hat{\theta}'_H(x) = \hat{\theta}_H(x) + \alpha$. With $\alpha = 2\pi$, $P'_j = P_j$ and $\hat{\theta}'_H = \hat{\theta}_H + 2\pi$. It implies that physics is invariant under

$$\theta_H \to \theta_H + 2\pi \quad . \tag{8}$$

4. Effective theory

Four-dimensional fluctuations of the AB phase θ_H correspond to the 4D neutral Higgs field $H(x)$. In the effective theory of the low energy fields the Higgs field $H(x)$ enters always in the combination of $\hat{\theta}_H(x)$ in (5). The effective Lagrangian must be invariant under $\hat{\theta}_H(x) \to \hat{\theta}_H(x) + 2\pi$. The effective Higgs interactions with the W, Z bosons, quarks and leptons at low energies are summarized as[4,14,16,17]

$$\mathcal{L}_{\text{eff}} = -V_{\text{eff}}(\hat{\theta}_H) - m_W^2(\hat{\theta}_H) W_\mu^\dagger W^\mu - \tfrac{1}{2} m_Z^2(\hat{\theta}_H) Z_\mu Z^\mu - \sum_{a,b} m_{ab}^F(\hat{\theta}_H) \overline{\psi}_a \psi_b \ . \quad (9)$$

$V_{\text{eff}}(\hat{\theta}_H)$ is the effective potential which arises at the one loop level. It is finite and independent of the cutoff.[1] From the second derivative at the global minimum a finite Higgs mass m_H is obtained. The finiteness of m_H^2 gives a solution to the gauge-hierarchy problem.[18]

The mass functions $m_W(\hat{\theta}_H)$, $m_Z(\hat{\theta}_H)$ and $m_{ab}^F(\hat{\theta}_H)$ arise at the tree level. It is essential to include contributions coming from KK excited states in intermedium states. The effective Lagrangian (9) is obtained after integrating out all heavy KK modes.

In the $SO(5) \times U(1)$ model under consideration these mass functions are found to be, to good accuracy in the warped space,[10,11,14]

	gauge-Higgs	**[SM]**
$m_W(\hat{\theta}_H) \sim$	$\dfrac{1}{2} g f_H \sin\hat{\theta}_H$,	$\left[\dfrac{1}{2} g(v+H) \right]$,
$m_Z(\hat{\theta}_H) \sim$	$\dfrac{1}{2\cos\theta_W} g f_H \sin\hat{\theta}_H$,	$\left[\dfrac{1}{2\cos\theta_W} g(v+H) \right]$,
$m_{ab}^F(\hat{\theta}_H) \sim$	$y_{ab}^F f_H \sin\hat{\theta}_H$,	$\left[y_{ab}^F (v+H) \right]$. (10)

We have listed the formulas in the standard model in brackets on the right. It is seen that $v + H$ in the standard model is replaced approximately by $f_H \sin\hat{\theta}_H$ in the gauge-Higgs unification. In the standard model the mass functions are linear in the Higgs field H, whereas they become periodic, nonlinear functions of H in the gauge-Higgs unification. It is a consequence of the phase nature of θ_H. In other words the gauge invariance in the warped space naturally leads to the nonlinear behavior.

The masses of the W, Z bosons and fermions are given by $m_W = \frac{1}{2} g f_H \sin\theta_H$, $m_Z = m_W / \cos\theta_W$, and $m_{ab}^F = y_{ab}^F f_H \sin\theta_H$. An immediate consequence is that the Higgs couplings to W, Z, and fermions deviate from those in the standard model. In particular,

$$\begin{pmatrix} WWH \\ ZZH \\ \text{Yukawa} \end{pmatrix} = \text{SM} \times \cos\theta_H \ ,$$

$$\begin{pmatrix} WWHH \\ ZZHH \end{pmatrix} = \text{SM} \times \cos 2\theta_H \quad . \tag{11}$$

As we shall see below, the effective potential is minimized at $\theta_H = \frac{1}{2}\pi$ so that the WWZ, ZZH and Yukawa couplings vanish.

5. Dynamical EW symmetry breaking

The value of θ_H in the vacuum is determined by the location of the global minimum of the effective potential $V_{\rm eff}(\theta_H)$, which is given at the one loop level by

$$V_{\rm eff}(\theta_H) = \sum_{\rm particles} \pm\frac{1}{2} \int \frac{d^4p}{(2\pi)^4} \sum_n \ln\left(p^2 + m_n(\theta_H)^2\right) \tag{12}$$

where $\{m_n(\theta_H)\}$ is a 4D mass spectrum in each KK tower with the AB phase θ_H. As originally shown in ref. 1 the θ_H-dependent part of $V_{\rm eff}(\theta_H)$ is finite, independent of how the theory is regularized. Evaluation of $V_{\rm eff}(\theta_H)$ in the RS warped space was initiated by Oda and Weiler.[19] Since then a powerful method for evaluation has been developed by Falkowski.[20]

In the model under consideration the electroweak symmetry $SU(2)_L \times U(1)_Y$ remains unbroken for $\theta_H = 0, \pi$. Otherwise the symmetry is broken to $U(1)_{\rm EM}$. If there were no fermions, the symmetry is unbroken. Among the fermions, multiplets in the bulk containing the top quark gives a dominant contribution to $V_{\rm eff}(\theta_H)$. The fact that the top quark mass (172 GeV) is larger than m_W is relevant. Contributions from other fermions, whose masses are much smaller than m_W, are negligible in the warped space. Contributions fom charm quarks, for instance, are suppressed by a factor 10^{-5}. The existence of the top quark triggers the electroweak symmetry breaking. $U(\theta_H) = (16\pi^4/m_{\rm KK}^4)\, V_{\rm eff}(\theta_H)$ is plotted with the warp factor $z_L = 10^{10}$ in fig. 4.

It is seen that the effective potential is minimized at $\theta_H = \pm\frac{1}{2}\pi$ so that the electroweak symmetry is dynamically broken. It follows from (10) that $m_W = \frac{1}{2} g f_H$ and $f_H \sim 256\,\text{GeV}$.

As the warp factor $z_L = e^{kL}$ is decreased, there appear two kinds of phase transitions. For $z_L < z_{L1} \sim 2200$ the top quark mass $m_t \sim 171\,\text{GeV}$ cannot be achieved. It necessarily becomes smaller than the observed value. One can set the bulk mass parameter to be zero, and further decrease the value of z_L. Then, for $z_L \sim z_{L2} \sim 1.67$ there appears a weakly-first-order phase transition. For $z_L < z_{L2}$ the global minima move to $\theta_H = 0$ and π so that the symmetry is unbroken.[a]

The curvature of the $V_{\rm eff}$ at the minimum is related to the Higgs mass m_H by

$$m_H^2 = \frac{\pi^2 g^2 kL}{4\, m_{\rm KK}^2} \frac{d^2 V_{\rm eff}}{d\theta_H^2}\bigg|_{\rm min} \quad , \quad m_{\rm KK} = \pi k z_L^{-1} \quad . \tag{13}$$

[a]There was a numerical error in the evaluation of $V_{\rm eff}(\theta_H)$ in Ref. 13.

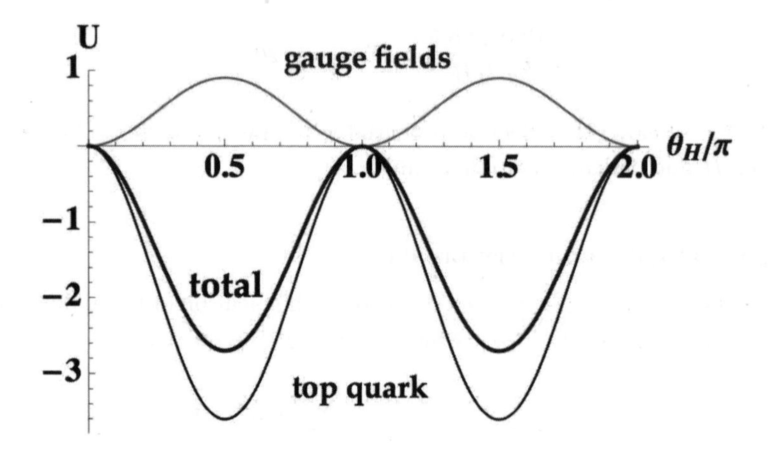

Fig. 4. Effective potential $V_{\rm eff}(\theta_H) = m_{\rm KK}^4/(16\pi^4)U(\theta_H)$ with $z_L = 10^{10}$.

For $z_L = 10^{10}$, $m_{\rm KK} = 1.2\,\text{TeV}$ and $m_H = 108\,\text{GeV}$. At first sight one might think that this value for m_H contradicts with the LEP2 bound $m_H > 114\,\text{GeV}$. However, the effective potential is minimized at $\theta_H = \pm\frac{1}{2}\pi$ where the ZZH coupling vanishes. See (11). The LEP2 bound is evaded as the process $e^+e^- \to Z^* \to ZH$ does not take place in the present model.

6. The absolute stability of the Higgs bosons

In the present gauge-Higgs unification model the AB phase is dynamically chosen to be $\theta_H = \frac{1}{2}\pi$ at the one loop level. It implies that all ZZH, WWH and Yukawa couplings of quarks and leptons vanish so that the Higgs boson cannot decay. Can the Higgs boson decay at higher orders? We show that the Higgs boson becomes absolutely stable in a class of the gauge-Higgs unification models including the present model.[4]

(i) Mirror reflection symmetry

The theory is invariant under the mirror reflection in the fifth dimension;

$$\begin{aligned}
(x^\mu, y) &\to (x^{\mu\prime}, y') = (x^\mu, -y) \,, \\
A_M(x,y) &\to A'_M(x',y') = (A_\mu, -A_y)(x,y) \,, \\
\Psi_a(x,y) &\to \Psi'_a(x',y') = \pm\gamma^5 \Psi_a(x,y) \,.
\end{aligned} \tag{14}$$

It implies that physics is invariant under

$$\hat\theta_H(x) = \theta_H + \frac{H(x)}{f_H} \to \hat\theta'_H(x') = -\hat\theta_H(x) \tag{15}$$

while all other SM particles remain unchanged.

(ii) Enhanced gauge invariance

One may notice that the effective potential $V_{\rm eff}(\theta_H)$ at the one loop level depicted in fig. 4 has periodicity π. Indeed, this remains true to all order. Gauge invariance is enhanced.

Consider a lare gauge transformation (7) with $\alpha = \pi$. With $\Omega^{\rm large}(y;\pi)$ the boundary conditions change to $(P_0'^{\rm vec}, P_1'^{\rm vec}) = (P_0{}^{\rm vec}, P_1{}^{\rm vec})$ in the vectorial representation and $(P_0'^{\rm sp}, P_1'^{\rm sp}) = (P_0{}^{\rm sp}, -P_1{}^{\rm sp})$ in the spinorial representation. The AB phase changes to $\theta_H' = \theta_H + \pi$. In the present model all bulk fermions are in the vector representation. The brane fermions and scalar field on the Planck brane are not affected by this transformation as $\Omega^{\rm large}(0;\pi) = 1$. Hence the theory is invariant under $\theta_H \to \theta_H + \pi$. All physical quantities become periodic in θ_H with a reduced period π.

(iii) H-parity

In a class of the $SO(5) \times U(1)$ gauge-Higgs unification models in the warped space which contains bulk fermions only in tensorial representations of $SO(5)$ and brane fermions only on the Planck brane, the enhanced gauge symmetry with the mirror reflection symmetry leads to

$$
\begin{aligned}
&V_{\rm eff}(\hat\theta_H + \pi) = V_{\rm eff}(\hat\theta_H) = V_{\rm eff}(-\hat\theta_H) ~, \\
&m_{W,Z}^2(\hat\theta_H + \pi) = m_{W,Z}^2(\hat\theta_H) = m_{W,Z}^2(-\hat\theta_H) ~, \\
&m_{ab}^F(\hat\theta_H + \pi) = -m_{ab}^F(\hat\theta_H) = m_{ab}^F(-\hat\theta_H) ~.
\end{aligned}
\tag{16}
$$

$m_{W,Z}(0) = m_{ab}^F(0) = 0$ as the EW symmetry is recovered at $\theta_H = 0$.

In the previous section we have seen that $V_{\rm eff}(\theta_H)$ is minimized at $\theta_H = \frac{1}{2}\pi$. It follows then from (16) that all of $V_{\rm eff}(\hat\theta_H)$, $m_{W,Z}^2(\hat\theta_H)$ and $m_{ab}^F(\hat\theta_H)$ satisfy a relation $F(\frac{1}{2}\pi + f_H^{-1}H) = F(\frac{1}{2}\pi - f_H^{-1}H)$. They are even functions of H when expanded around $\theta_H = \pm\frac{1}{2}\pi$. All odd-power Higgs couplings $H^{2\ell+1}$, $H^{2\ell+1}W_\mu^\dagger W^\mu$, $H^{2\ell+1}Z_\mu Z^\mu$, and $H^{2\ell+1}\overline{\psi}_a\psi_b$, vanish.

The effective interactions at low energies are invariant under $H(x) \to -H(x)$ with all other fields kept intact at $\theta_H = \pm\frac{1}{2}\pi$. There arises the H-parity invariance. Among low energy fields only the Higgs field is H-parity odd. The Higgs boson becomes absolutely stable, protected by the H-parity conservation.

7. Stable Higgs bosons as cold dark matter

Abosolutely stable Higgs bosons become cold dark matter (CDM) in the present universe. They are copiously produced in the very early universe. As the annihilation rate of Higgs bosons falls below the expansion rate of the universe, the annihilation processes get effectively frozen and the remnant Higgs bosons become dark matter.[4]

The annihilation rates can be estimated from the effective Lagrangian (9) with (10). At $\theta_H = \frac{1}{2}\pi$ one has

$$\mathcal{L}_{\rm eff} \sim -\Big\{ m_W^2 W_\mu^\dagger W^\mu + \frac{1}{2} m_Z^2 Z_\mu Z^\mu \Big\} \cos^2 \frac{H}{f_H} - \sum_a m_a \overline{\psi}_a \psi_a \cos \frac{H}{f_H} \ . \tag{17}$$

As $m_W \sim \frac{1}{2} g f_H$, f_H is determined to be $\sim 246\,\mathrm{GeV}$. Although the Yukawa coupling $\overline{\psi}_a \psi_a H$ vanishes, the $\overline{\psi}_a \psi_a H^2$ coupling is nonvanishing, given by $(m_a/2f_H^2)\overline{\psi}_a\psi_a H^2$. It is generated by two vertices $\psi_a \psi_a^{(n)} H$ where $\psi_a^{(n)}$ is the n-th KK excited state of ψ_a.

The Higgs mass m_H is predicted in the present model in the range of 50 GeV to 130 GeV, depending on the value of the warp factor. If $m_H > m_W$, the dominant annihilation modes are $HH \to WW$, ZZ. The rate is large so that the resultant relic abundance becomes very small. For $m_H < m_W$ the relevant annihilation modes are $HH \to W^*W^*$, Z^*Z^*, $b\bar{b}, c\bar{c}, \tau\bar{\tau}$, and gg. Here W^* (Z^*) indicates virtual W (Z) which subsequently decays into a fermion pair. Annihilation into a gluon (g) pair takes place through a top quark loop.

The relic abundance of Higgs bosons is evaluated as a function of m_H. It is depicted in fig. 5. It is seen that $\Omega_H h^2$ determined from the WMAP data is reproduced in the gauge-Higgs unification with $m_H \sim 70\,\mathrm{GeV}$. It is remarkable that the gauge-Higgs unification scenario gives $\Omega_H h^2$ in the just right order of magnitude for $20\,\mathrm{GeV} < m_H < 75\,\mathrm{GeV}$. With $m_H = 70\,\mathrm{GeV}$, the freeze-out temperature is $T_f \sim 3\,\mathrm{GeV}$. The relative contributions of the $HH \to W^*W^*$ and $b\bar{b}$ modes are 61% and 34%, respectively.

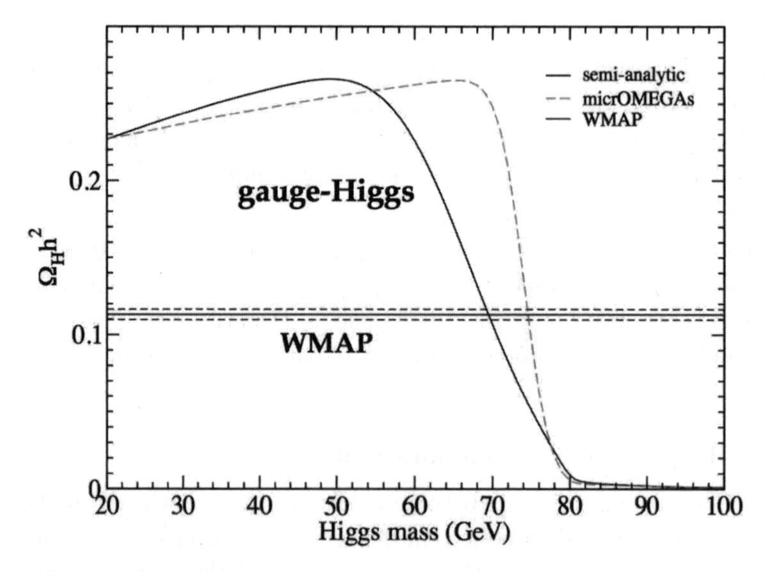

Fig. 5. Thermal relic density of Higgs boson DM with $f_H = 246\,\mathrm{GeV}$. The solid curve is obtained by the semi-analytic formulae. The horizontal band is the WMAP data $\Omega_{\rm CDM} h^2 = 0.1131 \pm 0.0034$. The value in the WMAP data is obtained with $m_H \sim 70\,\mathrm{GeV}$.

If Higgs bosons constitute the cold dark matter of the universe, they can be detected by observing Higgs-nucleon elastic scattering process, $HN \to HN$. The relevant part of the effective interaction (17) is $\mathcal{L}_{\text{eff}} = (H^2/2f_H^2)\sum_f m_f \bar{f} f$. With QCD corrections taken into account the effective interaction for the Higgs-nucleon coupling becomes

$$\mathcal{L}_{HN} \simeq \frac{2 + 7f_N}{9} \frac{m_N}{2f_H^2} H^2 \overline{N} N , \tag{18}$$

where $f_N = \sum_{q=u,d,s} f_q^N$ and $\langle N|m_q \bar{q} q|N\rangle = m_N f_q^N$. With this coupling (18) the spin-independent (SI) Higgs-nucleon scattering cross section is found to be

$$\sigma_{\text{SI}} \simeq \frac{1}{4\pi}\left(\frac{2 + 7f_N}{9}\right)^2 \frac{m_N^4}{f_H^4 (m_H + m_N)^2} , \tag{19}$$

in the non-relativistic limit. There is ambiguity in the value of f_N. Our prediction is $\sigma_{SI} \simeq (1.2 - 2.7) \times 10^{-43}\,\text{cm}^2$ for $f_N = (0.1 - 0.3)$. The present experimental upper bounds for the spin-independent WIMP-nucleon cross sections come from CDMS II[21] and XENON10.[22] From the recent CDMS II data $\sigma_{\text{SI}} \lesssim 7 \times 10^{-44}\,\text{cm}^2$ at 90 % CL with the WIMP mass 70 GeV. With many uncertainties and ambiguity in the analysis taken into account, this does not necessarily mean that the present gauge-Higgs unification model is excluded. The next generation experiments for the direct detection of WIMP-nucleon scattering are awaited to pin down the rate.

8. Stable Higgs bosons at Tevatron/LHC/ILC

In all of the experiments performed so far, Higgs bosons are searched by trying to identify their decay products. If Higgs bosons are stable, however, this way of doing experiments becomes a vain effort.

Higgs bosons are produced in pairs. Typical processes are $Z^* \to ZHH$, $W^* \to WHH$, $WW \to HH$, $ZZ \to HH$, and $gg \to HH$. Higgs bosons are stable so that they appear as missing energies and momenta in collider experiments. The appearance of two particles of missing energies and momenta in the final state makes experiments hard, but not impossible. These events must be distinguished from those involving neutrinos. Since $m_H \sim 70\,\text{GeV}$, Higgs bosons can be copiously produced at LHC. It is a challenging task to identify stable Higgs bosons at colliders.

The effects of the vanishing WWH and ZZH couplings can be seen in WW, WZ, and ZZ elastic scattering, too. These scattering amplitudes become large as the energy is increased, much faster than in the standard model.[23]

9. Gauge couplings of quarks and leptons

We have seen large deviations in the Higgs couplings from the standard model. There arise deviations in the gauge couplings of quarks and leptons as well, which have to be scrutinized with the electroweak precision measuremens.[5,8,9]

5D gauge couplings of fermions are universal. However, 4D gauge couplings of quarks and leptons are obtained by inserting their wave functions and W (Z) wave function into 5D Lagrangian and integrating over the fifth coordinate. Since the wave functions depend on quarks and leptons, slight variations appear in 4D gauge couplings, resulting in violation of the universality. The wave functions of W and Z bosons are determined in refs. 10 and 11, whereas those of quarks and leptons are determined in refs. 5 and 14.

There arise deviations in the W boson couplings from the standard model. For the t and b quarks their interactions are given by

$$\frac{1}{2}g_{tb,L}^{(W)}(W_\mu \bar{b}_L \gamma^\mu t_L + W_\mu^\dagger \bar{t}_L \gamma^\mu b_L) + \frac{1}{2}g_{tb,R}^{(W)}(W_\mu \bar{b}_R \gamma^\mu t_R + W_\mu^\dagger \bar{t}_R \gamma^\mu b_R). \tag{20}$$

Not only left-handed components but also right-handed components couple to W in general. The W coupling defined experimentally is the coupling between left-handed e and ν_e, $g_{e\nu,L}^{(W)}$. In the standard model the couplings are universal; $g_{f,L}^{(W)} = g_{e\nu,L}^{(W)}$ and $g_{f,R}^{(W)} = 0$. The violation of the universality for the left-handed quarks and leptons, $g_{f,L}^{(W)}/g_{e\nu,L}^{(W)} - 1$, is summarized in Table 1. The W couplings to the right-handed quarks and leptons are summarized in Table 2. One can see that the deviation from the standard model is extremely tiny except for top-bottom, well below the current limits. For the left-handed top-bottom quarks the deviation is about 2%.

Similarly the Z boson couplings deviate from the standard model. The couplings of t and b quarks take the form

$$\frac{1}{\cos\theta_W} Z_\mu \left\{ g_{tL}^{(Z)} \bar{t}_L \gamma^\mu t_L + g_{tR}^{(Z)} \bar{t}_R \gamma^\mu t_R + g_{bL}^{(Z)} \bar{b}_L \gamma^\mu b_L + g_{bR}^{(Z)} \bar{b}_R \gamma^\mu b_R \right\} . \tag{21}$$

The relevant couplings are $g_{f,LR}^{(Z)}/g_{e\nu,L}^{(W)}$. They are summarized in Table 3, in which the couplings in the standard model are also listed for comparison. It is seen from the table, the deviation from the standard model is rather small (0.1% to 1%)

Table 1. The deviation of the W couplings for left-handed leptons and quarks from the standard model for the warp factor $z_L = 10^{15}$.

$g_{fL}^{(W)}/g_{e\nu,L}^{(W)} - 1$

ν_μ, μ	ν_τ, τ	u, d	c, s	t, b
-1.022×10^{-8}	-2.736×10^{-6}	-2.778×10^{-11}	-1.415×10^{-6}	-0.02329

Table 2. The W couplings for right-handed leptons and quarks for the warp factor $z_L = 10^{15}$. They are all tiny.

$g_{fR}^{(W)}/g_{e\nu,L}^{(W)}$

ν_e, e	ν_μ, μ	ν_τ, τ
-4.761×10^{-21}	-4.141×10^{-16}	-1.133×10^{-13}

u, d	c, s	t, b
-1.392×10^{-11}	-1.896×10^{-7}	-0.001264

except for the top quark coupling for $z_L = 10^{15}$. The Z couplings of the left- and right-handed top quark deviate from those in the standard model by −7% and 18%, respectively. The deviations in the $Zb_L\bar{b}_L$ and $Zb_R\bar{b}_R$ couplings are 0.3% and 0.9%, respectively.

Table 3. The deviation of the Z couplings for leptons and quarks from the standard model for the warp factor $z_L = 10^{15}$. In the column SM the values in the standard model are quoted.

$$\tilde{g}^{(Z)}_{f,LR} = g^{(Z)}_{f,LR}/g^{(W)}_{e\nu,L}$$

f	ν_e	ν_μ	ν_τ	SM
$\tilde{g}^{(Z)}_{f,L}$	0.500822	0.500822	0.500822	0.5
$\tilde{g}^{(Z)}_{fR}$	-5.754×10^{-31}	-5.392×10^{-29}	-1.840×10^{-27}	0
	e	μ	τ	
$\tilde{g}^{(Z)}_{fL}$	−0.2692	−0.2692	−0.2692	−0.2688
$\tilde{g}^{(Z)}_{fR}$	0.2334	0.2333	0.2333	0.2312
	u	c	t	
$\tilde{g}^{(Z)}_{fL}$	0.3464	0.3464	0.3204	0.3459
$\tilde{g}^{(Z)}_{fR}$	−0.1556	−0.1556	−0.1823	−0.1541
	d	s	b	
$\tilde{g}^{(Z)}_{fL}$	−0.4236	−0.4236	−0.4241	−0.4229
$\tilde{g}^{(Z)}_{fR}$	0.07779	0.07777	0.07774	0.07707

There results an important prediction for the forward-backward asymmetry in the e^+e^- collisions on the Z pole as pointed out by Uekusa.[24] The asymmetry is given by

$$A^f_{FB} = \frac{3}{4} A^e_{LR} A^f_{LR} \ , \quad A^f_{LR} = \frac{\left(g^{(Z)}_{fL}\right)^2 - \left(g^{(Z)}_{fR}\right)^2}{\left(g^{(Z)}_{fL}\right)^2 + \left(g^{(Z)}_{fR}\right)^2} \ . \tag{22}$$

The predictions are summarized in Table 4. It is exciting that the gauge-Higgs unification model gives better fits to the observed data for the asymmetry in $b\bar{b}$ and $c\bar{c}$ than the standard model. CP violation, anomalous magnetic moment, and electric dipole moment in gauge-Higgs unification have been also discussed.[25]

10. Summary

The $SO(5) \times U(1)$ gauge-Higgs unification in the Randall-Sundrum warped spacetime is promising. The Higgs boson becomes a part of the gauge fields in higher dimensions, and is unified with gauge bosons. The resultant 4D gauge couplings of

Table 4. The forward-backward asymmetry on the Z pole in e^+e^- collisions. The numbers in the gauge-Higgs unification scenario with the warp factor $z_L = 10^{15}$ are quoted from Uekusa, ref. 24.

A^f_{FB}

f	Experiment	Gauge-Higgs	Standard Model
c	0.0707 ± 0.0035	0.07073	0.0738 ± 0.0006
s	0.0976 ± 0.0114	0.09950	0.1034 ± 0.0007
b	0.0992 ± 0.0016	0.09952	0.1033 ± 0.0007
e	0.0145 ± 0.0025	0.01511	0.01627 ± 0.00023
μ	0.0169 ± 0.0013	0.01513	
τ	0.0188 ± 0.0017	0.01515	

quarks and leptons are close to those in the standard model. The Higgs couplings, on the other hand, deviate significantly from those in the standard model.

In a large class of the $SO(5) \times U(1)$ gauge-Higgs unification models the Higgs boson becomes absolutely stable to all order in perturbation theory. In the evolution of the universe Higgs bosons become cold dark matter. From the WMAP data the Higgs boson mass is determined to be around 70 GeV.

In collider experiments Higgs bosons are produced in pairs. They appear as missing energies and momenta. The way of searching for Higgs bosons must be altered.

Acknowledgment

This work was supported in part by Scientific Grants from the Ministry of Education and Science, Grant No. 20244028, Grant No. 20025004, and Grant No. 50324744.

References

1. Y. Hosotani, Phys. Lett. B **126**, 309 (1983).
2. Y. Hosotani, Annals Phys. **190**, 233 (1989).
3. A. T. Davies and A. McLachlan, Phys. Lett. B **200**, 305 (1988); Nucl. Phys. B **317**, 237 (1989).
4. Y. Hosotani, P. Ko and M. Tanaka, Phys. Lett. B **680**, 179 (2009) [arXiv:0908.0212 [hep-ph]].
5. Y. Hosotani, S. Noda, and N. Uekusa, arXiv:0912.1173 [hep-ph] (to appear in Prog. Theoret. Phys.).
6. R. Contino, Y. Nomura and A. Pomarol, Nucl. Phys. B **671**, 148 (2003) [arXiv:hep-ph/0306259].
7. Y. Hosotani and M. Mabe, Phys. Lett. B **615**, 257 (2005) [arXiv:hep-ph/0503020].
8. K. Agashe, R. Contino and A. Pomarol, Nucl. Phys. B **719**, 165 (2005) [arXiv:hep-ph/0412089].
9. Y. Hosotani, S. Noda, Y. Sakamura and S. Shimasaki, Phys. Rev. D **73**, 096006 (2006) [arXiv:hep-ph/0601241].

10. Y. Sakamura and Y. Hosotani, Phys. Lett. B **645**, 442 (2007) [arXiv:hep-ph/0607236].
11. Y. Hosotani and Y. Sakamura, Prog. Theor. Phys. **118**, 935 (2007) [arXiv:hep-ph/0703212].
12. A. D. Medina, N. R. Shah and C. E. M. Wagner, Phys. Rev. D **76**, 095010 (2007) [arXiv:0706.1281 [hep-ph]].
13. Y. Hosotani, K. Oda, T. Ohnuma and Y. Sakamura, Phys. Rev. D **78**, 096002 (2008) [Erratum-ibid. D **79**, 079902 (2009)] [arXiv:0806.0480 [hep-ph]].
14. Y. Hosotani and Y. Kobayashi, Phys. Lett. B **674**, 192 (2009) [arXiv:0812.4782 [hep-ph]].
15. N. Haba, M. Harada, Y. Hosotani and Y. Kawamura, Nucl. Phys. B **657**, 169 (2003) [Erratum-ibid. B **669**, 381 (2003)] [arXiv:hep-ph/0212035].
16. Y. Sakamura, Phys. Rev. D **76**, 065002 (2007) [arXiv:0705.1334 [hep-ph]].
17. G. F. Giudice, C. Grojean, A. Pomarol and R. Rattazzi, JHEP **0706**, 045 (2007) [arXiv:hep-ph/0703164].
18. H. Hatanaka, T. Inami and C. S. Lim, Mod. Phys. Lett. A **13**, 2601 (1998) [arXiv:hep-th/9805067].
19. K. Oda and A. Weiler, Phys. Lett. B **606**, 408 (2005).
20. A. Falkowski, Phys. Rev. D **75**, 025017(2007).
21. Z. Ahmed *et al.* [CDMS Collaboration], Phys. Rev. Lett. **102**, 011301(2009) [arXiv:0802.3530 [astro-ph]]; arXiv:0912.3592 [astro-ph.CO].
22. J. Angle *et al.* [XENON Collaboration], Phys. Rev. Lett. **100**, 021303(2008) [arXiv:0706.0039 [astro-ph]].
23. N. Haba, Y. Sakamura and T. Yamashita, arXiv:0908.1042 [hep-ph].
24. N. Uekusa, arXiv:0912.1218 [hep-ph].
25. Y. Adachi, C. S. Lim and N. Maru, Phys. Rev. D **76**, 075009 (2007) [arXiv:0707.1735 [hep-ph]]; Phys. Rev. D **80**, 055025 (2009) [arXiv:0905.1022 [hep-ph]].

The Limits of Custodial Symmetry*

R. Sekhar Chivukula[a], Roshan Foadi[b] and Elizabeth H. Simmons[c]

Department of Physics and Astronomy
Michigan State University, East Lansing, MI 48824, USA
E-mails: [a]sekhar@msu.edu, [b]foadiros@msu.edu, [c]esimmons@msu.edu

Stefano Di Chiara

CP3-Origins, Campusvej 55, DK-5230 Odense M, Denmark
E-mail: dichiara@cp3.sdu.dk

We introduce a toy model implementing the proposal of using a custodial symmetry to protect the $Zb_L\bar{b}_L$ coupling from large corrections. This "doublet-extended standard model" adds a weak doublet of fermions (including a heavy partner of the top quark) to the particle content of the standard model in order to implement an $O(4) \times U(1)_X \sim SU(2)_L \times SU(2)_R \times P_{LR} \times U(1)_X$ symmetry in the top-quark mass generating sector. This symmetry is softly broken to the gauged $SU(2)_L \times U(1)_Y$ electroweak symmetry by a Dirac mass M for the new doublet; adjusting the value of M allows us to explore the range of possibilities between the $O(4)$-symmetric ($M \to 0$) and standard-model-like ($M \to \infty$) limits.

1. Introduction

Agashe[2] et al. have shown that the constraints on beyond the standard model physics related to the $Zb_L\bar{b}_L$ coupling can, in principle, be loosened if the global $SU(2)_L \times SU(2)_R$ symmetry of the electroweak symmetry breaking sector is actually a subgroup of a larger global symmetry of both the symmetry breaking and top quark mass generating sectors of the theory. In particular, they propose that these interactions preserve an $O(4) \sim SU(2)_L \times SU(2)_R \times P_{LR}$ symmetry, where P_{LR} is a parity interchanging $L \leftrightarrow R$. The $O(4)$ symmetry is then spontaneously broken to $O(3) \sim SU(2)_V \times P_{LR}$, breaking the elecroweak interactions but protecting g_{Lb} from radiative corrections, so long as the left-handed bottom quark is a P_{LR} eigenstate.

In this talk we report on the construction of the simplest $O(4)$-symmetric extension of the SM. For reasons that will shortly become clear, we call this model the doublet-extended standard model or DESM. Because the DESM is minimal, it displays the essential ingredients protecting g_{Lb} without the burden of additional states, interactions, or symmetry patterns that might otherwise obscure the role

*Speaker at conference: R. Sekhar Chivukula. This report is a shortened version of previously published work.[1]

played by custodial $O(3)$. Because it is concrete, it also enables us to explore how the new symmetry impacts the model's ability to conform with the constraints imposed by other precision electroweak data.

In our model, all operators of dimension-4 in the Higgs potential and the sector generating the top quark mass respect a global $O(4) \times U(1)_X$ symmetry; the $U(1)_X$ enables the SM-like fermions to obtain the appropriate electric charges and hypercharges. In addition to the particle content of the SM, we introduce a new weak doublet of Dirac fermions, $\Psi = (\Theta, T')$, and combine Ψ_L with the left-handed top-bottom doublet (t'_L, b_L) to form a $(2, 2^*)$ under the global $SU(2)_L \times SU(2)_R$ symmetry. The b_L state is thereby endowed with identical charges under the two global $SU(2)$ groups, $T^3_L = T^3_R$, making it a parity eigenstate, as desired. We also find that the T' mixes with t' to form a SM-like top quark and a heavy partner. The $O(4) \times U(1)_X$ symmetric Yukawa interaction can, of course, be extended to the bottom quark and the remaining electroweak doublets, by adding further spectator fermions; here we focus exclusively on the partners of the top quark since they give the dominant contribution to g_{Lb}.

To enable electroweak symmetry breaking and fermion mass generation to proceed, the global symmetry is explicitly broken to $SU(2)_L \times U(1)_Y$ by a dimension-three Dirac mass M for Ψ. As $M \to \infty$ the ordinary SM top Yukawa interaction is recovered; as $M \to 0$ the model becomes exactly $O(4) \times U(1)_X$ symmetric; adjusting the value of M allows us to interpolate between these extremes and to investigate the limits to which the custodial symmetry of the top-quark mass generating sector can be enhanced. When we calculate the dominant one-loop corrections to g_{Lb} in our model we find, consistent with previous results,[2] that because b_L is a P_{LR} eigenstate, g_{Lb} is protected from radiative corrections in the $M \to 0$ limit and these corrections return as M is switched on. However, when we study the behavior of oblique radiative corrections as M is varied, we find that in the small-M limit where g_{Lb} is closer to the experimental value,[3] the oblique corrections become unacceptably large. In particular, in the $M \to 0$ limit the enhanced custodial symmetry produces a potentially sizable negative contribution to αT.

2. Doublet-Extended Standard Model

2.1. *Custodial Symmetry and Z Coupling*

The tree-level coupling of a SM fermion ψ to the Z boson is,

$$\frac{e}{c_w s_w} \left(T^3_L - Q \sin^2\theta_W\right) Z^\mu \bar{\psi} \gamma_\mu \psi \ , \tag{1}$$

where T^3_L and Q are, respectively, the weak isospin and electromagnetic charges of fermion ψ, e is the electromagnetic coupling; c_w and s_w are the cosine and sine of the weak mixing angle. Because the electromagnetic charge is conserved, loop corrections to the $Z\bar{\psi}\psi$ coupling do not alter it; however, the weak symmetry $SU(2)_L$ is broken at low energies, and radiative corrections to the T^3_L coupling are present in the SM.

Following Agashe[2] *et. al.*, we wish to construct a scenario in which the T_L^3 coupling is not subject to flavor-dependent radiative corrections. To start, we note that the accidental custodial symmetry of the SM implies that the vectorial charge $T_V^3 \equiv T_L^3 + T_R^3$ is conserved

$$\delta T_V^3 = \delta T_L^3 + \delta T_R^3 = 0 \ . \tag{2}$$

This suggests a way to evade flavor-dependent corrections to T_L^3 itself, by adding a parity symmetry P_{LR} that exchanges $L \leftrightarrow R$. If ψ is an eigenstate of this parity symmetry and the symmetry persists at the energies of interest, then

$$\delta T_L^3 = \delta T_R^3 \ . \tag{3}$$

Now, we see that Eq. (2) is satisfied by having the two terms on the right hand side vanish separately, rather than remaining non-zero and canceling one another. In other words, $\delta T_L^3 = 0$ and the $Z\bar{\psi}\psi$ coupling remains fixed even to higher-order in this scenario. We will now show how to implement this idea for the b-quark in a toy model and examine the phenomenological consequences.

2.2. *The Model*

Let us construct a simple extension of the SM that implements this parity idea for the third-generation quarks, in order to suppress radiative corrections to the $Zb\bar{b}$ vertex. We extend the global $SU(2)_L \times SU(2)_R$ symmetry of the Higgs sector of the SM to an $O(4) \times U(1)_X \sim SU(2)_L \times SU(2)_R \times P_{LR} \times U(1)_X$ for both the symmetry breaking and top quark mass generating sectors of the theory. As usual, only the electroweak subgroup, $SU(2)_L \times U(1)_Y$, of this global symmetry is gauged; our model does not include additional electroweak gauge bosons. The global $O(4)$ spontaneously breaks to $O(3) \sim SU(2)_V \times P_{LR}$ which will protect g_{Lb} from radiative corrections, as above, provided that the left-handed bottom quark is a parity eigenstate: $P_{LR} b_L = \pm b_L$. The additional global $U(1)_X$ group is included to ensure that the light t and b eigenstates, the ordinary top and bottom quarks, obtain the correct hypercharges.

In light of the extended symmetry group, the relationships between electromagnetic charge Q, hypercharge Y, the left- and right-handed T^3 charges, and the new charge Q_X associated with $U(1)_X$ are as follows:

$$Y = T_R^3 + Q_X \ , \tag{4}$$

$$Q = T_L^3 + Y = T_L^3 + T_R^3 + Q_X \ . \tag{5}$$

Since the b_L state is supposed to correspond to the familiar bottom-quark, it has the familiar SM charges $T_L^3(b_L) = -1/2$, and $Q(b_L) = -1/3$, and $Y(b_L) = 1/6$. Because b_L must be an eigenstate under P_{LR}, we deduce that $T_R^3(b_L) = T_L^3(b_L) = -1/2$. Then to be consistent with Eqs. (4) and (5), its charge under the new global $U(1)_X$ must be $Q_X(b_L) = 2/3$. Moreover, since the left-handed b quark is an $SU(2)_L$ partner of the left-handed t quark, the full left-handed top-bottom doublet must have the charges $T_R^3 = -1/2$ and $Q_X = 2/3$, just as the full doublet has hypercharge $Y = 1/6$. Finally, the non-zero T_3^R charge of the top-bottom doublet tells us that this doublet forms part of a larger multiplet under the $SU(2)_L \times SU(2)_R$ symmetry and it will be necessary to introduce some new fermions with $T_R^3 = 1/2$ to complete the multiplet.

We therefore introduce a new doublet of fermions $\Psi \equiv (\Omega, T')$. The left-handed component, Ψ_L joins with the top-bottom doublet $q_L \equiv (t'_L, b_L)$ to form an $O(4) \times U(1)_X$ multiplet

$$\mathcal{Q}_L = \begin{pmatrix} t'_L & \Omega_L \\ b_L & T'_L \end{pmatrix} \equiv \begin{pmatrix} q_L & \Psi_L \end{pmatrix} , \tag{6}$$

which transforms as a $(2, 2^*)_{2/3}$ under $SU(2)_L \times SU(2)_R \times U(1)_X$. The parity operation P_{LR}, which exchanges the $SU(2)_L$ and $SU(2)_R$ transformation properties of the fields, acts on $\mathcal{Q}_L$ as:

$$P_{LR}\mathcal{Q}_L = -\left[(i\sigma_2)\,\mathcal{Q}_L\,(i\sigma_2)\right]^T = \begin{pmatrix} T'_L & -\Omega_L \\ -b_L & t'_L \end{pmatrix} \tag{7}$$

exchanging the diagonal components, while reversing the signs of the off-diagonal components. Thus t'_L and T'_L are constrained to share the same electromagnetic charge, in order to satisfy Eq. (5). In fact, we will later see that the t' and T' states mix to form mass eigenstates corresponding to the top quark (t) and a heavy partner (T). The charges of the components of $\mathcal{Q}_L$ are listed in Table 1.

Table 1. Charges of the fermions under the various symmetry groups in the model. Note that, as discussed in the text, other T_R^3 and Q_X assignments for the Ω_R and T'_R states are possible.

	t'_L	b_L	Ω_L	T'_L	t'_R	b_R	Ω_R	T'_R
T_L^3	$\frac{1}{2}$	$-\frac{1}{2}$	$\frac{1}{2}$	$-\frac{1}{2}$	0	0	$\frac{1}{2}$	$-\frac{1}{2}$
T_R^3	$-\frac{1}{2}$	$-\frac{1}{2}$	$\frac{1}{2}$	$\frac{1}{2}$	0	-1	0	0
Q	$\frac{2}{3}$	$-\frac{1}{3}$	$\frac{5}{3}$	$\frac{2}{3}$	$\frac{2}{3}$	$-\frac{1}{3}$	$\frac{5}{3}$	$\frac{2}{3}$
Y	$\frac{1}{6}$	$\frac{1}{6}$	$\frac{7}{6}$	$\frac{7}{6}$	$\frac{2}{3}$	$-\frac{1}{3}$	$\frac{7}{6}$	$\frac{7}{6}$
Q_X	$\frac{2}{3}$	$\frac{2}{3}$	$\frac{2}{3}$	$\frac{2}{3}$	$\frac{2}{3}$	$\frac{2}{3}$	$\frac{7}{6}$	$\frac{7}{6}$

We assign the minimal right-handed fermions charges that accord with the symmetry-breaking pattern we envision: the top and bottom quarks will receive mass via Yukawa terms that respect the full $O(4) \times U(1)_X$ symmetry, while the exotic states will have a dimension-three mass term that explicitly breaks the large symmetry to $SU(2)_L \times U(1)$. Moreover, to accord with experiment, the t'_R and b_R must have $T_L^3 = 0$ and share the electric charges of their left-handed counterparts. The top and bottom quarks will receive mass through a Yukawa interaction with a SM-like Higgs multiplet that breaks the electroweak symmetry. The simplest choice is to assign the Higgs multiplet to be neutral under $U(1)_X$; in this case, both t'_R and b_R share the $Q_X = 2/3$ charge of t'_L and b_L. Therefore, from Eqs. (4) and (5), we find $T_R^3(t'_R) = 0$ (meaning that t'_R can be chosen to be an $SU(2)_R$ singlet) and $T_R^3(b_R) = -1$ (so that b_R is, minimally, part of an $SU(2)_R$ triplet if we extend the symmetry to the bottom quark mass generating sector). Turning now to the T'_R and Ω_R states, we see that they must form an $SU(2)_L$ doublet with hypercharge 7/6 so that the Dirac mass term for Ψ preserves the electroweak symmetry as desired.[a] Finally, we choose $T_R^3(\Omega_R) = T_R^3(T'_R) = 0$, which implies $Q_X = 7/6$ for both states, as the minimal choice satisfying the constraint imposed by Eq. (4); other choices of T_R^3 charge would involve adding additional fermions to form complete $SU(2)_R$ multiplets. The charges of the fermions are listed in Table 1.

Now, let us describe the symmetry-breaking pattern and fermion mass terms explicitly. Spontaneous electroweak symmetry breaking proceeds through a Higgs multiplet that transforms as a $(2, 2^*)_0$ under $SU(2)_L \times SU(2)_R \times U(1)_X$:

$$\Phi = \frac{1}{\sqrt{2}} \begin{pmatrix} v + h + i\phi^0 & i\sqrt{2}\,\phi^+ \\ i\sqrt{2}\,\phi^- & v + h - i\phi^0 \end{pmatrix} . \tag{8}$$

Again, the parity operator P_{LR} exchanges the diagonal fields and reverses the signs of the off-diagonal elements. When the Higgs acquires a vacuum expectation value, the longitudinal W and Z bosons acquire mass and a single Higgs boson remains in the low-energy spectrum. The Higgs multiplet has an $O(4) \times U(1)_X$ symmetric Yukawa interaction with the top quark:

$$\mathcal{L}_{\text{Yukawa}} = -\lambda_t \text{Tr}\left(\overline{\mathcal{Q}}_L \cdot \Phi\right) t'_R \; + \text{h.c.} \tag{9}$$

that contributes to generating a top quark mass. In principle, the same Higgs multiplet can also contribute to the bottom quark mass through a separate, and similarly $O(4) \times U(1)_X$ symmetric, Yukawa interaction involving the $SU(2)_R$ triplet to which b_R belongs. Since the phenomenological issues that concern us in this paper are affected far more strongly by m_t than by the far-smaller m_b, we will neglect this and any other Yukawa interaction.

Next we break the full $O(4) \times U(1)_X$ symmetry to its electroweak subgroup. We do so first by gauging $SU(2)_L \times U(1)_Y$. In addition, we wish to preserve the $O(4)$ symmetry of the top quark mass generating sector in all dimension-4 terms, but break it softly by introducing a dimension-3 Dirac mass term for Ψ,

$$\mathcal{L}_{\text{mass}} = -M \; \bar{\Psi}_L \cdot \Psi_R + h.c. \tag{10}$$

that explicitly breaks the global symmetry to $SU(2)_L \times U(1)_Y$. Note that we therefore expect that any flavor-dependent radiative corrections to the $Zb_L\bar{b}_L$ coupling will vanish in the limit $M \to 0$, as the protective parity symmetry is restored; alternatively, as $M \to \infty$, the larger symmetry is pushed off to such high energies that the resulting theory looks more and more like the SM.

[a]This means that the Ω_R and T'_R states do not fill out the $SU(2)_R$ triplet to which b_R belongs – which is uncharged under $SU(2)_L$ and carries hypercharge 2/3; other exotic fermions must play that role if we wish to extend the symmetry to the bottom quark mass generating sector.

In addition to the fermions explicitly described above, a more complete version of this toy model must contain several other fermions to fill out the $SU(2)_R$ multiplet to which the b_R belongs and also some spectator fermions that cancel $U(1)_Y$ anomalies. However, the toy model suffices for exploration of the issues related to the $Zb_L\bar{b}_L$ coupling that is the focus of this paper.

2.3. *Mass Matrices and Eigenstates*

When the Higgs multiplet acquires a vacuum expectation value and breaks the electroweak symmetry, masses are generated for the top quark, its heavy partner T and the exotic fermion Ω through the mass matrix:

$$\mathcal{L}_{\text{mass}} = -\begin{pmatrix} t'_L & T'_L \end{pmatrix} \begin{pmatrix} m & 0 \\ m & M \end{pmatrix} \begin{pmatrix} t'_R \\ T'_R \end{pmatrix} - M\bar{\Omega}_L\Omega_R + \text{h.c} , \tag{11}$$

where

$$m = \frac{\lambda_t v}{\sqrt{2}} . \tag{12}$$

Note that the Ω field is decoupled from the SM sector, and its mass is simply $m_\Omega = M$. The bottom quark remains massless because we have ignored its Yukawa coupling.

Diagonalizing the top quark mass matrix yields mass eigenstates t (corresponding to the SM top quark) and T (a heavy partner quark), with corresponding eigenvalues

$$m_t^2 = \frac{1}{2}\left[1-\sqrt{1+\frac{4m^4}{M^4}}\right] M^2+m^2 , \qquad m_T^2 = \frac{1}{2}\left[1+\sqrt{1+\frac{4m^4}{M^4}}\right] M^2+m^2 . \tag{13}$$

The mass eigenstates are related to the original gauge eigenstates through the rotations:

$$\begin{pmatrix} t'_R \\ T'_R \end{pmatrix} = \begin{pmatrix} \cos\theta_R & \sin\theta_R \\ -\sin\theta_R & \cos\theta_R \end{pmatrix} \begin{pmatrix} t_R \\ T_R \end{pmatrix} , \quad \begin{pmatrix} t'_L \\ T'_L \end{pmatrix} = \begin{pmatrix} \cos\theta_L & \sin\theta_L \\ -\sin\theta_L & \cos\theta_L \end{pmatrix} \begin{pmatrix} t_L \\ T_L \end{pmatrix} , \tag{14}$$

whose mixing angles are given by

$$\sin\theta_R = \frac{1}{\sqrt{2}}\sqrt{1-\frac{1-2m^2/M^2}{\sqrt{1+4m^4/M^4}}} , \quad \sin\theta_L = \frac{1}{\sqrt{2}}\sqrt{1-\frac{1}{\sqrt{1+4m^4/M^4}}} . \tag{15}$$

From these equations the decoupling limit $M \to \infty$ is evident: m_t approaches its SM value as in Eq. (12), the $t-T$ mixing goes to zero, and T becomes degenerate with Ω. Conversely, in the limit $M \to 0$, the full $O(4)\times U(1)_X$ symmetry is restored and only the combination $T'_L + t'_L$ couples to t_R with mass m.

For phenomenological discussion, it will be convenient to fix m_t at its experimental value and express the other masses in terms of m_t and the ratio $\mu \equiv M/m$. Fig. 1 shows how m, M, and m_T, vary with μ; the horizontal line represents m_t which is being held fixed at 172 GeV. In the limit as μ becomes large, $m \to m_t$, $m_T \sim M$ grows steadily, and the mixing angles decline toward zero; this is a physically-sensible limit that ultimately leads back to the SM. However we see that the opposite limit, where $\mu \to 0$ can only be achieved for $m \to \infty$, which is not physically reasonable since it corresponds to taking $\lambda_t \to \infty$. Hence, we will need to take care in talking about the case of small μ.

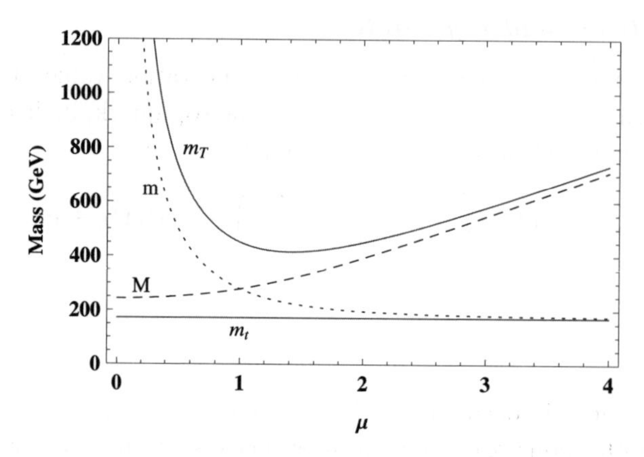

Fig. 1. The curves show the behaviors of m (dotted), M (dashed), and m_T (upper solid) as functions of $\mu \equiv M/m$ when m_t is held fixed. The solid horizontal line corresponds to $m_t \simeq$ 172 GeV.

3. Phenomenology

3.1. *Z Coupling to* $b_L\bar{b}_L$

We are now ready to study how the flavor-dependent corrections to the $Zb_L\bar{b}_L$ coupling behave in our toy model. Specifically, if we write the $Zb_L\bar{b}_L$ coupling as

$$\frac{e}{c_w s_w}\left(-\frac{1}{2}+\delta g_{Lb}+\frac{1}{3}\sin^2\theta_L\right)\,Z_\mu\,\bar{b}_L\gamma_\mu b_L\,, \tag{16}$$

then all the flavor-dependence is captured by δg_{Lb}. At tree-level, the $Zb_L\bar{b}_L$ coupling in our model has its SM value, with $\delta g_{Lb} = 0$, because the b_L has the same quantum numbers as in the SM. However, at one-loop, flavor-dependent vertex corrections arise and these give non-zero corrections to δg_{Lb}; these corrections differ from those in the SM due to the presence of vertex corrections involving exchange of T, the heavy partner of the top quark.

The calculation may be done conveniently in the "gaugeless" limit,[4–7] in which the Z boson is treated as a non-propagating external field coupled to the current $j^\mu_{3L} - j^\mu_Q \sin^2\theta_L$. Operationally, this involves replacing Z_μ with $\partial_\mu\phi^0/m_Z$ in the gauge current interaction, where ϕ^0 is the Goldstone boson eaten by the Z:

$$\begin{aligned}\frac{e}{c_w s_w}\,Z_\mu(j^\mu_{3L}-j^\mu_Q\sin^2\theta_L) \quad &\to \quad \frac{e}{c_w s_w m_Z}\,\partial_\mu\phi^0(j^\mu_{3L}-j^\mu_Q\sin^2\theta_L)\\ &= \frac{2}{v}\,\partial_\mu\phi^0(j^\mu_{3L}-j^\mu_Q\sin^2\theta_L)\end{aligned} \tag{17}$$

The general vertex diagram shown in Fig. 2, will yield radiative corrections to the effective operator $\partial_\mu \phi^0 \, \bar{b}_L \gamma^\mu b_L$; that is, the expression for this diagram will include a term of the form

$$A \, \partial_\mu \phi^0 \, \bar{b}_L \gamma^\mu b_L \, . \tag{18}$$

Comparing the last three equations shows that the coefficient A is proportional to the quantity we are interested in:

$$\delta g_{Lb} = \frac{v}{2} \, A \, . \tag{19}$$

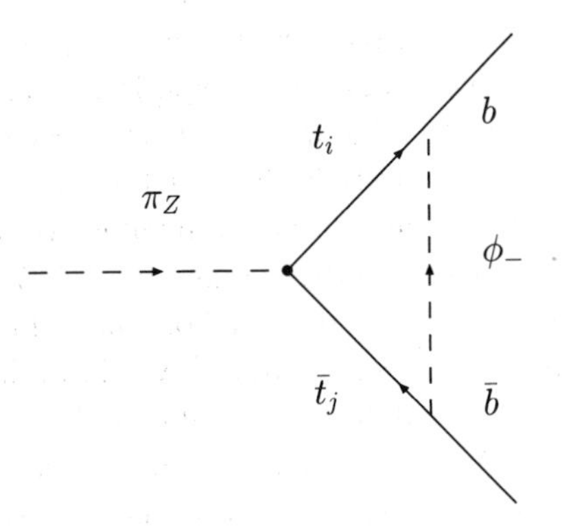

Fig. 2. One-loop vertex correction diagram for $\pi_Z \to b\bar{b}$ in our model. The $t_{i,j}$ may be either the top quark (t) or its heavy partner (T).

We have calculated the several loop diagrams represented by Fig. 2 and obtain the following expression for δg_{Lb}:

$$\begin{aligned} \delta g_{Lb} = & \frac{m_t^2}{16\pi^2 v^2} \left[\cos^2\theta_L \left(\cos 2\theta_L + \sin^2\theta_R \right) + \frac{m_T^2}{m_t^2} \sin^2\theta_L \left(\cos^2\theta_R - \cos 2\theta_L \right) \right. \\ & - \frac{m_T/m_t}{4} \sin 2\theta_L \sin 2\theta_R \\ & \left. - \frac{m_T/m_t}{2} \sin 2\theta_L \left(\frac{m_T^2/m_t^2 + 1}{4} \sin 2\theta_R - 2\frac{m_T}{m_t} \sin 2\theta_L \right) \frac{\log(m_T^2/m_t^2)}{m_T^2/m_t^2 - 1} \right] , \end{aligned} \tag{20}$$

where the prefactor proportional to m_t^2 is the SM result for this class of diagram. We expect to see δg_{Lb} vanish in the limit $M \to 0$ as the parity symmetry is restored; this expectation is fulfilled, since $m_t \to 0$ in this limit. At the other extreme, for large M, we expect to find δg_{Lb} take on its SM value by having the factor within square brackets approach one. This may be readily verified if we take the equivalent limit as $\mu \to \infty$ for fixed m_t:

$$\delta g_{Lb}(\mu \to \infty) \to \frac{m_t^2}{16\pi^2 v^2} \left[1 + \frac{\log(1/\mu^2)}{2\mu^2} + \mathcal{O}(1/\mu^4) \right] , \tag{21}$$

since in this limit $\sin\theta_L \to 1/\mu^2$, $\sin\theta_R \to 1/\mu$ and $m_T^2/m_t^2 \to \mu^2$. In other words, we find that adjusting the value of M allows us to interpolate between the SM value for δg_{Lb} at large M and the absence of a radiative correction at small M. While the limit of small μ is less useful, as we mentioned earlier, for completeness we note that

$$\delta g_{Lb}(\mu \to 0) \to \frac{m_t^2}{16\pi^2 v^2}\left[\frac{3\log(2/\mu)-1}{2} + \mu^2 \frac{6+\log(2/\mu)}{8} + \mathcal{O}(\mu^4)\right]. \tag{22}$$

since in this limit $\sin\theta_L \to (1/\sqrt{2})(1-\mu^2/4)$, $\sin\theta_R \to (1-\mu^2/8)$, and $m_T^2/m_t^2 \to 4/\mu^2$. This growth at small μ is visible in Fig. (3).

We now use our results to compare the value of g_{Lb} in our model (as a function of μ for fixed m_t) with the values given by experiment and the SM, as illustrated in Fig. (3). The experimental[8] value $g_{Lb}^{ex} = -0.4182 \pm 0.0015$ corresponds to the thick horizontal line; the thin (red) horizontal lines bordering the shaded band show the $\pm 1\sigma$ deviations from the experimental value. We calculated the SM value using ZFITTER[9,10] with a reference Higgs mass $m_h = 115$ GeV, and obtain $g_{Lb}^{SM} = -0.42114$. This is indicated by the dashed horizontal line, and may be seen to deviate from g_{Lb}^{ex} by 1.96σ. The (solid blue) curve shows how g_{Lb} varies with μ in our model; we required g_{Lb} to match the SM value with $m_t = 172$ GeV and $v = 246$ GeV as $\mu \to \infty$ and the shape of the curve reflects our results for δg_{Lb} in Eq. (21). We see that g_{Lb} in our model is slightly more negative than (i.e. slightly farther from the experimental value than) the SM value for $\mu > 1$, agrees with the SM value for $\mu = 1$, and comes within $\pm 1\sigma$ of the experimental value only for $\mu < 1$. Given the shortcomings of the small-μ limit, this is disappointing.

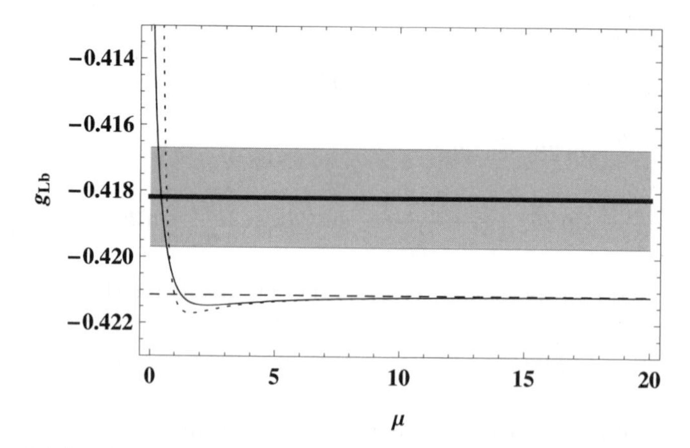

Fig. 3. The solid (blue) curve shows the DESM model's prediction for g_{Lb}, Eq. (21). The thick horizontal line corresponds to $g_{Lb}^{ex} = -0.4182$, while the two horizontal upper and lower solid lines bordering the shaded band correspond to the $\pm 1\sigma$ deviations.[8] The SM prediction is given by the dashed horizontal line. The leading-log contribution is shown by the dotted curve.

3.2. *Oblique Electroweak Parameters*

The flavor-universal corrections from new physics beyond the SM can be parametrized in a model independent way using the four oblique EW parameters αS, αT, $\alpha\delta$, $\Delta\rho$; the first two are the oblique paramters[11–13] for models without additional electroweak gauge bosons, while the other two incorporate the effects of an extended electroweak sector. In general, the oblique parameters are related as follows[14,15] to the neutral-current

$$-\mathcal{M}_{NC} = 4\pi\alpha\frac{QQ'}{P^2} + \frac{(T^3 - s_w^2 Q)(T'^3 - s^2 Q')}{\left(\frac{s_w^2 c_w^2}{4\pi\alpha} - \frac{S}{16\pi}\right)P^2 + \frac{1}{4\sqrt{2}G_F}\left(1 - \alpha T + \frac{\alpha\delta}{4 s_w^2 c_w^2}\right)} \tag{23}$$
$$+ \sqrt{2}G_F\,\frac{\alpha\delta}{s_w^2 c_w^2}\,T^3 T'^3 + 4\sqrt{2}G_F\,(\Delta\rho - \alpha T)\,(Q - T^3)(Q' - T'^3),$$

and charged-current electroweak scattering amplitudes

$$-\mathcal{M}_{CC} = \frac{(T^+T'^- + T^-T'^+)/2}{\left(\frac{s_w^2}{4\pi\alpha} - \frac{S}{16\pi}\right)P^2 + \frac{1}{4\sqrt{2}G_F}\left(1 + \frac{\alpha\delta}{4 s_w^2 c_w^2}\right)} + \sqrt{2}G_F\,\frac{\alpha\delta}{s_w^2 c_w^2}\,\frac{(T^+T'^- + T^-T'^+)}{2}, \tag{24}$$

with P^2 a Euclidean momentum-squared. In the DESM we may set $\Delta\rho = \alpha T$, because the model contains no extra $U(1)$ gauge group, and $\delta = 0$, because there is no extra $SU(2)$ gauge group. We therefore work purely in terms of αS and αT from here on. We take the origin of the $\alpha S, \alpha T$ parameter space to correspond to the SM with $m_H = 115$ GeV; this ensures that any non-zero prediction for the oblique parameters for a Higgs of this mass arises from physics beyond the SM. At the one-loop level, the only new contributions to αS and αT in the DESM come from heavy fermion loops in the vacuum polarization diagrams indicated in Figure 4. We therefore expect αS and αT to be of order a few percent.[b]

In this section, we will first separately derive expressions for αT and αS in DESM and see how each compares to current constraints from.[3] We then compare the DESM's joint prediction for αS and αT as a function of μ to the region of the $\alpha S - \alpha T$ plane that gives the best fit to existing data[3] and thereby derive a 95% confidence level lower bound on μ.

3.2.1. *Parameter αT*

The custodial-symmetry-breaking parameter αT is defined as[11]

$$\alpha T = \left[\frac{\Pi_{WW}(0)}{M_W^2} - \frac{\Pi_{ZZ}(0)}{M_Z^2}\right], \tag{25}$$

[b]There are, in principle, additional oblique parameters such as αU that arise at higher order. These will be suppressed relative to αS or αT by a factor of order m_Z^2/m_T^2; since we can see from Figure 1 that $m_T > 2m_t$, the suppression is by at least an order of magnitude and we shall neglect αU and its ilk from here on.

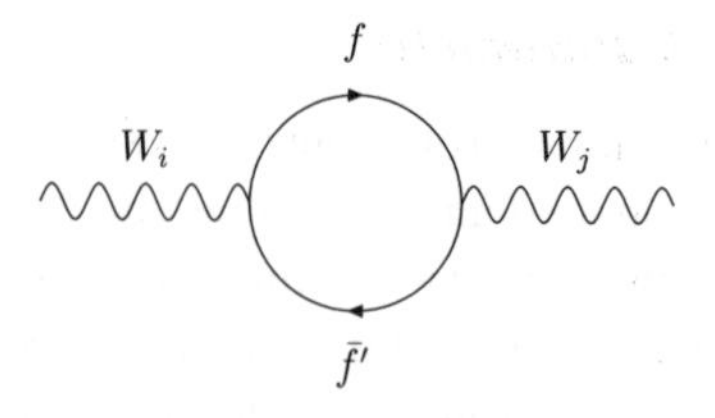

Fig. 4. Vacuum polarization diagram contributing to the oblique electroweak parameters. The indices $i,\ j = 1,2,3,Q$, refer to weak ($i = 1,2,3$) or electromagnetic (Q) generators, while f, f' run over the appropriate combinations of t, b, T and Ω.

where the contributions proportional to $g^{\mu\nu}$ in the vacuum polarization diagrams of Fig. (4) for the W and Z are labeled Π_{WW} and Π_{ZZ}, respectively. Each contribution sums over various $f\bar{f}'$ pairs – for W we have $f\bar{f}' = t\bar{b},\ T\bar{b},\ t\bar{\Omega},\ T\bar{\Omega}$; while for Z, we have $f\bar{f}' = t\bar{t},\ T\bar{T},\ t\bar{T}, \Omega\bar{\Omega}, b\bar{b}, b\bar{\Omega}$.

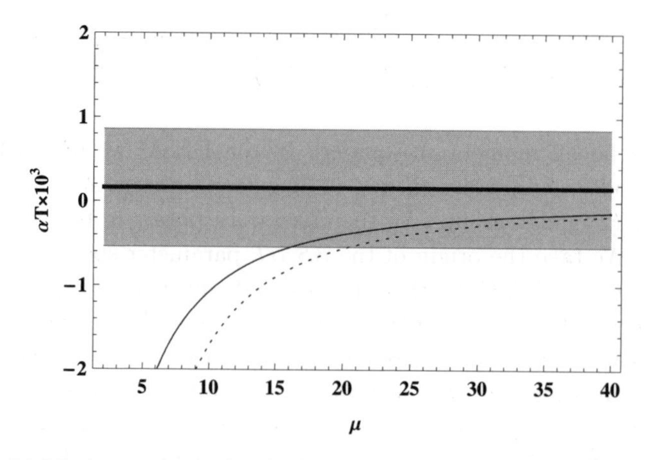

Fig. 5. The solid (blue) curve shows the DESM model's prediction for αT^{th} as a function of μ. The horizontal lines show the optimal fit value of $\alpha T = 0.16 \times 10^{-3}$ (thick solid line) and the relative $\pm 1\sigma$ deviations[3] (solid lines bordering the shaded band). The line $\alpha T = 0$ corresponds to the SM value (with $m_h = 115$ GeV), by definition. The leading-log contribution to αT is shown by the dotted curve.

The analytical result for αT^{DESM} cannot be written in compact form; the result[c] in the limit $\mu >> 1$ is:

$$\alpha T^{DESM} = \frac{3m_t^2}{16\pi^2 v^2}\left(1 - 4\frac{\ln \mu^2}{\mu^2} + \frac{22}{3\mu^2}\right). \tag{26}$$

One can see that, for $\mu \to \infty$, Eq. (26) reproduces the leading SM result $\alpha T^{SM}(m_t) = 3m_t^2/(4\pi v)^2$,[11] as expected. It interesting to note that the leading

[c]This is consistent with Eq. (33) in Pomarol[21] *et. al.*, when only the contributions from new fermions are included ($c_L = 0$).

log contribution arising from the heavy states *reduces*[d] the value of αT. This is to be expected, since the custodial symmetry is enhanced in the small-μ limit and αT measures the *change* in the amount of isospin violation relative to the standard model.

Subtracting the SM contribution from the top-quark, the numerical value of

$$\alpha T^{th} = \alpha T^{DESM} - \alpha T^{SM}(m_t), \tag{27}$$

as a function of μ is plotted as the solid blue curve in Fig. (5); the dotted curve shows just the leading-log term (second term of Eq. (26)). The thick solid horizontal line corresponds to the best-fit value of $\alpha T = 0.16 \times 10^{-3}$ obtained by ref.[3] when setting $U = 0$; the two horizontal solid lines bordering the shaded band show the relative $\pm 1\sigma$ deviations from that central fit value. Unlike the case of δg_{Lb}, the experimental constraints on αT clearly favor large values of μ, closer to the SM limit.

By way of comparison, it is interesting to note that in an extra-dimensional model[19,20] where a SM-like weak-singlet top quark was in the same $SO(5)$ multiplet as extra quarks forming a weak doublet, radiative corrections produced an experimentally-disfavored large negative contribution to αT at one loop. Given that the $SO(5)$ multiplet in 4D includes an $SO(4) = SU(2)_L \times SU(2)_R$ bi-doublet, our results are consistent with theirs.

3.2.2. *Parameter αS*

The parameter S is defined[11] as

$$\alpha S = 16\pi\alpha \left[\frac{d}{dq}\Pi_{33}(0) - \frac{d}{dq}\Pi_{3Q}(0)\right], \tag{28}$$

where q is the gauge boson momentum. The complete expression for αS^{DESM} cannot be written in compact form; the limiting case where $\mu >> 1$ is given by:

$$\alpha S^{DESM} = \frac{1}{6\pi}\left(3 + 2\ln\frac{m_b}{m_t} + \frac{8}{\mu^2}(2 - \ln\mu)\right), \tag{29}$$

where we reintroduce a non-zero mass for the b quark to cut off a divergence in the integral over the fermion loop momenta. One can check that Eq. (29) reproduces the SM result $\alpha S^{SM}(m_t, m_b)$[11] for $\mu \to \infty$. Defining

$$\alpha S^{th} = \alpha S^{DESM}(\mu) - \alpha S^{SM}(m_t, m_b), \tag{30}$$

we plot the result in Fig. (6), along with the value, $\alpha S = 0.31 \times 10^{-3}$, that provides an optimal fit to the data (for $U = 0$) and the $\pm 1\sigma$ relative deviations.[3] From Fig. (6) one can see that αS is within the $\pm 1\sigma$ bounds unless $\mu < 3$; as with αT, smaller values of μ are disfavored, though the constraint in this case is less severe.

---

[d]This does not violate the theorem[16,17] stating that $\Delta\rho \geq 0$ when mixing occurs only between particles of the same T^3 and hypercharge. In the DESM, there is significant mixing between the t'_L and T'_L which have different T^3 and hypercharge values. As a result, we also expect significant GIM violation in the third generation.

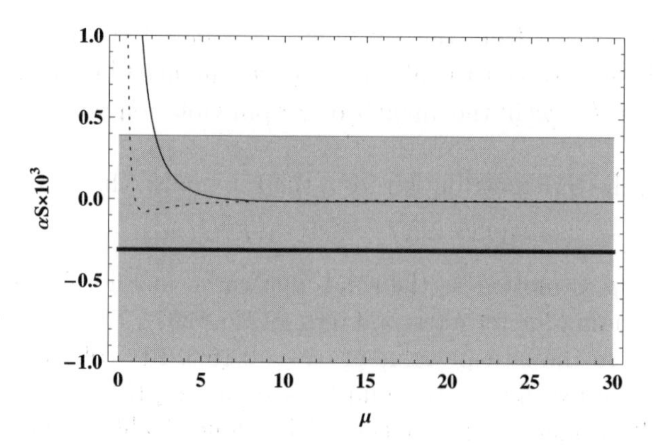

Fig. 6. The solid curve shows the DESM model's prediction for αS^{th} as a function of μ. The horizontal lines show the optimal fit value of $\alpha S = -0.31 \times 10^{-3}$ (thick solid line) and the relative $\pm 1\sigma$ deviations (solid lines bordering the shaded band) from.[3] The line $\alpha S = 0$ corresponds to the SM value (with $m_h = 115$ GeV), by definition. The leading-log contribution to αS is shown by the dotted curve.

3.3. *The αS– αT Plane*

In Figure 7 we show the DESM predictions for $[\alpha S^{th}(\mu), \alpha T^{th}(\mu)]$ from Eqs. (30, 26) using $m_h = 115$ GeV, and illustrating the successive mass-ratio values $\mu = 3, 4, ..., 20, \infty$; the point $\mu = \infty$ corresponds to the SM limit of the DESM and therefore lies at the origin of the αS - αT plane. On the same plane we also plot the elliptical curves that define the 95% confidence level (CL) bounds on the αS - αT plane, relative to the optimal values of αS and αT found in.[3] Using the best-fit values[3] and corresponding $\pm 1\sigma$ deviations for $m_h = 115$ GeV, 300 GeV, along with the correlation matrix, we obtained the approximate values appropriate to $m_h = 1$ TeV by extrapolating based on the logarithmic dependence of αS and αT on m_h. To calculate the 95% CL ellipses, we solved the equation $\Delta\chi^2 = \chi^2 - \chi^2_{min} = 5.99$, as appropriate to the χ^2 probability distribution for two degrees of freedom.

From this figure, we observe directly that the 95%CL lower limit on μ for $m_h = 115$ GeV is about 20, while for any larger value of m_h the DESM with $\mu \leq 20$ is excluded at 95%CL. In other words, the fact that a heavier m_h tends to worsen the fit of even the SM ($\mu \to \infty$) to the electroweak data is exacerbated by the new physics contributions within the DESM. The bound $\mu \geq 20$ corresponding to a DESM with a 115 GeV Higgs boson also implies, at 95%CL, that $m_T \geq \mu\ m_t \cong 3.4$ TeV, so that the heavy partners of the top quark would likely be too heavy for detection at LHC.

4. Conclusions

We have introduced the doublet-extended standard model (DESM) as a simple realization of the idea[2] of using custodial symmetry to protect the $Zb_L\bar{b}_L$ coupling (g_{Lb}) from receiving large radiative corrections. In this toy model, all terms of dimension-

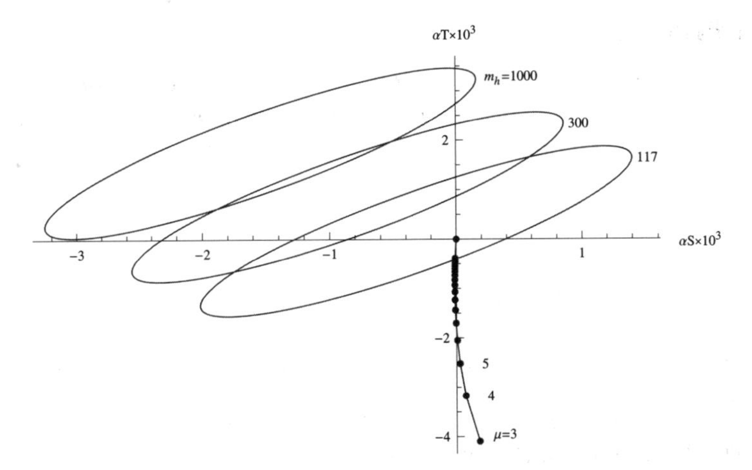

Fig. 7. The dots represent the theoretical predictions of the DESM (with m_h set to the reference value 115 GeV), showing how the values of αS and αT change as μ successively takes on the values 3, 4, 5, ..., 20, ∞. The three ellipses enclose the 95%CL regions of the αS - αT plane for the fit to the experimental data performed in;[3] they correspond to Higgs boson mass values of $m_h = 115\,\text{GeV}$, 300 GeV, and 1 TeV. Comparing the theoretical curve with the ellipses shows that the minimum allowed value of μ is 20, for $m_h = 115$ GeV.

4 in the top-quark mass generating sector obey a global $O(4) \times U(1)_X$ symmetry, which includes a parity symmetry protecting g_{Lb} from radiative corrections. That global symmetry is softly broken to its $SU(2)_L \times U(1)_Y$ subgroup by a Dirac mass term for the extra fermion doublet that incorporates the heavy partner of the top quark. Varying the size of this Dirac mass M allows the model to interpolate between the $O(4) \times U(1)_X$-symmetric case ($M = 0$) in which $\delta g_{Lb} = 0$ and the SM-like case ($M \to \infty$) in which the one-loop corrections to g_{Lb} are as in the SM, and enabled us to investigate the degree to which the custodial symmetry of the top-quark mass generating sector can be enhanced. By comparing the predictions of the DESM with experimental constraints on the oblique parameters αS and αT from,[3] we found the DESM to be consistent with experiment only for $\mu > 20$ at 95%CL, with a Higgs mass $m_h = 115$ GeV. The bound on μ translates into the 95% CL lower bound of 3.4 TeV on the masses of the extra quarks – placing them out of reach of the LHC. This result demonstrates that electroweak data strongly limits the amount by which the custodial symmetry of the top-quark mass generating sector can be enhanced relative to the standard model.

While we cannot discuss this here, the toy model reproduces the behavior of extra-dimensional models in which the extended custodial symmetry is invoked to control the size of additional contributions to αT and the $Zb_L\bar{b}_L$ coupling,[19,20] while leaving the standard model contributions essentially unchanged. Finally, we also note that our results are also consistent with a general effective field-theory analysis,[21] which confirms that the toy DEWSB model illustrates the electroweak physics operative in a broad class of models.

Acknowledgments

This work was supported in part by the US National Science Foundation under grant PHY-0354226. RSC and EHS acknowledge the support of the Aspen Center for Physics and the Institute for Advanced Study where part of this work was completed.

References

1. R. Sekhar Chivukula, S. Di Chiara, R. Foadi and E. H. Simmons, Phys. Rev. D **80**, 095001 (2009) [arXiv:0908.1079 [hep-ph]].
2. K. Agashe, R. Contino, L. Da Rold and A. Pomarol, Phys. Lett. B **641**, 62 (2006) [arXiv:hep-ph/0605341].
3. C. Amsler *et al.* [Particle Data Group], Phys. Lett. B **667**, 1 (2008).
4. R. Barbieri, M. Beccaria, P. Ciafaloni, G. Curci and A. Vicere, Phys. Lett. B **288**, 95 (1992) [Erratum-ibid. B **312**, 511 (1993)] [arXiv:hep-ph/9205238].
5. R. Barbieri, M. Beccaria, P. Ciafaloni, G. Curci and A. Vicere, Nucl. Phys. B **409**, 105 (1993).
6. J. F. Oliver, J. Papavassiliou and A. Santamaria, Phys. Rev. D **67**, 056002 (2003) [arXiv:hep-ph/0212391].
7. T. Abe, R. S. Chivukula, N. D. Christensen, K. Hsieh, S. Matsuzaki, E. H. Simmons and M. Tanabashi, arXiv:0902.3910 [hep-ph].
8. [ALEPH Collaboration and DELPHI Collaboration and L3 Collaboration and], Phys. Rept. **427**, 257 (2006) [arXiv:hep-ex/0509008].
9. D. Y. Bardin, P. Christova, M. Jack, L. Kalinovskaya, A. Olchevski, S. Riemann and T. Riemann, Comput. Phys. Commun. **133**, 229 (2001) [arXiv:hep-ph/9908433].
10. A. B. Arbuzov *et al.*, Comput. Phys. Commun. **174**, 728 (2006) [arXiv:hep-ph/0507146].
11. M. E. Peskin and T. Takeuchi, Phys. Rev. D **46**, 381 (1992).
12. G. Altarelli and R. Barbieri, Phys. Lett. B **253**, 161 (1991).
13. G. Altarelli, R. Barbieri and S. Jadach, Nucl. Phys. B **369**, 3 (1992) [Erratum-ibid. B **376**, 444 (1992)].
14. R. S. Chivukula, E. H. Simmons, H. J. He, M. Kurachi and M. Tanabashi, Phys. Lett. B **603**, 210 (2004) [arXiv:hep-ph/0408262].
15. R. Barbieri, A. Pomarol, R. Rattazzi and A. Strumia, Nucl. Phys. B **703**, 127 (2004) [arXiv:hep-ph/0405040].
16. M. B. Einhorn, D. R. T. Jones and M. J. G. Veltman, Nucl. Phys. B **191**, 146 (1981).
17. A. G. Cohen, H. Georgi and B. Grinstein, Nucl. Phys. B **232**, 61 (1984).
18. F. Braam, M. Flossdorf, R. S. Chivukula, S. Di Chiara and E. H. Simmons, Phys. Rev. D **77**, 055005 (2008) [arXiv:0711.1127 [hep-ph]].
19. M. S. Carena, E. Ponton, J. Santiago and C. E. M. Wagner, Nucl. Phys. B **759**, 202 (2006) [hep-ph/0607106].
20. M. S. Carena, E. Ponton, J. Santiago and C. E. M. Wagner, Phys. Rev. D **76**, 035006 (2007) [arXiv:hep-ph/0701055].
21. A. Pomarol and J. Serra, Phys. Rev. D **78**, 074026 (2008) [arXiv:0806.3247 [hep-ph]].

Higgs Searches at the Tevatron

Kazuhiro Yamamoto [for the CDF and DØ Collaborations]
Department of Physics, Osaka City University
Osaka 558-8585, Japan

We present the latest results on searches for the standard and beyond-the-standard model Higgs bosons in proton-antiproton collisions at $\sqrt{s}$ = 1.96 TeV by the CDF and DØ experiments at the Fermilab Tevatron. No significant excess is observed above the expected background, and the cross section limits for the Higgs bosons are calculated. It is noticed that the standard model Higgs boson in the mass range 163 – 166 GeV/c^2 is excluded at the 95% C.L.

1 Introduction

The Higgs mechanism plays a role of a cornerstone in the modern particle physics. It was originally proposed to explain the non-zero masses of the weak bosons while keeping the theory gauge-invariant in the standard model (SM). As a result of the spontaneous electroweak symmetry braking, the weak bosons acquire the masses and simultaneously a new fundamental scalar particle, the Higgs boson (H), appears. On the other hand in the minimal supersymmetric standard model (MSSM) which is one of the simplest extensions beyond the SM, two Higgs doublet fields result in five Higgs bosons denoted as h, H, A, $H^{\pm}$. In spite of many experiments searching for these predicted particles, they have yet to be discovered. For the SM Higgs boson, the direct searches at the CERN LEP collider have set the lower limit on the Higgs boson mass (M_H) to be 114.4 GeV/c^2 at the 95% C.L. [1]. Also combining this limit with precision electroweak measurements provides the constraint on M_H to be less than 186 GeV/c^2 at the 95% C.L. [2]. The Tevatron experiments, CDF and DØ, are probing the Higgs boson in the most probable region, $100 < M_H < 200$ GeV/c^2.

2 Experimental Apparatus

The Fermilab Tevatron collider is providing proton-antiproton collisions at $\sqrt{s}$ = 1.96 TeV. The CDF and DØ detectors are general-purpose cylindrically symmetric detectors put into place at the Tevatron. Both detectors consist of a superconducting solenoid magnet, a precision tracking system, electromagnetic and hadronic calorimeters and muon spectrometers. The detailed description can be found elsewhere [3][4]. The accelerators and detectors keep running very well in stable condition. As of December 2009, the typical peak luminosity is 3×10^{32} cm^{-1}s^{-1} with the week integration of 50 ~ 60 pb^{-1}. The delivered luminosity to each detector is accumulated to be 7.4 fb^{-1}, among of which 6.1 fb^{-1} is recorded on tape as the collision data.

3 Search for the Standard Model Higgs Boson

There are three dominant processes for producing the SM Higgs boson at the Tevatron energy: gluon fusion ($gg \rightarrow H$), associated production with a vector boson ($q\bar{q}^{(\prime)} \rightarrow VH$, $V = W$ or Z), and vector boson fusion ($q\bar{q}^{(\prime)} \rightarrow q\bar{q}^{(\prime)}H$). For the Higgs mass of 100 ~ 200 GeV/c^2, the cross section of each process is predicted to be 0.2 ~ 1 pb, 0.01 ~ 0.3 pb, and 0.02 ~ 0.1 pb respectively [5]. As for the decay mode, $H \rightarrow b\bar{b}$ is dominant in the low mass case (M_H < 140 GeV/c^2), while $H \rightarrow WW$ becomes dominant for the high mass Higgs (M_H > 140 GeV/c^2). The Higgs searches are performed with the proper combinations of the above production and decay modes. Considering the huge backgrounds from QCD multijet production, it is appropriate to use VH production with $H \rightarrow b\bar{b}$ for the low mass Higgs because a high-P_T lepton and/or missing E_T coming from the vector boson decay allows us to trigger the events efficiently. On the other hand for the high mass Higgs, we take advantage of the large cross section of $gg \rightarrow H$ followed by $H \rightarrow WW$. The analyses are optimized for every channel of the different topological signatures in terms of the decay mode and jet multiplicity. The event selections and analysis strategies of each Higgs search channel are described below. The data used here correspond to an integrated luminosity of 2.0 ~ 5.4 fb^{-1}. The obtained results on the cross section upper limit are summarized in Table I. The further details can be found in Refs. [6] – [15].

We search for $ZH \rightarrow \nu\bar{\nu}b\bar{b}$ by investigating the missing E_T + $b\bar{b}$ signature. Considering the charged leptons which escape from the detector, this signature is also sensitive to a portion of $WH \rightarrow \ell\nu b\bar{b}$ and $ZH \rightarrow \ell^+\ell^-b\bar{b}$. The event selections used in the CDF and DØ analyses are similar. They are performed by vetoing events with isolated charged leptons (e or μ) and also by requiring large missing E_T and 2 or 3 jets containing at least one b-tagged jet. For b-tagging, the secondary vertex tagger (SECVTX) and the jet probability tagger (JP) are utilized at CDF, while the neural network (NN) tagger is developed and used at DØ. The dominant backgrounds are W/Z + jets, $t\bar{t}$, diboson (WW, WZ, ZZ), and QCD multi-jets. The multivariate discriminant technique is applied in order to effectively separate signal from background; the CDF uses a NN while the DØ uses a boosted decision tree (BDT). The event separation is performed in two steps; the first discriminant is trained to remove the multi-jet background, and the second discriminant separates the expected signal from the remaining SM background.

The process $WH \rightarrow \ell\nu b\bar{b}$ is also a very sensitive channel in the low mass region as well as $ZH \rightarrow \nu\bar{\nu}b\bar{b}$. The events are selected by requiring a single isolated high-p_T lepton, large missing E_T and 2 or 3 energetic jets with at least one b-tagged jet. The dominant background in the selected events is $Wb\bar{b}$ for the 2 jets sample and $t\bar{t}$ for the 3 jets sample. The other backgrounds include W/Z + jets, single top, diboson and non-W QCD jets. For the DØ analysis, a NN is employed to discriminate signal from background after the base selection. On the other hand for the CDF analysis, efforts to achieve better sensitivity are made by two analysis groups each of which employs a matrix element method (ME) and a NN respectively. The discriminant is optimized for each b-tag category.

Another important channel in the low mass region is $ZH \rightarrow \ell^+\ell^-b\bar{b}$. In the base selection, events containing Z candidates reconstructed from e^+e^- and $\mu^+\mu^-$ pairs are selected at first. Then, two or more energetic jets at least one of which is b-tagged are required. Backgrounds of this analysis are dominated by Z + jets, $t\bar{t}$, WZ, ZZ, and QCD events where a hadron is misidentified as a lepton. In order to discriminate signal from background, a BDT is used in the DØ analysis, while a two dimensional NN is implemented in the CDF analysis, where Z + jets and $t\bar{t}$ backgrounds are separated out from signal simultaneously.

The Higgs events of which final state includes a $\tau^+\tau^-$ pair also become significant by combining all relevant channels. For the DØ analysis, the following five channels: $ZH(Z \to \tau^+\tau^-, H \to q\bar{q})$, $HZ(H \to \tau^+\tau^-, Z \to q\bar{q})$, $HW(H \to \tau^+\tau^-, W \to q\bar{q}')$, $q\bar{q}^{(\prime)} \to Hq\bar{q}^{(\prime)}(H \to \tau^+\tau^-$, vector boson fusion), and $gg \to H \to \tau^+\tau^-$ (gluon fusion) are considered. Identification of $\tau^+\tau^-$ decay is essential for this analysis, and events in which one τ decays leptonically and the other τ decays hadronically are selected in order to increase the signal to noise ratio. An isolated object consisting of one or three charged particles in a narrow cone is required as a signature of a hadronically decaying τ, and a NN is used for further τ identification. After selecting the τ rich sample, an analysis using a BDT is performed to separate the expected signal from the remaining backgrounds which are mainly $t\bar{t}$, W/Z + jets, and QCD multi-jets.

In the high mass region, $H \to WW^* \to \ell^+\nu\ell^-\bar{\nu}$ is the most sensitive channel at the Tevatron. The most dominant process for producing the Higgs boson in this channel is gluon fusion ($gg \to H$). The other contribution comes from associated production with a vector boson (WH, ZH) and vector boson fusion ($q\bar{q}^{(\prime)}H$). The event selection is performed by requiring high-P_T opposite-sign dilepton with large missing E_T. The main backgrounds are $t\bar{t}$, Drell-Yan, $W\gamma$, and W + jets. Considering the scalar nature of the Higgs boson and the parity violation in W decays, two charged leptons from the Higgs decay tend to go in the same direction, which is different from the SM backgrounds. The final discriminants are NN outputs which are obtained by feeding a number of kinematic variables as inputs. The opening angle between two leptons mentioned above is one of the effective input variables. The NN training is carefully performed for the separated subsamples; the CDF analysis separates the sample by the jet multiplicity (0 jet, 1 jet, 2 or more jets), while the DØ analysis makes classification by dilepton flavor (ee, $\mu\mu$, $e\mu$). Also the low dilepton mass region ($M_{\ell\ell}$ < 16 GeV/c^2) and the same-sign dilepton + jets sample ($WH \to WWW^*$, $ZH \to ZWW^*$) are added to increase sensitivity. Due to the different background composition, the dedicated NN training is performed separately.

All the available results on the Higgs boson searches are combined into a single limit on the Higgs boson production cross section at the 95% C.L. for each Higgs boson mass ranging from 100 GeV/c^2 to 200 GeV/c^2. The combination process is made in two steps; the results of various analysis channels are combined in each experiment of CDF and DØ separately at first [15][16], and then the both results are combined into the single Tevatron combination [17]. In calculating the combined result, systematic correlations among the analysis channels and also those between experiments are carefully taken into account. The obtained limit is presented as a ratio to the predicted cross section by the SM in Fig. 1. We obtain the observed (expected) values of 2.70 (1.78) at M_H = 115 GeV/c^2 and 0.94 (0.89) at M_H = 165 GeV/c^2. We exclude the SM Higgs boson in the mass range 163 GeV/c^2 to 166 GeV/c^2, while the corresponding expected excluded region is 159 GeV/c^2 to 169 GeV/c^2.

TABLE I: Observed (Expected) limits at the 95% C.L. on the production cross section as a ratio to the SM Higgs cross section. The limits at $M_H = 115$ GeV/c^2 are presented for the first four processes, while 165 GeV/c^2 for $H \rightarrow WW$. Two analysis results are shown for the CDF $WH \rightarrow \ell\nu b\bar{b}$ channel; one is a result by using a neural network (NN) and the other is obtained with a matrix element method (ME).

Analysis channel	M_H (GeV/c^2)	CDF Observed (Expected) limit [σ/σ_{SM}]	CDF $\mathcal{L}$ (fb^{-1})	DØ Observed (Expected) limit [σ/σ_{SM}]	DØ $\mathcal{L}$ (fb^{-1})
$VH \rightarrow \not{E}_T b\bar{b}$	115	6.1 (4.2) [6]	3.6	3.7 (4.6) [7]	5.2
$WH \rightarrow \ell\nu b\bar{b}$	115	5.3 (4.0), NN [8] 6.6 (4.1), ME [9]	4.3	6.9 (5.1) [10]	5.0
$ZH \rightarrow \ell^+\ell^- b\bar{b}$	115	5.9 (6.8) [11]	4.1	9.1 (8.0) [12]	4.2
$\tau^+\tau^- q\bar{q}$ final state	115	Update in progress		27.0 (15.9) [13]	4.9
$H \rightarrow WW$	165	1.23 (1.21) [14]	4.8	1.55 (1.36) [15]	5.4

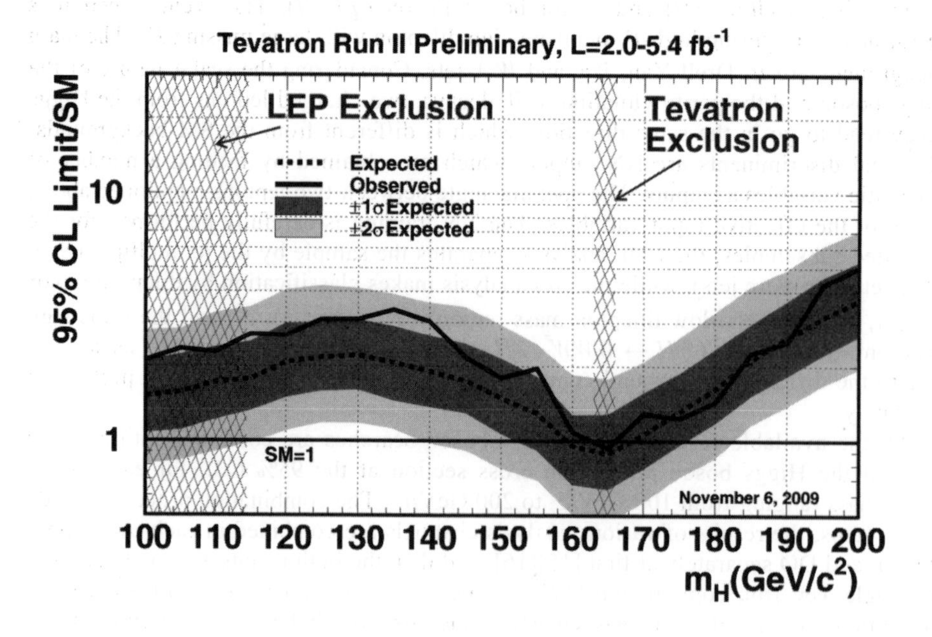

FIG. 1: Observed (solid line) and expected (dashed line) 95% C.L. upper limits on the ratio to the SM Higgs cross section as functions of the Higgs boson mass for the combined CDF and DØ analyses. The bands indicate the 68% and 95% probability regions where the limit is expected to fluctuate, in the absence of signal.

4 Search for the Higgs Bosons Beyond the SM

The minimal supersymmetric standard model (MSSM) is one of the simplest extensions beyond the SM. The production cross section of the neutral Higgs bosons ($\phi = h, H, A$) is enhanced by $\tan^2\beta$. At the Tevatron energy, the production cross section is 10 ~ 50 pb at $\tan\beta = 40$ [5], which is one order of magnitude larger than the SM. The main decay modes of ϕ are $b\bar{b}$ (~90%) and $\tau^+\tau^-$ (~10%) in a wide mass range (100 ~ 200 GeV/c^2).

One of the most promising channels to search for the neutral MSSM Higgs bosons at the Tevatron is $\phi \rightarrow \tau^+\tau^-$ in combination with the Higgs production through gluon fusion ($gg \rightarrow \phi$) and $b\bar{b}$ fusion ($b\bar{b} \rightarrow \phi$). Tau pairs are identified in decay modes of $e\mu$, e + hadrons and μ + hadrons, where the hadronic τ decay is identified with the similar method as the SM Higgs search described in the previous section. The background is dominated by $Z \rightarrow \tau^+\tau^-$, diboson, $t\bar{t}$ and W + jets. Also certain portion of the background arises from $Z \rightarrow ee/\mu\mu$ by misidentifying a lepton as a hadron. Figure 2 shows the combined result of the CDF and DØ data corresponding to an integrated luminosity of 1.0 ~ 2.2 fb^{-1}. It presents the 95% C.L. upper limit on $\tan\beta$ as a function of the A boson mass (M_A) with an assumption of the typical MSSM scenario of the maximal stop mixing and $\mu = -200$ GeV [19].

Another explorable channel of ϕ is $gb \rightarrow b\phi \rightarrow bb\bar{b}$. Three b-tagged jets are required in the analysis, and a peak structure is searched for in the dijet mass spectrum above the large multi-jet background. Figure 3 shows the results released with 1.9 fb^{-1} at CDF [20] and 2.6 fb^{-1} at DØ [21].

The charged MSSM Higgs ($H^{\pm}$) bosons can be investigated in top quark decays; a top quark decays into not only Wb but also $H^{\pm}b$ when $H^{\pm}$ is lighter than t. In this case, there would be some deviation from the SM prediction for the final states of $t\bar{t}$ decay. However, the results are consistent with the SM prediction at both CDF and DØ, and hence the upper limits are obtained as shown in Fig. 4 [22][23].

In some models beyond the SM, the neutral Higgs couplings to fermions are suppressed due to some specific mechanisms, thus the Higgs boson decay into vector bosons are significantly increased. This kind of "fermiophobic" Higgs boson mainly decays into $\gamma\gamma$ for the low mass region, while WW and ZZ become the main decay mode in the intermediate and high mass region. For simplicity, the benchmark scenario is

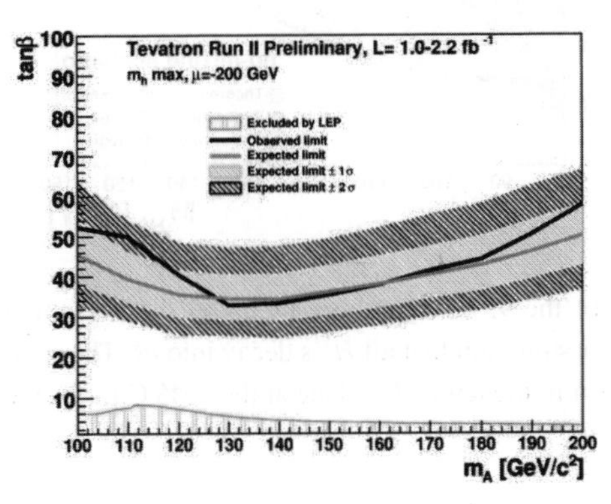

FIG. 2: The 95% C.L. limits in the $\tan\beta$-M_A plane for the typical MSSM scenario of the maximal stop mixing and $\mu = -200$ GeV. The black line denotes the observed limit, the gray line denotes the expected limit, and the hatched regions indicate the ±1σ and ±2σ bands around the expectation.

considered by assuming that the Higgs does not couple to any fermions but couples to electroweak bosons with the SM coupling strength. Figure 5 shows the results from CDF with 3.0 fb^{-1} and DØ with 4.2 fb^{-1} in $\gamma\gamma$ decay mode [24][25]. Both experiments exclude the fermiophobic Higgs boson with masses below ~110 GeV/c^2.

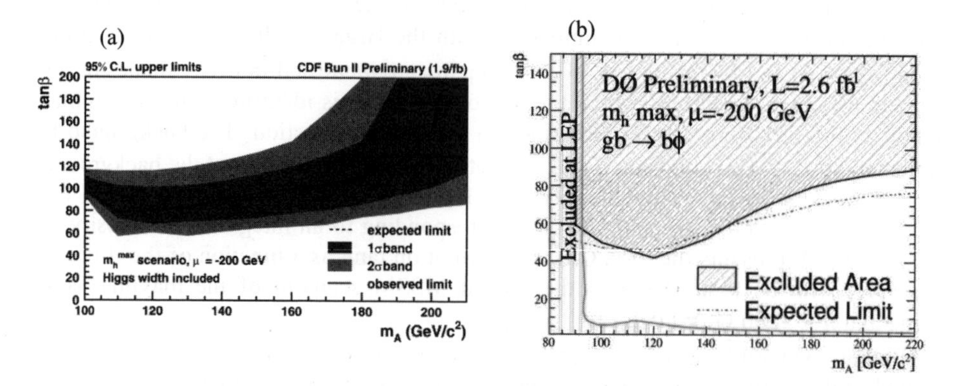

FIG. 3: The 95% C.L. upper limits in the tanβ-M_A plane for the MSSM scenario of maximal stop mixing and μ = –200 GeV, which is obtained from the analyses of $gb \rightarrow b\phi \rightarrow bb\bar{b}$ channel. The left plot (a) is a result of the CDF analysis, and the right plot (b) is that of the DØ analysis.

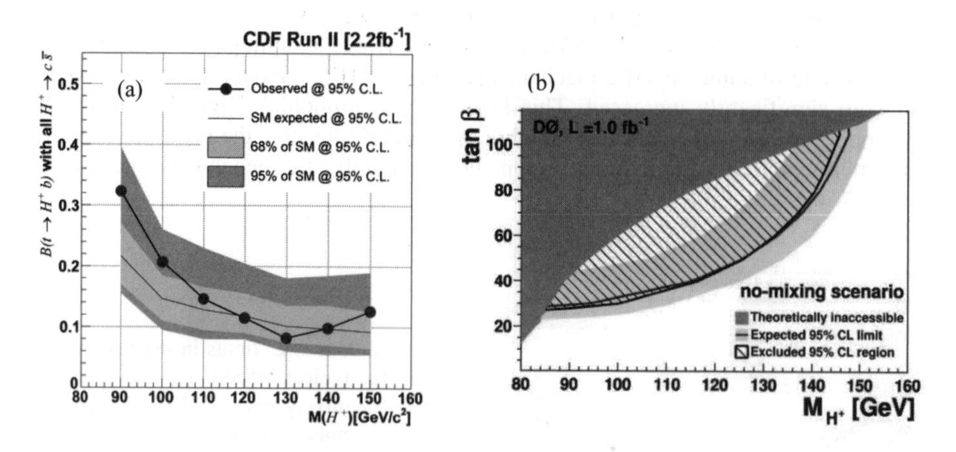

FIG. 4: The left plot (a) shows the CDF result of the 95% C.L. upper limits on the branching fraction of $t \rightarrow H^+b$ as a function of M_{H^+} with an assumption that all H^+'s decay into $c\bar{s}$. The right plot (b) shows the DØ result of the excluded region in the tanβ-M_{H^+} plane at the 95% C.L. for the no-mixing scenario.

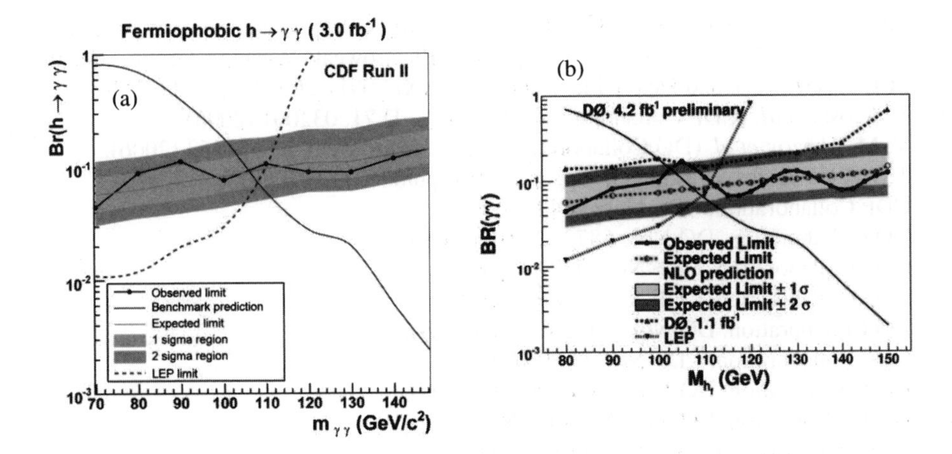

FIG. 5: The 95% C.L. upper limits on the branching fraction for the fermiophobic Higgs boson decay to diphotons, as a function of the fermiophobic Higgs mass. The result on the left plot (a) is obtained at CDF and that on the right plot (b) is obtained at DØ.

5 Conclusions and Future Prospects

The Tevatron accelerators and the collider detectors, CDF and DØ, are performing very well. The integrated delivered luminosity as of December 2009 is 7.4 fb^{-1}, while 6.1 fb^{-1} is acquired as the collision data on tape. The Higgs boson searches are in progress in the various production and decay channels. The preliminary results with up to 5.4 fb^{-1} of data are presented in this report. For the SM Higgs boson, the combined results of CDF and DØ exclude the mass range from 163 to 166 GeV/c^2 at the 95% C.L., while the expected exclusion range is 159 to 168 GeV/c^2. On the other hand for the Higgs bosons beyond the SM, no sign of discovery has been observed yet. However, the sensitivity is steadily increasing. By the end of FY2010, the luminosity is expected to be accumulated up to 9 fb^{-1}, and the sensitivity for seeing the Higgs boson will significantly increase.

Acknowledgements

I would like to thank the members of the CDF and DØ collaborations for their work and effort in achieving the results presented in this report.

References

1. The ALEPH, DELPHI, L3 and OPAL Collaborations, and the LEP Working Group for Higgs Boson Searches, Phys. Lett. B **565**, 61 (2003).

2. The ALEPH, CDF, DØ, DELPHI, L3, OPAL, SLD Collaborations, the LEP Electroweak Working Group, the Tevatron Electroweak Working Group, and the SLD Electroweak and Heavy Flavor Groups, arXiv:0911.2604 [hep-ex] (2009).
3. T. Acosta *et al.* (CDF Collaboration), Phys. Rev. D **71**, 032001 (2005).
4. V. M. Abazov *et al.* (DØ Collaboration), Nucl. Instr. Meth. **A565**, 463 (2006).
5. TeV4LHC Higgs Working Group, arXiv:hep-ph/0612172 (2007).
6. CDF Collaboration, CDF Note 9642 (2009).
7. DØ Collaboration, DØ Note 5872-CONF (2009).
8. CDF Collaboration, CDF Note 9997 (2009).
9. CDF Collaboration, CDF Note 9985 (2009).
10. DØ Collaboration, DØ Note 5972-CONF (2009).
11. CDF Collaboration, CDF Note 9889 (2009).
12. DØ Collaboration, DØ Note 5876-CONF (2009).
13. DØ Collaboration, DØ Note 5845-CONF (2009).
14. CDF Collaboration, CDF Note 9887 (2009).
15. DØ Collaboration, DØ Note 6006-CONF (2009).
16. CDF Collaboration, CDF Note 9999 (2009).
17. DØ Collaboration, DØ Note 6008-CONF (2009).
18. The TEVNPH Working Group, arXiv:0911.3930 [hep-ex] (2009).
19. The TEVNPH Working Group, FERMILAB-PUB-09-394-E, CDF Note 9888, DØ Note 5980-CONF (2009).
20. CDF Collaboration, CDF Note 9284 (2008).
21. DØ Collaboration, DØ Note 5726-CONF (2008).
22. T. Aaltonen *et al.* (CDF Collaboration), Phys. Rev. Lett. **103**, 101803 (2009).
23. DØ Collaboration, DØ Note 5715-CONF (2008).
24. T. Aaltonen *et al.* (CDF Collaboration), Phys. Rev. Lett. **103**, 061803 (2009).
25. DØ Collaboration, DØ Note 5880-CONF (2009).

The Top Triangle Moose*

R. S. Chivukula, N. D. Christensen, B. Coleppa and E. H. Simmons†

Department of Physics and Astronomy, Michigan State University
East Lansing, Michigan, 48824, USA
† *E-mail: esimmons@msu.edu*
http://www.pa.msu.edu/hep/hept/index.php

We introduce a deconstructed model that incorporates both Higgsless and top-color mechanisms. The model alleviates the typical tension in Higgsless models between obtaining the correct top quark mass and keeping $\Delta\rho$ small. It does so by singling out the top quark mass generation as arising from a Yukawa coupling to an effective top-Higgs which develops a small vacuum expectation value, while electroweak symmetry breaking results largely from a Higgsless mechanism. As a result, the heavy partners of the SM fermions can be light enough to be seen at the LHC.

Keywords: Top quark, electroweak symmetry breaking, new strong dynamics.

1. Introduction

Higgsless models[1] have recently emerged as a novel way of understanding the mechanism of electroweak symmetry breaking (EWSB) without the presence of a scalar particle in the spectrum. In an extra dimensional context, these can be understood in terms of a $SU(2) \times SU(2) \times U(1)$ gauge theory in the bulk of a finite AdS spacetime,[2–5] with symmetry breaking encoded in the boundary conditions of the gauge fields. One can understand the low energy properties of such theories in a purely four dimensional picture by invoking the idea of deconstruction.[6,7] The "bulk" of the extra dimension is replaced by a chain of gauge groups strung together by non linear sigma model fields. The spectrum typically includes extra sets of charged and neutral vector bosons and heavy fermions. A general analysis of Higgsless models[8–13] suggests that to satisfy precision electroweak constraints, the standard model (SM) fermions must be 'delocalized' into the bulk. A useful realization of this idea, "ideal fermion delocalization"',[14] dictates that the light fermions be delocalized so as not to couple to the heavy charged gauge bosons. The simplest framework capturing these ideas is the "three site Higgsless model",[15] with just one gauge group in the bulk and correspondingly, only one set of heavy vector bosons. The twin constraints of getting the correct value of the top quark mass and having an admissible ρ parameter push the heavy fermion masses into the TeV regime[15] in that model.

*Speaker at SCGT09: E. H. Simmons.

This presentation summarizes Ref. 16, in which we seek to decouple these constraints by separating the mechanisms that break the electroweak symmetry and generate the masses of the third family of fermions. In this way, one can obtain a massive top quark and heavy fermions in the sub TeV region, without altering tree level electroweak predictions. To present a minimal model with these features, we modify the three site model by adding a "top Higgs" field, Φ, that couples preferentially to the top quark. The resulting model is shown in Moose notation[22] in Figure 1; we will refer to it as the "top triangle moose."

The idea of a top Higgs is motivated by top condensation models (see references in Ref. 16), and the specific framework shown here is most closely aligned with topcolor assisted technicolor theories first proposed in Ref. 17, in which EWSB occurs via technicolor[18,19] interactions while the top mass has a dynamical component arising from topcolor[20,21] interactions and a small component generated by an extended technicolor mechanism. The dynamical bound state arising from topcolor dynamics can be identified as a composite top Higgs field, and the low-energy spectrum includes a top Higgs boson. The extra link in our triangle moose that corresponds to the top Higgs field results in the presence of uneaten Goldstone bosons, the top pions, which couple preferentially to the third generation. The model can thus be thought of as the deconstructed version of a topcolor assisted technicolor model.

2. The Model

We now introduce the essential features of the model, which are required in order to understand the LHC phenomenology. Full details are presented in Ref. 16.

The electroweak gauge structure of our model is $SU(2)_0 \times SU(2)_1 \times U(1)_2$. This is shown using Moose notation[22] in Figure 1, in which the $SU(2)$ groups are associated with sites 0 and 1, and the $U(1)$ group is associated with site 2. The SM fermions deriving their $SU(2)$ charges mostly from site 0 (which is most closely associated with the SM $SU(2)$) and the bulk fermions mostly from site 1. The extended electroweak gauge structure of the theory is the same as that of the BESS models,[23,24] motivated by models of hidden local symmetry.[25–29]

The non linear sigma field Σ_{01} breaks the $SU(2)_0 \times SU(2)_1$ gauge symmetry down to $SU(2)$, and field Σ_{12} breaks $SU(2)_1 \times U(1)_2$ down to $U(1)$. The left handed fermions are $SU(2)$ doublets residing at sites 0 (ψ_{L0}) and 1 (ψ_{L1}), while the right handed fermions are a doublet under $SU(2)_1(\psi_{R1})$ and two $SU(2)$-singlet fermions at site 2 (u_{R2} and d_{R2}). The fermions ψ_{L0}, ψ_{L1}, and ψ_{R1} have SM-like $U(1)$ charges (Y): $+1/6$ for quarks and $-1/2$ for leptons. Similarly, the fermion u_{R2} (d_{R2}) has an SM-like $U(1)$ charge of $+2/3$ ($-1/3$); the right-handed leptons, likewise, have $U(1)$ charges corresponding to their SM hypercharge values. The third component of isospin, T_3, takes values $+1/2$ for "up" type fermions and $-1/2$ for "down" type fermions, just like in the SM. The electric charges satisfy $Q = T_3 + Y$.

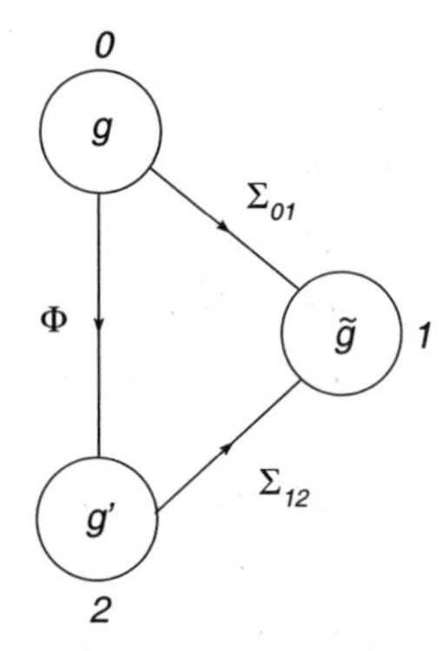

Fig. 1. The $SU(2) \times SU(2) \times U(1)$ gauge structure of the model in Moose notation.[22] The $SU(2)$ coupling g and $U(1)$ coupling g' of sites 0 and 2 are approximately the SM $SU(2)$ and hypercharge gauge couplings, while the $SU(2)$ coupling $\tilde{g}$ represents the 'bulk' gauge coupling.

We add a 'top-Higgs' link to separate top quark mass generation from EWSB. The top quark couple preferentially to the top Higgs link via the Largangian:

$$\mathcal{L}_{top} = -\lambda_t \bar{\psi}_{L0}\, \Phi\, t_R + h.c. \tag{1}$$

When the field Φ develops a non zero vacuum expectation value, Eqn.(1) generates a top quark mass term. Since we want most EWSB to come from the Higgsless side, we choose the vacuum expectation values of Σ_{01} and Σ_{02} to be $F = \sqrt{2}\, v\, \cos\omega$ and the one associated with the top Higgs sector to be $f = \langle\Phi\rangle = v\, \sin\omega$ (where ω is small). The top Higgs sector also includes the uneaten Goldstone bosons, the top pions; we assume they are heavy enough not to affect electroweak phenomenology.

The mass terms for the light fermions arise from Yukawa couplings of the fermionic fields with the non linear sigma fields

$$\mathcal{L} = M_D \left[\epsilon_L \bar{\psi}_{L0} \Sigma_{01} \psi_{R1} + \bar{\psi}_{R1} \psi_{L1} + \bar{\psi}_{L1} \Sigma_{12} \begin{pmatrix} \epsilon_{uR} & 0 \\ 0 & \epsilon_{dR} \end{pmatrix} \begin{pmatrix} u_{R2} \\ d_{R2} \end{pmatrix} \right]. \tag{2}$$

We denote the Dirac mass setting the scale of the heavy fermion masses as M_D. Here, ϵ_L is a flavor-universal parameter describing delocalization of the left handed fermions. All the flavor violation for the light fermions is encoded in the last term; the delocalization parameters for the right handed fermions, ϵ_{fR}, can be adjusted to realize the masses and mixings of the up and down type fermions. For our phenomenological study, we will, for the most part, assume that all the fermions, except the top, are massless and hence will set these ϵ_{fR} parameters to zero.

The *tree level* contributions to precision measurements in Higgsless models come from the coupling of standard model fermions to the heavy gauge bosons. Choosing the profile of a light fermion bilinear along the Moose to be proportional to the profile of the light W boson makes the fermion current's coupling to the W' vanish because the W and W' fields are mutually orthogonal. This procedure (called ideal fermion delocalization[14]) keeps deviations from the SM values of all electroweak quantities at a phenomenologically acceptable level. We find that the ideal delocalization condition in this model is $\epsilon_L^2 = M_W^2/2M_{W'}^2$, as in the three-site model.

The top quark mass matrix may be read from Eqns. (1) and (2) and is given by:

$$\begin{pmatrix} M_D\epsilon_{tL} & \lambda_t v \sin\omega \\ M_D & M_D\epsilon_{tR} \end{pmatrix}. \tag{3}$$

Diagonalizing the top quark mass matrix perturbatively in ϵ_{tL} and ϵ_{tR}, we find the mass of the top quark is:

$$m_t = \lambda_t v \sin\omega \left[1 + \frac{\epsilon_{tL}^2 + \epsilon_{tR}^2 + \frac{2}{a}\epsilon_{tL}\epsilon_{tR}}{2(-1+a^2)}\right], \qquad a \equiv \frac{\lambda_t\, v \sin\omega}{M_D}, \tag{4}$$

Thus, we see that m_t depends mainly on v and only slightly on ϵ_{tR}, in contrast to the situation in the three-site model where $m_t \propto M_D\epsilon_L\epsilon_{tR}$.

Since the b_L is the $SU(2)$ partner of the t_L, its delocalization is (to the extent that $\epsilon_{bR} \simeq 0$) also determined by ϵ_{tL}. Thus, the tree level value of the $Zb_L\bar{b}_L$coupling can be used to constrain ϵ_{tL}. We find g_L^{Zbb} equals its tree-level SM value if the left-handed top quark is delocalized exactly as the light fermions are: $\epsilon_{tL} = \epsilon_L$.

Finally, the contribution of the heavy top-bottom doublet to $\Delta\rho$ is of the same form as in the three-site model:[15] $\Delta\rho = M_D^2\,\epsilon_{tR}^4/16\,\pi^2\,v^2$. The key difference is that, since the top quark mass is dominated by the vev of the top Higgs instead of M_D, ϵ_{tR} can be as small as the ϵ_R of any light fermion. There is no conflict between the twin goals of a large top quark mass and a small value of $\Delta\rho$. Thus, the heavy fermions in the top triangle moose can be light enough to be seen at the LHC.

3. Heavy Quarks at the LHC

We now summarize our analysis[16] of the possible discovery modes of the heavy quarks at the LHC; this work employed the CalcHEP package.[30]

3.1. *Pair production: $pp \to Q\bar{Q} \to WZqq \to ll l\nu jj$*

Pair production of heavy quarks occurs at LHC via gluon fusion and quark annihilation processes, with the former dominating for smaller M_D. Each heavy quark decays to a vector boson and a light fermion. For $M_D < M_{W',Z'}$, the decay is purely to the standard model gauge bosons. We study the case where one heavy quark decays to $Z + j$ and the other decays to $W + j$, with the gauge bosons subsequently decaying leptonically. Thus, the final state is $lll\nu jj$.

To enhance the signal to background ratio, we have imposed a variety of cuts, as shown in Table 1. We note that the the two jets in the signal should have a high p_T ($\sim M_D/2$), since they each come from the 2-body decay of a heavy fermion. Thus, imposing strong p_T cuts on the outgoing jets can eliminate much of the SM background without affecting the signal too much. We also expect the η distribution of the jets to be largely central, which suggests an η cut: $|\eta| \leq 2.5$. We impose standard separation cuts between the two jets and between jets and leptons to ensure that they are observed as distinct final state particles. We also impose basic identification cuts on the leptons and missing transverse energy.

Table 1. Cuts employed in the pair (left) and single (right) production channels for the heavy quarks. $\Delta R_{jj} = \sqrt{\Delta\eta_{jj} + \Delta\phi_{jj}}$ refers to the separation between the two jets; ΔR_{jl} refers to the angular separation between a lepton and a jet.

Variable	Cut
p_{Tj}	>100 GeV
p_{Tl}	>15 GeV
Missing E_T	>15 GeV
$\lvert\eta_j\rvert$	< 2.5
$\lvert\eta_l\rvert$	< 2.5
ΔR_{jj}	>0.4
ΔR_{jl}	>0.4
M_{ll}	89 GeV$< M_{ll} <$ 93 GeV

Variable	Cut
p_{Tj} hard	>200 GeV
p_{Tj} soft	>15 GeV
p_{Tl}	>15 GeV
Missing E_T	>15 GeV
$\lvert\eta_j$ hard$\rvert$	< 2.5
$\lvert\eta_j$ soft$\rvert$	$2< \lvert\eta\rvert < 4$
$\lvert\eta_l\rvert$	< 2.5
ΔR_{jj}	>0.4
ΔR_{jl}	>0.4

We identify the leptons that came from the Z by imposing the invariant mass cut $(M_Z - 2\,\text{GeV}) < M_{ll} < (M_Z + 2\,\text{GeV})$. We then combine this lepton pair with a leading-p_T light jet to reconstruct the heavy fermion mass. Because one cannot know which light jet came from the Q, we actually combine the lepton pair first with the light jet of largest p_T and then, separately, with the light jet of next-largest p_T, and include both reconstructed versions of each event in our analysis. This yields[16] an invariant mass distribution with a narrow signal peak standing out cleanly at M_D above a tiny "background" from the wrongly-reconstructed signal events.

When generating the signal events, we included the four flavors of heavy quarks, U, D, C, S, that should have similar phenomenology. We estimate the size of the peak by counting the signal events in the invariant mass window: $(M_D - 10\,\text{GeV}) < M_{jll} < (M_D + 10\,\text{GeV})$. To analyze the SM background, we fully calculated the irreducible $pp \to ZWjj$ process and subsequently decayed the W and Z leptonically. Imposing the full set of cuts on the final state $lll\nu jj$ entirely eliminates the SM background.

We find there are many signal events in the region of parameter space where $Q \to Vq$ decays are allowed but $Q \to V'q$ decays are kinematically forbidden. The precise number is controlled by the branching ratio of the heavy fermion into the SM vector bosons. Since the SM background is negligible, if we assume the signal events are Poisson-distributed, then we can take 10 events to represent a 5σ signal at 95% c.l. Figure 2 shows the results, recast in the form of the integrated luminosity required to achieve a 5σ discovery signal. We see that the pair-production process we have studied spans almost the entire parameter space. However, in the region where $M_D \geq 900$ GeV and $M_{W'} \leq M_D$ there will not be enough signal events for the discovery of the heavy quark since the decay channel $Q \to W'q$ becomes significant. To explore this region, we now investigate the single production channel where the heavy quark decays to a heavy gauge boson.

3.2. *Single production: $pp \to Qq \to W'qq' \to WZqq'$*

While the single production channel is electroweak, the smaller cross section is compensated by the fact that the u and d are valence quarks, and their parton

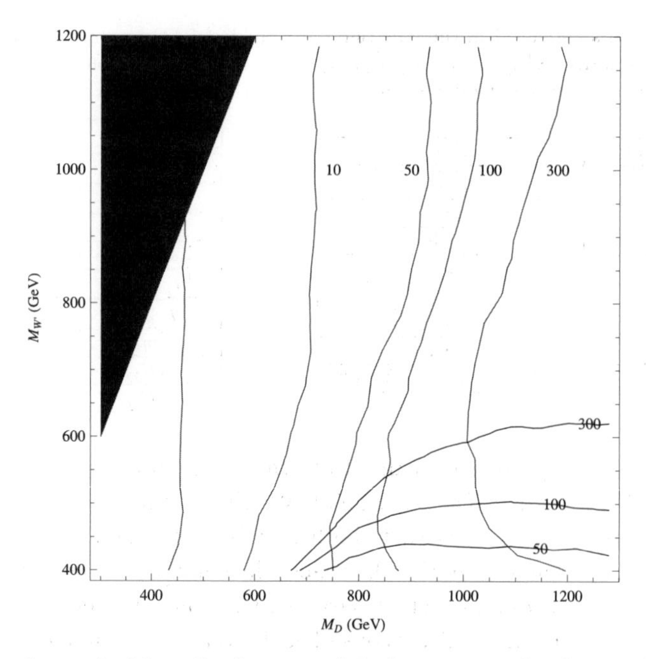

Fig. 2. Luminosity required for a 5σ discovery of the heavy vector fermions at the LHC in the single (blue curves, nearly horizontal) and pair (red curves, nearly vertical) production channels. The shaded area is non perturbative and not included in the study. The two channels are complementary and allow almost the entire region to be covered in 300 fb^{-1}.

distribution functions fall less sharply than the gluon's. Also, there is less phase space suppression in the single production channel than in the pair production case. We analyze the processes $[u, u \to u, U]$, $[d, d \to d, D]$ and $[u, d \to u, D$ or $U, d]$, which occur through a t channel exchange of a Z and Z'. Since we want to look at the region of parameter space where $M_{W'}$ is smaller than M_D, we let the heavy quark decay to a W'. The W' decays 100% of the time to a W and Z, because its coupling to two SM fermions is zero in the limit of ideal fermion delocalization. We constrain both the Z and W to decay leptonically so the final state is $lll\nu jj$.

Again, we expect the jet from the decay of the heavy quark to have a large p_T, and we impose a strong p_T cut on this "hard jet". As before, this jet should be central, so we impose the same η cut on the hard jet. We expect the η distribution of the "soft jet" arising from the light quark in the production process to be in the forward region, $2 < |\eta| < 4$. We impose the same ΔR jet separation and jet-lepton separation cuts as before, along with basic identification cuts on the leptons and missing transverse energy. The complete set of cuts is in the right side of Table 1.

The leptonic W decay introduces the usual two fold ambiguity in determining the neutrino momentum and hence, we have performed a transverse mass analysis of the process, defining the transverse mass variable[31] of interest as:

$$M_T^2 = \left(\sqrt{M^2(lllj) + p_T^2(lllj)} + |p_T(missing)| \right)^2 - |\overrightarrow{p_T}(lllj) + \overrightarrow{p_T}(missing)|^2 \quad (5)$$

We expect the distribution to fall sharply at M_D in the narrow width approximation, and indeed we find that there are typically few or no events beyond $M_D + 20$ GeV in the distributions. Thus, we take the signal events to be those in the transverse mass window: $(M_D - 200\,\text{GeV}) < M_T < (M_D + 20\,\text{GeV})$.

The SM background for this process, $pp \rightarrow WZjj \rightarrow jjl\nu ll$, was calculated summing over the u, d, c, s and gluon jets and the first two families of leptons. Since we apply a strong p_T cut on only one of the jets (unlike in the pair production case), there is a non zero SM background, as plotted in Ref. 16. The luminosity necessary for a 5σ discovery at 95% c.l. can be calculated by requiring $(N_{signal}/\sqrt{N_{bkrnd}}) \geq 5$, as per a Gaussian distribution. Figure 2) shows the results, again recast in the form of the integrated luminosity required to achieve a 5σ discovery signal. Almost the entire parameter space is covered, with the pair and single production channels nicely complementing each other.

References

1. C. Csaki et al., Phys. Rev. D **69**, 055006 (2004).
2. K. Agashe et al., JHEP **0308**, 050 (2003).
3. C. Csaki et al., Phys. Rev. Lett. **92** (2004) 101802.
4. G. Burdman and Y. Nomura, Phys. Rev. D **69,** 115013 (2004).
5. G. Cacciapaglia et al., Phys. Rev. D **70**, (2004) 075014.
6. N. Arkani-Hamed, A. G. Cohen and H. Georgi, Phys. Rev. Lett. **86** (2001) 4757-4761.
7. C. T. Hill, S. Pokorski and J. Wang, Phys. Rev. D **64** (2001) 105005.
8. R. S.Chivukula et al., Phys. Rev. D **71** (2005) 035007.
9. G. Cacciapaglia et al., Phys. Rev. D 71, (2005) 035015.
10. G. Cacciapaglia et al., Phys. Rev. D 72, (2005) 095018.
11. R. Foadi, S. Gopalakrishna and C. Schmidt, Phys. Lett. B **606** (2005) 157.
12. R. Casalbuoni et al., Phys. Rev. D **71**, 075015 (2005).
13. R. Foadi, and C. Schmidt, Phys. Rev. D **73** (2006) 075011.
14. R. S. Chivukula et al., Phys. Rev. D **72** (2005) 015008.
15. R. S. Chivukula et al., Phys. Rev. D **74** (2006) 075011.
16. R. Sekhar Chivukula et al., Phys. Rev. D **80**, 035011 (2009).
17. C. T. Hill, **B 345**: 483-489 (1995).
18. E. Eichten and K. D. Lane, Phys. Lett. B **90**, 125 (1980).
19. S. Dimopoulos and L. Susskind, Nucl. Phys. B **155**, 237 (1979).
20. C. T. Hill, Phys. Lett. **B 266** , 419-424 (1991)
21. C. T. Hill and S. Parke, Phys. Rev. **D49**, 4454 (1994).
22. H. Georgi, Nucl. Phys. **B266** (1986) 274.
23. R. Casalbuoni, S. De Curtis, D. Dominici, and R. Gatto, *Phys. Lett.* **B155** (1985) 95.
24. R. Casalbuoni *et. al., Phys. Rev.* **D53** (1996) 5201-5221.
25. M. Bando et al., Phys. Rev. Lett. **54** (1985) 1215.
26. M. Bando, T. Kugo, and K. Yamawaki, Nucl. Phys. **B259** (1985) 493.
27. M. Bando, T. Kugo, and K. Yamawaki, Phys. Rept. **164** (1988) 217-314.
28. M. Bando, T. Fujiwara, and K. Yamawaki, Prog. Theor. Phys. **79** (1988) 1140.
29. M. Harada and K. Yamawaki, Phys. Rept. **381** (2003) 1-233.
30. A. Pukhov, arXiv: hep-ph/0412191.
31. J. Bagger, *et al.,* Phys. Rev. D **52**, 3878 (1995).

Conformal Phase Transition in QCD Like Theories and Beyond

V. A. Miransky

Department of Applied Mathematics, University of Western Ontario
London, Ontario N6A 5B7, Canada
E-mail: vmiransk@uwo.ca

The dynamics with an infrared stable fixed point in the conformal window in QCD like theories with a relatively large number of fermion flavors is reviewed. The emphasis is on the description of a clear signature for the conformal window, which in particular can be useful for lattice computer simulations of these gauge theories.

Keywords: Conformal phase transition, infrared fixed point, scaling law for hadron masses, light glueballs.

1. Introduction

The Landau, or σ-model-like, phase transition[1] is characterized by the following basic feature. Around the critical point $z = z_c$ (where z is a generic notation for parameters of a theory, as the coupling constant α, number of particle flavors N_f, etc.), an order parameter X is

$$X = \Lambda f(z), \tag{1}$$

where Λ is an ultraviolet cutoff and the function $f(z)$ has such a non-essential singularity at $z = z_c$ that $\lim f(z) = 0$ as z goes to z_c both in symmetric and non-symmetric phases. The standard form for $f(z)$ is $f(z) \sim (z - z_c)^\nu$, $\nu > 0$, around $z = z_c$ [for convenience, we assume that $z > z_c$ ($z < z_c$) in the nonsymmetric (symmetric) phase].[a] The conformal phase transition (CPhT), whose conception was introduced in Ref. 3, is a very different continuous phase transition. It is defined as a phase transition in which an order parameter X is given by Eq. (1) where $f(z)$ has such an *essential* singularity at $z = z_c$ that while

$$\lim_{z \to z_c} f(z) = 0 \tag{2}$$

as z goes to z_c from the side of the non-symmetric phase, $\lim f(z) \neq 0$ as $z \to z_c$ from the side of the symmetric phase (where $X \equiv 0$). Notice that since the relation (2) ensures that the order parameter $X \to 0$ as $z \to z_c$, the phase transition is continuous.

[a]Strictly speaking, Landau considered the mean-field phase transition. By the Landau phase transition, we understand a more general class, when fields may have anomalous dimensions.[2]

There are the following basic differences between the Landau phase transition (LPhT) and the CPhT one:[3]

(1) In the case of the LPhT, masses of light excitations are continuous functions of the parameters z around the critical point $z = z_c$ (though they are non-analytic at $z = z_c$). In the case of the CPhT, the situation is different: there is an abrupt change of the spectrum of light excitations, as the critical point $z = z_c$ is crossed. This implies that the effective actions describing low energy dynamics in the phases with $z < z_c$ and $z > z_c$ are different in a system with CPhT.
(2) Unlike the LPhT, the parameter z governing the CPhT is connected with a marginal operator [in the LPhT phase transition, such a parameter is connected with a relevant operator; it is usually a mass term].
(3) The fact that the parameter z is connected with a marginal operator in the CPhT implies that in the continuum limit, when $z \to z_c + 0$, the conformal symmetry is broken by a marginal operator in nonsymmetric phase, i.e., there is a conformal anomaly.
(4) Unlike the LPhT, in the case of CPhT, the structures of renormalizations (i.e., the renormalization group at high momenta) are different in symmetric phase and nonsymmetric one.

In relativistic field theory, the CPhT is realized in the two dimensional Gross-Neveu (GN) model[4] at the critical coupling constant $g_c = 0$, reduced (or defect) QED,[5,6] and quenched QED.[7–10] It was suggested that the chiral phase transition with respect to the number of fermion flavors N_f in QCD is a CPhT one.[3,11] In condensed matter physics, a CPhT like phase transition is realized in the Berezinskii-Kosterlitz-Thouless (BKT) model[12] and, possibly, graphene.[13]

Recently, the interest to the dynamics with the CPht phase transition has essentially increased. It is in particular connected with a progress in numerical lattice studies of gauge theories with a varied number of fermion flavors (for a recent review, see Ref. 14), the revival of the interest to the electroweak symmetry breaking based on the walking technicolor like dynamics[15,16] (for a recent review, see Ref.[17]), and intensive studies of graphene, a single atomic layer of graphite (for a review, see Ref. 18).

2. Dynamics in the conformal window in QCD-like theories

2.1. *General description*

In this section, we will consider the problem of the existence of a nontrivial conformal dynamics in 3+1 dimensional non-supersymmetric vector like gauge theories, with a relatively large number of fermion flavors N_f. We will discuss their phase diagram in the $(\alpha^{(0)}, N_f)$ plane, where $\alpha^{(0)}$ is the bare coupling constant. We also discuss a clear signature for the conformal window in lattice computer simulations of these theories suggested quite time ago in Ref. 19.

The roots of this problem go back to a work of Banks and Zaks[20] who were first to discuss the consequences of the existence of an infrared-stable fixed point $\alpha = \alpha^*$ for $N_f > N_f^*$ in vector-like gauge theories.[21] The value N_f^* depends on the gauge group: in the case of SU(3) gauge group, $N_f^* = 8$ in the two-loop approximation. In Nineties, a new insight in this problem[3,11] was, on the one hand, connected with using the results of the analysis of the Schwinger-Dyson (SD) equations describing chiral symmetry breaking in quenched QED[7–10] and, on the other hand, with the discovery of the conformal window in $N = 1$ supersymmetric QCD.[22]

In particular, Appelquist, Terning, and Wijewardhana[11] suggested that, in the case of the gauge group SU(N_c), the critical value $N_f^{cr} \simeq 4N_c$ separates a phase with no confinement and chiral symmetry breaking ($N_f > N_f^{cr}$) and a phase with confinement and with chiral symmetry breaking ($N_f < N_f^{cr}$). The basic point for this suggestion was the observation that at $N_f > N_f^{cr}$ the value of the infrared fixed point α^* is smaller than a critical value $\alpha_{cr} \simeq \frac{2N_c}{N_c^2-1}\frac{\pi}{3}$, presumably needed to generate the chiral condensate.[7–10]

The authors of Ref. 11 considered only the case when the running coupling constant $\alpha(\mu)$ is less than the fixed point α^*. In this case the dynamics is asymptotically free (at short distances) both at $N_f < N_f^{cr}$ and $N_f^{cr} < N_f < N_f^{**} \equiv \frac{11N_c}{2}$. Yamawaki and the author[3] analyzed the dynamics in the whole ($\alpha^{(0)}, N_f$) plane and suggested the ($\alpha^{(0)}, N_f$)-phase diagram of the SU(N_c) theory, where $\alpha^{(0)}$ is the bare coupling constant (see Fig 1 below).[b] In particular, it was pointed out that one can get an interesting non-asymptotically free dynamics when the bare coupling constant $\alpha^{(0)}$ is *larger* than α^*, though not very large.

The dynamics with $\alpha^{(0)} > \alpha^*$ admits a continuum limit and is interesting in itself. Also, its better understanding can be important for establishing the conformal window in lattice computer simulations of the SU(N_c) theory with such large values of N_f. In order to illustrate this, let us consider the following example. For $N_c = 3$ and $N_f = 16$, the value of the infrared fixed point α^* calculated in the two-loop approximation is small: $\alpha^* \simeq 0.04$. To reach the asymptotically free phase, one needs to take the bare coupling $\alpha^{(0)}$ less than this value of α^*. However, because of large finite size effects, the lattice computer simulations of the SU(3) theory with such a small $\alpha^{(0)}$ would be unreliable. Therefore, in this case, it is necessary to consider the dynamics with $\alpha(\mu) > \alpha^*$.

In Ref. 19, this author suggested a clear signature of the existence of the infrared fixed point α^*, which in particular can be useful for lattice computer simulations. The signature is based on two characteristic features of the the spectrum of low energy excitations in the presence of a bare fermion mass in the conformal window: a) a strong (and simple) dependence of the masses of all the colorless bound states (including glueballs) on the bare fermion mass, and b) unlike QCD with a small N_f ($N_f = 2$ or 3), glueballs are lighter than bound states composed of fermions, if the value of the infrared fixed point is not too large.

---

[b]This phase diagram is different from the original Banks-Zaks diagram.[20]

2.2. *Phase diagram*

The phase diagram in the $(\alpha^{(0)}, N_f)$-plane in the $SU(N_c)$ gauge theory is shown in Fig. 1. The left-hand portion of the curve in this figure coincides with the line of the infrared-stable fixed points $\alpha^*(N_f)$:[21]

$$\alpha^{(0)} = \alpha^* = -\frac{b}{c}, \tag{3}$$

where

$$b = \frac{1}{6\pi}(11N_c - 2N_f), \tag{4}$$

$$c = \frac{1}{24\pi^2}(34N_c^2 - 10N_cN_f - 3\frac{N_c^2-1}{N_c}N_f). \tag{5}$$

It separates two symmetric phases, S_1 and S_2, with $\alpha^{(0)} < \alpha^*$ and $\alpha^{(0)} > \alpha*$, respectively. Its lower end is $N_f = N_f^{cr}$ (with $N_f^{cr} \simeq 4N_c$ if $\alpha_{cr} \simeq \frac{2N_c}{N_c^2-1}\frac{\pi}{3}$): at $N_f^* < N_f < N_f^{cr}$ the infrared fixed point is washed out by generating a dynamical fermion mass (here N_f^* is the value of N_f at which the coefficient c in Eq. (5) becomes positive and the fixed point disappears).

The horizontal, $N_f = N_f^{cr}$, line describes a phase transition between the symmetric phase S_1 and the phase with confinement and chiral symmetry breaking. As it was suggested in Ref.11, based on a similarity of this phase transition with that in quenched QED_4,[8,9] there is the following scaling law for m_{dyn}^2:

$$m_{dyn}^2 \sim \Lambda_{cr}^2 \exp\left(-\frac{C}{\sqrt{\frac{\alpha^*(N_f)}{\alpha_{cr}} - 1}}\right), \tag{6}$$

where the constant C is of order one and Λ_{cr} is a scale at which the running coupling is of order α_{cr}. It is a CPhT phase transition with an essential singularity at $N_f = N_f^{cr}$.

At last, the right-hand portion of the curve on the diagram occurs because at large enough values of the bare coupling, spontaneous chiral symmetry breaking takes place for any number N_f of fermion flavors. This portion describes a phase transition called a bulk phase transition in the literature, and it is presumably a first order phase transition. [c] The vertical line ends above N_f=0 since in pure gluodynamics there is apparently no phase transition between weak-coupling and strong-coupling phases.

2.3. *Signature for the conformal window*

Up to now we have considered the case of a chiral invariant action. But how will the dynamics change if a bare fermion mass term is added in the action? This question

[c] The fact that spontaneous chiral symmetry breaking takes place for any number of fermion flavors, if $\alpha^{(0)}$ is large enough, is valid at least for lattice theories with Kogut-Susskind fermions. Notice however that since the bulk phase transition is a lattice artifact, the form of this portion of the curve can depend on the type of fermions used in simulations.

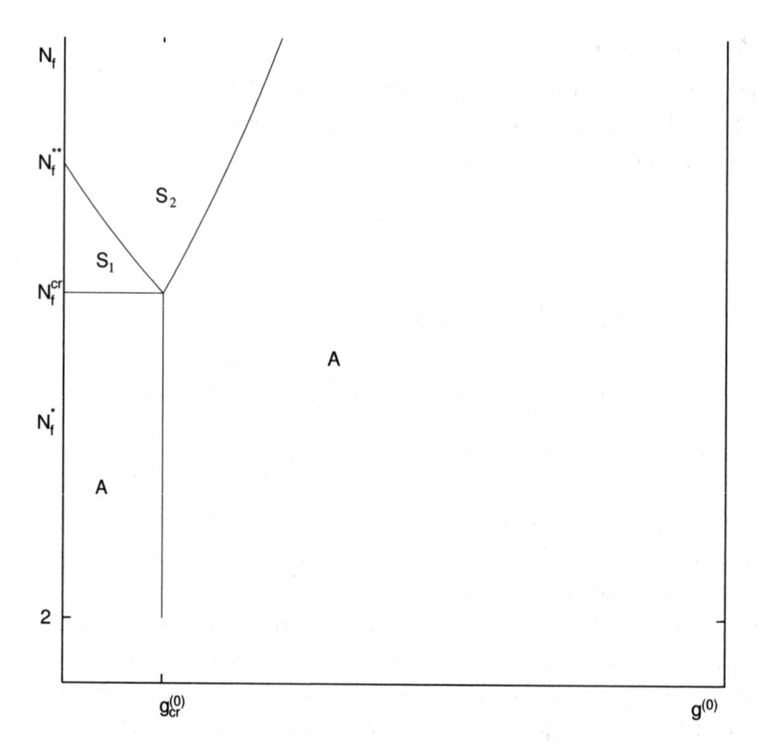

Fig. 1. The phase diagram in an SU(N_c) gauge model. The coupling constant $g^{(0)} = \sqrt{4\pi\alpha^{(0)}}$ and S and A denote symmetric and asymmetric phases, respectively.

is in particular relevant for lattice computer simulations: for studying a chiral phase transition on a finite lattice, it is necessary to introduce a bare fermion mass. As was pointed out in Ref.,[19] adding even an arbitrary small bare fermion mass results in a dramatic changing the dynamics both in the S_1 and S_2 phases.

Recall that in the case of confinement SU(N_c) theories, with a small, $N_f < N_f^{cr}$, number of fermion flavors, the role of a bare fermion mass $m^{(0)}$ is minor if $m^{(0)} << \Lambda_{QCD}$ (where Λ_{QCD} is a confinement scale). The only relevant consequence is that massless Nambu-Goldstone pseudoscalars get a small mass (the PCAC dynamics).

The reason for that is the fact that the scale Λ_{QCD}, connected with a conformal anomaly, describes the breakdown of the conformal symmetry connected *both* with perturbative and nonperturbative dynamics: the running coupling and the formation of bound state. Certainly, a small bare mass $m^{(0)} << \Lambda_{QCD}$ is irrelevant for the dynamics of those bound states.

Now let us turn to the phases S_1 and S_2, with $N_f > N_f^{cr}$. There is still the conformal anomaly in these phases: because of the running of the effective coupling constant, the conformal symmetry is broken. It is restored only if $\alpha^{(0)}$ is equal to the infrared fixed point α^*. However, the essential difference with respect to confinement theories is that this conformal anomaly have nothing to do with the dynamics forming bound states: Since at $N_f > N_f^{cr}$ the effective coupling is relatively weak, it is impossible to form bound states from *massless* fermions and gluons (recall that the S_1 and S_2 phases are chiral invariant).

Therefore the absence of a mass for fermions and gluons is a key point for *not* creating bound states in those phases. The situation changes dramatically if a bare fermion mass is introduced: indeed, even weak gauge, Coulomb-like, interactions can easily produce bound states composed of massive constituents, as it happens, for example, in QED, where electron-positron (positronium) bound states are present. To be concrete, let us consider the case when all fermions have the same bare mass $m^{(0)}$. It leads to a mass function $m(q^2) \equiv B(q^2)/A(q^2)$ in the fermion propagator $G(q) = (\hat{q}A(q^2) - B(q^2))^{-1}$. The current fermion mass m is given by the relation

$$m(q^2)|_{q^2=m^2} = m. \tag{7}$$

For the clearest exposition, let us consider a particular theory with a finite cutoff Λ and the bare coupling constant $\alpha^{(0)} = \alpha(q)|_{q=\Lambda}$ being not far away from the fixed point α^*. Then, the mass function is changing in the "walking" regime[16] with $\alpha(q^2) \simeq \alpha^*$. It is

$$m(q^2) \simeq m^{(0)} \left(\frac{M}{q}\right)^{\gamma_m} \tag{8}$$

where γ_m is the anomalous dimension of the operator $\bar{\psi}\psi$: $\gamma_m = 3 - d_{\bar{\psi}\psi}$ with $d_{\bar{\psi}\psi}$ being the dynamical dimension of this operator. In the walking regime, $\gamma_m \simeq 1 - (1 - \frac{\alpha^*}{\alpha_{cr}})^{1/2}$ (see Refs. 9,16).

Eqs.(7) and (8) imply that

$$m \simeq \Lambda \left(\frac{m^{(0)}}{\Lambda}\right)^{\frac{1}{1+\gamma_m}}. \tag{9}$$

Recall that the anomalous dimension $\gamma_m \geq 0$, and $\gamma_m \lesssim 2$ in the "walking" regime.

There are two main consequences of the presence of the bare mass:

(a) bound states, composed of fermions, occur in the spectrum of the theory. The mass of a n-body bound state is $M^{(n)} \simeq nm$. Therefore they satisfy the scaling

$$M^{(n)} \simeq nm \sim n \left(m^{(0)}\right)^{\frac{1}{1+\gamma_m}}. \tag{10}$$

(b) At low momenta, $q < m$, fermions and their bound states decouple. There is a pure SU(N_c) Yang-Mills theory with confinement. Its spectrum contains glueballs.

To estimate glueball masses, notice that at momenta $q < m$, the running of the coupling is defined by the parameter $\bar{b}$ of the Yang-Mills theory,

$$\bar{b} = \frac{11}{6\pi} N_c. \tag{11}$$

Therefore the glueball masses M_{gl} are of order

$$\Lambda_{YM} \simeq m \exp(-\frac{1}{\bar{b}\alpha^*}). \tag{12}$$

For $N_c = 3$, we find from Eqs.(4), (5), and (11) that $\exp(-\frac{1}{b\alpha^*})$ is 6×10^{-7}, 2×10^{-2}, 10^{-1}, and 3×10^{-1} for N_f=16, 15, 14, and 13, respectively. Therefore at N_f=16, 15 and 14, the glueball masses are essentially lighter than the masses of the bound states composed of fermions.

The situation is similar to that in confinement QCD with heavy (nonrelativistic) quarks, $m >> \Lambda_{QCD}$. However, there is now a new important point. In the conformal window, *any* value of $m^{(0)}$ (and therefore m) is "heavy": the fermion mass m sets a new scale in the theory, and the confinement scale Λ_{YM} (12) is less, and rather often much less, than this scale m. One could say that the latter plays a role of a dynamical ultraviolet cutoff for the pure YM theory.

This leads to a spectacular "experimental" signature of the conformal window in lattice computer simulations: the masses of all colorless bound states, including glueballs, decrease as $(m^{(0)})^{\frac{1}{1+\gamma_m}}$ with the bare fermion mass $m^{(0)}$ for *all* values of $m^{(0)}$ less than cutoff Λ. Moreover, one should expect that glueball masses are lighter than the masses of the bound states composed of quarks.

Few comments are in order:

(1) The phases S_1 and S_2 have essentially the same long distance dynamics. They are distinguished only by their dynamics at short distances: while the dynamics of the phase S_1 is asymptotically free, that of the phase S_2 is not. Also, while around the infrared fixed point α^* the sign of the beta function is negative in S_1, it is positive in S_2.[3] When all fermions are massive (with the current mass m), the continuum limit $\Lambda \to \infty$ of the S_2-theory is a non-asymptotically free confinement theory. Its spectrum includes colorless bound states composed of fermions and gluons. For $q < m$ the running coupling $\alpha(q)$ is the same as in pure SU(N_c) Yang-Mills theory, and for all $q \gg m$ $\alpha(q)$ is very close to α^* ("walking", actually, "standing" dynamics). For those values N_f for which α^* is small (as N_f=16, 15 and 14 at N_c=3), glueballs are much lighter than the bound states composed of fermions. Notice that unlike the case with $m = 0$, corresponding to the unparticle dynamics,[23] there exists a conventional S-matrix in this theory.

(2) In order to get the clearest exposition, we assumed such estimates as $N_f^{cr} \simeq 4N_c$ for N_f^{cr} and $\gamma_m = 1 - \sqrt{1 - \frac{\alpha^*}{\alpha_{cr}}}$ for the anomalous dimension γ_m. While the latter should be reasonable for $\alpha^* < \alpha_{cr}$ (and especially for $\alpha^* << \alpha_{cr}$),[9] the former is based on the assumption that $\alpha_{cr} \simeq \frac{2N_c}{N_c^2-1}\frac{\pi}{3}$ which, though seems reasonable, might be crude for some values of N_c. It is clear however that the dynamical picture presented above is essentially independent of those assumptions.

2.4. *Lattice computer simulations*

During last two years, there has been an essential progress in the lattice computer simulations of gauge theories with a varied number of fermion flavors.[d] For a recent review, see Ref. 14 and the papers of Tom Appelquist, George Fleming, Kieran Holland, Julius Kuti, Maria Lombardo, and Donald Sinclair in this volume.

[d]For pioneer papers in this area, see Refs. 24–26.

This author is certainly not an expert in lattice computer simulations. Here I would like to discuss this topic only in the connection with the phase diagram and the signature of the conformal window considered in the Secs. 2.2 and 2.3 above.

In Ref. 27, based on the fact that the sign of the beta function changes from negative to positive when the line between the S_1 and S_2 phases is crossed, the existence of the conformal window in QCD with $N_f = 12$ was studied by using the measurements of the chiral condensate and the mass spectrum. The analysis supports the existence of the conformal window in this theory.

In Ref. 28, the scaling law (10) was rediscovered and applied to the study of the conformal window in the SU(3) lattice gauge theory with two flavors of color sextet fermions (the parameter y_m in Ref. 28 is connected with the anomalous dimension γ_m as $y_m = 1+\gamma_m$). The main conclusion of that study was that $y_m \sim 1.5$ ($\gamma_m \sim 0.5$). This value is smaller than $\gamma_m \simeq 1$ in walking technicolor and at this moment it is unclear whether this theory contains an infrared-stable fixed point.

The authors of Ref. 29 studied the spectrum of mesons and glueballs in the SU(2) lattice gauge theory with adjoint fermions. They found that for light constituent fermions the lightest glueballs are lighter than the lightest mesons. It is tempting to speculate that in accordance with the signature for the conformal window discussed above, this fact indicates on the existence of a infrared fixed point in this theory. However, as the authors point out, a lot of issues should still be clarified in order to reach a solid conclusion.

It is clear that lattice simulations of gauge theories with varied numbers of fermion flavors are crucial for further progress in our understanding of such dynamics. The important point is that CPhT is a long range interactions phenomenon, which is very sensitive to any screening and finite-size effects. The progress made in this area during last two years is certainly encouraging.

Acknowledgments

I am grateful to the organizers of SCGT09 Workshop, in particular Koichi Yamawaki, for their warm hospitality. This work was supported by the Natural Sciences and Engineering Research Council of Canada.

References

1. L. D. Landau, Phys. Z. Sowjet. **11**, 26 (1937).
2. K. G. Wilson, Phys. Rev. **B4** (1971)3174; K. G. Wilson and J. B. Kogut, Phys. Rep. **12** (1974)75.
3. V. A. Miransky and K. Yamawaki, Phys. Rev. D **55**, 5051 (1997).
4. D. J. Gross and A. Neveu, Phys. Rev. D **10**, 3235 (1974).
5. E. V. Gorbar, V. P. Gusynin and V. A. Miransky, Phys. Rev. D **64**, 105028 (2001); S.-J. Rey, Prog. Theor. Phys. Suppl. No. **177**, 128 (2009).
6. D. B. Kaplan, J. W. Lee, D. T. Son and M. A. Stephanov, Phys. Rev. D **80**, 125005 (2009).
7. T. Maskawa and H. Nakajima, Prog. Theor. Phys. **52** (1974)1326; R. Fukuda and T. Kugo, Nucl. Phys. **B117** (1976)250.

8. P. I. Fomin and V. A. Miransky, Phys. Lett. B **64**, 166 (1976); P. I. Fomin, V. P. Gusynin and V. A. Miransky, Phys. Lett. B **78**, 136 (1978); P. I. Fomin, V. P. Gusynin, V. A. Miransky and Yu. A. Sitenko, Riv. Nuovo Cim. **6N5**, 1 (1983).
9. V. A. Miransky, Phys. Lett. B **91**, 421 (1980); Nuovo Cim. A **90**, 149 (1985).
10. W. A. Bardeen, C. N. Leung and S. T. Love, Phys. Rev. Lett. **56**, 1230 (1986); C. N. Leung, S. T. Love and W. A. Bardeen, Nucl. Phys. B **273**, 649 (1986).
11. T. Appelquist, J. Terning, and L. C. R. Wijewardhana, Phys. Rev. Lett. **75**, 2081 (1995).
12. V. L. Berezinskii, Zh. Eksp. Teor. Fiz. **59**, 907 (1970); J. M. Kosterlitz and D. J. Thouless, J. Phys. C **6**, 1181 (1973).
13. E. V. Gorbar, V. P. Gusynin, V. A. Miransky and I. A. Shovkovy, Phys. Rev. B **66**, 045108 (2002); H. Leal and D. V. Khveshchenko, Nucl. Phys. B **687**, 323 (2004).
14. E. Pallante, arXiv:0912.5188 [hep-lat].
15. B. Holdom, Phys. Rev. D **24**, 1441 (1981).
16. K. Yamawaki, M. Bando and K. i. Matumoto, Phys. Rev. Lett. **56**, 1335 (1986); T. W. Appelquist, D. Karabali and L. C. R. Wijewardhana, Phys. Rev. Lett. **57**, 957 (1986); T. Akiba and T. Yanagida, Phys. Lett. B **169**, 432 (1986).
17. F. Sannino, arXiv:0911.0931 [hep-ph].
18. A. H. Castro Neto, F. Guinea, N. M. R. Peres, K. S. Novoselov, and A. K. Geim, Rev. Mod. Phys. **81**, 109 (2009).
19. V. A. Miransky, Phys. Rev. D **59**, 105003 (1999).
20. T. Banks and A. Zaks, Nucl. Phys. B **196**, 189 (1982).
21. D. R. T. Jones, Nucl. Phys. **B75**, 531 (1974); W. E. Caswell, Phys. Rev. Lett. **33**, 244 (1974).
22. N. Seiberg, Phys. Rev. **D49**, 6857 (1994); K. Intriligator and N. Seiberg, Nucl. Phys. B (Proc. Suppl.) **45 B and C**, 1 (1996).
23. H. Georgi, Phys. Rev. Lett. **98**, 221601 (2007)
24. J. B. Kogut and D. R. Sinclair, Nucl. Phys. **B295**, 465 (1988); F. Brown et al., Phys. Rev. D **46**, 5655 (1992).
25. Y. Iwasaki, K. Kanaya, S. Kaya, S. Sakai, and T. Yoshie, Nucl Phys. B (Proc. Suppl.) **53**, 449 (1997); D. Chen and R. D. Mawhinney, ibid. **53**, 216 (1997).
26. P. H. Damgaard, U. M. Heller, A. Krasnitz and P. Olesen, Phys. Lett. B **400**, 169 (1997).
27. A. Deuzeman, M. P. Lombardo and E. Pallante, arXiv:0904.4662 [hep-ph].
28. T. DeGrand and A. Hasenfratz, Phys. Rev. D **80**, 034506; T. DeGrand, arXiv:0910.3072.
29. L. Del Debbio, B. Lucini, A. Patella, C. Pica and A. Rago, Phys. Rev. D **80**, 074507 (2009); B. Lucini, arXiv:0911.0020 [hep-ph].

Gauge-Higgs Unification at LHC

Nobuhito Maru*

Department of Physics, Chuo University
Tokyo, 112-8551, Japan
E-mail: maru@phys.chuo-u.ac.jp

Nobuchika Okada

Department of Physics and Astronomy, University of Alabama
Tuscaloosa, AL 35487, USA
E-mail: okadan@ua.edu

Higgs boson production by the gluon fusion and its decay into two photons at the LHC are investigated in the context of the gauge-Higgs unification scenario. The qualitative behaviors for these processes in the scenario are quite distinguishable from those of the Standard Model and the universal extra dimension scenario because of the overall sign difference for the effective couplings induced by one-loop corrections through Kaluza-Klein (KK) modes.

Keywords: Gauge-Higgs unification, Higgs production and decay at LHC.

1. Introduction

Gauge-Higgs unification (GHU) is a fascinating scenario solving the hierarchy problem without invoking supersymmetry.[1] In this scenario, Higgs scalar field in the Standard Model (SM) is identified with extra components of higher dimensional gauge field. The remarkable thing in this scenario is that the quantum correction to Higgs mass is finite due to the higher dimensional gauge symmetry regardless of the nonrenormalizability of the theory[2] (See also[3]). Such a UV insensitivity in other physical observables has been also investigated for S and T parameters,[4] g-2,[5] the violation of gauge-Yukawa universality[6] and the gluon fusion (two photon decay) of Higgs boson.[7] The last one is the issue discussed in this talk.

The Large Hadron Collider (LHC) started its operation again and the collider signatures of various new physics models beyond the SM have been extensively studied. The GHU shares the similar structure with the universal extra dimension (UED) scenario,[8] namely, Kaluza-Klein (KK) states of the SM particles appear. The collider phenomenology on the KK particles will be quite similar to the one in the UED scenario. A crucial difference should lie in the Higgs sector, because the Higgs

*Speaker of the conference.

doublet originates from the higher dimensional gauge field and its interactions are controlled by the higher dimensional gauge invariance. The discovery of Higgs boson is expected at the LHC, by which the origin of the electroweak symmetry breaking and the mechanism responsible for generating fermion masses will be revealed. Precise measurements of Higgs boson properties will provide us the information of a new physics relevant to the Higgs sector

In this talk, we investigate the effect of GHU on Higgs boson phenomenology at the LHC, namely, the production and decay processes of Higgs boson[9] (See also.[10]). At the LHC, the gluon fusion is the dominant Higgs boson production process and for light Higgs boson with mass $m_h < 150$ GeV, and two photon decay mode of Higgs boson becomes the primary discovery mode[11] nevertheless its branching ratio is $\mathcal{O}(10^{-3})$. The coupling between Higgs boson and these gauge bosons are induced through quantum corrections at one-loop level even in the SM. Therefore, we can expect a sizable effect from new particles if they contribute to the coupling at one-loop level. In a five dimensional GHU model, we calculate one-loop diagrams with KK fermions for the effective couplings between Higgs boson and the gauge bosons (gluons and photons). If the KK mass scale is small enough, we can see a sizable deviation from the SM couplings and as a result, the number of signal events from Higgs production at the LHC can be altered from the SM one. Interestingly, reflecting the special structure of Higgs sector in the GHU, there is a clear qualitative difference from the UED scenario, the signs of the effective couplings are opposite to those in the UED scenario.

2. Model

Let us consider a toy model of five dimensional (5D) $SU(3)$ GHU with an orbifold S^1/Z_2 compactification, in order to avoid unnecessary complications for our discussion. Although the predicted Weinberg angle in this toy model is unrealistic, $\sin^2\theta_W = \frac{3}{4}$, this does not affect our analysis. We introduce an $SU(3)$ triplet fermion as a matter field, which is identified with top and bottom quarks and their KK excited states, although the top quark mass vanishes and the bottom quark mass $m_b = M_W$ in this simple toy model. In our analysis, we will take into account a situation where a realistic top quark mass is realized and the bottom quark contributions are negligible comparing to the top quark ones.

The $SU(3)$ gauge symmetry is broken to $SU(2) \times U(1)$ by the orbifolding on S^1/Z_2. The remaining gauge symmetry $SU(2) \times U(1)$ is supposed to be broken by the vacuum expectation value (VEV) of the zero-mode of A_5, the extra space component of the gauge field identified with the SM Higgs doublet. We do not address the origin of $SU(2) \times U(1)$ gauge symmetry breaking and the resultant Higgs boson mass in the one-loop effective Higgs potential, which is highly model-dependent and out of our scope of this work.

The Lagrangian is simply given by

$$\mathcal{L} = -\frac{1}{2}\mathrm{Tr}(F_{MN}F^{MN}) + i\bar{\Psi}\not{D}\Psi. \tag{1}$$

The periodic boundary conditions are imposed along S^1 for all fields. The non-trivial Z_2 parities are assigned for each field as follows,

$$A_\mu(y_i - y) = \mathcal{P} A_\mu(y_i + y)\mathcal{P}^\dagger, \tag{2}$$

$$A_y(y_i - y) = -\mathcal{P} A_y(y_i + y)\mathcal{P}^\dagger, \tag{3}$$

$$\Psi(y_i - y) = \mathcal{P}\gamma^5\Psi(y_i + y) \tag{4}$$

where $\gamma^5\psi_L = \psi_L$, $\mathcal{P} = \text{diag}(+,+,-)$ at fixed points $y_i = 0, \pi R$. By this Z_2 parity assignment, $SU(3)$ is explicitly broken to $SU(2) \times U(1)$. Higgs scalar field is identified with the off-diagonal block of zero mode $A_y^{(0)}$.

After the electroweak gauge symmetry breaking, 4D effective Lagrangian among KK fermions, the SM gauge boson and Higgs boson (h) defined as $h^0 = (v+h)/\sqrt{2}$ can be derived from the term $\mathcal{L}_{\text{fermion}} = i\bar{\Psi}\not{D}\Psi$ in Eq. (1). Integrating over the fifth dimensional coordinate, we obtain a relevant 4D effective Lagrangian in the mass eigenstate of nonzero KK modes:

$$\begin{aligned}\mathcal{L}^{(4D)}_{\text{fermion}} &= \sum_{n=1}^{\infty} \Big\{ (\bar{\psi}_1^{(n)}, \bar{\tilde{\psi}}_2^{(n)}, \bar{\tilde{\psi}}_3^{(n)}) \\ &\times \begin{pmatrix} i\gamma^\mu\partial_\mu - m_n & 0 & 0 \\ 0 & i\gamma^\mu\partial_\mu - \left(m_+^{(n)} + \frac{m}{v}h\right) & 0 \\ 0 & 0 & i\gamma^\mu\partial_\mu - \left(m_-^{(n)} - \frac{m}{v}h\right) \end{pmatrix} \begin{pmatrix} \psi_1^{(n)} \\ \tilde{\psi}_2^{(n)} \\ \tilde{\psi}_3^{(n)} \end{pmatrix} \\ &+ \text{gauge interaction part} + \text{ zero-mode part.}\end{aligned} \tag{5}$$

where $m = \frac{gv}{2}(= M_W)$ is the bottom quark mass in this toy model, $g = \frac{g_5}{\sqrt{2\pi R}}$ is the 4D gauge coupling. Note that the mass splitting $m_\pm^{(n)} \equiv m_n \pm m = \frac{n}{R} \pm m$ occurs associated with a mixing between the $SU(2)$ doublet component and singlet component. Note that the mass eigenstate for $m_\pm^{(n)}$ has the Yukawa coupling $\mp m/v$, which is exactly the same as the one for the zero mode. In UED, however, the KK mode mass spectrum and Yukawa couplings are given by $M_n = \sqrt{m_n^2 + m_t^2}$ without mass splitting and $-(m_t/v) \times (m_t/M_n)$, respectively.[8] Together with the mass splitting of KK modes, this property is a general one realized in any GHU model and leads to a clear qualitative difference of the GHU from the UED scenario, as we will see.

3. Effective couplings between Higgs boson and gauge bosons

Before calculating KK fermion contributions to one-loop effective couplings between Higgs boson and gauge bosons (gluons and photons), it is instructive to recall the SM result. We parameterize the effective coupling between Higgs boson and gluons or photons as

$$\mathcal{L}_{\text{eff}} = C_g^{SM}\, h\, G^{a\mu\nu}G^a_{\mu\nu}, \tag{6}$$

$$\mathcal{L}_{\text{eff}} = C_\gamma^{SM} hF^{\mu\nu}F_{\mu\nu}, \tag{7}$$

where $G^a_{\mu\nu}(F_{\mu\nu})$ is a gluon (photon) field strength tensor. This coupling is generated by one-loop corrections (triangle diagram) on which quarks are running. The top quark loop diagram gives the dominant contribution and the coupling C_g^{SM} and C_γ^{SM} is described in the following form:

$$C_g^{SM} = -\frac{m_t}{v} \times \frac{\alpha_s F_{1/2}(4m_t^2/m_h^2)}{8\pi m_t} \times \frac{1}{2} \simeq \frac{\alpha_s}{12\pi v}, \tag{8}$$

$$\begin{aligned} C_\gamma^{SM} &= -\frac{m_t}{v} \times \frac{\alpha_{em} F_{1/2}(4m_t^2/m_h^2)}{8\pi m_t} \times \frac{4}{3} - \frac{m_W^2}{v} \times \frac{\alpha_{em} F_1(4m_W^2/m_h^2)}{8\pi m_W^2} \\ &\simeq -\frac{47\alpha_{em}}{72\pi v} \end{aligned} \tag{9}$$

where the first (second) term in C_γ^{SM} is KK top quark (KK W-boson) contributions, respectively, $\alpha_{s,em}$ is the fine structure constant of QCD, electromagnetic coupling, the loop function $F_{1/2,1}(\tau)$ given by

$$F_{1/2}(\tau) = -2\tau\left(1 + (1-\tau)\left[\sin^{-1}(1/\sqrt{\tau})\right]^2\right) \to -\frac{4}{3} \quad \text{for } \tau \gg 1, \tag{10}$$

$$F_1(\tau) = 2 + 3\tau + 3\tau(2-\tau)[\sin^{-1}(1/\sqrt{\tau})]^2 \to 7 \text{ for } \tau \gg 1. \tag{11}$$

It is well-known that in the top quark decoupling limit $m_t \gg m_h$, $F_{1/2,1}$ becomes a constant and the resultant effective coupling becomes independent of m_t, m_W and m_h.

Calculations of KK mode contributions are completely analogous to the top loop correction. The structure described in our toy model is common in any GHU model, we will have KK modes of top quark with mass eigenvalue $m_\pm^{(n)} = m_n \pm m_t$ and Yukawa couplings $\mp m_t/v$, respectively. The KK mode contributions are found to be

$$\begin{aligned} \mathcal{L}_{\text{eff}} &= C_g^{KK(GH)}\, h\, G^{a\mu\nu} G^a_{\mu\nu}, \\ C_g^{KK(GH)} &= -\sum_{n=1}^{\infty}\left[\frac{m_t}{v} \times \frac{\alpha_s F_{1/2}(4m_+^{(n)2}/m_h^2)}{8\pi m_+^{(n)}} \times \frac{1}{2}\right] + (m_+^{(n)} \leftrightarrow m_-^{(n)}) \\ &\simeq \frac{m_t \alpha_s}{12\pi v}\sum_{n=1}^{\infty}\left[\frac{1}{m_+^{(n)}} - \frac{1}{m_-^{(n)}}\right] \simeq -\frac{\alpha_s}{6\pi v}\sum_{n=1}^{\infty}\frac{m_t^2}{m_n^2} \end{aligned} \tag{12}$$

where we have taken the limit m_h^2, $m_t^2 \ll m_n^2$ to simplify the results. Note that this result is finite due to the cancellation between two divergent corrections with opposite signs. Also, note that the KK mode contribution is subtractive against the top quark contribution in the SM. The results are depicted in Fig. 1 as a function of the mass of the lightest KK mode (diagonal) mass eigenvalue (m_1). For the bulk fermion with the (half-)periodic boundary condition $m_1 = 1/R(1/(2R))$. In this analysis, we take $m_h = 120$ GeV. The result is not sensitive to the Higgs boson mass if $m_h < 2m_t$. For reference, the result in the UED scenario[8] is also shown, for which only the periodic fermion has been considered. The KK fermion contribution is subtractive and the Higgs production cross section is reduced in the GHU, while

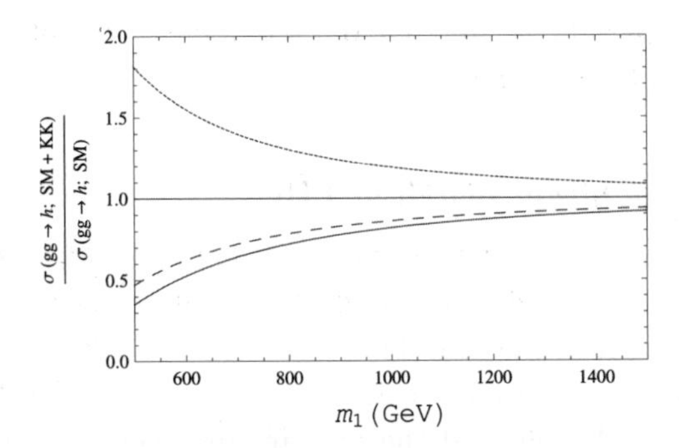

Fig. 1. The ratio of the Higgs boson production cross sections in the GHU and in the SM as a function of the KK mode mass m_1. The solid (dashed) line corresponds to the result including the (half-)periodic fermion contributions, respectively. As a reference, the result in the UED scenario with top quark KK modes is also shown (dotted line). We have taken $m_h = 120$ GeV.

it is increased in the UED scenario. This is a crucial point to distinguish the GHU from the UED scenario.

The contribution of top quark KK modes to the effective coupling between Higgs boson and photons are calculated similarly.

$$\begin{aligned}\mathcal{L}_{\text{eff}} &= C_\gamma^{KK(GH)}\, h\, F^{\mu\nu}F_{\mu\nu}, \\ C_\gamma^{KK(GH)} &= -\sum_{n=1}^{\infty}\left[\frac{m_t}{v}\times\frac{\alpha_{em}F_{1/2}(4m_+^{(n)2}/m_h^2)}{8\pi m_+^{(n)}}\times\frac{4}{3}\right] + (m_+^{(n)} \leftrightarrow m_-^{(n)}) \\ &\simeq \frac{2m_t\alpha_{em}}{9\pi v}\sum_{n=1}^{\infty}\left[\frac{1}{m_+^{(n)}} - \frac{1}{m_-^{(n)}}\right] \simeq -\frac{4\alpha_{em}}{9\pi v}\sum_{n=1}^{\infty}\frac{m_t^2}{m_n^2}. \end{aligned} \tag{13}$$

For the effective coupling with photons, in addition to the KK fermion contributions, we also have the KK W-boson loop corrections as in the SM. However, we neglect such contributions compared to those from the KK top quark ones by the following plausible reasons.

- KK top (KK W-boson) contributions are decoupling effects, and are proportional to mass squared of top (W-boson), respectively. This indicates that KK top quark contributions are likely to be dominant.
- In the GHU model on the flat space, a large dimensional representation in which the SM top quark is embedded must be introduced to reproduce a realistic top Yukawa coupling.[13] Therefore, the effective 4D theory includes extra vector-like top-like quarks and its KK modes. Thus, KK top quark contributions are enhanced by a number of extra top-like quarks.
- In some GHU models, bulk top-like quarks with the half-periodic boundary condition are often introduced to realize the correct electroweak symmetry breaking and a viable Higgs boson mass. The lowest KK mass of the half-periodic fermions is half of the lowest KK mass of periodic ones, so that their loop contributions can dominate over those by periodic KK mode fields.

After these considerations, we see that the KK mode contributions to two photon decay is the same sign as those of the SM.

4. Effects on Higgs boson search at LHC

As we have shown, the KK mode loop contribution to the effective coupling between Higgs boson and gluons (photons) is subtractive (slightly constructive) to the top quark loop contribution in the SM. This fact leads to remarkable effects on Higgs boson search at the LHC. Since the main production process of Higgs boson at the LHC is through gluon fusion, and the primary discovery mode of Higgs boson is its two photon decay channel if Higgs boson is light $m_h < 150$ GeV. Therefore, the deviations of the effective coupling between Higgs boson and gluons or photons from the SM one give important effects on the Higgs boson production and the number of two photon events from Higgs boson decay.

We show the ratio of the number of two photon events from Higgs decay produced through gluon fusion at the LHC. As a good approximation, this ratio is described as

$$\frac{\sigma(gg \to h;\ \mathrm{SM+KK}) \times BR(h \to \gamma\gamma;\ \mathrm{SM+KK})}{\sigma(gg \to h;\ \mathrm{SM}) \times BR(h \to \gamma\gamma;\ \mathrm{SM})} \simeq \left(1 + \frac{C_g^{KK(GH)}}{C_g^{SM}}\right)^2 \left(1 + \frac{C_\gamma^{KK(GH)}}{C_\gamma^{SM}}\right)^2 \qquad (14)$$

where σ is Higgs boson production cross section, BR denotes the branching ratio of two photon decay of Higgs boson. Fig. 2 shows the results for the periodic and half-periodic KK modes as a function of m_1 for the case of n_t =1, 3 and 5 extra top-like quark fermions. Even for $m_1 = 1$ TeV and $n_t = 1$, the deviation is sizable $\simeq$14%. When m_1 is small and n_t is large, the new physics contribution can dominate.

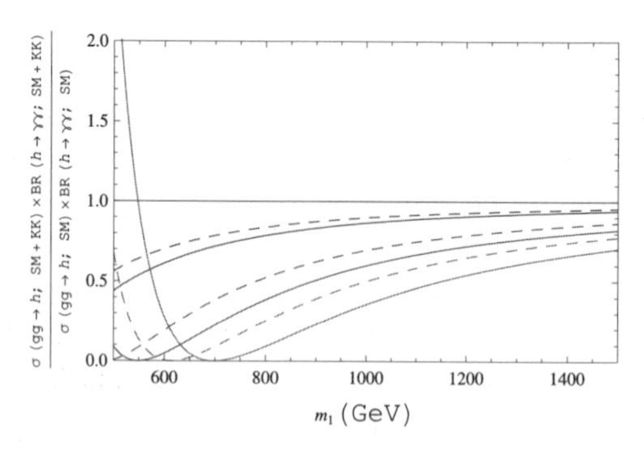

Fig. 2. The ratio of the number of two photon events in the GHU scenario to those in the SM as a function of the lowest KK mass m_1. The solid (dashed) lines represent the results including the n_t (half-)periodic KK fermion contributions. $n_t = 1, 3, 5$ are shown from the top to the bottom at $m_1 = 1500$ GeV. Higgs mass is taken to be 120 GeV.

5. Summary

We have calculated the one-loop KK fermion contributions to the Higgs effective couplings between Higgs boson and gluons or photons and found them to be finite. This finiteness is achieved by a non-trivial cancellations between two KK mass eigenstates, although each contribution is divergent. The overall sign of the contributions is opposite compared to the SM result by top quark loop corrections and the similar result in the UED scenario. Therefore, this feature is a clue to distinguish the GHU from the UED scenario. Our analysis have shown that even with the KK mode mass is around 1 TeV, the KK mode loop corrections provide $\mathcal{O}(10\%)$ deviations from the SM results in Higgs boson phenomenology at the LHC. In a realistic GHU model, some extra top-like quarks would be introduced to reproduce the top Yukawa coupling in the SM. In such a case, the KK mode contributions are enhanced and the signal events of Higgs boson production at the LHC are quite different from those in the SM.

Acknowledgments

The work of authors was supported in part by the Grant-in-Aid for Scientific Research of the Ministry of Education, Science and Culture, No.18204024 and No.20025005.

References

1. N. S. Manton, Nucl. Phys. B **158**, 141 (1979); D. B. Fairlie, Phys. Lett. B **82**, 97 (1979); J. Phys. G **5**, L55 (1979); Y. Hosotani, Phys. Lett. B **126**, 309 (1983); Phys. Lett. B **129**, 193 (1983); Annals Phys. **190**, 233 (1989).
2. H. Hatanaka, T. Inami and C. S. Lim, Mod. Phys. Lett. A **13**, 2601 (1998).
3. I. Antoniadis, K. Benakli and M. Quiros, New J. Phys. **3**, 20 (2001); G. von Gersdorff, N. Irges and M. Quiros, Nucl. Phys. B **635**, 127 (2002); C. S. Lim, N. Maru and K. Hasegawa, J. Phys. Soc. Jap. **77**, 074101 (2008); K. Hasegawa, C. S. Lim and N. Maru, Phys. Lett. B **604**, 133 (2004); N. Maru and T. Yamashita, Nucl. Phys. B **754**, 127 (2006); Y. Hosotani, N. Maru, K. Takenaga and T. Yamashita, Prog. Theor. Phys. **118**, 1053 (2007).
4. C. S. Lim and N. Maru, Phys. Rev. D **75**, 115011 (2007).
5. Y. Adachi, C. S. Lim and N. Maru, Phys. Rev. D **76**, 075009 (2007); Phys. Rev. D **79**, 075018 (2009).
6. C. S. Lim and N. Maru, arXiv:0904.0304 [hep-ph].
7. N. Maru, Mod. Phys. Lett. A **23**, 2737 (2008) [arXiv:0803.0380 [hep-ph]].
8. T. Appelquist, H. C. Cheng and B. A. Dobrescu, Phys. Rev. D **64**, 035002 (2001) .
9. N. Maru and N. Okada, Phys. Rev. D **77**, 055010 (2008).
10. A. Falkowski, Phys. Rev. D **77**, 055018 (2008); G. Cacciapaglia, A. Deandrea and J. Llodra-Perez, JHEP **0906**, 054 (2009); I. Low, R. Rattazzi and A. Vichi, arXiv:0907.5413 [hep-ph].
11. See, for example, A. Djouadi, arXiv:hep-ph/0503172, references therein.
12. F. J. Petriello, JHEP **0205**, 003 (2002).
13. G. Cacciapaglia, C. Csaki and S. C. Park, JHEP **0603**, 099 (2006).

$W_L W_L$ Scattering in Higgsless Models: Identifying Better Effective Theories

Alexander S. Belyaev[a], R. Sekhar Chivukula[b], Neil D. Christensen[b],
Hong-Jian He[c], Masafumi Kurachi[d,*], Elizabeth H. Simmons[b] and Masaharu Tanabashi[e]

[a]*School of Physics & Astronomy, University of Southampton*
Highfield, Southampton SO17 1BJ, UK
Particle Physics Department, Rutherford Appleton Laboratory
Chilton, Didcot, Oxon OX11 0QX, UK
[b]*Department of Physics and Astronomy, Michigan State University*
East Lansing, MI 48824, USA
[c]*Center for High Energy Physics, Tsinghua University, Beijing 100084, China*
[d]*Department of Physics, Tohoku University, Sendai 980-8578, Japan*
[e]*Department of Physics, Nagoya University, Nagoya 464-8602, Japan*

The three site Higgsless model has been offered as a benchmark for studying the collider phenomenology of Higgsless models. In this talk, we present how well the three site Higgsless model performs as a general representative of Higgsless models, in describing $W_L W_L$ scattering, and which modifications can make it more representative. We employ general sum rules relating the masses and couplings of the Kaluza-Klein (KK) modes of the gauge fields in continuum and deconstructed Higgsless models as a way to compare the different theories. After comparing the three site Higgsless model to flat and warped continuum Higgsless models, we analyze an extensions of the three site Higgsless model, namely, the Hidden Local Symmetry (HLS) Higgsless model. We demonstrate that $W_L W_L$ scattering in the HLS Higgsless model can very closely approximate scattering in the continuum models, provided that the parameter 'a' is chosen to mimic ρ-meson dominance of $\pi\pi$ scattering in QCD.

Keywords: Higgsless model, $W_L W_L$ scattering, sum rule, Hidden Local Symmetry.

1. Introduction

Higgsless model[1] is an attractive alternative to the Standard Model (SM) for describing the Electroweak symmetry breaking, in which the symmetry is broken by the boundary conditions of the five dimensional gauge theory. It turned out that dimensional deconstruction[2] is quite useful to understand the important nature of the model, such as the delay of perturbative unitarity violation, boundary conditions, etc. (See Refs.[3,4]) It was also extensively used to study the constraints from the electroweak precision measurements.[5]

The three site Higgsless model[6] was proposed as an extremely deconstructed version of five dimensional Higgsless models, in which only one copy of weak gauge

*Speaker, Email: kurachi@tuhep.phys.tohoku.ac.jp.

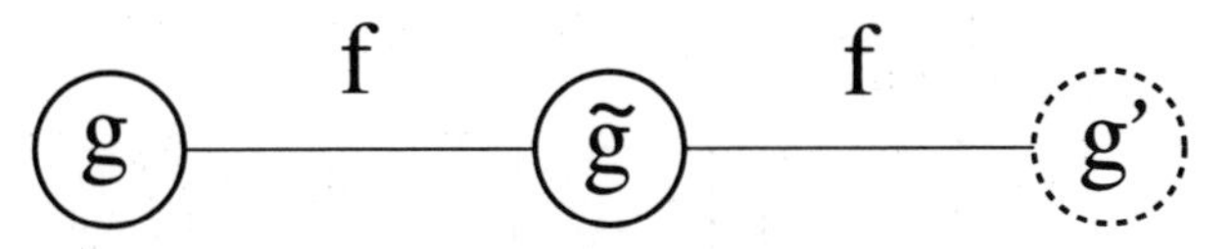

Fig. 1. Gauge sector of the three site Higgsless model. The solid circles represent $SU(2)$ gauge groups, with coupling strengths g_0 and g_1, and the dashed circle is a $U(1)$ gauge group with coupling g_2.

boson (W', Z') is introduced as a new resonance. Such a model contains sufficient complexity to incorporate interesting physics issues related to fermion masses and electroweak observables, yet remains simple enough that it could be encoded in a Matrix Element Generator program for use with Monte Carlo simulations. Such program was already done by several groups.[7] In this talk, we present how well the three site Higgsless model performs as a general representative of Higgsless models, in describing $W_L W_L$ scattering, and which modifications can make it more representative. After briefly reviewing the three site Higgsless model, we compare the three site Higgsless model to flat and warped continuum Higgsless models. Then, we analyze a Hidden Local Symmetry (HLS) generalization of the three site Higgsless model.[a]

2. The three site Higgsless model

In this section, we briefly review the three site Higgsless model.[6] The gauge sector of the three site Higgsless model is illustrated in Fig. 1 using "Moose notation".[9] The model incorporates an $SU(2) \times SU(2) \times U(1)$ gauge group (with couplings g_0, g_1 and g_2, respectively), and two nonlinear $(SU(2) \times SU(2))/SU(2)$ sigma models in which the global symmetry groups in adjacent sigma models are identified with the corresponding factors of the gauge group. The symmetry breaking between the middle $SU(2)$ and the $U(1)$ follows an $SU(2)_L \times SU(2)_R/SU(2)_V$ symmetry breaking pattern with the $U(1)$ embedded as the T_3-generator of $SU(2)_R$. This extended electroweak gauge sector is in the same class as models of extended electroweak gauge symmetries,[10] which are considered as an application of the hidden local symmetry[11] to the electroweak sector.[b] The decay constants, f_1 and f_2, of two nonlinear sigma models can be different in general, however, we take $f_1 = f_2 = f(= \sqrt{2}v)$ for simplicity. Also, we work in the limit $x \equiv g_0/g_1 \ll 1$, $y \equiv g_2/g_1 \ll 1$, in which case we expect a massless photon, light W and Z bosons, and a heavy set of bosons W' and Z'. Numerically, then, $g_{0,2}$ are approximately equal to the standard model $SU(2)_W$ and $U(1)_Y$ couplings, and we therefore denote $g_0 \equiv g$ and $g_2 \equiv g'$, and define an angle θ such that $\frac{g'}{g} = \frac{\sin\theta}{\cos\theta} \equiv \frac{s}{c} (\equiv t)$. In addition, we denote $g_1 \equiv \tilde{g}$.

[a]This talk is based on the work done in Ref.[8]

[b]The new physics discussed in Ref.[6] is related to the fermion sector.

Table 1. Leading expressions of $Z'WW$ coupling and $ZW'W$ coupling in each models.

	Three Site	5D Flat	5D Warped
$g_{Z'WW}$	$-\frac{1}{2}\frac{e}{s}\left(\frac{M_W}{M_{W'}}\right)$	$-\frac{4\sqrt{2}}{\pi^2}\frac{e}{s}\left(\frac{M_W}{M_{W'}}\right)$	$-0.36\left(\frac{M_W}{M_{W'}}\right)$
$g_{ZW'W}$	$-\frac{1}{2}\frac{e}{sc}\left(\frac{M_W}{M_{W'}}\right)$	$-\frac{4\sqrt{2}}{\pi^2}\frac{e}{sc}\left(\frac{M_W}{M_{W'}}\right)$	$-0.36\frac{1}{c}\left(\frac{M_W}{M_{W'}}\right)$

3. Comparison between continuum and the three site Higgsless models

The three site Higgsless model can be viewed as an extremely deconstructed version of 5-dimensional $SU(2)\times SU(2)$ Higgsless model in which $SU(2)\times SU(2)$ symmetry is broken down to its diagonal $SU(2)$ by the boundary condition (BC) at one end of the extra dimension, while one of $SU(2)$ is broken down to its $U(1)$ subgroup by the BC at the other end. Thus, it is tempting to investigate how well the three site Higgsless model performs as a low energy effective theory of continuum Higgsless models. For this purpose, we consider $SU(2)\times SU(2)$ Higgsless model in the flat and the warped extra dimension as example models to compare.

The electroweak gauge sector of $SU(2)\times SU(2)$ Higgsless models in both the flat and the warped extra dimension can be characterized (with certain simplifications) by four free parameters. In flat case, those are R (the size of the extra dimension), g_5 (the bulk gauge coupling), g_0 and g_Y (couplings for brane localized kinetic terms of $SU(2)\times U(1)$ gauge bosons). In warped case, those are R' (the seize of extra dimension), b (warp factor), g_5 and g_Y. (See Ref.[12] for detailed descriptions of the models.) Since the gauge sector of the three site Higgsless model is also characterized by four free parameter (see, Fig. 1), these three models can be compared by choosing four parameters so that they reproduce same values of four physical quantities. Three of four parameters should be chosen in a way that electroweak observables (e, M_Z and M_W, for example) are correctly reproduced. Then, there is still one free parameter to fix. Here, we use that parameter to fix the scale of KK mode (the mass of W' boson, for example). Now, there are no free parameter left in each model, and any physical quantities other than those four quantities can be calculated as predictions in each models, which should be compared among three models to see how much the three site Higgsless model approximates continuum Higgsless models.

In the present analysis, we focus on the values of triple gauge boson couplings each model predicts. As for triple gauge boson couplings which involve only SM gauge bosons, it can be shown that those are of the same form as in the SM at the leading order in the expansion of $M_W^2/M_{W'}^2$, and we require that the deviation from the SM value, which appears as next to leading order, should be within experimental bound (which put a lower bound on $M_{W'}$).

In Table 1, we listed the leading expressions of $Z'WW$ coupling and $ZW'W$ coupling in each models. Making the numerical approximations $\frac{4\sqrt{2}}{\pi^2}\frac{e}{s}\simeq 0.36$ and $\frac{4\sqrt{2}}{\pi^2}\frac{e}{sc}\simeq 0.41$, we find, $\frac{g_{Z'WW}|_{\text{warped}-5\text{d}}}{g_{Z'WW}|_{\text{flat}-5\text{d}}}\simeq\frac{g_{ZW'W}|_{\text{warped}-5\text{d}}}{g_{ZW'W}|_{\text{flat}-5\text{d}}}\simeq 1$. In other words, the values of $g_{Z'WW}$ and $g_{ZW'W}$ in these continuum models are essentially independent of the geometry of the extra dimension to leading order. Then we can compare the couplings in continuum models to those in the three site Higgsless model, assuming a common value for $M_W/M_{W'}$: $\frac{g_{Z'WW}|_{\rm three-site}}{g_{Z'WW}|_{\rm flat-5d}} \simeq \frac{g_{ZW'W}|_{\rm three-site}}{g_{ZW'W}|_{\rm flat-5d}} \simeq \frac{\pi^2}{8\sqrt{2}} \simeq 0.87$. The values of $g_{Z'WW}$ and $g_{ZW'W}$ in the three site Higgsless model are about 13% smaller than those values in 5-dimensional $SU(2) \times SU(2)$ Higgsless models. Then, the question which naturally arises is — "Why do $g_{Z'WW}$ and $g_{ZW'W}$ take similar values in different models?" There are two keywords to be addressed to answer to this question: sum rules and the lowest KK mode dominance.

In any continuum five-dimensional gauge theory, the sum rules that guarantee the absence, respectively, of $\mathcal{O}(E^4)$ and $\mathcal{O}(E^2)$ growth in the amplitude for $W_L^+W_L^- \to W_L^+W_L^-$ elastic scattering have the following form,[1,13]

$$\sum_{i=1}^{\infty} g_{Z_iWW}^2 = g_{WWWW} - g_{ZWW}^2 - g_{\gamma WW}^2 \,, \tag{1}$$

$$3\sum_{i=1}^{\infty} g_{Z_iWW}^2 M_{Z_i}^2 = 4 g_{WWWW} M_W^2 - 3 g_{ZWW}^2 M_Z^2 \,, \tag{2}$$

where Z_i represents the i-th KK mode of the neutral gauge boson. (Z_1 is identified as Z'.) We focuses on the degree to which the first KK mode saturates the sum on the LHS of the identities (1) and (2). Suppose that we form the ratio of the $n = 1$ term in the sum on the LHS to the full combination of terms on the RHS, evaluated to leading order in $(M_W/M_{W'})^2$. The ratios derived from Eqs. (1) and (2) are:

$$\frac{g_{Z'WW}^2}{g_{WWWW} - g_{ZWW}^2 - g_{\gamma WW}^2} \,, \tag{3}$$

$$\frac{3 g_{Z'WW}^2 M_{Z'}^2}{4 g_{WWWW} M_W^2 - 3 g_{ZWW}^2 M_Z^2} \,. \tag{4}$$

If the $n = 1$ KK mode saturates the identity, then the related ratio will be 1.0; ratio values less than 1.0 reflect contributions from higher KK modes. We see from Table 2 that each of these ratios is nearly 1.0 in both the $SU(2) \times SU(2)$ flat and warped Higgsless models, confirming that the first KK mode nearly saturates the sum rules in these continuum models. The similar behavior of the two extra dimensional models is consistent with our finding that the $g_{Z'WW}$ coupling is relatively independent of geometry.

Because the first KK mode nearly saturates the identities (1) and (2) in these continuum models, the ratios (3) and (4) should be useful for drawing comparisons with the three site Higgsless model, which only possesses a single KK gauge mode. As shown in the 3rd column of Table 2, the first ratio has the value one in the three site Higgsless model, meaning that the identity (1) is still satisfied. The ratio related to identity (2), however, has the value 3/4 for the three site Higgsless model, meaning that the second identity is not satisfied; the longitudinal gauge boson scattering amplitude continues to grow as E^2 due to the underlying non-renormalizable interactions in the three site Higgsless model. Since the value of the denominator in

Table 2. Ratios relevant to evaluating the degree of cancellation of growth in the $W_L W_L$ scattering amplitude from the lowest lying KK resonance at order E^4 (top row, from Eq. (3)), and at order E^2 (second row, from Eq. (4)). A value close to one indicates a high degree of cancellation from the lowest lying resonance. Shown in successive columns for the $SU(2) \times SU(2)$ flat and warped continuum models discussed in the text, and the three site Higgsless model.

	5d 2×2 Flat	5d 2×2 Warped	Three-site
$\frac{g^2_{Z'WW}}{g_{WWWW} - g^2_{ZWW} - g^2_{\gamma WW}}$	$\frac{960}{\pi^6} \simeq 0.999$	0.992	1
$\frac{3g^2_{Z'WW} M^2_{Z'}}{4g_{WWWW} M^2_W - 3g^2_{ZWW} M^2_Z}$	$\frac{96}{\pi^4} \simeq 0.986$	0.986	3/4

Eq. (4) has not changed appreciably, this indicates a difference between the values of the $g_{Z'WW}$ couplings in the continuum and three site Higgsless models. This is the reason why $g_{Z'WW}$ in the three site Higgsless model is 13% (or 25% in $g^2_{Z'WW}$) smaller than the value of $g_{Z'WW}$ in continuum models as we have shown in the previous section.

4. Hidden Local Symmetry generalization of the three site Higgsless models

In this section, we consider an Hidden Local Symmetry (HLS) generalization of the three site Higgsless model, in which an extra parameter a is introduced and the Lagrangian takes the following form:[11]

$$\begin{aligned}
\mathcal{L}^{\rm HLS} &= -\frac{v^2}{4}\mathrm{Tr}\left[(D^\mu\Sigma_1^\dagger)\Sigma_1 - (D^\mu\Sigma_2)\Sigma_2^\dagger\right]^2 - a\frac{v^2}{4}\mathrm{Tr}\left[(D^\mu\Sigma_1^\dagger)\Sigma_1 + (D^\mu\Sigma_2)\Sigma_2^\dagger\right]^2, \\
&= \frac{v^2}{4}(1+a)\mathrm{Tr}\left[(D_\mu\Sigma_1)^\dagger(D^\mu\Sigma_1) + (D_\mu\Sigma_2)^\dagger(D^\mu\Sigma_2)\right] \\
&\quad + \frac{v^2}{2}(1-a)\mathrm{Tr}\left[(D_\mu\Sigma_1)^\dagger\Sigma_1(D_\mu\Sigma_2)\Sigma_2^\dagger\right], \\
&= \frac{av^2}{2}\mathrm{Tr}\left[(D_\mu\Sigma_1)^\dagger(D^\mu\Sigma_1) + (D_\mu\Sigma_2)^\dagger(D^\mu\Sigma_2)\right] \\
&\quad + \frac{v^2}{4}(1-a)\,\mathrm{Tr}\left[(D_\mu(\Sigma_1\Sigma_2))^\dagger D^\mu(\Sigma_1\Sigma_2)\right]. \qquad (5)
\end{aligned}$$

Note that taking $a = 1$ reproduce the Lagrangian of the three site Higgsless model. Fermion sector of the model is also discussed in detail in Ref.,[8] in which we showed that the model can accommodate the ideal fermion delocalization which is needed for the consistency with the precision EW experiments.

It is straightforward to calculate triple gauge boson couplings of this model, from which we can evaluate the values in Eqs. (3) and (4):

$$\frac{g^2_{Z'WW}}{g_{WWWW} - g^2_{ZWW} - g^2_{\gamma WW}} = 1, \qquad \frac{3g^2_{Z'WW} M^2_{Z'}}{4g_{WWWW} M^2_W - 3g^2_{ZWW} M^2_Z} = \frac{3}{4}a\,. \qquad (6)$$

By comparing this result with the ones shown in Table 2, we see that the HLS Higgsless model can very closely approximate scattering in the continuum models if we take $a = \frac{4}{3}$.

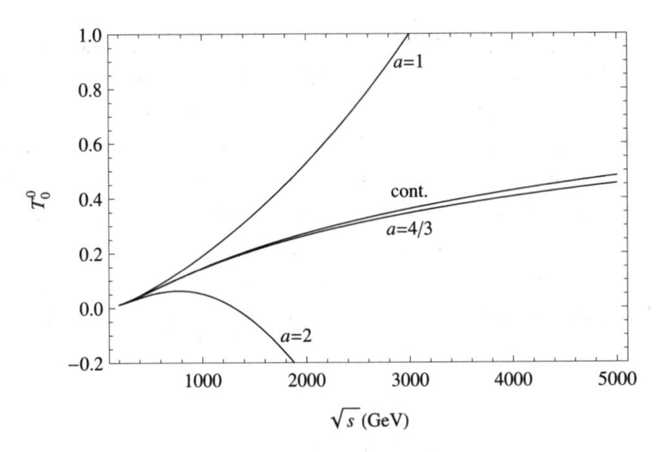

Fig. 2. Behavior of the partial wave amplitude T_0^0 for NG boson scattering in the triangular moose model with various a. The values $v = 250$ GeV, $M_1 = 500$ GeV are assumed. The curve labeled "cont." shows T_0^0 in the continuum flat $SU(2) \times SU(2)$ model for $M_1 = 500$ GeV.

It is interesting to note that $a = \frac{4}{3}$ is the choice in which the four-point Nambu-Goldstone (NG) boson coupling, $g_{\pi\pi\pi\pi} = 1 - \frac{3}{4}a$, vanishes in this model.[14] Considering that, in the language of dimensional deconstruction,[2] NG bosons are identified as a fifth-component of the gauge boson field (A_5) in a discretized five-dimensional gauge theory, and also considering the fact that there is no four-point A_5 coupling in five dimensional gauge theories, it is natural that the parameter choice in which the four-point NG boson coupling vanishes well approximates the continuum models.

Fig. 2 shows the partial wave amplitude T_0^0 for NG boson scattering in the global (which means we set $g = g' = 0$ for simplicity) continuum flat $SU(2) \times SU(2)$ model (with the mass of the first KK mode, M_1, taken to be $M_1 = 500$ GeV) compared with T_0^0 in the HLS Higgsless model for several values of the parameter a. The result in the globalthree site Higgsless model is shown by the curve labeled $a = 1$; the value $a = 2$ is motivated by the phenomenological KSRF relation.[15] This plot ties our results together quite neatly: while the curves with three different values of a all give a reasonable description of T_0^0 at very low energies, the best approximation to the continuum behavior of T_0^0 over a wide range of energies is given by the HLS Higgsless model curve with $a = 4/3$. At low energies, the fact that the three-site and the HLS Higgsless models both prevent E^4 growth of the amplitude suffices; but at higher energies, the fact that the HLS Higgsless model with $a = 4/3$ has $g_{\pi\pi\pi\pi} = 0$ and enables it to cut off the E^2 growth of the amplitude as well, as consistent with the behavior in the continuum model.

5. Conclusions

We have considered how well the three site Higgsless model performs as a general representative of Higgsless models, and have studied HLS generalization have the potential to improve upon its performance. Our comparisons have employed sum rules relating the masses and couplings of the gauge field KK modes. We find that the tendency of the sum rules to be saturated by contributions from the lowest-lying KK resonances suggests a way to quantify the extent to which a highly-deconstructed theory like the three site Higgsless model can accurately describe the low-energy physics. We demonstrated that $W_L W_L$ scattering in the HLS Higgsless model can very closely approximate scattering in the continuum models, provided that the HLS parameter a is chosen appropriately. This observation confirms that the collider phenomenology studied, such as in Ref.[7] , are applicable not only to the three site Higgsless model, but also to extra-dimensional Higgsless models.

References

1. C. Csaki, C. Grojean, H. Murayama, L. Pilo and J. Terning, Phys. Rev. D **69**, 055006 (2004); C. Csaki, C. Grojean, L. Pilo and J. Terning, Phys. Rev. Lett. **92**, 101802 (2004).
2. N. Arkani-Hamed, A. G. Cohen and H. Georgi, Phys. Rev. Lett. **86**, 4757 (2001); C. T. Hill, S. Pokorski and J. Wang, Phys. Rev. D **64**, 105005 (2001).
3. R. S. Chivukula, D. A. Dicus and H. J. He, Phys. Lett. B **525**, 175 (2002); R. S. Chivukula and H. J. He, Phys. Lett. B **532**, 121 (2002).
4. H. J. He, Int. J. Mod. Phys. A **20** (2005) 3362.
5. R. S. Chivukula, E. H. Simmons, H. J. He, M. Kurachi and M. Tanabashi, Phys. Rev. D **70**, 075008 (2004); Phys. Lett. B **603**, 210 (2004); Phys. Rev. D **71**, 035007 (2005); Phys. Rev. D **71**, 115001 (2005); Phys. Rev. D **72**, 015008 (2005); Phys. Rev. D **72**, 095013 (2005).
6. R. Sekhar Chivukula, B. Coleppa, S. Di Chiara, E. H. Simmons, H. J. He, M. Kurachi and M. Tanabashi, Phys. Rev. D **74**, 075011 (2006).
7. H. J. He *et al.*, Phys. Rev. D **78**, 031701 (2008); J. G. Bian *et al.*, Nucl. Phys. B **819**, 201 (2009).
8. A. S. Belyaev, R. Sekhar Chivukula, N. D. Christensen, H. J. He, M. Kurachi, E. H. Simmons and M. Tanabashi, Phys. Rev. D **80**, 055022 (2009).
9. H. Georgi, Nucl. Phys. B **266**, 274 (1986).
10. R. Casalbuoni, S. De Curtis, D. Dominici and R. Gatto, Phys. Lett. B **155**, 95 (1985); R. Casalbuoni, A. Deandrea, S. De Curtis, D. Dominici, R. Gatto and M. Grazzini, Phys. Rev. D **53**, 5201 (1996).
11. M. Bando, T. Kugo, S. Uehara, K. Yamawaki and T. Yanagida, Phys. Rev. Lett. **54**, 1215 (1985); M. Bando, T. Kugo and K. Yamawaki, Nucl. Phys. B **259**, 493 (1985); M. Bando, T. Fujiwara and K. Yamawaki, Prog. Theor. Phys. **79**, 1140 (1988); M. Bando, T. Kugo and K. Yamawaki, Phys. Rept. **164**, 217 (1988).
12. R. S. Chivukula, E. H. Simmons, H. J. He, M. Kurachi and M. Tanabashi, Phys. Rev. D **72**, 075012 (2005).
13. R. S. Chivukula, H. J. He, M. Kurachi, E. H. Simmons and M. Tanabashi, Phys. Rev. D **78**, 095003 (2008).
14. See section 3 in M. Harada and K. Yamawaki, Phys. Rept. **381**, 1 (2003); The relation between $a = 4/3$ and ρ meson dominance was also pointed out in M. Harada, S. Matsuzaki and K. Yamawaki, Phys. Rev. D **74**, 076004 (2006).
15. K. Kawarabayashi and M. Suzuki, Phys. Rev. Lett. **16**, 255 (1966); Riazuddin and Fayyazuddin, Phys. Rev. **147**, 1071 (1966).

Holographic Estimate of Muon $g - 2$

Deog Ki Hong
Department of Physics, Pusan National University
Busan 609-735, Korea
E-mail: dkhong@pusan.ac.kr

I present recent calculations of the hadronic contributions to muon anomalous magnetic moment in holographic QCD, based on gauge/gravity duality. The holographic estimates are compared well with the analysis based on recently revised BaBar measurements of $e^+e^- \to \pi^+\pi^-$ cross-sections and also with other model calculations for the light-by-light scattering contributions.

Keywords: Anomalous magnetic moment, muon, gauge/gravity duality.

1. Introduction

The successful prediction of anomalous magnetic moments in early days of particle physics was one of the first triumphs of quantum field theory.[1] Since then it has served as a critical test of the standard model of particle physics, which is based on quantum field theory, guided by gauge principle. Recent measurement of anomalous magnetic moment of muon (E821 at BNL)[2] provides the most stringent test so far, at the level of sub parts per million, of the standard model,

$$a_\mu^{\rm exp} = \frac{g_\mu - 2}{2} = 11659208.0(5.4)(3.3) \times 10^{-10}\,, \tag{1}$$

which exhibits currently 3.2σ deviation from the standard model estimate:

$$\Delta a_\mu = a_\mu^{\rm exp} - a_\mu^{\rm SM} = 302(63)(61) \times 10^{-11}\,. \tag{2}$$

If the discrepancy persists in more improved measurements, such as the E969 experiment, planned at BNL, or other experiments at the Fermi Lab, it will certainly indicate a hint of new physics. Therefore it is quite necessary to understand the standard model predictions more precisely to pin down the possible hint of new physics at the level of 5σ or more.

The theoretical prediction of muon $g-2$ in the standard model consists of three different contributions:

$$a_\mu^{\rm th} = a_\mu^{\rm QED} + a_\mu^{\rm weak} + a_\mu^{\rm had}\,.$$

Among them the QED contribution is most dominant one and has been calculated to be $a_\mu^{\rm QED} = 116584718.10(0.16) \times 10^{-11}$ at 4.5 loops by Kinoshita et al,[3] while the weak interaction corrections are found to be $a_\mu^{\rm weak} = 154(2) \times 10^{-11}$ at the two-loop level.[4]

As hadrons (or quarks) contribute to the anomalous magnetic moment of muon only through quantum fluctuations, the strong interaction contribution is suppressed, compared to that of QED. The current estimate of the hadronic contributions is

$$a_\mu^{\rm had} = 116591778(2)(46)(40) \times 10^{-11}\,. \tag{3}$$

The most uncertainty in the SM estimate is however coming from the hadronic contributions, since we poorly understand strong dynamics, while the electroweak corrections can be calculated accurately in perturbation. Direct calculation of hadronic contributions from QCD requires lattice calculations, which are currently not accurate enough to give a meaningful result due to large systematic uncertainties.[5]

In this talk I present new estimates of the hadronic contributions, based on a holographic model of QCD. Holographic models have been proposed recently for QCD,[6,7] inspired by the gauge/gravity duality, found in string theory,[8] which states that certain strongly coupled gauge theories are equivalent to weakly coupled gravity in one-higher dimension, the extra dimension being the renormalization scale. The holographic models of QCD were found to be quite successful to account for the properties of hadrons at low energy and give relations to their couplings and also new sum rules,[6,7,9,10] offering theoretical understanding of phenomenological rules for hadrons found in 60's such as vector meson dominance, proposed by Sakurai.[11]

2. Holographic QCD

Holographic QCD (hQCD) is a model for 5D gravity dual theory of QCD in the large N_c and large 't Hooft coupling limit ($\lambda \equiv g_s^2 N_c \gg 1$), describing QCD directly with hadrons. As a gravity dual of QCD with three light flavors we consider $U(3)_L \times U(3)_R$ flavor gauge theory in a slice of AdS_5,

$$S = \int d^5x \sqrt{g}\, {\rm Tr}\left\{ |DX|^2 + 3|X|^2 - \frac{1}{4g_5^2}({F_L}^2 + {F_R}^2) \right\} + S_Y + S_{CS} \tag{4}$$

where the metric is given as, taking the AdS radius $R = 1$,

$$ds^2 = \frac{1}{z^2}(dx^\mu dx_\mu - dz^2), \quad \epsilon \le z \le z_m\,. \tag{5}$$

We take the ultraviolet (UV) regulator $\epsilon \to 0$ and introduce an infra-red (IR) brane at z_m to implement the confinement of QCD, breaking the conformal symmetry.[7] The bulk scalar (X) and the bulk gauge fields (A) are dual to $\bar{q}q$ and $\bar{q}\gamma^\mu T^a q$, respectively. Following Katz and Schwartz,[12] we have introduced a flavor-singlet bulk scalar (Y) for the η' meson, which is dual to F^2 ($F\tilde{F}$) and is described by

$$S_Y = \int d^5x \sqrt{g} \left[\frac{1}{2}|DY|^2 - \frac{\kappa}{2}(Y^{N_f} \det(X) + {\rm h.c.}) \right]\,. \tag{6}$$

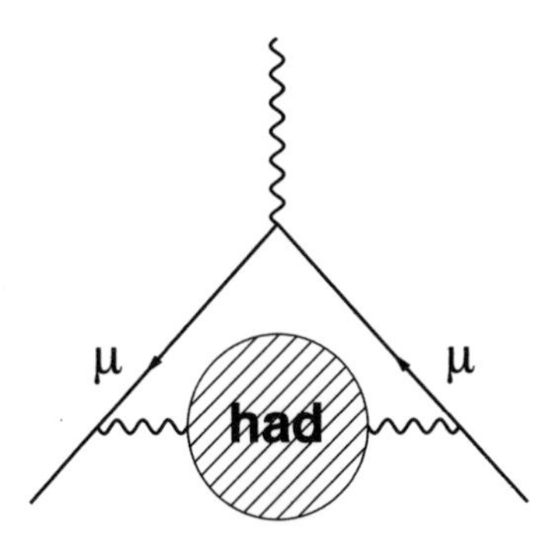

Fig. 1. Leading hadronic-vacuum-polarization contribution to muon $g-2$.

Finally we introduce a Chern-Simons term to reproduce the QCD flavor anomaly,[13]

$$S_{CS} = \frac{N_c}{24\pi^2}\int \left[\omega_5(A_L) - \omega_5(A_R)\right] , \tag{7}$$

where $d\,\omega_5(A) = \mathrm{Tr}\,F^3$. Our calculations can be easily applied to other holographic models like Sakai-Sugimoto model, which will give similar results.

Gauge/gravity duality at large N_c and large λ implies that the generating functional of one-particle irreducible Green's functions is given by the classical gravity action of the dual theory:

$$W_{4D}[\phi_0(x)] = S_{5D\mathrm{eff}}[\phi(x,\epsilon)] \quad \text{with} \quad \phi(x,\epsilon) = \phi_0(x)\,. \tag{8}$$

Using this equality we will be able to calculate the hadronic contributions at the leading order in $1/N_c$ and $1/\lambda$ expansion.

3. Holographic calculations of hadronic contributions

The strong interaction contributions to the muon magnetic moment consist of three pieces; the hadronic vacuum polarization (HLO), the higher-order hadronic vacuum-polarization effect, and the hadronic light-by-light (LBL) scattering.

The HLO contribution (see Fig. 1) is given as[5]

$$a_\mu^{\mathrm{HLO}} = 4\pi^2\left(\frac{\alpha}{\pi}\right)^2 \int_0^\infty dQ^2 f(Q^2)\,\bar{\Pi}_{\mathrm{em}}^{\mathrm{had}}(Q^2)\,, \tag{9}$$

where the vacuum polarization is defined as $\bar{\Pi}_{\mathrm{em}}^{\mathrm{had}}(Q^2) = \Pi_{\mathrm{em}}^{\mathrm{had}}(Q^2) - \Pi_{\mathrm{em}}^{\mathrm{had}}(0)$ and the kernel is

$$f(Q^2) = \frac{m_\mu^2 Q^2 Z^3(1-Q^2 Z)}{1+m_\mu^2 Q^2 Z^2} \quad \text{with } Z = -\frac{Q^2 - \sqrt{Q^4 + 4m_\mu^2 Q^2}}{2m_\mu^2 Q^2}\,. \tag{10}$$

From the AdS/CFT formula (8) the vector current correlator $\Pi_V(q^2)$ can then be expressed in terms of the infinite set of the vector meson wave-functions $\psi_{V_n}(z)$, as

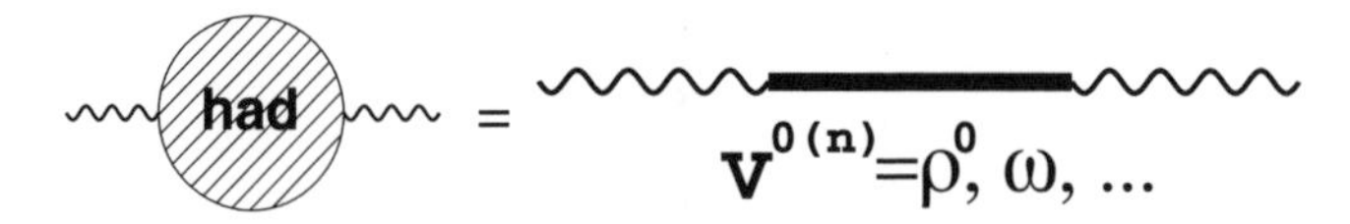

Fig. 2. A diagram illustrating neutral vector meson exchanges giving dominant contributions to Π_V at the large N_c limit.

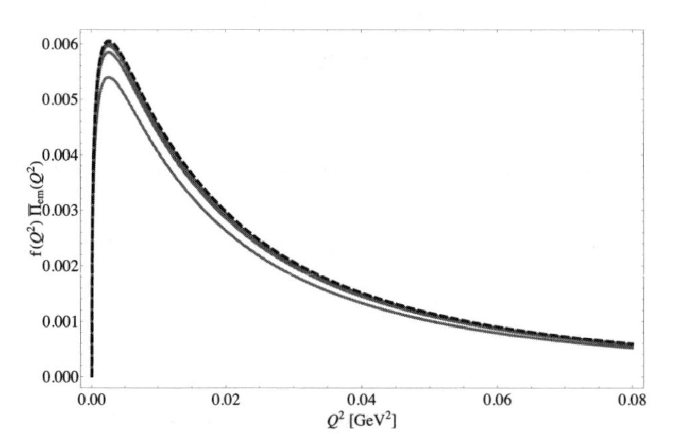

Fig. 3. Comparison of the integral kernel $f(Q^2)\bar{\Pi}_{\rm em}(Q^2)$ which has a peak around $Q^2 = m_\mu^2$: The dashed curve corresponds to the result including full contributions from the infinite tower of vector mesons, while four bold curves are obtained by integrating out the infinite tower at the levels of $n = 1, 2, 3, 4$. The dashed curve is almost (within about 1% deviation) reproduced when $n = 4$. In the plot the number of flavors N_f is taken to be 2.

shown in Fig. 2, which is the holographic realization of the vector meson dominance proposed by Sakurai,

$$\Pi_V(q^2) = \frac{1}{g_5^2}\sum_{n=1}^{\infty}\frac{[\dot{\psi}_{V_n}(\epsilon)/\epsilon]^2}{(q^2 - M_{V_n}^2)M_{V_n}^2}\,, \tag{11}$$

where the dot denotes the derivative with respect to z, $\dot{\psi} \equiv \partial_z\psi$. Using the holographic renormalization to take care of the UV divergence and keeping only the first four low-lying states, we find for Euclidean momentum $Q^2 = -q^2 \geq 0$[14]

$$\bar{\Pi}_V(Q^2) \simeq \sum_{n=1}^{4}\frac{Q^2F_{V_n}^2}{(Q^2 + M_{V_n}^2)M_{V_n}^4} + \mathcal{O}(Q^2/(M_{V_5}^2)) \tag{12}$$

where F_{V_n} is the decay constant of n-th excited vector mesons. Plugging the holographic vacuum polarization (12) into the formula (9), we obtain[14]

$$a_\mu^{\rm HLO}|_{\rm AdS/QCD}^{N_f=2} = 470.5 \times 10^{-10}, \tag{13}$$

which agrees, within 30% errors, with the currently updated value,[15] estimated from new 2009 BaBar data[16]

$$a_\mu^{\rm HLO}[\pi\pi]|_{\rm BABAR} = (514.1 \pm 3.8) \times 10^{-10}\,. \tag{14}$$

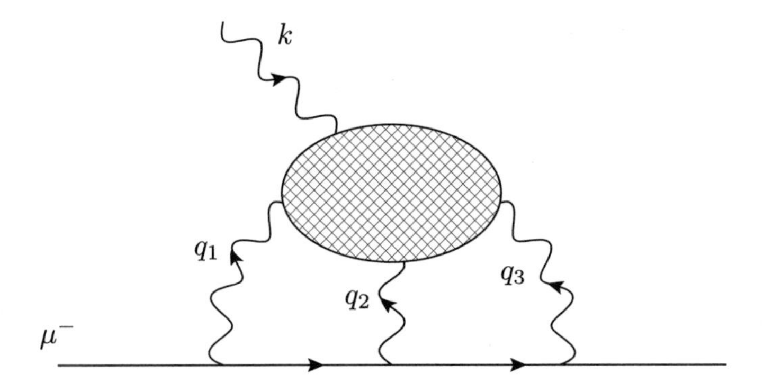

Fig. 4. Light-by-light corrections to muon $g-2$.

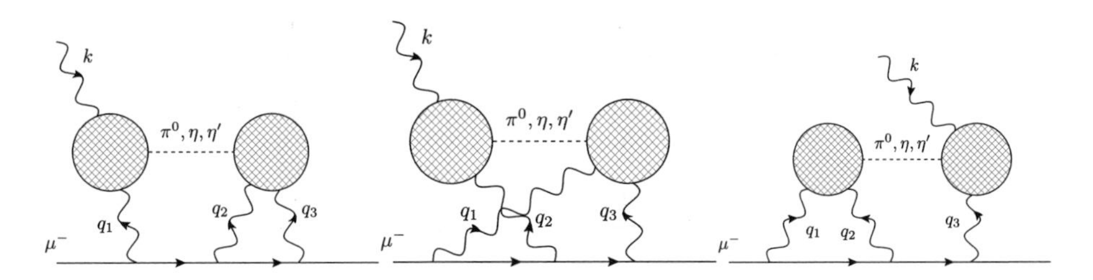

Fig. 5. Light-by-light correction is dominated by the pseudo scalar mesons exchange.

For the hadronic light-by-light corrections, shown in Fig. 4, we need to calculate 4-point functions of flavor currents. Since there is no quartic term for $A_{Q_{\rm em}}$ ($Q_{\rm em} = 1/2 + I_3$), there is no 1PI 4-point function for the EM currents in hQCD, because higher order terms like F^4 or F^2X^2 terms are suppressed. In hQCD the LBL diagram is therefore dominated by VVA or VVP diagrams, which come from the Chern-Simons term, Eq. 7:

$$F_{\gamma^*\gamma^*P(A)}(q_1, q_2) = \frac{\delta^3}{\delta V(q_1)\delta V(q_2)\delta A(-q_1-q_2)} S_{CS} \tag{15}$$

where the gauge fields satisfy in the axial gauge, $V_5 = 0 = A_5$,

$$\left[\partial_z\left(\frac{1}{z}\partial_z V_\mu^{\hat{a}}(q,z)\right) + \frac{q^2}{z}V_\mu^{\hat{a}}(q,z)\right]_\perp = 0\,, \tag{16}$$

$$\left[\partial_z\left(\frac{1}{z}\partial_z A_\mu^{\hat{a}}\right) + \frac{q^2}{z}A_\mu^{\hat{a}} - \frac{g_5^2v^2}{z^3}A_\mu^{\hat{a}}\right]_\perp = 0\,, \tag{17}$$

where $\perp$ denotes the projection onto the transversal components. For two flavors the longitudinal components, $A_{\mu\|}^a = \partial_\mu\phi^a$, and the phase of bulk scalar X are related by EOM as

$$\partial_z\left(\frac{1}{z}\partial_z\phi^a\right) + \frac{g_5^2v^2}{z^3}(\pi^a - \phi^a) = 0\,, \tag{18}$$

$$-q^2\partial_z\phi^a + \frac{g_5^2v^2}{z^2}\partial_z\pi^a = 0\,. \tag{19}$$

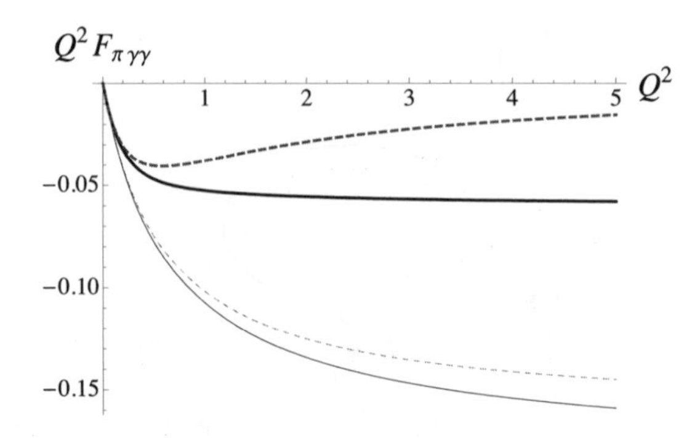

Fig. 6. $F_{\pi\gamma^*\gamma}(Q^2, 0)$ for lower part; $F_{\pi\gamma^*\gamma^*}(Q^2, Q^2)$ for upper part (Brodsky-Lepage).

Table 1. Muon $g-2$ in unit of 10^{-10} from AdS/QCD.

Vector modes	$a_\mu^{\pi^0}$	a_μ^{η}	$a_\mu^{\eta'}$	$a_\mu^{\rm PS}$
4	7.5	2.1	1.0	10.6
6	7.1	2.5	0.9	10.5
8	6.9	2.7	1.1	10.7

The anomalous form factor is then given as,[17] with $\psi^a(z) = \phi^a - \pi^a$ and $J_q = V(iq, z)$

$$F_{\pi\gamma^*\gamma^*} = \frac{N_c}{12\pi^2}\left[\psi(z_m)J(Q_1, z_m)J(Q_2, z_m) - \int_z \partial_z\psi J_{Q_1} J_{Q_2}\right], \tag{20}$$

where we take as boundary conditions $\psi(\epsilon) = 1$ and $\partial_z\psi(z)|_{z_m} = 0$.

To calculate the hadronic LBL contribution to the muon anomalous magnetic moment we expand the photon line in the anomalous form factor (20) as

$$J(-iQ, z) = V(q, z) = \sum_\rho \frac{-g_5 f_\rho \psi_\rho(z)}{q^2 - m_\rho^2 + i\epsilon}$$

The result is shown in Table 1 for several choices of vector-mode truncations.[18] Our results are comparable with recent results by A. Nyffeler,[19] obtained in the LMD+V model;

$$a_\mu^{\rm PS} = 9.9(1.6) \times 10^{-10}\,. \tag{21}$$

4. Discussions

In the era of electroweak precision, it becomes more important to understand precisely QCD corrections to the electroweak processes. As being non-perturbative strong dynamics, QCD corrections are often difficult to estimate and one resorts to lattice calculations, which are, however, not precise enough for certain measurements such as muon anomalous magnetic moment. Recent development in gauge/gravity duality shows that in the large N_c and large λ limit the estimate of QCD contributions can be made precisely in holographic QCD, thus may be useful in assessing the new physics effects. I present recent estimates[14,18] of anomalous magnetic moment of muon in holographic QCD, which are found to be in consistent with other calculations.

Acknowledgments

D.K.H. thanks Doyoun Kim and Shinya Matsuzaki for the collaborations upon which this talk is based on. This work is supported by the Korea Research Foundation Grant funded by the Korean Government (MOEHRD, Basic Research Promotion Fund) (KRF-2007-314- C00052).

References

1. J. S. Schwinger, Phys. Rev. **82**, 664 (1951).
2. G. W. Bennett *et al.* [Muon G-2 Collaboration], Phys. Rev. D **73**, 072003 (2006).
3. For recent five-loop calculations, see T. Aoyama, M. Hayakawa, T. Kinoshita and M. Nio, Phys. Rev. D **78**, 113006 (2008).
4. A. Czarnecki, B. Krause and W. J. Marciano, Phys. Rev. Lett. **76**, 3267 (1996).
5. T. Blum, Phys. Rev. Lett. **91**, 052001 (2003).
6. T. Sakai and S. Sugimoto, Prog. Theor. Phys. **113**, 843 (2005).
7. J. Erlich, E. Katz, D. T. Son and M. A. Stephanov, Phys. Rev. Lett. **95**, 261602 (2005); L. Da Rold and A. Pomarol, Nucl. Phys. B **721**, 79 (2005).
8. J. M. Maldacena, Adv. Theor. Math. Phys. **2**, 231 (1998) [Int. J. Theor. Phys. **38**, 1113 (1999)]; S. S. Gubser, I. R. Klebanov and A. M. Polyakov, Phys. Lett. B **428**, 105 (1998); E. Witten, Adv. Theor. Math. Phys. **2**, 253 (1998).
9. D. K. Hong, T. Inami and H. U. Yee, Phys. Lett. B **646**, 165 (2007).
10. D. K. Hong, M. Rho, H. U. Yee and P. Yi, Phys. Rev. D **76**, 061901 (2007); JHEP **0709**, 063 (2007); Phys. Rev. D **77**, 014030 (2008).
11. J. J. Sakurai, Phys. Rev. Lett. **22**, 981 (1969).
12. E. Katz and M. D. Schwartz, JHEP **0708**, 077 (2007).
13. S. K. Domokos and J. A. Harvey, Phys. Rev. Lett. **99**, 141602 (2007); C. T. Hill, Phys. Rev. D **73**, 085001 (2006).
14. D. K. Hong, D. Kim and S. Matsuzaki, arXiv:0911.0560 [hep-ph].
15. M. Davier, A. Hoecker, B. Malaescu, C. Z. Yuan and Z. Zhang, Eur. Phys. J. C **66**, 1 (2010).
16. B. Aubert *et al.* [BABAR Collaboration], Phys. Rev. Lett. **103**, 231801 (2009).
17. H. R. Grigoryan and A. V. Radyushkin, Phys. Rev. D **77**, 115024 (2008).
18. D. K. Hong and D. Kim, Phys. Lett. B **680**, 480 (2009).
19. A. Nyffeler, Phys. Rev. D **79**, 073012 (2009).

Gauge-Higgs Dark Matter

T. Yamashita

Department of Physics, Nagoya University
Nagoya 464-8602, Japan
E-mail: yamasita@eken.phys.nagoya-u.ac.jp

We discuss a dark matter candidate peculiar in the gauge-Higgs unification scenario*.

1. Introduction

The standard model (SM) of the particle physics successfully explains the experimental results with energies up to the electroweak scale. Nevertheless, we have several motivations to explore the physics beyond the SM. One of them is the hierarchy problem, and another one is the absence of dark matter (DM) candidate. In this talk, we discuss a possible solution to these two issues.

Among the proposals for the former problem, we examine the so-called gauge-Higgs unification scenario, in which the Higgs field is identified with an extra-dimensional component of a higher-dimensional gauge field. This scenario, thus, works in higher-dimensional gauge theories, for example a five dimensional model with the S^1/Z_2 orbifold that we discuss here. To be more concrete, we write the bulk gauge symmetry as G, which is broken at the two endpoints down to H_1 and H_2, respectively. This model can be viewed as four-dimensional theories with infinite copies of the hidden local symmetry G and one H_1 and one H_2, written in a open moose diagram with H_1 (H_2) resides on a one (the other) boundary, via the deconstruction method [2] (latticization). In this view, the extra-dimensional component of the gauge field appears as the "link" field which is the pseudo-Nambu-Goldstone (pNG) modes originated from the process of the "deconstruction" [2]. The diagonalization of the mass matrix for these pNG modes leads to a massless modes on $G/H_1 \cap G/H_2$ (at the tree-level). For instance, if we set G =SO(5) and $H_1 = H_2$ =SO(4) [a], the remaining symmetry is SO(4)$\approx$SU(2)$_L\times$SU(2)$_R$, and the massless pNG appears as a vector of SO(4), that is, a bi-doublet of SU(2)$_L\times$SU(2)$_R$ which can be identified with our Higgs field. Since this pNG mode is for the collective breaking of the infinite symmetry groups, the Higgs mass is protected from ultraviolet divergences, solving the hierarchy problem.

---

*This proceeding is based on our work Ref. [1], to appear in JHEP, in collaboration with Naoyuki Haba, Shigeki Matsumoto and Nobuchika Okada.

[a]In our analysis, we add a U(1) symmetry to reproduce a correct Weinberg angle, in addition to the color SU(3) symmetry.

So far, the latter issue has not been investigated in this scenario, except for a few literatures [3, 4], and so we examine a DM candidate peculiar in this scenario.

We discuss first a new candidate of the DM in general higher-dimensional models, in Sec. 2. Then, we apply this to the gauge-Higgs unification scenario and show that a weakly interacting massive particle (WIMP) can be realized, in Sec. 3.

2. A New Candidate of the Dark Matter

In this section, we claim that the lightest mode among the anti-periodic (AP) fields, if exist, can be stable and thus be a good DM candidate.

In models with the toroidal compactification, no matter with or without further orbifolding, the Lagrangian should be invariant under a discrete coordinate shift along the compactified direction, for example as

$$\mathcal{L}(x, y + 2\pi R) = \mathcal{L}(x, y), \tag{1}$$

where x (y) is the non-compact four (fifth) dimensional coordinate and R is radius of the toroidal compactification. This invariance, or periodicity, represents that the fundamental region of this direction is finite and thus the direction is actually compactified.

A point is that we can impose the AP boundary condition under the shift to a field Φ:

$$\Phi(x, y + 2\pi R) = -\Phi(x, y), \tag{2}$$

Then, they can not appear alone, but always in pairs, in the Lagrangian so as to make it periodic. This means that this Lagrangian has an accidental Z_2 symmetry, under which the AP fields are odd while the periodic fields are even. This accidental symmetry forbids the lightest mode among the AP fields from decaying, and thus the lightest mode can be a good candidate of the DM if it is colorless and neutral.

In this way, if we introduce AP fields in higher-dimensional models, a DM candidate can be accommodated. There are, however, in general no motivations to introduce AP fields, other than serving the DM candidate. It is, of course, not bad, but it is more desirable if there are. In the next section, we discuss a model with such a motivation.

3. Gauge-Higgs Dark Matter

We examine the above new DM candidate in a model of the gauge-Higgs unification.

It is known that the Kaluza-Klein (KK) scale, the top mass and the Higgs mass tend to be too small in this scenario with flat metric, while these can be made heavy enough easily in models with warped metric. In the latter models, as a general property of the warped models, the contributions to the electroweak parameters tend to be too large, and it is claimed in Ref. [5] that the Wilson line phase which corresponds to the Higgs vacuum expectation value (VEV) in this scenario should be small to suppress the contributions. In Ref. [6], we have shown that if we introduce

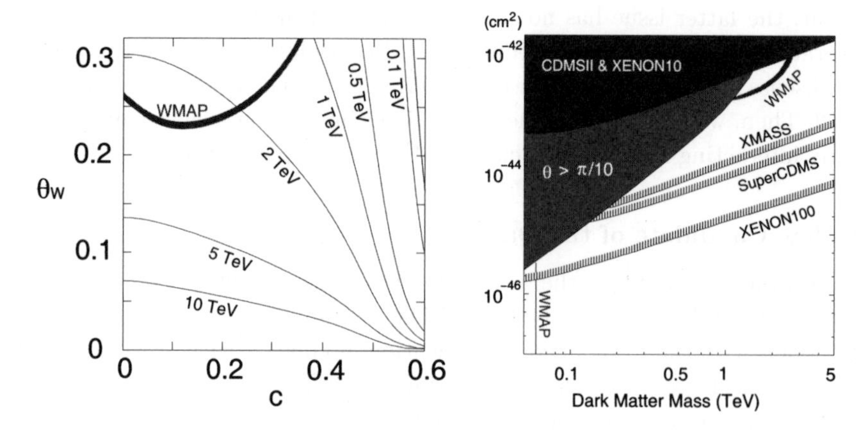

Fig. 1. The relic abundance (left) and the direct detection (right).

AP fermions, a small VEV can be obtained. This is the motivation we mentioned above[b]. The lightest mode among these AP fermions introduced to suppress the contributions is stable and thus a good candidate of the DM, as discussed in Sec. 2.

A bonus by applying this new way in the gauge-Higgs unification scenario is that the relevant interactions of the fermionic DM are the gauge interactions and those to the Higgs field, and both of them are controlled by the gauge symmetry since the Higgs field is a part of the gauge field. This leads to a peculiar predictions [1], and we call this candidate gauge-Higgs dark matter.

We investigate this possibility in a model with the SO(5)×U(1) gauge symmetry which is broken down to SO(4)×U(1) by the boundary conditions, further down to SU(2)×U(1) by a boundary Higgs mechanism and finally to U(1) symmetry by the VEV of the Wilson line phase θ_W [1]. We do not specify the SM fermion sector, just by supposing it goes well, and treat the Higgs mass m_h and θ_W, which should be determined by the loop induced effective potential which is calculable if the fermion sector is also fixed, as free parameters. We introduce a AP fermion with $\mathbf{5}_0$ quantum number of SO(5)×U(1), in addition to the SM fermion sector, as the DM candidate. For this fermion multiplet, we can introduce a parity odd bulk mass term [7], c, which is our third free parameter.

Then, we derive the four-dimensional effective Lagrangian for light KK modes, as function of the three free parameters (m_h, θ_W and c) [1]. Using the Lagrangian, we calculate the coannihilation cross sections, and then solve the Boltzmann equation to predict the thermal relic abundance of the DM candidate. In a similar way, we also evaluate the elastic scattering cross sections between the dark matter particle and nucleon to compare with the bounds from the direct detection experiments. The results are depicted in Figure 1, where the Higgs boson mass is set as $m_h = 120$ GeV. The relic abundances consistent with the observations are drawn in the red

[b]In Ref. [3], with a similar purpose, a similar Z_2 symmetry is imposed by hand.

regions. There are two allowed regions: the very narrow region in upper-right and a band in upper-left of the left figure, where the contours of fixed dark matter masses are also shown. In the former region, the right relic abundance is achieved by the enhancement of the annihilation cross section through the s-channel Higgs boson resonance, so that the dark matter mass is $m_{\rm DM} \simeq m_h/2 = 60$ GeV there. In the other one, the DM mass is around a few TeV, in accordance with the discussion of the WINP scenario. From the right figure, where the present (future expected) bound is shown in brown region (blue lines), we find that a part of the band has been excluded already and will be covered by the coming experiments.

References

1. N. Haba S. Matsumoto, N. Okada and T. Yamashita,
2. deconstruction
3. M. Regis, M. Serone and P. Ullio, JHEP **0703** (2007) 084;
   G. Panico, E. Ponton, J. Santiago and M. Serone, Phys. Rev. D **77** (2008) 115012;
   M. Carena, A. D. Medina, N. R. Shah and C. E. M. Wagner, Phys. Rev. D **79** (2009) 096010.
4. Y. Hosotani, P. Ko and M. Tanaka, Phys. Lett. B **680** (2009) 179.
5. K. Agashe and R. Contino, Nucl. Phys. B **742** (2006) 59.
6. N. Haba, Y. Hosotani, Y. Kawamura and T. Yamashita, Phys. Rev. D **70** (2004) 015010.
7. T. Gherghetta and A. Pomarol, Nucl. Phys. B **586** (2000) 141.

Topological and Curvature Effects in a Multi-Fermion Interaction Model

T. Inagaki and M. Hayashi

Department of Physics, Hiroshima University
Higashi-Hiroshima, Hiroshima 739-8526, Japan

A multi-fermion interaction model is investigated in a compact spacetime with non-trivial topology and in a weakly curved spacetime. Evaluating the effective potential in the leading order of the $1/N$ expansion, we show the phase boundary for a discrete chiral symmetry in an arbitrary dimensions, $2 \leq D < 4$.

Keywords: Eight-fermion interaction, dynamical symmetry breaking.

1. Introduction

It is believed that a fundamental theory with higher symmetry was realized at the early stage of our universe and the symmetry was broken down to the observed symmetry in the present universe. The mechanism of the symmetry breaking may be found in a dynamics of the fundamental theory. Thus the origin of symmetry breaking is quite important to find a consequence of the fundamental theory in cosmological and astrophysical phenomena.

In QCD the chiral symmetry is dynamically broken according to a non-vanishing expectation value for a composite operator constructed by a quark and an anti-quark field, $\bar{q}q$. The chiral symmetry restoration is theoretically predicted in extreme conditions at the QCD scale. The symmetry restoration at high density may be found in phenomena of dense stars. The heavy ion collision at RIHC and LHC provide experimental data for the symmetry restoration at high temperature.

We can apply the dynamical mechanism to symmetry breaking at GUT era. It is one of candidates to describe symmetry breaking of the fundamental symmetry in GUT. It is natural to expect that the broken symmetry is restored in extreme conditions at GUT scale. Thus we have launched a plan to study the symmetry restoration at high temperature, high density and strong curvature. A topological effect is also interesting before inflationally expansion of our spacetime. In this paper we focus on the topological and the curvature effects.

A variety of works has been done in a simple four-fermion interaction model. The topological effect has been investigated in the spacetime with one compactified dimension, S^1,[1] and the torus universe.[2–6] It has been found that the finite size effect restore the broken symmetry if we adopt the anti-periodic boundary condition to the fermion fields. On the other hand, the fermion fields which possess the periodic boundary condition contribute to break the symmetry. The curvature effects has been studied in two,[7,8] three,[9] four[10] and arbitrary dimensions.[11] The broken symmetry is restored if the spacetime curvature is positive and strong enough. However, the symmetry is always broken in a negative curvature spacetime.[12] A combined effect has been also discussed in a weakly curved spacetime,[13] the maximally symmetric spacetime[14,15] and Einstein space.[16,17] For a review, see for example Ref. 18.

In these works the four-fermion interaction model is considered to have something essential as a low energy effective model of a fundamental theory. To discuss the model dependence or independence of above results we have to extend the four-fermion interaction model. Here we consider a multi-fermion interaction model[19–23] as a simple extension of the four-fermion interaction model and study the contribution from a higher dimensional operator. In Sec. 2 we introduce a multi-fermion interaction model which is considered in this paper. We show an explicit expression of the effective potential in an arbitrary dimensions, $2 \leq D < 4$. We consider a spacetime $R^{D-1} \otimes S^1$ in Sec. 3. Evaluating the effective potential, we study the topological effect. In Sec. 4 we assume that the spacetime curves slowly and investigate the curvature effect. In Sec. 5 we give some concluding remarks.

2. Multi-Fermion Interaction Model

As in well-known, the chiral symmetry is dynamically broken by a simple scalar type four-fermion interaction model.[24,25] It is a useful low energy effective theory of QCD to describe meson properties. The four-fermion interaction model is also useful as a simple toy model in the study of low energy phenomena of the strong coupling gauge theory at high energy scale in various environments. But there is no reason to neglect higher dimensional operators in extreme conditions at the early universe.

In the present paper we extend the model to include scalar type multi-fermion interactions,

$$S = \int d^D x \sqrt{-g} \left[\sum_{l=1}^{N} \bar{\psi}_l i \gamma^\mu (x) \nabla_\mu \psi_l + \sum_{k=1}^{n} \frac{G_k}{N^{2k-1}} (\sum_{l=1}^{N} \bar{\psi}_l \psi_l)^{2k} \right], \tag{1}$$

where index l represents flavors of the fermion field ψ, N is the number of fermion flavors. We neglect the flavor index below. The multi-fermion interaction is unrenormalizable in four spacetime dimensions. The model depends on regularization methods. In this paper we adopt the dimensional regularization and regards the spacetime dimension, D, for the integration of internal fermion lines as one of parameters in the effective theory.[26–30] In QCD it can be fixed to reproduce meson properties. Here we leave it as an arbitrary parameter to be fixed phenomenologically.

The action (1) possesses the discrete chiral symmetry, $\bar{\psi}\psi \to -\bar{\psi}\psi$, and the global $SU(N)$ flavor symmetry, $\psi \to e^{i \sum_a \theta_a T_a} \psi$. The discrete chiral symmetry prohibits the fermion mass term. We can adopt the $1/N$ expansion as a non-perturbative approach to investigate the dynamical symmetry breaking under the global $SU(N)$ symmetry.

For practical calculations it is more convenient to introduce the auxiliary fields[31] and start from the action,

$$S_y = \int d^D x \sqrt{-g} \left[\bar{\psi} i \gamma^\mu(x) \nabla_\mu \psi + \sum_{k=1}^{n} \frac{N G_k \sigma^{2k}}{(2G_1)^{2k}} - \frac{N}{2G_1} s \left(\sigma + \frac{2G_1}{N} \bar{\psi}\psi \right) \right] . \quad (2)$$

The multi-fermion interactions in the original action are replaced by the auxiliary fields σ and s. If the auxiliary field s develops a non-vanishing expectation value, the fermion field acquires a mass term and the chiral symmetry is eventually broken.

In this paper we only consider the case $n = 2$ for simplicity and concentrate on the contribution from the eight-fermion interaction. To study the phase structure of the model we calculate the expectation value for the auxiliary field, s. It is obtained by observing the minimum of the effective potential. It should be noted that the expectation value for the composite operator $\bar{\psi}\psi$ is given by the value of the auxiliary field σ at the minimum of the effective potential. In the leading order of the $1/N$ expansion we can analytically integrate out the fermion field and get the effective potential,

$$V(s, \sigma) = -\frac{N}{4G_1}(\sigma - s)^2 + \frac{1}{4G_1} s^2 - \frac{N G_2}{16 G_1^4} \sigma^4 + i \mathrm{Tr} \ln \langle x | \left[i \gamma^\mu(x) \nabla_\mu - s \right] | x \rangle . \quad (3)$$

The trace for the Dirac operator in Eq.(3) depends on the spacetime structure. The minimum of the effective potential satisfies the gap equation,

$$\left. \frac{\partial V}{\partial \sigma} \right|_s = \left. \frac{\partial V}{\partial s} \right|_\sigma = 0. \quad (4)$$

If we consider the four-fermion interaction model, $G_2 = 0$, in the Minkowski spacetime, R^D, the gap equation allows a nontrivial solution only for a negative G_1. The solution is give by

$$\sigma = s = m_0 \equiv \left(\frac{(4\pi)^{D/2}}{\mathrm{tr} 1 \Gamma\left(1 - D/2\right)} \frac{1}{2G_1} \right)^{1/(D-2)} . \quad (5)$$

Before investigating topological and curvature effects, we numerically calculate the effective potential in the Minkowski spacetime. We are interested in the model where the primordial symmetry is broken down at low energy scale. Thus we confine ourselves to a case of a negative G_1. In this case the effective potential (3) only depends on m_0 and the rate of the coupling constants, G_2/G_1^3. We normalize all the mass scales by m_0 and set $g = G_2 m_0^2 / G_1^3$. As is seen in Fig. 1, a positive g suppresses the symmetry breaking, while a negative g enhances it. A new local minimum appears for a negative g.

3. Phase Structure in a Spacetime with Non-Trivial Topology

One of extreme conditions we have to consider at the early universe is the topological effect. It is expected that the boundary condition for matter fields restricts how to

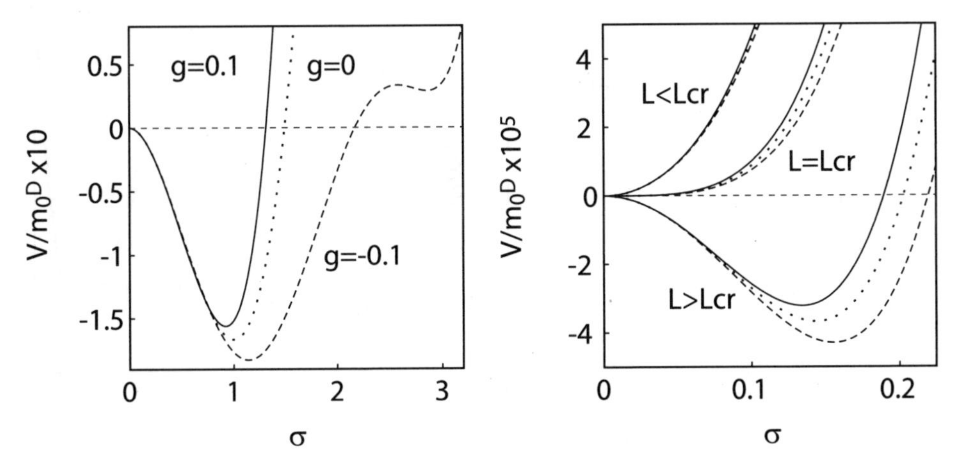

Fig. 1. Behavior of the effective potential in the tree dimensional Minkowski spacetime R^3. We set $g = 0.1, 0$ and -0.1 and draw the solid, dotted and dashed lines, respectively.

Fig. 2. Behavior of the effective potential on $R^2 \otimes S^1$ for $\delta_{p,1} = 1$, as L varies. We set $g = 0.1, 0$ and -0.1 and draw the solid, dotted and dashed lines, respectively.

compactify the spacetime. In this section we assume that one of space directions is compactified and investigate the multi-fermion interaction model on the cylindrical spacetime, $R^{D-1} \otimes S^1$. It is a flat spacetime with a non-trivial topology.

On $R^{D-1} \otimes S^1$ the effective potential, (3) is given by

$$\frac{V(s,\sigma)}{Nm_0^D} = -\frac{\mathrm{tr}1}{2(4\pi)^{D/2}}\Gamma\left(1-\frac{D}{2}\right)\left(\frac{(\sigma-s)^2}{m_0^2} - \frac{s^2}{m_0^2} + \frac{g}{4}\frac{\sigma^4}{m_0^4}\right)$$
$$+\frac{\mathrm{tr}1}{2(4\pi)^{(D-1)/2}}\Gamma\left(\frac{1-D}{2}\right)\frac{1}{Lm_0}\sum_{n=-\infty}^{\infty}\left[\left(\frac{(2n+\delta_{p,1})\pi}{Lm_0}\right)^2 + \frac{s^2}{m_0^2}\right]^{(D-1)/2}, \quad (6)$$

where we set $\delta_{p,1} = 0$ for the periodic and $\delta_{p,1} = 1$ for the anti-periodic boundary conditions. In Fig. 2 the effective potential is shown for $\delta_{p,1} = 1$, as L varies. It is clearly seen that the broken symmetry is restored through the second order phase transition as L is decreased. On the other hand only the broken phase is realized for the periodic boundary condition.

To see the situation more precisely we perform a rigorous analysis on the critical length L_{cr} for $\delta_{p,1} = 1$ and find the boundary which divides the symmetric and the broken phases. For a non-negative g only the second order phase transition is realized. In this case we can find the explicit expression for the critical length,[18]

$$L_{cr}m_0 = 2\pi\left[\frac{2\Gamma((3-D)/2)}{\sqrt{\pi}\Gamma(1-D/2)}(2^{3-D}-1)\zeta(3-D)\right]^{1/(D-2)}. \quad (7)$$

It is illustrated by the solid line in Fig. 3. The second local minimum appears for a negative g. The dashed line in Fig. 3 shows the length L where the effective potential satisfies $V = 0$ at the second local minimum for $g = -0.1$. We plot the effective potential on this dashed line in Fig. 4. The broken symmetry is restored below both the solid and the dashed lines on $D-g$ plane for $g = -0.1$.

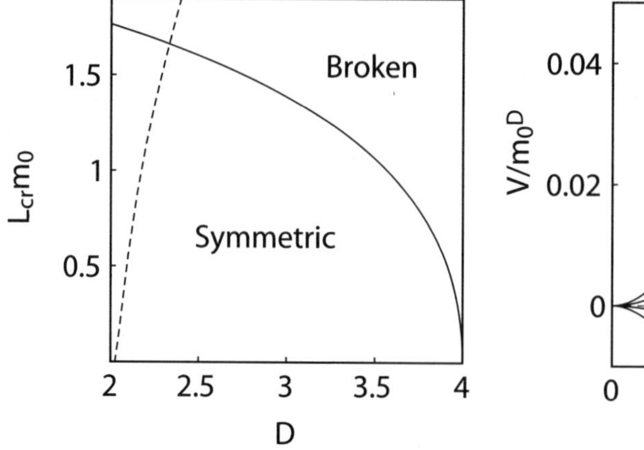

Fig. 3. Critical length. The dashed line shows the length to satisfies $V = 0$ at the second local minimum for $g = -0.1$.

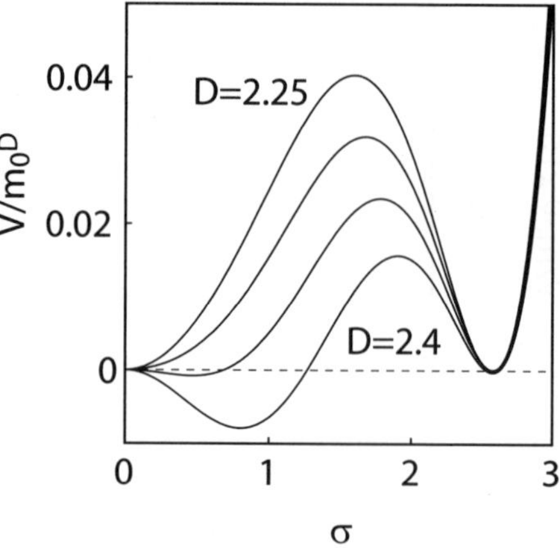

Fig. 4. Behavior of the effective potential on the dashed line in Fig. 3 for $D = 2.25, 2.3, 2.35$ and 2.4.

4. Phase Structure in a Weakly Curved Spacetime

At the early universe the symmetry breaking may be induced under the influence of the spacetime curvature. In this section we assume that the spacetime curved slowly and keep only terms independent of the curvature R and terms linear in R. We discuss the curvature induced phase transition by observing the minimum of the effective potential. Following the procedure developed in Ref. 11, we obtain the effective potential up to linear in R,

$$\frac{V(s,\sigma)}{Nm_0^D} = -\frac{\text{tr}1}{2(4\pi)^{D/2}}\Gamma\left(1-\frac{D}{2}\right)\left(\frac{(\sigma-s)^2}{m_0^2}-\frac{s^2}{m_0^2}+\frac{g}{4}\frac{\sigma^4}{m_0^4}\right)$$
$$-\frac{\text{tr}1}{(4\pi)^{D/2}D}\Gamma\left(1-\frac{D}{2}\right)\frac{|s|^D}{m_0^D}-\frac{\text{tr}1}{(4\pi)^{D/2}}\frac{R}{24m_0^2}\Gamma\left(1-\frac{D}{2}\right)\frac{|s|^{D-2}}{m_0^{D-2}}. \tag{8}$$

In the four-fermion interaction model, $g = 0$, the phase transition takes place by varying the curvature R. The broken symmetry is restored for a large positive curvature $R > R_{cr} \geq 0$. The phase transition is of the first order for $2 < D < 4$. We evaluate the solution of the gap equation and find the analytic expression for the critical curvature,[11]

$$\frac{R_{cr}}{m_0^2} = 6(D-2)\left(\frac{(4-D)D}{4}\right)^{(4-D)/(D-2)}. \tag{9}$$

It is drawn as a function of the spacetime curvature in Fig. 5.

The critical curvature R_{cr} is modified by the eight-fermion interaction. For a positive g we find a smaller R_{cr} except for two and four dimensions. The dashed line shows the curvature to satisfy $V = 0$ at the second minimum for $g = -0.1$. Thus the symmetric phase is realized above both the solid and the dashed lines for $g = -0.1$. In Fig. 6 we show the behavior of the effective potential near the point A

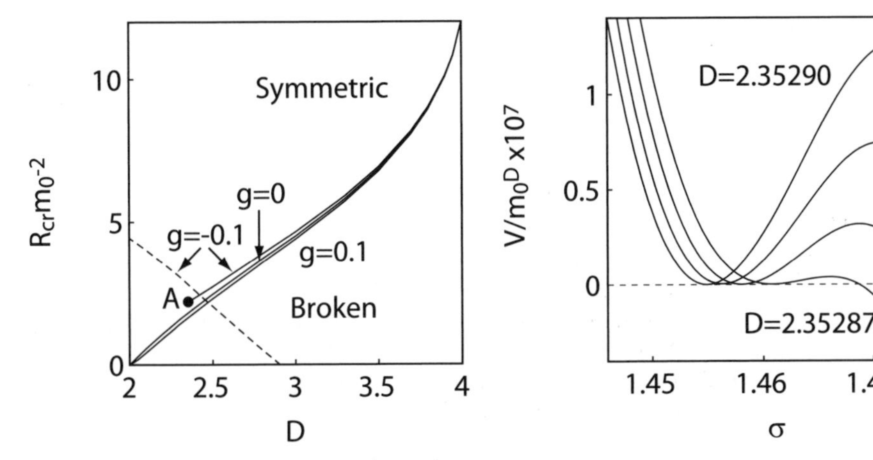

Fig. 5. Critical curvature for $g = -0.1, 0$ and 0.1. The dashed line shows the curvature to satisfies $V = 0$ at the second local minimum.

Fig. 6. Behavior of the first local minimum of the effective potential on the solid line for $g = -0.1$ near the point A in Fig.5.

on the solid line in Fig. 5. It is observed that the first local minimum disappears at the point A. Finally we noted that only the broken phase is realized in a spacetime with a negative curvature.[12]

5. Conclusion

We have investigated the multi-fermion interaction model under the influence of the spacetime topology and curvature. In the practical calculation the four- and the eight-fermion interactions is studied. Evaluating the effective potential in the leading order of the $1/N$ expansion, we show the phase structure of the model in an arbitrary dimension, $2 \leq D < 4$.

On $R^{D-1} \otimes S^1$ the finite size effect restores the broken chiral symmetry for fermion fields with the anti-periodic boundary condition. In this case the theory is equivalent to the finite temperature field theory. The phase transition is of the second order for a non-positive g. We found that the phase boundary is not modified by the eight-fermion interaction with a positive g. For a negative g an additional local minimum contributes to the phase boundary. A new boundary appears in lower dimensions and the first order phase transition takes place.

The spacetime curvature also contributes to the phase structure of the theory. For $2 \leq D < 4$ the broken symmetry is restored through the first order phase transition as increasing the curvature R. The eight-fermion interaction with a positive g suppresses the chiral symmetry breaking and changes the phase boundary. For a negative g we observe a contribution from an additional local minimum to enhance the broken symmetry in lower dimensions.

In the present paper our study is restricted on the analysis of the phase structure of the theory. We are interested in applying our results to critical phenomena at the early universe. We will continue our work and hope to report on these problems.

Acknowledgment

The authors would like to thank H. Takata, D. Kimura, Y. Kitadono and Y. Mizutani for fruitful discussions. T. I. is supported by the Ministry of Education, Science, Sports and Culture, Grant-in-Aid for Scientific Research (C), No. 18540276, 2009.

References

1. A. S. Vshivtsev, K. G. Klimenko and B. V. Magnitsky, Phys. Atom. Nucl. **59**, 529 (1996) [Yad. Fiz. **59**, 557 (1996)].
2. S. K. Kim, W. Namgung, K. S. Soh and J. H. Yee, Phys. Rev. D **36**, 3172 (1987).
3. D. Y. Song and J. K. Kim, Phys. Rev. D **41**, 3165 (1990).
4. D. K. Kim, Y. D. Han and I. G. Koh, Phys. Rev. D **49**, 6943 (1994).
5. K. Ishikawa, T. Inagaki, K. Yamamoto and K. Fukazawa, Prog. Theor. Phys. **99**, 237 (1998).
6. L. M. Abreu, M. Gomes and A. J. da Silva, Phys. Lett. B **642**, 551 (2006).
7. H. Itoyama, Prog. Theor. Phys. **64**, 1886 (1980).
8. I. L. Buchbinder and E. N. Kirillova, Int. J. of Mod. Phys. A4, 143 (1989).
9. E. Elizalde, S. D. Odintsov and Yu. I. Shil'nov, Mod. Phys. Lett. **A9**, 913 (1994).
10. T. Inagaki, T. Muta and S. D. Odintsov, Mod. Phys. Lett. **A8**, 2117 (1993).
11. T. Inagaki, Int. J. of Mod. Phys. **A11**, 4561 (1996).
12. E. V. Gorbar, Phys. Rev. D **61**, 024013 (2000).
13. E. Elizalde, S. Leseduarte and S. D. Odintsov, Phys. Rev. **D49**, 5551 (1994).
14. T. Inagaki, S. Mukaigawa, T. Muta, Phys. Rev. **D52**, 4267 (1995).
15. E. Elizalde, S. Leseduarte, S. D. Odintsov and Yu. I. Shilnov, Phys. Rev. **D53**, 1917 (1996).
16. K. Ishikawa, T. Inagaki, T. Muta, Mod. Phys. Lett. **A11**, 939 (1996).
17. D. Ebert, K. G. Klimenko, A. V. Tyukov and V. C. Zhukovsky, Eur. Phys. J. C **58**, 57 (2008).
18. T. Inagaki, T. Muta and S. D. Odintsov, Prog. Theor. Phys. Suppl. **127**, 93 (1997).
19. G. 't Hooft, Phys. Rev. **D14**, 3432 (1976); **D18**, 2199 (E) (1978); Phys. Rep. **142**, 357 (1986).
20. R. Alkofer and I. Zahed, Phys. Lett. B **238**, 149 (1990).
21. J. Moreira, A. A. Osipov, B. Hiller, A. H. Blin, J. Providencia, Annals of Physics **322**, 2021 (2007); A. A. Osipov, B. Hiller, A. H. Blin, J. da Providencia, Phys. Lett. **B650**, 262 (2007); A. A. Osipov, B. Hiller, J. Moreira, A. H. Blin, J. da Providencia, Phys. Lett. **B646**, 91 (2007).
22. M. Hayashi, T. Inagaki and H. Takata, to be published in Int. J. Mod. Phys. **A**, arXiv:0812.0900 (hep-ph).
23. T. Inagaki, *The Problems of Modern Cosmology*, ed. P. M. Lavrov, (Tomsk Pedagogical University Press, 2009), pp. 214-221.
24. Y. Nambu and G. Jona-Lasinio, Phys. Rev. **124**, 246 (1961).
25. D. J. Gross and A. Neveu, Phys. Rev. **D10**, 3235 (1974).
26. S. Krewald, K. Nakayama, Ann. Phys. **216**, 201 (1992).
27. H.-J. He, Y.-P. Kuang, Q. Wang, Y.-P. Yi, Phys. Rev. **D45**, 4610 (1992).
28. T. Inagaki, T. Kouno and T. Muta, Int. J. Mod. Phys. **A10**, 2241 (1995).
29. R. G. Jafarov, V. E. Rochev, Russ. Phys. J. **49**, 364 (2006).
30. T. Inagaki, D. Kimura and A. Kvinikhidze, Phys. Rev. **D77**, 116004 (2008); T. Fujihara, D. Kimura, T. Inagaki and A. Kvinikhidze, Phys. Rev. **D79**, 096008 (2009).
31. H. Reinhardt, R. Alkofer, Phys. Lett. **B207**, 482 (1988).

A Model of Soft Mass Generation

J. Hošek

Department of Theoretical Physics, Nuclear Physics Institute
Academy of Sciences of the Czech Republic
25068 Řež (Prague), Czech Republic
E-mail: hosek@ujf.cas.cz

We replace the Higgs sector of the electroweak gauge $SU(2)_L \times U(1)_Y$ model of three fermion families with its numerous free parameters by a horizontal gauge $SU(3)_F$ quantum flavor dynamics with eight flavor gluon fields C_a^μ and one coupling constant h. We suggest that the new strong low momentum dynamics generates spontaneously the masses of its eight flavor gluons, of leptons and quarks, and of the intermediate W and Z bosons. Absence of axial anomalies requires neutrino right-handed electroweak singlets and brings into the model a new non-Abelian global symmetry. Because the crucial test of the model, the reliable computation of the fermion mass spectrum is not available, we briefly discuss the experimental implications of the resulting neutrino sector.

Keywords: Gauge principle, dynamical symmetry breaking, Majorana neutrinos.

1. Introduction

Spontaneous (i.e. soft) fermion and intermediate boson mass generation in the electroweak gauge $SU(2)_L \times U(1)_Y$ interactions is a necessity: Hard fermion and W and Z masses ruin the unitary behavior of scattering amplitudes with longitudinally polarized intermediate vector bosons. Some interactions/concepts responsible for soft mass generation are necessary. With the advent of the LHC these will be harshly tested experimentally.

We consider the glorious, weakly coupled Higgs sector responsible for the spontaneous mass generation in the Standard model beautiful but phenomenological by construction: (1) It contains too many theoretically arbitrary and numerically vastly different parameters. (2) The elementary scalar fields are unnatural. (3) It is a tree-level effect. (4) The fermion and the intermediate boson masses are not related.

Phenomenological interpretation of the Higgs mechanism means that the massive spinless particle of the elementary real scalar field with properties given by the Standard model Lagrangian does not exist.

Alternatives to the SM Higgs mechanism are numerous.[1] If weakly coupled like SUSY extensions they parameterize the data with too many parameters. If strongly coupled like technicolor-like extensions they do not allow for quantitative computations of their properties and, frankly, have also many parameters. Fortunately, life with QCD taught us to be modest. New concepts like those referring to extra dimensions do not provide, to the best of our knowledge, ordinary operational framework. Our suggestion[2] is conservative and nasty: We gauge the $SU(3)$ flavor (family) index and argue that the resulting non-Abelian quantum $SU(3)_F$ flavor dynamics (QFD) is a viable[3] candidate for spontaneous electroweak symmetry breaking i.e., for the dynamical generation of both the fermion (lepton and quark) and intermediate vector boson masses.

2. Basic reasoning

(i) Because 'in QCD we trust' the $SU(3)_F$ must not be vector-like. For if it were it would be confining. Instead, referring to[4] we argue that all flavor gluons acquire masses by Schwinger mechanism[5] i.e., by their self-interactions and by interactions with fermions. Operationally, in the transverse flavor gluon polarization tensor

$$\Pi^{\mu\nu}_{ab}(q) \equiv (q^2 g^{\mu\nu} - q^\mu q^\nu)\Pi_{ab}(q^2) \tag{1}$$

the scalars Π_{ab} develop dynamically the massless poles. Mediating the flavor changing electric charge conserving processes the flavor gluons have to be rather heavy. For definiteness we take $M_a \sim 10^6$GeV.
(ii) The flavor gluon exchanges, strong at low momenta, build up the bridges (see Fig.1.) between the left- and right-handed fermion fields i.e., the fermion masses. Operationally, they are described by the chiral symmetry changing fermion proper self energies $\hat{\Sigma}$ in the inverse fermion propagators $S(p)^{-1} = \not{p} - \hat{\Sigma}(p^2)$ where $\hat{\Sigma} = \Sigma P_L + \Sigma^+ P_R$ and $P_{L,R} = \frac{1}{2}(1 \mp \gamma_5)$. Such matrices have an inverse

$$S(p) = (\not{p} + \Sigma^+)(p^2 - \Sigma\Sigma^+)^{-1}P_L + (\not{p} + \Sigma)(p^2 - \Sigma^+\Sigma)^{-1}P_R \tag{2}$$

Basically, the charged lepton and quark masses of three electroweakly identical fermion families differ due to a unique assignment of the the chiral fermion multiplets to triplet and antitriplet representations of QFD. The choice is in fact unique: Take q_L as an $SU(3)_F$ triplet (i.e. both u_L and d_L are triplets). Then (u_R, d_R) can be either $(3, \bar{3})$ or $(\bar{3}, 3)$, since for the choices $(3, 3)$ and $(\bar{3}, \bar{3})$ the mass matrices of the u- and d-type quarks would come out equal. Without lack of generality choose $(u_R, d_R) = (3, \bar{3})$. Then l_L (i.e. both ν_L and e_L) cannot be a triplet. For if it were the charged lepton mass matrix would be equal either to the u-type or the d-type quark matrix. Hence, l_L (i.e. both ν_L and e_L) must be an antitriplet. Let e_R be a triplet (case I). At this point we impose the the field theory restriction of the the absence of axial anomalies.[6] It requires introduction of *three neutrino right-handed flavor triplets*, ν_{NR}, $N = 1, 2, 3$. For definiteness we conside this case here. (e_R can also be an antitriplet (case II). Anomaly freedom requires introduction of *five neutrino right-handed flavor triplets*, ν_{NR}, $N = 1, ..., 5$. Knowledge of the charged lepton mass spectrum would select uniquely one of the two possibilities.)

We will show that within the given electric charge the fermion masses differ due to the low-momentum effective sliding coupling depending upon the flavor gluon mass matrix.

(iii) *Dynamically generated fermion masses break spontaneously the electroweak $SU(2)_L \times U(1)_Y$ symmetry down to $U(1)_{em}$.* The resulting three multicomponent 'would-be' composite Nambu-Goldstone bosons become the longitudinal polarization states of the intermediate bosons W and Z. Because of this realization of the general Schwinger mechanism[5] the intermediate boson masses are expressed in terms of the fermion self-energies by sum rules.
(iv) Neutrinos are peculiar in our model.[7] Dynamical generation of both Dirac and Majorana mass terms results in 12 massive Majorana neutrinos and in 9 massless composite Majorons.[8]

3. Flavor gluon mass generation

Here we follow the analysis of Ward identities with flavor gluons in accordance with:[4] Divergences of the full vertices $\Gamma^{\mu\nu\lambda}_{abc}(p+q,p)$ (three-flavor-gluon vertex) and $\Gamma^{f;\mu}_{ij;c}(p+q,p)$ (fermion-flavor-gluon vertex) at vanishing momenta are expressed in terms of the full inverse flavor gluon and fermion propagators i.e., in terms of the self energies Π and Σ. We assume that the ghost propagators do not play any dynamical role in the generically nonperturbative reasoning. This assumption is manifest in the 'pinch technique'.[9] If the symmetry is unbroken the Ward identities are fulfilled trivially. *If Π_{ab} and Σs develop the symmetry breaking parts the validity of the Ward identities requires the massless poles in the vertices themselves.* They correspond to the 'would-be' NG bosons composed by construction from both flavor gluons and from all fermion species in the world:

$$\Gamma^{\mu\nu\lambda}_{abc}(p+q,p)|_{pole} = P^{\nu\lambda}_{bc;d}(p+q,p)\frac{i}{q^2}h(-iq^\mu)\Lambda_{da}(q^2) \tag{3}$$

$$\Gamma^{f;\mu}_{ij;a}(p+q,p)|_{pole} = P^{f}_{ij;d}(p+q,p)\frac{i}{q^2}h(-iq^\mu)\Lambda_{da}(q^2) \tag{4}$$

where

$$-iq^\mu\Lambda_{da}(q^2) \equiv [I^\mu_{C;da}(q) + \sum_f I^\mu_{f;da}(q)] \tag{5}$$

Physical interpretation of this decomposition should be clear: (1) There are eight 'would-be' NG bosons composed both of the flavor gluons and of all fermions in the model. (2) $P^{\nu\lambda}_{bc;d}$ is the effective coupling of the NG boson with flavor gluons. (3) $P^f_{ij;d}$ is the effective coupling of the NG boson with the fermion f. (4) $I^\mu_{C;da}(q)$ and $I^\mu_{f;da}(q)$ are the vectorial tadpole UV finite loop integrals. They convert in terms of the effective vertices P, the elementary vertices and the full flavor gluon and fermion propagators both the flavor gluon components and the fermion components of the 'would-be' NG bosons to the flavor gluons. (5) The crucial effective bilinear derivative vertex between the flavor gluon octet and the 'would-be' NG boson octet is given by (5).

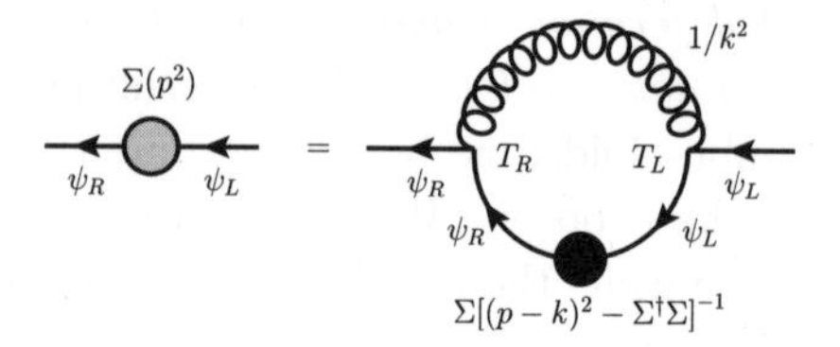

Fig. 1. Structure of the SD equation for chirality changing Σ of a fermion (charged lepton or quark) ψ. $T_{L,R}$ are the triplet $(\frac{1}{2}\lambda)$ or antitriplet $(-\frac{1}{2}\lambda^*)$ generators of a given chiral fermion $\psi_{L,R}$.

The vertex (5) gives rise to the massless pole in the longitudinal part of the flavor gluon polarization tensor (1). Although its transversality is saved by contributions which we cannot compute explicitly, it follows from it that the flavor gluon mass matrix $M^2_{ab}(q^2)$ is given by the formula[4]

$$-q^2\Pi_{ab}(q^2) \equiv M^2_{ab}(q^2) = \sum_d \Lambda_{ad}(q^2)\Lambda_{bd}(q^2) \tag{6}$$

Practical applications will demand diagonalization of the mass matrix $M^2_{ab}(0)$ and introduction of the flavor gluon mass eigenstates.

4. Fermion mass generation

Structure of the SD equations for the chiral symmetry changing fermion proper self energies $\Sigma(p^2)$ (NJL type self-consistency condition[10]) of electrically charged fermions is easily read off the interaction Lagrangian of QFD using the general form of the massive fermion propagator. It is shown in Fig.1. The (bare) flavor gluon propagator is taken in the Feynman gauge as suggested by the pinch technique.[9] The integration in Fig.1. extends over all momenta and the SD equations *must*[3] be improved by taking into account the momentum-dependent sliding coupling $\bar{h}^2_{ab}(k^2)$ vanishing at high momenta. We know no way of knowing $\bar{h}^2_{ab}(k^2)$ at low momenta other than solving the theory.[11] It is likely that it is dominated by the exchanges of the composite 'would-be' NG bosons with the effective vertices $P^{\nu\lambda}_{bc;d}(p+q,p)$ and $P^f_{ij;d}(p+q,p)$ to both gluons and fermions, respectively.

To proceed we write

$$\frac{1}{k^2} = \frac{1}{k^2}\{[1+\Pi(k^2)]^{-1} + \Pi(k^2)[1+\Pi(k^2)]^{-1}\} \tag{7}$$

and argue as follows:

(1) At high momenta Π is given by perturbation theory. The first term in (7) when used in Fig.1. gives rise to flavor insensitive Σ due to massless flavor gluon exchange with asymptotically free flavor insensitive interaction strength $\bar{h}^2(k^2) = h^2[1+\Pi(k^2)]^{-1}$. The corresponding SD equation giving the high-momentum asymptotics of Σ was studied in QCD in detail in.[12] The second term in (7) corresponding to massless gluon exchange with the bare charge should be ignored.

(2) At low momenta the non-perturbative Π is given by (6). The first term in (7) corresponds to the massive gluon exchanges with a bare charge ($\frac{1}{k^2}(1+\Pi)^{-1} = (k^2 - M^2)^{-1}$) and in Fig.1. it should be ignored. The second term in (7) when used in Fig.1. gives rise to Σs due to massless gluon exchange with the low-momentum $\bar{h}^2_{ab}(k^2)$ running to a non-perturbative IR stable fixed point:

$$\bar{h}^2_{ab}(k^2) = h^2_*[\Pi(k^2)(1+\Pi(k^2))^{-1}]_{ab}\,. \tag{8}$$

The corresponding matrix SD equations, still rather schematic, hopefully illustrate the main point: The symmetry-breaking form of Π_{ab} implies that *at low momenta* the fermion self energies Σ differ in different flavor channels by the low-momentum flavor sensitive interaction strengths (8), basically due to the low momentum symmetry breaking flavor gluon self energy.

At this exploratory stage we are merely able to illustrate that the low momentum Ansatz (8) for the sliding coupling is bona fide responsible for strong suppression of the fermion mass with respect to the huge flavor gluon mass. We replace both $M^2_{ab}(k^2)$ and the fermion self energies $\Sigma_{ij}(p^2) = \Sigma_{ij}(0) = m_{ij}$ by real numbers M^2 and m, respectively. The SD equation in Fig.1. turns into an algebraic equation

$$m = \frac{h^2_*}{16\pi^2}\int_0^\infty dk^2 \frac{M^2}{k^2+M^2}\frac{m}{k^2+m^2} \tag{9}$$

with solution $m = M\exp[-8\pi^2/h^{*2}]$. Indeed, for $M = 10^6$ GeV the "neutrino" mass $m_\nu = 10^{-9}$ GeV is obtained with an effective interaction strength $h^2_\nu/4\pi = 2\pi/15\ln 10$, and the "top quark" mass $m_t = 10^2$ GeV with $h^2_t/4\pi = 2\pi/4\ln 10$.

5. Intermediate boson mass generation

Masses of the W and Z vector bosons are the necessary consequence of the dynamically generated fermion masses: The fermion proper self-energies $\Sigma(p^2)$ generated by strong QFD break spontaneously also the 'vertical' $SU(2)_L \times U(1)_Y$ symmetry down to $U(1)_{em}$. The properties of three composite 'would-be' NG bosons can be extracted from the $SU(2)_L \times U(1)_Y$ Ward identities.[13,14] For simplicity we consider the fermion proper self energies diagonal.

$$\Gamma^\alpha_W(p+q,p) = \frac{g}{2\sqrt{2}}\{\gamma^\alpha(1-\gamma_5) - \frac{q^\alpha}{q^2}[(1-\gamma_5)\Sigma_U(p+q) - (1+\gamma_5)\Sigma_D(p)]\},$$

$$\Gamma^\alpha_Z(p+q,p) = \frac{g}{2\cos\theta_W}\{t_3\gamma^\alpha(1-\gamma_5) - 2Q\gamma^\alpha\sin^2\theta_W - \frac{q^\alpha}{q^2}t_3[\Sigma(p+q)+\Sigma(p)]\gamma_5\}.$$

When the electroweak gauge interactions are switched on as weak external perturbations the W and Z bosons dynamically acquire masses. Their squares are defined as the residues at single massless poles of the W and Z polarization tensors:

$$m_W^2 = \frac{1}{4}g^2 \sum (m_U^2 I_{U;D}(0) + m_D^2 I_{D;U}(0)) \tag{10}$$

$$m_Z^2 = \frac{1}{4}(g^2 + g'^2) \sum (m_U^2 I_{U;U}(0) + m_D^2 I_{D;D}(0)) \tag{11}$$

In the formulas above U and D abbreviate upper and nether fermions in electroweak doublets, respectively. The neutrinos are considered as massive Dirac fermions for simplicity. The quantities I in (10, 11) defined in[14] are the UV finite loop integrals depending upon Σs. The Weinberg relation $m_W^2/m_Z^2 \cos^2\theta_W = 1$ is tolerably violated.[14]

6. Neutrino mass generation[7]

In the case I there is one $SU(3)_F$ antitriplet of left-handed neutrinos ν_{fL}, three $SU(3)_F$ triplets of the right-handed neutrinos ν_{NfR}, and the global sterility symmetry $U(3)$. Building the neutrino left-right mass matrix bridge in accord with Fig.1. proceeds in steps: (1) It is convenient to collect all neutrino fields into one left-handed multiplet n_L, $n_L^T = (\nu_{fL}, (\nu_{NfR})^{\mathcal{C}})$ and its corresponding right-handed charge conjugate, $n_R = (n_L)^{\mathcal{C}}$ $((\psi)^{\mathcal{C}} = C\bar{\psi}^T, C^+ = C^T = C^{-1} = -C)$. (2) Rewrite the known vertices of neutrinos with flavor gluons in terms of n_L and n_R resulting in the corresponding neutrino n vertices T_L and T_R. (3) The most general bilinear neutrino Lagrangian[15] which defines the chirality changing neutrino proper self energy $\hat{\Sigma}$ and the neutrino propagator S has in the momentum space the form $\mathcal{L}'_0 = \overline{\nu_L}\not p\nu_L + \overline{\nu_R}\not p\nu_R - \overline{\nu_L}\,\Sigma_D^+\,\nu_R - \overline{\nu_R}\,\Sigma_D\,\nu_L - \frac{1}{2}\overline{\nu_L}\,\Sigma_L^+\,(\nu_L)^{\mathcal{C}} - \frac{1}{2}\overline{(\nu_L)^{\mathcal{C}}}\,\Sigma_L\,\nu_L - \frac{1}{2}\overline{\nu_R}\,\Sigma_R\,(\nu_R)^{\mathcal{C}} - \frac{1}{2}\overline{(\nu_R)^{\mathcal{C}}}\,\Sigma_R^+\,\nu_R = \frac{1}{2}\bar{n}(\not p - \hat{\Sigma})\,n \equiv \bar{n}S^{-1}(p)\,n$. Here $n = n_L + (n_L)^{\mathcal{C}}$ is the column of 12 Majorana fields and Σ in $\hat{\Sigma} = \Sigma P_L + \Sigma^+ P_R$ $(P_{L,R} = \frac{1}{2}(1 \mp \gamma_5))$is a general complex symmetric momentum dependent[16] matrix. It can be cast into the real diagonal form with positive entries Σ_d by the transformation $U^*\Sigma U^+ = \Sigma_d$ where U is a unitary matrix. Upon elimination of unphysical phases the lepton sector of the model has the fixed CP properties.[7] (4) By construction the structure of the schematic SD equation for Σ is identical to that depicted in Fig.1. Identical is, unfortunately, also our inability to find reliable solutions.

While computation of the charged fermion masses would "merely" be a postdiction, a little is known at present experimentally about the neutrino sector. Because the model is uniquely defined, we conclude that the nontrivial non-degenerate solutions of the neutrino SD equations, if found, describe twelve massive Majorana neutrinos, cause complete spontaneous breakdown of the global $U(3)$ sterility symmetry manifested by nine massless composite Majorons, and yield definite CP properties of the leptonic sector.

7. Conclusion

Present model of soft generation of masses of the Standard model particles by a strong-coupling dynamics, having just one unknown parameter h, is either right or plainly wrong. Reliable computation of the fermion mass spectrum is, however, far away. Ultimately, masses should be related. (1) One elaborated example of mass relations is the sum rules for the intermediate boson masses m_W (10) and m_Z (11). The implication is interesting: *There is no generic Fermi scale in the model.* The intermediate boson masses are merely a manifestation of the large top quark mass. (2) Detailed analysis of the uniquely defined neutrino sector is a challenge. The very existence of almost sterile neutrinos introduced for anomaly freedom should have experimental consequences in neutrino oscillations and in astrophysics.[17] Another low energy manifestation of the model are the elusive Majorons. (3) The fermion SD equations can fix also the fermion mixing parameters. (4) It is natural to expect that the unitarization of the scattering amplitudes with longitudinal polarization states of massive spin one particles proceeds in the present model via the massive composite 'cousins' of the composite 'would-be' NG bosons. Its practical implementation is obscured, however, by our ignorance of the detailed properties of the spectrum of strongly coupled $SU(3)_F$.

Acknowledgements

I thank Petr Beneš and Adam Smetana for help with the manuscript and for collaboration. Financial support of this work by the Committee for the CR-CERN collaboration is gratefully acknowledged.

References

1. E. Accomando *et al.*, *Workshop on CP studies and non-standard Higgs physics*, arXiv:hep-ph/0608079.
2. J. Hošek, arXiv:0909.0629.
3. H. Pagels, *Phys. Rev.* **D21**, 2336 (1980).
4. E.J.Eichten and F.L.Feinberg, *Phys.Rev.***D10**,3254(1974).
5. J. Schwinger, *Phys. Rev.* **125**, 397 (1962).
6. D. J. Gross and R. Jackiw, *Phys. Rev.* **D6**,477(1972).
7. P. Beneš, J. Hošek and A. Smetana, *Spontaneous breakdown of sterility*, to be published.
8. Y. Chikashige, R. N. Mohapatra and R. D. Peccei, *Phys. Lett.* **B98**, 26 (1981).
9. D.Binosi and J. Papavassiliou, *Phys.Reports* **479**:1-152(2009).
10. Y. Nambu and G. Jona-Lasinio, *Phys. Rev.* **122**, 345 (1961).
11. K. G. Wilson, *Phys. Rev.* **D3**, 1818(1970).
12. H. Pagels, Phys. Rev.**D19**,3080(1979).
13. J. Hošek, preprint JINR E2-82-542 (1982);
    J. Hošek, preprint CERN-TH.4104/85(1985).
14. J. Hošek, Phys. Rev. **D36**,2093(1987).
15. S. M. Bilenky and S. T. Petcov, *Rev. Mod. Phys.***59**,671(1987).
16. P. Beneš, arXiv:0904.0139.
17. A. Kusenko, New J. Phys.**11**: 105007, 2009.

TeV Physics and Conformality

Thomas Appelquist

Department of Physics, Sloane Laboratory, Yale University
New Haven, Connecticut, 06520, USA

1. Introduction

Lattice gauge theory has been very successful in deepening our understanding of the strong nuclear interactions. During the past two years, stimulated to some extent by the start-up of the Large Hadron Collider, interest is growing in applying lattice methods to new, strongly interacting theories that could play a role in extending the standard model. I will focus here on the use of lattice methods to study strongly coupled gauge theories with some application to models of dynamical electroweak symmetry breaking.

2. The conformal window and walking

2.1. *Perturbative RG flow in generalized Yang-Mills theories*

Consider a Yang-Mills theory with local gauge symmetry group $SU(N_c)$, coupled to N_f massless Dirac fermion flavors:

$$\mathcal{L}_{YM} = -\frac{1}{4g^2}\sum_{a=1}^{N_c} F^a_{\mu\nu}F^{a,\mu\nu} + \sum_{i=1}^{N_f} \bar{\psi}_i(i\not{D})\psi_i, \tag{1}$$

With the fermions in a representation R of the gauge group. The scale dependence of the renormalized coupling $g = g(\mu)$ is determined by the β-function, which we can expand perturbatively:

$$\beta(\alpha) \equiv \frac{\partial\alpha}{\partial(\log\mu^2)} = -\beta_0\alpha^2 - \beta_1\alpha^3 - \beta_2\alpha^4 - \ldots \tag{2}$$

with $\alpha(\mu) \equiv g(\mu)^2/4\pi$. The universal values for the first two coefficients are

$$\beta_0 = \frac{1}{4\pi}\left(\frac{11}{3}N_c - \frac{4}{3}T(R)N_f\right), \tag{3}$$

$$\beta_1 = \frac{1}{(4\pi)^2}\left[\frac{34}{3}N_c^2 - \left(4C_2(R) + \frac{20}{3}N_c\right)T(R)N_f\right], \tag{4}$$

Table 1. Casimir invariants and dimensions of some common representations of $SU(N)$: fundamental (F), two-index symmetric (S_2), two-index antisymmetric (A_2), and adjoint (G).

Representation	$\dim(R)$	$T(R)$	$C_2(R)$
F	N	$\frac{1}{2}$	$\frac{N^2-1}{2N}$
S_2	$\frac{N(N+1)}{2}$	$\frac{N+2}{2}$	$\frac{(N+2)(N-1)}{N}$
A_2	$\frac{N(N-1)}{2}$	$\frac{N-2}{2}$	$\frac{(N-2)(N+1)}{N}$
G	N^2-1	N	N

where $T(R)$ and $C_2(R)$ are the trace normalization and quadratic Casimir invariant of the representation R, respectively. The Casimir invariants for a few commonly-used representations of $SU(N)$ are shown in Table 1. So long as $N_f/N_c < 11/(4T(R))$ so that $\beta_0 > 0$, the theory is asymptotically free. One may continue on to higher order in this expansion, at the cost of specifying a renormalization scheme; in the commonly used $\overline{\text{MS}}$ scheme, the next two coefficients are known.[1]

For present purposes, a key step in studying the RG flow of the coupling constant is to identify any fixed points. Keeping the first two terms in the β-function, we see that in addition to the trivial ultraviolet fixed point $\alpha = 0$, there is a second solution located at

$$\alpha_\star^{(2L)} = -\frac{\beta_0}{\beta_1}. \tag{5}$$

Since $\beta < 0$ for all $0 < \alpha < \alpha_\star^{(2L)}$, this solution is an infrared-stable fixed point, describing the limit $\mu \to 0$. If the fixed-point coupling is sufficiently weak, the theory is perturbative at all scales. This condition will be satisfied if N_f is near the value $11N_c/4T(R)$ at which asymptotic freedom is lost.[2,3] Since confinement of color charges and spontaneous breaking of chiral symmetry, are strong-coupling effects, they are absent in a theory that is perturbative at all scales.

As N_f is decreased, the value of the fixed-point coupling increases, at some point reaching a critical value N_f^c at which the infrared behavior changes from conformal to confining. Theories within the range

$$N_f^c < N_f < \frac{11N_c}{4T(R)} \tag{6}$$

are said to lie in the conformal window, due to the approximate restoration of conformal symmetry in the infrared. Perturbation theory is a-priori unreliable to describe physics in the vicinity of the infrared fixed point as N_f approaches the transition point N_f^c, so in order to determine the location of the transition, some non-perturbative estimate is required.

2.2. *Infrared conformality and walking behavior*

Theories that lie inside the conformal window, although interesting in their own right, are not generally useful in describing electroweak symmetry breaking due to the lack of chiral symmetry breaking (although it is possible to force trigger symmetry breaking even in the conformal window by explicit construction.[4]) However, the basic picture outlined above suggests that a theory outside the window but very close to the transition could also have a novel structure while still breaking chiral symmetry spontaneously.

The idea is as follows:[5–7] suppose that there is some (scheme-dependent) critical coupling α_c, which when exceeded will trigger the spontaneous breaking of chiral symmetry. Now consider a theory with a beta-function such that the coupling is approaching a somewhat supercritical fixed point $\alpha_\star > \alpha_c$. When α_c is exceeded, confinement and chiral symmetry breaking set in. The fermions which were responsible for the existence of the fixed point develop masses and are screened out of the theory, causing the coupling to run as in the $N_f = 0$ theory below the generated mass scale.

This idea, known as walking, results in a separation of scale between the UV physics where the coupling runs perturbatively and the IR scale at which confinement sets in. This dynamical scale separation is what one needs to address the FCNC problem in extended technicolor. The conflict there was between trying to simultaneously match the standard model particle masses and suppress FCNC-generating effects, both of which are tied to the same ETC scale Λ_i. With walking, the ultraviolet-sensitive condensate can pick up a large additional contribution from the scales between Λ_{TC} and Λ_{ETC}, allowing recovery of standard model masses without violation of precision electroweak experimental bounds.

While walking technicolor offers a solution to the difficulties with technicolor models, the existence of such a theory is speculative. The onset of walking is a strong-coupling effect, so that perturbative methods are unlikely to be useful in a walking theory. The search for walking is closely linked to more general questions about the location and nature of the conformal transition at $N_f = N_f^c$.

3. Lattice studies of the conformal transition

3.1. *Overview*

Lattice field theory provides an ideal way to study the conformal transition, and more generally the properties of Yang-Mills theories as we vary N_f. Lattice simulations are truly non-perturbative, although the continuum limit must be taken carefully to recover information about continuum physics (the ability to take this limit being made possible by the asymptotic freedom of the theories in which we are interested.) Lattice simulations allow broad investigation; a large number of different observables can be computed simultaneously on a single set of gauge configurations.

3.2. *Running coupling*

A natural application of lattice simulation to investigate the conformal window is by the direct computation of a running coupling constant.

The first step is to select a non-perturbative definition of a running coupling which we can measure on the lattice. There are a number of such choices possible, including the standard extraction of the static potential from Wilson loops, the Schrödinger functional,[8–12] the twisted Polyakov loop scheme,[13] and constructions using ratios of Wilson loops.[14,15]

Regardless of the definition of the running coupling, the goal is to map out its evolution over a large range of distance scales R. If we work at a fixed lattice spacing a, then the range of available R at which we can measure the coupling is quite small, with the computational expense quickly becoming prohibitive even in QCD. The problem is exacerbated in a theory with large N_f, where the size of the β-function is small; to go from weak to strong coupling, a change in scale of many orders of magnitude is often required. To achieve our goal, then, we must find some way to match together lattice measurements of the running coupling taken at different lattice spacings and combine them into an overall measurement of continuum evolution. A technique known as *step scaling*[16,17] provides a systematic approach.

Step scaling is a recursive procedure which describes the evolution of the coupling constant $g(R)$ as the scale changes from $R \to sR$, where s is a numerical scaling factor known as the *step size.* We define the relation between the coupling at these two scales in the continuum through the *step-scaling function,*

$$\sigma(s, g^2(R)) \equiv g^2(sR). \tag{7}$$

The step-scaling function σ is nothing more than a discrete version of the usual continuum β-function, both of which describe the evolution of the coupling as a function of the coupling strength. In a lattice calculation, the step-scaling function which we extract also contains lattice artifacts in the form of a/R corrections. We denote the lattice version of the step-scaling function by Σ, and it is related to the continuum σ by extrapolation of the lattice spacing a to zero:

$$\sigma(s, g^2(R)) = \lim_{a \to 0} \Sigma(s, g^2(R), a/R). \tag{8}$$

Generically, the implementation of step scaling begins with the choice of some initial value for $g^2(R)$. Several ensembles at different a/R are then generated, tuning the lattice bare coupling $\beta = 2N_c/g_0^2$ so that on each ensemble we measure the chosen value of the renormalized coupling, $g^2(R)$. Then we generate a second ensemble at each β, but measure the coupling at a longer scale $R \to sR$. The value of the coupling measured on the second lattice is exactly $\Sigma(s, g^2(R), a/R)$. One can then extrapolate $a/R \to 0$ and recover the continuum value $\sigma(s, g^2(R))$. Taking $\sigma(s, g^2(R))$ to be the new starting value, we repeat the procedure, mapping $g^2(R) \to g^2(sR) ... \to g^2(s^n R)$ until we have sampled the coupling over a large range of R values.

An efficient approach to step scaling is to measure $g^2(R)$ for a wide range of values in β and R/a, and then to generate an interpolating function. Step scaling may then be done analytically using the interpolated values. Such an interpolating function should reproduce the perturbative relation $g^2(R) = g_0^2 + \mathcal{O}(g_0^4)$ at weak coupling, but otherwise its form is not strongly constrained. One possible choice which has worked quite well in some studies is an expansion of the inverse coupling $1/g^2(\beta, R/a)$ as a set of polynomial series in the bare coupling $g_0^2 = 2N_c/\beta$ at each R/a:

$$\frac{1}{g^2(\beta, R/a)} = \frac{\beta}{2N_c}\left[1 - \sum_{i=1}^{n} c_{i,R/a}\left(\frac{2N_c}{\beta}\right)^i\right]. \tag{9}$$

The order n of the polynomial is arbitrary, and can be varied as a function of R/a to achieve the optimal fit to the available data.

There is a natural caveat on the step-scaling procedure, especially in the context of studying theories with infrared fixed points. The procedure as outlined above depends crucially on the ability to take the limit $a/R \to 0$. If we hold $g^2(R)$ fixed and take the limit $a/R \to 0$, it is important that the bare coupling $g_0^2(a/R)$, which depends on the short-distance behavior of the theory, does not become strong enough to trigger a bulk phase transition. This is satisfied automatically if the short-distance behavior is determined by asymptotic freedom, in which case $g_0^2(a/R)$ vanishes as $1/\log(R/a)$. However, if we work in a theory with an infrared fixed point and measure values of $g^2(R)$ lying above the fixed point at $g_\star^2$, then $g_0^2(a/R)$ will increase as $a \to 0$, with no evidence that it remains bounded and therefore that a continuum limit exists. Even so, we can extrapolate to small enough values of a/R to render lattice artifact corrections negligible, providing that $g_0^2(a/R)$ is kept small enough to avoid triggering a bulk transition into a strong-coupling phase.

Our work has relied on one definition of a running coupling based on the Schrödinger functional (SF). The SF running coupling is defined through the response of a system to variation in strength of a background chromoelectric field. It is a finite-volume method, with the coupling strength defined at the spatial box size L, so that we identify $R = L$ and can discard finite-volume corrections. Formally, the Schrödinger functional describes the quantum mechanical evolution of some system from a given state at time $t = 0$ to another given state at time $t = T$, in a spatial box of size L with periodic boundary conditions.[8,9,18] The temporal extent T is fixed proportional to L, so that the Euclidean box size depends only on a single parameter. The initial and final states are described as Dirichlet boundary conditions which are imposed at $t = 0$ and $t = T$, and for measurement of the coupling constant are chosen such that the minimum-action configuration is a constant chromo-electric background field of strength $O(1/L)$. This can be implemented both in the continuum[8] and on the lattice.[10]

One can represent the Schrödinger functional as the path integral

$$\mathcal{Z}[W, \zeta, \overline{\zeta}; W', \zeta', \overline{\zeta}'] = \int [DAD\psi D\overline{\psi}] e^{-S_G(W,W') - S_F(W,W',\zeta,\overline{\zeta},\zeta',\overline{\zeta}')}, \tag{10}$$

where A is the gauge field and ψ, $\overline{\psi}$ are the fermion fields. W and W' are the boundary values of the gauge fields, and $\zeta, \overline{\zeta}, \zeta', \overline{\zeta}'$ are the boundary values of the fermion fields at $t = 0$ and $t = T$, respectively. The fermionic boundary values are subject only to multiplicative renormalization,[19] and as such are generally taken to be zero in order to simplify the calculation.

The gauge boundary fields W, W' are chosen to given a constant chromo-electric field in the bulk, whose strength is of order $1/L$ and controlled by a dimensionless parameter η.[20] The Schrödinger functional (SF) running coupling is then defined by the response of the action to variation of η:

$$\frac{k}{\overline{g}^2(L,T)} = -\frac{\partial}{\partial \eta} \log \mathcal{Z}\bigg|_{\eta=0} , \tag{11}$$

where (with the standard choice of gauge boundary fields for $SU(3)$), the normalization factor k is

$$k = 12\left(\frac{L}{a}\right)^2 \left[\sin\left(\frac{2\pi a^2}{3LT}\right) + \sin\left(\frac{\pi a^2}{3LT}\right)\right] . \tag{12}$$

The presence of k ensures that $\overline{g}^2(L,T)$ is equal to the bare coupling g_0^2 at tree level in perturbation theory. In general, $\overline{g}^2(L,T)$ can be thought of as the response of the system to small variations in the background chromo-electric field.

For most fermion discretizations, at this point we can take $T = L$ in order to define the running coupling as a function of a single scale, $\overline{g}^2(L)$. However, if staggered fermions are used (as they are often in order to offset the cost of simulating additional fermion flavors), then an additional complication arises which can be envisioned geometrically. The staggered approach to fermion discretization can be formulated as splitting the 16 spinor degrees of freedom available up over a 2^4 hypercubic sublattice. Clearly, such a framework requires an even number of lattice sites in all directions. If all boundaries are periodic or anti-periodic, then setting $T = L$ can be done so long as L is even. However, with Dirichlet boundaries in the time direction, the site $t = T$ is no longer identified with $t = 0$, so that a total of $T/a + 1$ lattice sites must exist. In order to accommodate staggered fermions, T/a must be odd.

Thus when using staggered fermions, the closest we can come to our desired choice of T is $T = L \pm a$. In the continuum limit, the desired relation $T = L$ is recovered. However, at finite lattice spacing $O(a)$ lattice artifacts are introduced into observables. This is undesirable, since staggered fermion simulations contain bulk artifacts only at $O(a^2)$ and above. Fortunately, there is a solution: simulating at both choices $T = L \pm a$ and averaging over the results has been shown to eliminate the induced $O(a)$ bulk artifact in the running coupling.[21] We define $\overline{g}^2(L)$ through the average:

$$\frac{1}{\overline{g}^2(L)} = \frac{1}{2}\left[\frac{1}{\overline{g}^2(L, L-a)} + \frac{1}{\overline{g}^2(L, L+a)}\right] . \tag{13}$$

I will now discuss simulation data and results from a Schrödinger functional running coupling study of the $SU(3)$ fundamental, $N_f = 8$ and 12 theories, using staggered fermions.[12] I begin with the $N_f = 8$ theory, for which data was gathered in the range $4.55 \leq \beta \leq 192$ on lattice volumes given by $L/a = 6, 8, 12, 16$. The lower limit on β was determined to keep the lattice coupling too weak to trigger a bulk phase transition. A selection of the data, together with interpolating function fits of the form Eq. (9), are shown in Fig. 1. Note that at any fixed value of β, the coupling strength $\overline{g}^2(L)$ increases with L/a, showing no evidence of the "backwards" running that we would expect to observe in a theory with an infrared fixed point.

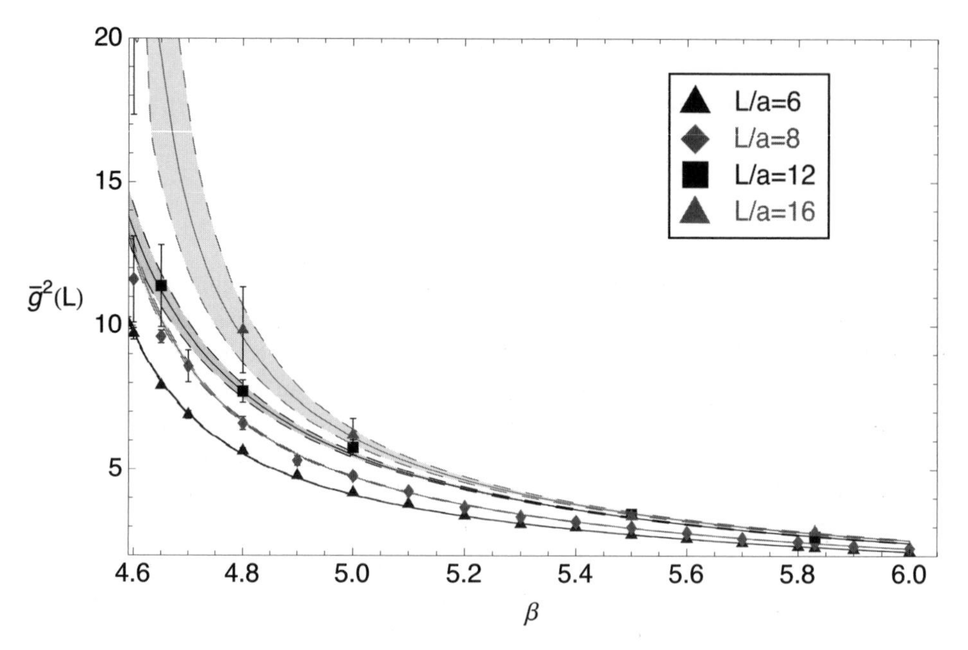

Fig. 1. Measured values $\overline{g}^2(L)$ versus β for $N_f = 8$. The interpolating curves shown represent the best fit to the data, using the functional form Eq. (9). The errors are statistical.

Although the study of Fig. 1 is indicative that the $N_f = 8$ theory lies outside the conformal window, it is possible for results at fixed β to be misleading; we must take the continuum limit in order to recover information about the continuum theory. We apply the step-scaling procedure detailed above in order to extract the continuum step-scaling function $\sigma(2, g^2(R))$, by extrapolation of $a/L \to 0$ with each doubling of the scale L. Our results for $\sigma(2, g^2(R))$ will depend on the choice of continuum extrapolation, i.e. the model function for a/L dependence of $\Sigma(2, g^2(R), a/L)$. As this is a staggered fermion study, the leading bulk lattice artifacts are expected to be of $O(a^2)$, but there are additional boundary artifacts of $O(a)$ which are only partially cancelled off by subtraction of their perturbative values. However, in this case the a/L dependence is weak, with the associated systematic error dominated by the statistical errors on the points, so that a constant extrapolation (i.e. weighted average of the two points) is used to extract $\sigma(2, g^2(R))$ here.

The resulting continuum running of $\overline{g}^2(L)$ for $N_f = 8$ is shown in Fig. 2. L_0 is an arbitrary length scale here defined by the condition $\overline{g}^2(L) = 1.6$, anchoring the step-scaling curve at a relatively weak value. The points shown correspond to repeated doubling of the scale L relative to L_0. Derivation of statistical errors uses a bootstrap technique.[12] Perturbative running at two and three loops is also shown for comparison up through $\overline{g}^2(L) \approx 10$, beyond which the accuracy of perturbation theory is expected to degrade. The coupling measured in this simulation follows the perturbative curve closely up through $\overline{g}^2(L) \approx 4$, and then begins to increase more rapidly, reaching values that exceed typical estimates of the coupling strength needed to induce spontaneous chiral symmetry breaking. As there is no evidence for an infrared fixed point, or even for an inflection point in the running of $\overline{g}^2(L)$, this study supports the assertion that the $N_f = 8$ theory lies outside the conformal window.

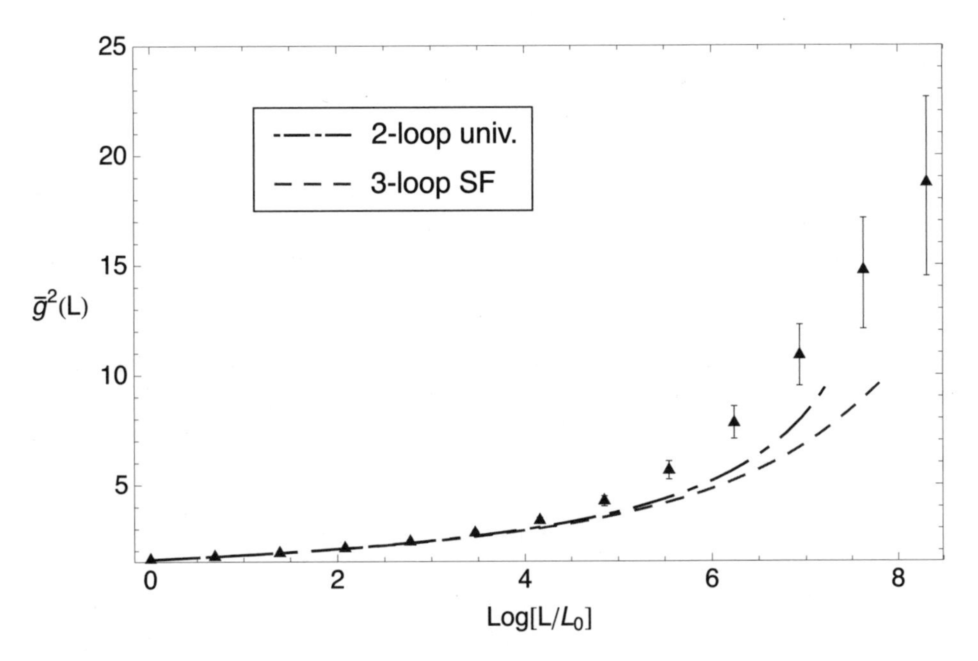

Fig. 2. Continuum running for $N_f = 8$. Purple points are derived by step-scaling using the constant continuum-extrapolation. The error bars shown are purely statistical. Two-loop and three-loop perturbation theory curves are shown for comparison.

Data and interpolating fits for the $N_f = 12$ theory are shown in Fig. 3. Here, simulations were performed on lattice extents of $L/a = 10, 20$, in addition to the values of $L/a = 6, 8, 12, 16$ of the $N_f = 8$ case. The data and fits shown in Fig. 3 show striking qualitative differences with their counterparts at $N_f = 8$ (Fig. 1).

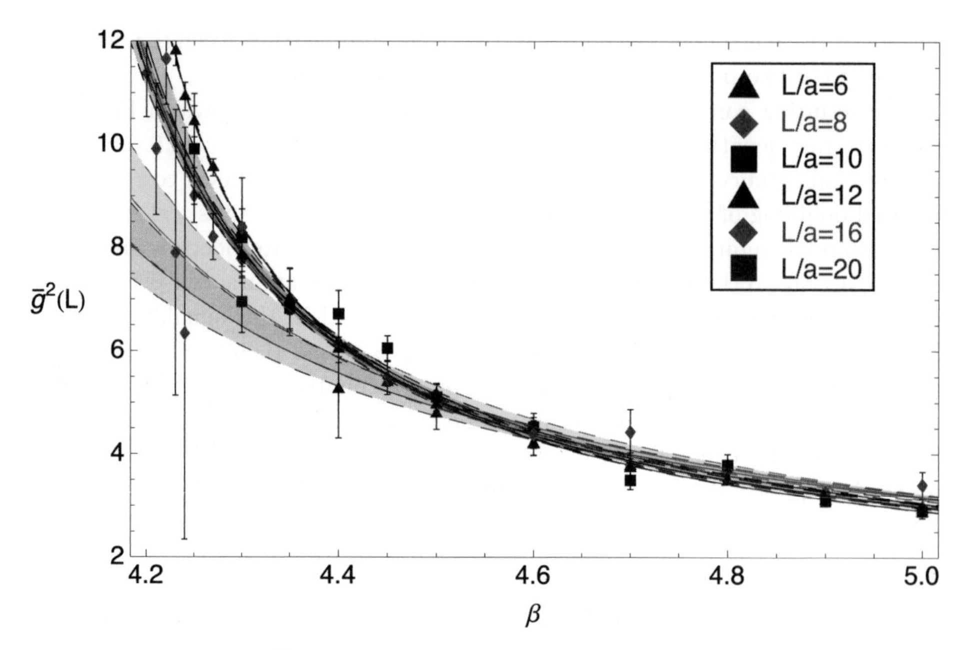

Fig. 3. Measured values $\bar{g}^2(L)$ versus β, $N_f = 12$. The interpolating curves shown represent the best fit to the data, using the functional form of Eq. (9).

In particular, there are some hints of a "crossover" phenomenon taking place, in which the order of the curves in L/a from weak to strong coupling is inverted at small β. Such a crossover is indicative of a region in which the coupling decreases towards the infrared, and is thus a signature of an infrared fixed point. However, it should be emphasized again that it is important to go through the full step-scaling procedure in order to extract meaningful continuum physics, and any result indicated by working at fixed β should not be taken as definitive.

As above, we choose a constant continuum extrapolation, i.e. weighted average of the three points.

Results for continuum running, again from the starting value $\bar{g}^2(L_0) = 1.6$, are shown in Fig. 4. Two-loop and three-loop perturbative curves are shown again for reference. The figure clearly shows the running coupling tracking towards an infrared fixed point, whose exact value lies within the statistical error band and which is consistent with the value predicted by three-loop perturbation theory. It should be noted that the error bars of Fig. 4 are highly correlated, with correlation approaching 100% near the fixed point, due to the use of an underlying interpolating function. This causes the error bars to approach a stable value asymptotically, even as we increase the number of steps towards infinity.

The infrared fixed point here also governs the infrared behavior of the theory for values of $\bar{g}^2(L)$ which lie above the fixed point. As discussed previously, we cannot naively apply the step-scaling procedure in this region, since we can no longer approach the ultraviolet fixed point at zero coupling strength in order to

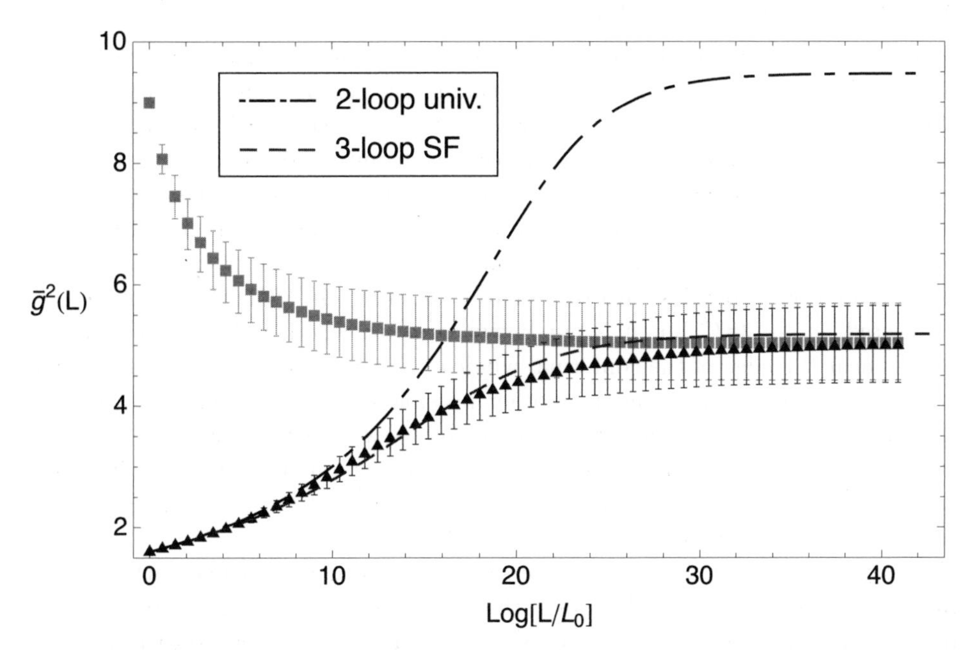

Fig. 4. Continuum running for $N_f = 12$. Results shown for running from below the infrared fixed point (purple triangles) are based on $\bar{g}^2(L_0) \equiv 1.6$. Also shown is continuum backwards running from above the fixed point (light blue squares), based on $\bar{g}^2(L_0) \equiv 9.0$. Error bars are again purely statistical, although strongly correlated due to the underlying interpolating functions. Two-loop and three-loop perturbation theory curves are shown for comparison.

take the continuum limit. Instead, we can restrict our attention to finite but small value of a/L, small enough to keep lattice artifacts small and yet large enough so that $g_0^2(a/L)$ does not trigger a bulk phase transition for $\bar{g}^2(L)$ near (but above) the fixed point. With these caveats in mind, the step-scaling procedure can then be applied and leads to the running from above the fixed point shown in Fig. 4. The observation of this "backwards-running" region is crucial to distinguishing theories with true infrared fixed points from walking theories, in which the β-function may become vanishingly small before turning over and confining.

Having shown evidence for the existence of an infrared fixed point in the $N_f = 12$ theory and demonstrated its absence up to strong coupling at $N_f = 8$, we have constrained the edge of the conformal window for the case of $N_c = 3$ with fermions in the fundamental rep, $8 < N_f^c < 12$. Similar measurements at other values of N_f can allow us to further constrain N_f^c. Furthermore, if a walking theory exists just below the transition value, a lattice measurement of the scale dependence of the coupling could directly reveal the expected plateau behavior and resulting separation between infrared and ultraviolet scales. In addition, the non-perturbative β function can be used in conjunction with additional lattice measurements to extract the anomalous dimension γ_m of the mass operator.[22]

3.3. *Spectral and chiral properties*

I will next discuss the evolution with N_f of various observables on the broken side of the conformal transition. In order to meaningfully compare any quantity between theories with different N_f, we must first identify a physical scale to hold fixed. A natural choice is the Goldstone-boson decay constant F. However, the extraction of F from lattice simulations can be challenging. The rho meson mass m_ρ is much more easily determined, due to the lack of a chiral logarithm at next-to-leading order (NLO) in a χPT-derived fit.[25] However, in the end we are more interested in the evolution of physics with respect to F than m_ρ. In QCD the two scales are connected, $m_\rho \sim 2\pi F$, but it is not known *a priori* whether this connection will persist near the edge of the conformal window. The Sommer scale r_0,[26] associated with the scale of confinement, is another possible choice with similar advantages and drawbacks to m_ρ. In the present discussion we will assume that these scales do not evolve with respect to each other, so that holding any one constant with N_f is sufficient. This assumption is supported by available data, but the choice of scale is an open question going forward.

As these lattice simulations are necessarily performed at finite mass, while we are interested in the behavior of theories in the chiral limit, extrapolation of results $m \to 0$ is crucial. Chiral perturbation theory provides a consistent way to carry out this extrapolation. The familiar expressions for the Goldstone boson mass M_m, decay constant F_m and chiral condensate $\langle\overline{\psi}\psi\rangle_m$ (with the subscript denoting evaluation at finite quark mass m) are easily generalized to theories with arbitrary $N_f \geq 2$ by inclusion of known counting factors. The next-to-leading order (NLO) expressions for a theory with 3 colors are:[27]

$$M_m^2 = \frac{2m\langle\overline{\psi}\psi\rangle}{F^2}\left\{1 + zm\left[\alpha_M + \frac{1}{N_f}\log(zm)\right]\right\}, \tag{14}$$

$$F_m = F\left\{1 + zm\left[\alpha_F - \frac{N_f}{2}\log(zm)\right]\right\}, \tag{15}$$

$$\langle\overline{\psi}\psi\rangle_m = \langle\overline{\psi}\psi\rangle\left\{1 + zm\left[\alpha_C - \frac{N_f^2 - 1}{N_f}\log(zm)\right]\right\}, \tag{16}$$

where $z = 2\langle\overline{\psi}\psi\rangle/(4\pi)^2F^4$. These expressions have also been computed at next-to-next-to-leading order (NNLO) and for fermions in the adjoint representation and the pseudo-real representations of the 2-color theory.[28]

Each of the unknown coefficients $\alpha_M, \alpha_F, \alpha_C$ also contain terms that grow linearly with N_f. α_C also contains a unique, N_f-independent "contact term" which remains even in the absence of spontaneous chiral symmetry breaking. This contribution is linear in m, quadratically sensitive to the ultraviolet cutoff (here the lattice spacing a^{-1}.) This term dominates the chiral expansion of $\langle\overline{\psi}\psi\rangle_m$, making

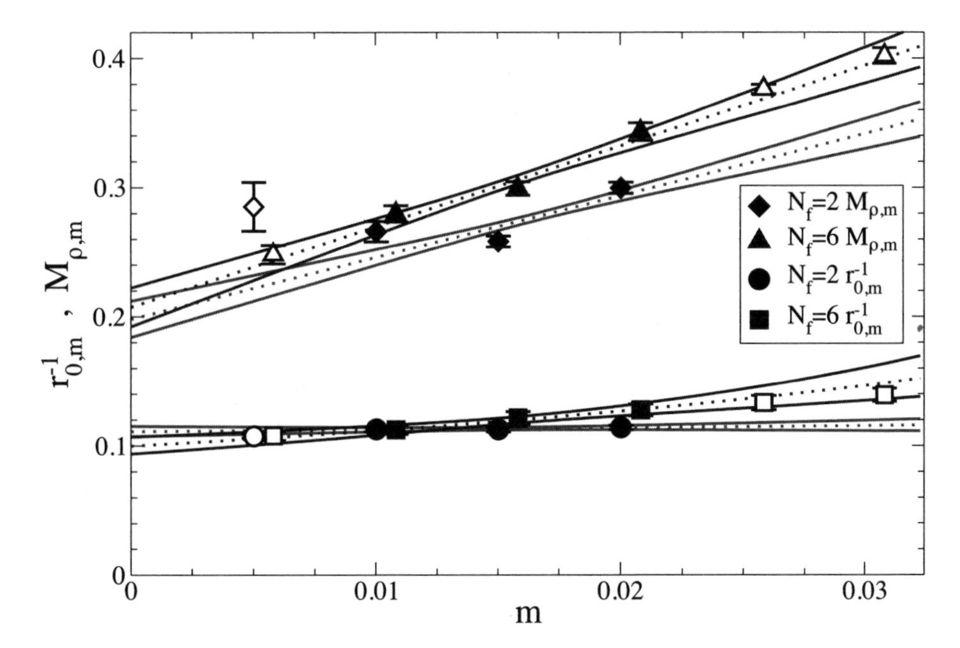

Fig. 5. From.[29] Linear chiral extrapolations of $M_{\rho,m}$ and the Sommer scale $r_{0,m}^{-1}$, in lattice units, based on the (solid) points at $m_f = 0.01-0.02$. Both show agreement within error between $N_f = 2$ and $N_f = 6$ in the chiral limit.

numerically accurate extrapolation of the condensate more difficult. Finally, the growth of the chiral log term in F_m with N_f forces the use of increasingly smaller fermion masses m as N_f is increased, in order to keep the NLO terms small enough relative to the leading order so that χPT is trustworthy.

The goal here is to search for enhancement of the condensate relative to the scale F. One way to proceed is to construct the ratio $\langle\overline{\psi}\psi\rangle_m/F_m^3$, and extrapolate directly $m \to 0$; however, as noted above the presence of the contact term can make such an extrapolation difficult to carry out precisely. By making use of the additional quantity M_m^2 and the Gell-Mann-Oakes-Renner (GMOR) relation $M_m^2 F_m^2 = 2m\langle\overline{\psi}\psi\rangle_m$, incorporated into the NLO formulas shown above, we can construct other ratios at finite m which will also extrapolate to $\langle\overline{\psi}\psi\rangle/F^3$ in the chiral limit: the other two possibilities are $M_m^2/(2mF_m)$, and $(M_m^2/2m)^{3/2}/\langle\overline{\psi}\psi\rangle_m^{1/2}$. Due to the contact term in $\langle\overline{\psi}\psi\rangle_m$, $M_m^2/(2mF_m)$ should have the mildest chiral extrapolation of the three ratios.

A lattice study of the type outlined here, investigating the evolution from $N_f = 2$ to $N_f = 6$ in the $SU(3)$ fundamental case, has been carried out by the Lattice Strong Dynamics (LSD) collaboration.[29] I refer to this reference for details of the simulation and analysis, and here give only selected details. The physical scales chosen to be matched are the rho mass m_ρ and the Sommer scale r_0^{-1}; each observable was first measured in the $N_f = 6$ case, and then matched by tuning the bare lattice coupling

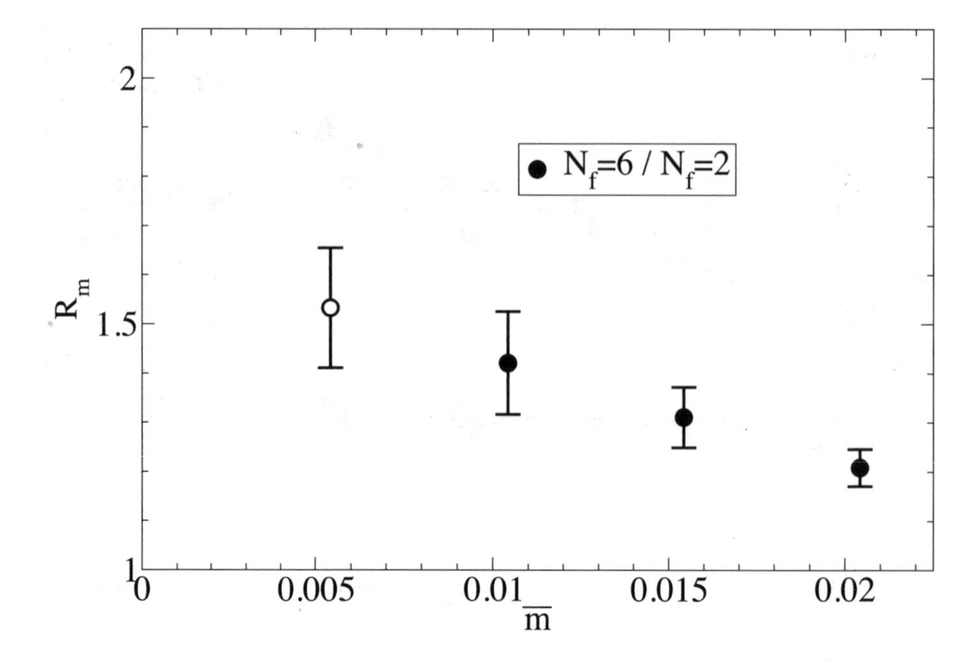

Fig. 6. From.[29] $R_m \equiv [M_m^2/2mF_m]_{6f}/[M_m^2/2mF_m]_{2f}$, versus $\overline{m} \equiv (m(2f)+m(6f))/2$, showing enhancement of $\langle\overline{\psi}\psi\rangle/F^3$ at $N_f = 6$ relative to $N_f = 2$. The open symbol at $m = 0.005$ denotes the presence of possible systematic errors.

at $N_f = 2$. The resulting chiral extrapolation of these quantities is shown in Fig. 5, and shows good agreement, so that the lattice cutoffs are well matched between $N_f = 2$ and $N_f = 6$.

Determination of the presence or absence of condensate enhancement is done through comparison of the quantity $\langle\overline{\psi}\psi\rangle/F^3$ between the $N_f = 6$ and $N_f = 2$ theories, by way of the equivalent ratio $M_m^2/(2mF_m)$. We can directly construct a "ratio of ratios"

$$R_m \equiv \frac{\left[M_m^2/2mF_m\right]_{N_f=6}}{\left[M_m^2/2mF_m\right]_{N_f=2}}, \tag{17}$$

A value of $R_m > 1$ then implies enhancement of the condensate as N_f increases. The result is shown in Fig. 6, and indicates that $R_m \gtrsim 1.5$ in the chiral limit, barring a downturn in R_m - an unlikely outcome, as the curvature of the NLO logarithm is naturally upwards in the chiral expansion of R_m itself. The magnitude of R_m is significant and larger than expected; an $\overline{\text{MS}}$ perturbation theory estimate of the enhancement from $N_f = 2$ to 6 by integrating the anomalous dimension of the mass operator γ_m leads to an expected increase on the order of 5-10%. Some care must be taken in comparing this value to our lattice result, since the condensate $\langle\overline{\psi}\psi\rangle$ and by extension $\langle\overline{\psi}\psi\rangle/F^3$ depends on the renormalization scheme chosen. The conversion factor $Z^{\overline{\text{MS}}}$ between the lattice-cutoff scheme with domain wall fermions and $\overline{\text{MS}}$

is known from Ref.[30] From that reference, for this simulation the required factor to convert R_m is $Z_6^{\overline{\rm MS}}/Z_2^{\overline{\rm MS}} = 1.449(29)/1.227(11) = 1.18(3)$. This increases the perturbative estimate of expected enhancement to the order of $20 - 30\%$, so the observed $R_m \gtrsim 1.5$ is still significantly larger than anticipated.

A direct computation of the S-parameter is also important to the study of general Yang-Mills theories with a focus on technicolor, and is well within the reach of existing lattice techniques. Some results have been reported,[31,32] and more work is underway by the LSD collaboration.

References

1. T. van Ritbergen, J. A. M. Vermaseren and S. A. Larin, *Phys. Lett.* **B400**, 379 (1997).
2. W. E. Caswell, *Phys. Rev. Lett.* **33**, p. 244 (1974).
3. T. Banks and A. Zaks, *Nucl. Phys.* **B196**, p. 189 (1982).
4. M. A. Luty, *JHEP* **04**, p. 050 (2009).
5. B. Holdom, *Phys. Lett.* **B150**, p. 301 (1985).
6. K. Yamawaki, M. Bando and K.-i. Matumoto, *Phys. Rev. Lett.* **56**, p. 1335 (1986).
7. T. W. Appelquist, D. Karabali and L. C. R. Wijewardhana, *Phys. Rev. Lett.* **57**, p. 957 (1986).
8. M. Lüscher, R. Narayanan, P. Weisz and U. Wolff, *Nucl. Phys.* **B384**, 168 (1992).
9. S. Sint, *Nucl. Phys.* **B421**, 135 (1994).
10. A. Bode, P. Weisz and U. Wolff, *Nucl. Phys.* **B576**, 517 (2000), Erratum-ibid.B608:481,2001.
11. T. Appelquist, G. T. Fleming and E. T. Neil, *Phys. Rev. Lett.* **100**, p. 171607 (2008).
12. T. Appelquist, G. T. Fleming and E. T. Neil, *Phys. Rev.* **D79**, p. 076010 (2009).
13. E. Bilgici *et al.* (2009).
14. E. Bilgici *et al.*, *Phys. Rev.* **D80**, p. 034507 (2009).
15. Z. Fodor, K. Holland, J. Kuti, D. Nogradi and C. Schroeder, *Phys. Lett.* **B681**, 353 (2009).
16. M. Lüscher, P. Weisz and U. Wolff, *Nucl. Phys.* **B359**, 221 (1991).
17. S. Caracciolo, R. G. Edwards, S. J. Ferreira, A. Pelissetto and A. D. Sokal, *Phys. Rev. Lett.* **74**, 2969 (1995).
18. A. Bode *et al.*, *Phys. Lett.* **B515**, 49 (2001).
19. R. Sommer (2006).
20. M. Lüscher, R. Sommer, P. Weisz and U. Wolff, *Nucl. Phys.* **B413**, 481 (1994).
21. U. M. Heller, *Nucl. Phys.* **B504**, 435 (1997).
22. F. Bursa, L. Del Debbio, L. Keegan, C. Pica and T. Pickup, *Phys. Rev.* **D81**, p. 014505 (2010).
23. X.-Y. Jin and R. D. Mawhinney, *PoS* **LAT2009**, p. 049 (2009).
24. L. Del Debbio, B. Lucini, A. Patella, C. Pica and A. Rago, *Phys. Rev.* **D80**, p. 074507 (2009).
25. D. B. Leinweber, A. W. Thomas, K. Tsushima and S. V. Wright, *Phys. Rev.* **D64**, p. 094502 (2001).
26. R. Sommer, *Nucl. Phys.* **B411**, 839 (1994).
27. J. Gasser and H. Leutwyler, *Phys. Lett.* **B184**, p. 83 (1987).
28. J. Bijnens and J. Lu, *JHEP* **11**, p. 116 (2009).
29. T. Appelquist, A. Avakian, R. Babich, R. C. Brower, M. Cheng, M. A. Clark, S. D. Cohen, G. T. Fleming, J. Kiskis, E. T. Neil, J. C. Osborn, C. Rebbi, D. Schaich and P. Vranas, *Phys. Rev. Lett.* **104**, p. 071601 (2010).

30. S. Aoki, T. Izubuchi, Y. Kuramashi and Y. Taniguchi, *Phys. Rev.* **D67**, p. 094502 (2003).
31. E. Shintani *et al.*, *Phys. Rev. Lett.* **101**, p. 242001 (2008).
32. P. A. Boyle, L. Del Debbio, J. Wennekers and J. M. Zanotti, *Phys. Rev.* **D81**, p. 014504 (2010).

Conformal Higgs, or Techni-Dilaton — Composite Higgs Near Conformality

Koichi Yamawaki*
Department of Physics, Nagoya University
Nagoya, Japan
E-mail: yamawaki@kmi.nagoya-u.ac.jp

In contrast to the folklore that Technicolor (TC) is a "Higgsless theory", we shall discuss existence of a composite Higgs boson, Techni-Dilaton (TD), a pseudo-Nambu-Goldstone boson of the scale invariance in the Scale-invariant/Walking/Conformal TC (SWC TC) which generates a large anomalous dimension $\gamma_m \simeq 1$ in a wide region from the dynamical mass $m = \mathcal{O}$ (TeV) of the techni-fermion all the way up to the intrinsic scale $\Lambda_{\rm TC}$ of the SWC TC (analogue of $\Lambda_{\rm QCD}$), where $\Lambda_{\rm TC}$ is taken typically as the scale of the Extended TC scale $\Lambda_{\rm ETC}$: $\Lambda_{\rm TC} \simeq \Lambda_{\rm ETC} \sim 10^3$ TeV ($\gg m$). All the techni-hadrons have mass on the same order $\mathcal{O}(m)$, which in SWC TC is extremely smaller than the intrinsic scale $\Lambda_{\rm TC} \simeq \Lambda_{\rm ETC}$, in sharp contrast to QCD where both are of the same order. The mass of TD arises from the *non-perturbative scale anomaly* associated with the techni-fermion mass generation and is typically 500-600 GeV, even *smaller than other techni-hadrons* of the same order of $\mathcal{O}(m)$, in another contrast to QCD which is believed to have no scalar $\bar{q}q$ bound state lighter than other hadrons. We discuss the TD mass in various methods, Gauged NJL model via ladder Schwinger-Dyson (SD) equation, straightforward calculations in the ladder SD/ Bethe-Salpeter equation, and the holographic approach including techni-gluon condensate. The TD may be discovered in LHC.

Keywords: Walking technicolor, scale invariance, conformal symmetry, techni-dilaton, fixed point, composite Higgs, large anomalous dimension, holographic gauge theory.

1. Introduction

Toshihide Maskawa is famous for 2008 Nobel prize-winning paper with Makoto Kobayashi on CP violation but did also fundamental contributions particularly to the SCGT, the topics of this workshop: Back in 1974 he found with Hideo Nakajima[1] that spontaneous chiral symmetry breaking (SχSB) solution does exists for and only for the strong gauge coupling, with the critical coupling of order 1, based on the ladder Schwinger-Dyson (SD) equation with non-running (scale-invariant) coupling, namely the walking gauge dynamics what is called today. This turned out to be the origin of SCGT activities toward understanding the Origin of Mass. The present workshop SCGT 09 was held in honor of his 70th birthday on February 7, 2010

*Present Address: Kobayashi-Maskawa Institute for the Origin of Particles and the Universe (KMI), Nagoya University.

and the 35th anniversary of his crucial contributions to SCGT. I will later explain impact of Maskawa-Nakajima solution on the conformal gauge dynamics.

The Origin of Mass is the most urgent issue of the particle physics today and is to be resolved at the LHC experiments. In the standard model (SM), all masses are attributed to a single parameter of the vacuum expectation value (VEV), $\langle H \rangle$ of the hypothetical elementary particle, the Higgs boson. The VEV simply picks up the mass scale of the input parameter M_0 which is tuned to be tachyonic ($M_0^2 < 0$) in such a way as to tune $\langle H \rangle \simeq 246\,\text{GeV}$ ("naturalness problem"). As such SM does not explain the Origin of Mass.

Technicolor (TC)[2] is an attractive idea to account for the Origin of Mass without introducing ad hoc Higgs boson and tachyonic mass parameter: The mass arises *dynamically* from the condensate of the techni-fermion and the anti techni-fermion pair $\langle \bar{T}T \rangle$ which is triggered by the attractive gauge forces between the pair analogously to the quark-antiquark condensate $\langle \bar{q}q \rangle$ in QCD. For the TC with $SU(N_{\text{TC}})$ gauge symmetry and N_f flavors ($N_f/2$ weak doublets) of techni-fermions, the techni-pion decay constant $F_\pi = \langle H \rangle / \sqrt{N_f/2}$ corresponds to the pion decay constant $f_\pi \simeq 93\,\text{MeV}$ in QCD, and hence the TC may be a scale-up of QCD by the factor $F_\pi/f_\pi \simeq 2650/\sqrt{N_f/2}$. Then the mass scale of the condensate $\Lambda_\chi = (-\langle \bar{T}T \rangle / N_{\text{TC}})^{1/3}$ as the Origin of Mass may be estimated as

$$\Lambda_\chi \simeq \left(\frac{-\langle \bar{q}q \rangle}{N_c} \right)^{1/3} \cdot \frac{F_\pi/\sqrt{N_{\text{TC}}}}{f_\pi/\sqrt{N_c}} \simeq 450\,\text{GeV} \cdot \left(\frac{N_c/N_{\text{TC}}}{N_f/2} \right)^{1/2}, \tag{1}$$

where we have used a typical value $(-\langle \bar{q}q \rangle)^{1/3} \simeq 250\,\text{MeV}$ ($N_c = 3$).

The dynamically generated mass scale of the condensate Λ_χ, or the dynamical mass of the techni-fermion, $m\,(\sim \Lambda_\chi \sim F_\pi)$, in fact picks up the intrinsic mass scale Λ_{TC} of the theory (analogue of Λ_{QCD} in QCD) already generated by the scale anomaly through quantum effects ("dimensional transmutation") in the gauge theory which is scale-invariant at classical level (for massless flavors):

$$\Lambda_{\text{TC}} = \mu \cdot \exp\left(- \int^{\alpha(\mu)} \frac{d\alpha}{\beta(\alpha)} \right) = \Lambda_0 \cdot \exp\left(- \int^{\alpha(\Lambda_0)} \frac{d\alpha}{\beta(\alpha)} \right), \tag{2}$$

where the running of the coupling constant $\alpha(\mu)$, with non-vanishing beta function $\beta(\alpha) \equiv \mu \frac{d\alpha(\mu)}{d\mu} \neq 0$, is a manifestation of the scale anomaly and Λ_0 is a fundamental scale like Planck scale. Note that Λ_{TC} is independent of the renormalization point μ, $\frac{d\Lambda_{\text{TC}}}{d\mu} = 0$, and can largely be separated from Λ_0 through logarithmic running ("naturalness"). Thus the Origin of Mass is eventually the quantum effect in this picture: In the simple scale-up of QCD we would have

$$\text{Naturalness (QCD scale up)}: \qquad m \sim \Lambda_\chi \sim \Lambda_{\text{TC}} \ll \Lambda_0\,. \tag{3}$$

The original version of TC, just a simple scale-up of QCD, however, is plagued by the notorious problems: Excessive flavor-changing neutral currents (FCNCs), and excessive oblique corrections of $\mathcal{O}(1)$ to the Peskin-Takeuchi S parameter[3] compared with the typical experimental bound about 0.1.

The FCNC problem was resolved long time ago by the TC based on the near conformal gauge dynamics with $\gamma_m \simeq 1$,[4,5] initially dubbed "scale-invariant TC" and then "walking TC", with almost *non-running (conformal) gauge coupling*, based on the pioneering work by Maskawa and Nakajima[1] who discovered *non-zero critical coupling*, $\alpha_{\rm cr} (\neq 0)$, for the SχSB to occur. We may call it "Scale-invarinat/Walking/Conformal TC" (SWC TC) (For reviews see Ref.[6]).

In addition to solving the FCNC problem, the theory made a definite prediction of "Techni-dilaton (TD)",[4] a pseudo Nambu-Goldstone (NG) boson of the spontaneous breaking of the (approximate) scale invariance of the theory. This will be the main topics of this talk in the light of modern version of SWC TC.

The modern version[7–9] of SWC TC is based on the Caswell-Banks-Zaks (CBZ) infrared (IR) fixed point [10], $\alpha_* = \alpha_*(N_f, N_{\rm TC})$, which appears at two-loop beta function for the number of massless flavors $N_f (< 11 N_{\rm TC}/2)$ larger than a certain number $N_f^* (\gg N_{\rm TC})$. See Fig. 1 and later discussions. Due to the IR fixed point the

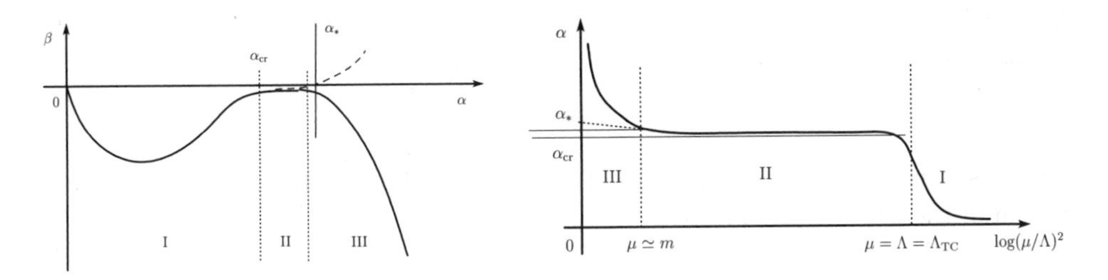

Fig. 1. The beta function and $\alpha(\mu)$ for SWC-TC.

coupling is almost non-running ("walking") all the way up to the intrinsic scale $\Lambda_{\rm TC}$ which is generated by the the scale anomaly associated with the (two-loop) running of the coupling analogously to QCD scale-up in Eq. (2). For $\mu > \Lambda_{\rm TC}$ (Region I of Fig. 1) the coupling no longer walks and runs similarly to that of QCD. When we set α_* slightly larger than $\alpha_{\rm cr}$, we have a condensate or the dynamical mass of the techni-fermion $m \, (\sim \Lambda_\chi)$, much smaller than the intrinsic scale of the theory $m \ll \Lambda_{\rm TC}$. The CBZ-IR fixed point α_* actually disappears (then becoming would-be IR fixed point) at the scale $\mu \lesssim m$ where the techni-fermions have acquired the mass m and get decoupled from the beta function for $\mu < m$ (Region III in Fig. 1). Nevertheless, the coupling is still walking due to the remnant of the CBZ-IR fixed point conformality in a wide region $m < \mu < \Lambda_{\rm TC}$ (Region II in Fig. 1). Thus the *symmetry responsible for the natural hierarchy $m \sim \Lambda_\chi \ll \Lambda_{\rm TC}$ is the (approximate) conformal symmetry*, while the naturalness for the hierarchy $\Lambda_{\rm TC} \ll \Lambda_0$ is the same as that of QCD scale-up in Eq.(3):

$$\text{Naturalness (SWC TC)}: \qquad m \sim \Lambda_\chi \ll \Lambda_{\rm TC} \quad (\ll \Lambda_0) \,. \tag{4}$$

The theory acts like the SWC-TC[4,5] : It develops a large anomalous dimension $\gamma_m \simeq 1$ for the almost non-running coupling in the Region II .[8,9] Here $\Lambda_{\rm TC}$ plays a role of cutoff Λ identified with the ETC scale: $\Lambda_{\rm TC} = \Lambda = \Lambda_{\rm ETC}$.

Moreover, there also exists a possibility[11,12] that the S parameter can be reduced in the case of SWC-TC.

In this talk I will argue[13] [a] that in contrast to the simple QCD scale-up which is widely believed to have no composite Higgs particle ("higgsless"), a salient feature of SWC TC is the *conformality which manifests itself by the appearance of a composite Higgs boson ("conformal Higgs") as the Techni-dilaton (TD)*[4] with mass relatively lighter than other techni-hadrons: $M_{\rm TD} < M_\rho, M_{a_1} \cdots = \mathcal{O}(\Lambda_\chi) \ll \Lambda_{\rm TC} = \Lambda_{\rm ETC}$, where $M_\rho, M_{a_1} \cdots$ denote the mass of techni-ρ, techni-a_1, etc. This is contrasted to the QCD dynamics where there are no scalar bound states lighter than others. Note that there is no idealized limit where the TD becomes exactly massless to be a true NG boson, in sharp contrast to the chiral symmetry breaking. Scale symmetry is always broken explicitly as well as spontaneously [b] .

For the phenomenological purpose, I will argue through several different calculations[13,15,16] that the techni-dilaton mass in the typical SWC TC models will be in the range (see the footnote below Eq.(30), however):

$$m_{\rm TD} = 500 - 600\,{\rm GeV}, \tag{5}$$

which is definitely larger than the SM Higgs bound but still within the discovery region of the LHC experiments.

2. Scale-Invariant/Walking/Conformal Technicolor

Let us briefly review the SWC TC.

The FCNC problem is related with the mass generation of quarks/leptons mass. In order to communicate the techni-fermion condensate to the quarks/leptons masses $m_{q/l}$, we would need interactions between the quarks/leptons and the techni-fermions which are typically introduced through Extended TC (ETC)[17] [c] with much higher scale $\Lambda_{\rm ETC}(\gg \Lambda_\chi)$: $m_{q/l} \sim \frac{-1}{\Lambda_{\rm ETC}^2} \langle \bar{T}T \rangle_{\Lambda_{\rm ETC}}$, where $\langle \bar{T}T \rangle_{\Lambda_{\rm ETC}}$ is the condensate measured at the scale of $\Lambda_{\rm ETC}$. (We here do not refer to the origin of the mass scale $\Lambda_{\rm ETC}$ which should also be of dynamical origin such as the tumbling.) Since the newly introduced ETC interactions characterized by the same scale $\Lambda_{\rm ETC}$ should induce extra FCNC's, we should impose a constraint $\Lambda_{\rm ETC} > 10^6$GeV in

[a] Preliminary discussions on the revival of the techni-dilaton[4] were given in several talks[14] .

[b] The straightforward calculations near the conformal edge indicated[16] that there is no isolated massless spectrum: $M_{\rm TD}/F_\pi, M_{\rm TD}/M_\rho, \cdots \to$ const. $\neq 0$ even in the limit of $\alpha_* \to \alpha_{\rm cr}$ ($N_f \to N_f^{\rm crit}$) where $F_\pi/\Lambda_{\rm TC}, M_{\rm TD}/\Lambda_{\rm TC}, M_\rho/\Lambda_{\rm TC}, \cdots \to 0$. In the case of holographic TD,[13] this fact is realized in a different manner: Although there apparently exists an isolated massless spectrum, $M_{\rm TD}/F_\pi \to 0$ while $M_\rho/F_\pi, M_{a_1}/F_\pi \to$ const. $\neq 0$, the decay constant of the TD diverges $F_{\rm TD}/F_\pi \to \infty$ in that limit and hence it gets decoupled. See later discussions.

[c] The same can be done in a composite model where quarks/leptons and techni-fermions are composites on the same footing.[18]

order to avoid the excessive FCNC's (typically involving s quark). If we assume a simple QCD scale up, $\langle \bar{T}T \rangle_{\Lambda_{\rm ETC}} \simeq \langle \bar{T}T \rangle_{\Lambda_\chi} = -N_{\rm TC} \cdot \Lambda_\chi^3$, we would have

$$m_{q/l} \sim \frac{\Lambda_\chi^3}{\Lambda_{\rm ETC}^2} \cdot N_{\rm TC} < 0.1\,{\rm MeV} \cdot N_{\rm TC} \left(\frac{N_{\rm c}/N_{\rm TC}}{N_d} \right)^{3/2}, \tag{6}$$

which implies that the typical mass (s-quark mass) would be roughly 10^{-3} smaller than the reality. We would desperately need 10^3 times enhancement.

This was actually realized dynamically by the TC based on the near conformal gauge dynamics,[4,5] based on the Maskawa-Nakajima solution[1] of the (scale-invariant) ladder Schwinger-Dyson (SD) equation for fermion full propagator $S_F(p)$ parameterized as $iS_F^{-1}(p) = A(p^2)\not{p} - B(p^2)$ with *non-running* (conformal, an ideal limit of the "walking") gauge coupling, $\alpha(Q) \equiv \alpha =$ constant, with $Q^2 \equiv -p^2 > 0$. (See Fig. 2)

Maskawa and Nakajima discovered that the SχSB can only take place for strong coupling $\alpha > \alpha_{\rm cr} = \mathcal{O}(1)$, *non-zero critical coupling.*[d] The critical value reads:[20] $C_2(F)\alpha_{\rm cr} = \pi/3$, or

$$\alpha_{\rm cr} = (\pi/3) \cdot 2N_{\rm TC}/(N_{\rm TC}^2 - 1) \tag{7}$$

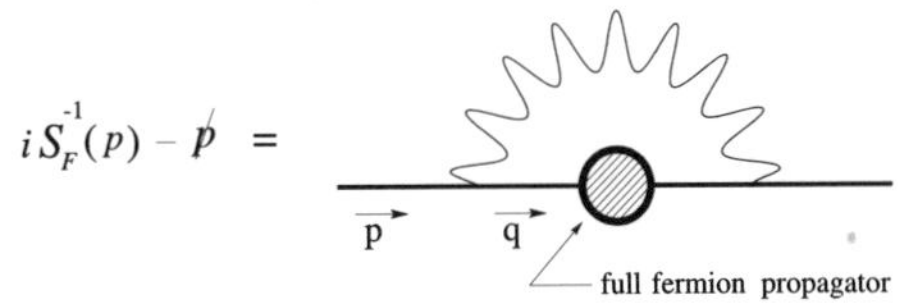

Fig. 2. Graphical expression of the SD equation in the ladder approximation.

in the $SU(N_{\rm TC})$ gauge theory, where $C_2(F)$ is the quadratic Casimir of the techni-fermion representation of the TC. The asymptotic form of the Maskawa-Nakajima SχSB solution of the fermion mass function $\Sigma(Q) = B(p^2)/A(p^2)$ in Landau gauge ($A(p^2) \equiv 1$) reads,[1,20]

$$\Sigma(Q) \sim 1/Q \quad (Q \gg \Lambda_\chi)\,. \tag{8}$$

We then proposed a "Scale-invariant TC" [4], based on the observation that Eq.(8) implies a special value of the anomalous dimension

$$\gamma_m = -\Lambda \frac{\partial \ln Z_m}{\partial \Lambda} = 1, \tag{9}$$

to be compared with the operator product expansion (OPE), $\Sigma(Q) \sim 1/Q^2 \cdot (Q/\Lambda_\chi)^{\gamma_m}$. Accordingly, we had an enhanced condensate $\langle \bar{T}T \rangle_{\Lambda_{\rm ETC}} = Z_m^{-1} \cdot \langle \bar{T}T \rangle_{\Lambda_\chi} \simeq -N_{\rm TC}(\Lambda_{\rm ETC}\Lambda_\chi^2)$, with the (inverse) mass renormalization constant being $Z_m^{-1} = (\Lambda_{\rm ETC}/\Lambda_\chi)^{\gamma_m} \simeq \Lambda_{\rm ETC}/\Lambda_\chi \simeq 10^3$, which in fact yields the desired enhancement. We actually obtained a different formula than Eq.(6):[4]

$$m_{q/l} \sim \frac{\Lambda_\chi^2}{\Lambda_{\rm ETC}} \cdot N_{\rm TC}\,. \tag{10}$$

[d] Earlier works[19] in the ladder SD equation with non-running coupling all confused explicit breaking solution with the SSB solution and thus implied $\alpha_{\rm cr} = 0$

The model[4] was formulated in terms of the Renormalization Group Equation (RGE) a la Miransky [21] for the Maskawa-Nakajima solution $\Sigma(m) = m$ which takes the form[20,21]

$$\Lambda_\chi \sim m \sim 4\Lambda \exp\left(-\pi/\sqrt{\alpha/\alpha_{\rm cr} - 1}\right), \tag{11}$$

where Λ is the cutoff of the SD equation. This has an *essential singularity* often called "Miransky scaling" and implies the *non-perturbative beta function* having a *multiple zero* [e] :

$$\beta(\alpha)_{\rm NP} = \Lambda \frac{\partial\alpha(\Lambda)}{\partial\Lambda} = -\frac{2}{3C_2(F)}\left(\frac{\alpha}{\alpha_{\rm cr}} - 1\right)^{3/2}, \tag{12}$$

with the critical coupling $\alpha_{\rm cr}$ identified with a nontrivial ultraviolet (UV) stable fixed point $\alpha = \alpha(\Lambda) \to \alpha_{\rm cr}$ as $\Lambda/m \to \infty$.

Subsequently, similar enhancement effects of the condensate were also studied[5] within the same framework of the ladder SD equation, without use of the RGE concepts of anomalous dimension and fixed point, rather emphasizing the asymptotic freedom of the TC theories with slowly-running (walking) coupling which was implemented into the ladder SD equation ("improved ladder SD equation").

Today the Scale-invariant/Walking/Conformal TC (SWC TC) is simply characterized by near conformal property with $\gamma_m \simeq 1$ (For a review see Ref.[6]). Such a theory should have an almost non-running and strong gauge coupling (larger than a certain non-zero critical coupling for SχSB) to be realized either at UV fixed point or IR fixed point, or both ("fusion" of the IR and UV fixed points), as was characterized by "*Conformal Phase Transition (CPT)*".[9]

The essential feature of the above is precisely what happens in the modern version [7–9] of the SWC TC based on the CBZ IR fixed point[10] of the large N_f QCD, the QCD-like theory with many flavors $N_f\,(\gg N_{\rm TC})$ of massless techni-fermions, [f] see Fig. 1. The two-loop beta function is given by $\beta(\alpha) = \mu\frac{d}{d\mu}\alpha(\mu) = -b\alpha^2(\mu) - c\alpha^3(\mu)$, where $b = (11N_{\rm TC} - 2N_f)/(6\pi)$, $c = \left[34N_{\rm TC}^2 - 10N_fN_{\rm TC} - 3N_f(N_{\rm TC}^2-1)/N_{\rm TC}\right]/(24\pi^2)$. When $b > 0$ and $c < 0$, i.e., $N_f^* < N_f < \frac{11}{2}N_{\rm TC}$ ($N_f^* \simeq 8.05$ for $N_{\rm TC} = 3$), there exists an IR fixed point (CBZ IR fixed point) at $\alpha = \alpha_*$, $\beta(\alpha_*) = 0$, where

$$\alpha_* = \alpha_*(N_{\rm TC}, N_f) = -b/c. \tag{13}$$

Note that $\alpha_* = \alpha_*(N_f, N_{\rm TC}) \to 0$ as $N_f \to 11N_{\rm TC}/2$ ($b \to 0$) and hence there exists a certain range $N_f^{\rm cr} < N_f < 11N_{\rm TC}/2$ ("Conformal Window") satisfying $\alpha_* < \alpha_{\rm cr}$, where the gauge coupling $\alpha(\mu)\,(< \alpha_*)$ gets so weak that attractive forces are no longer strong enough to trigger the SχSB as was demonstrated by Maskawa-

[e] Simple zero of the beta function, $\beta(\alpha) \sim (\alpha - \alpha_{\rm cr})^1$, never reproduces the essential singularity scaling, as is evident from Eq.(2).

[f] For SWC TC based on higher representation/other gauge groups see, e.g., Ref.[22]

Nakajima.[1] $N_f^{\rm cr}$ such that $\alpha_*(N_{\rm TC}, N_f^{\rm cr}) = \alpha_{\rm cr}$ may be evaluated by using the value of $\alpha_{\rm cr}$ from the ladder SD equation Eq.(7): [8] $N_f^{\rm cr} \simeq 4N_{\rm TC}$ ($= 12$ for $N_{\rm TC} = 3$) [g].

Here we are interested in the SχSB phase slightly off the conformal window, $0 < \alpha_* - \alpha_{\rm cr} \ll 1$ ($N_f \simeq N_f^{\rm cr}$). We may use the same equation as the ladder SD equation with $\alpha(\mu) \simeq$ const. $= \alpha_*$, yielding the same form as Eq.(11) :[8]

$$m \sim 4\Lambda_{\rm TC} \exp\left(-\pi/\sqrt{\alpha_*/\alpha_{\rm cr} - 1}\right) \ll \Lambda_{\rm TC} \quad (\alpha_* \simeq \alpha_{\rm cr}), \tag{14}$$

where the cutoff Λ was identified with $\Lambda_{\rm TC} (= \Lambda_{\rm ETC})$. We also have the same result as Eqs. (8),(9):

$$\Sigma(Q) \sim 1/Q, \qquad \gamma_m \simeq 1. \tag{15}$$

Hence it acts like SWC TC. Incidentally, Eq.(14) implies a *multiple zero* at $\alpha_* = \alpha_{\rm cr}$ in a non-perturbative beta function for $\alpha_* = \alpha_*(\Lambda)$ similar to Eq.(12), which would suggest "running" of the IR fixed point α_* with its UV fixed point $\alpha_* = \alpha_{\rm cr}$ in the limit $\Lambda_{\rm TC}/m \to \infty$.

The actual running of the coupling largely based on two-loop perturbation is already depicted in Fig. 1. The critical coupling $\alpha_{\rm cr}$ can be regarded as the UV fixed point viewed from the IR part of the Region II ($m < \mu < \mu_{\rm cr}$, with $\mu_{\rm cr}$ such that $\alpha(\mu_{\rm cr}) = \alpha_{\rm cr}$), while it is regarded as the IR fixed point from the UV part of the Region II ($\Lambda_{TC} > \mu > \mu_{\rm cr}$), with the Region II regarded as the *fusion of the IR and UV fixed points* in the idealized limit of non-running (perturbative) coupling in Region II (or $\Lambda_{\rm TC}/m \to \infty$). Although the perturbative (two-loop) beta function has a *simple zero*, which *never corresponds to the essential singularity* scaling as we noted before, the coupling near $\alpha_{\rm cr}$ should be sensitive to the non-perturbative effects in such as way that the beta function looks like the *multiple zero* as in Eq. (12) from both sides, corresponding to the *essential singularity scaling* as in Eq. (11). This should be tested by the fully non-perturbative studies like lattice simulations. A possible phase diagram (Fig 3 of Ref.[9]) of the large N_f QCD on the lattice is also waiting for the test by simulations.

3. Conformal Phase Transition[9,14]

Such an *essential singularity* scaling law like Eq.(11),(14), or equivalently the *multiple zero* of the non-perturbative beta function, characterizes an unusual phase transition, what we called "*Conformal Phase Transition (CPT)*", where the Ginzburg-Landau effective theory breaks down:[9] Although it is a second order (continuous) phase transition where the order parameter m ($\alpha_* > \alpha_{\rm cr}$) is continuously changed to $m = 0$ in the symmetric phase (conformal window, $\alpha_* < \alpha_{\rm cr}$), the spectra do not, i.e., while there exist light composite particles whose mass vanishes at the critical

[g]The value should not be taken seriously, since $\alpha_* = \alpha_{\rm cr}$ is of $\mathcal{O}(1)$ and the perturbative estimate of α_* is not so reliable there, although the chiral symmetry restoration in large N_f QCD has been supported by many other arguments, most notably the lattice QCD simulations,[23,24] which however suggest diverse results as to $N_f^{\rm cr}$; See e.g.,[25] for recent results.

point when approached from the side of the SSB phase, no isolated light particles do not exist in the conformal window, recently dubbed "unparticle".[26] This reflects the feature of the conformal symmetry in the conformal window. In fact explicit computations show no light (composite) spectra in the conformal window, in sharp contrast to the SχSB phase where light composite spectra do exist with mass of order $\mathcal{O}(m)$ which vanishes as we approach the conformal window $N_f \nearrow N_f^{\rm cr}$.[8,9,16]

The essence of CPT was illustrated [9] by a simpler model, 2-dimensional Gross-Neveu Model. This is the $D \to 2$ limit of the D-dimensional Gross-Neveu model ($2 < D < 4$) which has the beta function and the anomalous dimension:[27,28]

$$\beta(g) = -2g(g - g_*), \quad \gamma_m = 2g\,, \tag{16}$$

where $g = g_*(\equiv D/2 - 1) = g_{\rm cr}$ and $g = 0$ are respectively the UV and IR fixed points of the dimensionless four-fermion coupling, g, properly normalized (as $g_* = 1$ for the $D = 4$ NJL model). There exist light composites π, σ near the UV fixed point (phase boundary) $g \simeq g_*$ in both sides of symmetric ($0 < g < g_*$) and SSB ($g > g_*$) phases as in the NJL model.

Now we consider $D \to 2$ ($g_* \to 0$) where we have a well-known effective potential: $V(\sigma,\pi) \sim (1/g - 1)\rho^2 + \rho^2 \ln(\rho^2/\Lambda^2)$, or $\partial^2 V/\partial\rho^2|_{\rho=0} = -\infty$, where $\rho^2 = \pi^2 + \sigma^2$. This implies breakdown of the Ginzburg-Landau theory which distinguishes the SSB (< 0) and symmetric (> 0) phases by the signature of the finite $\partial^2 V/\partial\rho^2$ at the critical point $g = 0$. Eq. (16) now reads:

$$\beta(g) = -2g^2\,, \ \gamma_m|_{g=0} = 0 \qquad (D = 2)\,, \tag{17}$$

namely a *fusion of the UV and IR fixed points* at $g = 0$ as a result of *multiple zero* (not a simple zero) at $g = 0$. Now the symmetric phase is squeezed out to the region $g < 0$ (conformal phase) which corresponds to a *repulsive* four-fermion interaction and no composite states exist, while in the SSB phase ($g > 0$) there exists a composite state σ of mass $M_\sigma = 2m$ where the dynamical mass of the fermion is given by $m^2 \sim \Lambda^2 \exp(-1/g) \to 0$ ($g \to +0$), which shows an *essential singularity* scaling, in accord with the beta function with *multiple zero*, $\beta(g) = \Lambda\partial g/\partial\Lambda = -2g^2$. Note the would-be composite mass in the symmetric phase $|M|^2 \sim \Lambda^2 \exp(-1/g) \to \infty$ ($g \to -0$).

Now look at the SWC TC as modeled by the large N_f QCD: When the walking coupling $\alpha(Q) \simeq \alpha_*$ is close to the critical coupling, $\alpha_* \simeq \alpha_{\rm cr}$, we should include the *induced* four-fermion interaction, $(G/2)[(\bar\psi\psi)^2 + (\bar\psi i\gamma_5\psi)^2]$, which becomes relevant operator due to the anomalous dimension $\gamma_m = 1$, and the system becomes "gauged Nambu-Jona-Lasinio" model[29] whose solution in the full parameter space was obtained in Ref.[30]

Thus we may regard the *SWC TC as the gauged Nambu-Jona-Lasinio (NJL) model.* It was found[30] that SχSB solution exists for the parameter space $g > g_{(+)} = (1 + \sqrt{1 - \alpha_*/\alpha_{\rm cr}})^2/4$ ($\alpha_* < \alpha_{\rm cr}$) as well as the region $\alpha_* > \alpha_{\rm cr}$, where the dimensionless four-fermion coupling $g \equiv G\Lambda^2\,(N_{\rm TC}/4\pi^2)$ is normalized as $g = 1$ for $\alpha_* = 0$ (pure NJL model without gauge interaction). Based on the solution (including the running coupling case), the RGE flow in (α, g) space was found to be along the line of $\alpha = \alpha_*$ (α does not run), [h] on which the four-fermion coupling g runs, with the beta function and anomalous dimension given by [28,32,33]

$$\beta(g) = -2(g - g_{(+)})(g - g_{(-)}), \quad \gamma_m = 2g + \alpha_*/(2\alpha_{\rm cr}) \tag{18}$$

where $g = g_{(\pm)} \equiv (1 \pm \sqrt{1 - \alpha_*/\alpha_{\rm cr}})^2/4$ are regarded as the UV/IR fixed points (fixed lines) for $\alpha_* \leq \alpha_{\rm cr}$. The above anomalous dimension takes the values: $\gamma_m = 1 + \sqrt{1 - \alpha_*/\alpha_{\rm cr}}$ [31] at the UV fixed line while $\gamma_m = 1 - \sqrt{1 - \alpha_*/\alpha_{\rm cr}}$ at the IR fixed line. Light composite spectra only exist near the UV fixed line (phase boundary) $g \simeq g_{(+)}$ in both SSB ($g > g_{(+)}$) and symmetric ($g > g_{(+)}$) phases as in NJL model. Thus it follows that as $\alpha_* \to \alpha_{\rm cr}$ Eq. (18) takes the form

$$\beta(g) = -2(g - g_*)^2\,, \quad \gamma_m|_{g=g_*} = 1\,, \qquad (\alpha_* = \alpha_{\rm cr})\,, \tag{19}$$

with $g_{(\pm)} \to 1/4 \equiv g_*$, and hence we again got a *multiple zero* and *fusion of UV and IR fixed lines* [28,32,33] which corresponds to the *essential singularity* scaling;[30] $m^2 \sim \Lambda^2 \exp(-1/(g - g_*))$. A similar observation was also made recently.[34]

In passing, it should be stressed that the *anomalous dimension never changes discontinuously across the phase boundary* as is seen from Eq.(16) and Eq.(18).[27,28,32]

The scale anomaly in this case is given by:[9]

$$\begin{aligned}\langle \partial^\mu D_\mu \rangle = \langle \theta^\mu_\mu \rangle = 4\langle \theta^0_0 \rangle &= \frac{\beta(g)}{g} \cdot \frac{G}{2} \langle (\bar\psi\psi)^2 + (\bar\psi i\gamma_5\psi)^2 \rangle \\ &\simeq -m^4 \cdot (4N_f N_{\rm TC}/\pi^4) = \mathcal{O}(\Lambda_\chi^4)\,,\end{aligned} \tag{20}$$

where the second line was from the explicit computation[35] of the vacuum energy $\langle \theta^0_0 \rangle$ in the limit $\Lambda/m \to \infty$ ($g \to g_*$) at $\alpha \equiv \alpha_{\rm cr}$ (The result coincides with the one for $\alpha \to \alpha_{\rm cr}$ with $g \equiv 0$, see Eq.(24).[36]). Again there is a composite state this time having mass[15]

$$M_\sigma \to \sqrt{2} m\,, \tag{21}$$

as $g \to g_* + 0$, while there are no composites $|M|^2 \sim \Lambda^2 \exp(-1/(g - g_*)) \to \infty$ for $g \to g_* - 0$. Eq.(21) is compared with $M_\sigma = 2m$ in the pure NJL case with $\alpha \equiv 0$. This slightly lighter scalar may be identified with the techni-dilaton in the SWC TC. I will come back to this later.

The absence of the composites in the symmetric phase $g < g_*$ may be understood as in the 2-dimensional Gross-Neveu model for $g < 0$, namely the *repulsive* four-fermion interactions: From the analysis of the RG flow, it was argued[32] that the IR fixed line $g = g_{(-)}$ is due to the *induced* four-fermion interaction by the walking TC dynamics itself, while deviation from that line, $g - g_{(-)}$, is due to the *additional* four-fermion interactions, repulsive ($g < g_{(-)}$) and attractive ($g > g_{(-)}$), from UV

[h] The beta function in Eq.(12) may be regarded as an artificial one keeping $g \equiv$ const. which is not along the renormalized trajectory in the extended parameter space (α, g).

dynamics other than the TC (i.e., ETC). It is clear that no light composites exist for repulsive four-fermion interaction $g < g_{(-)}$, which becomes $g < g_*$ at $\alpha_* = \alpha_{cr}$.

4. *S* Parameter Constraint

Now we come to the next problem of TC, so-called S, T, U parameters[3] measuring possible new physics in terms of the deviation of the LEP precision experiments from the SM. In particular, S parameter excludes the TC as a simple scale-up of QCD which yields $S = (N_f/2) \cdot \hat{S}$ with $\hat{S}_{\rm QCD} = 0.32 \pm 0.04$. For a typical ETC model with one-family TC, $N_f = 8$,[2] we would get $S = \mathcal{O}(1)$ which is much larger than the experiments $S < 0.1$. This is the reason why many people believe that the TC is dead. However, since the simple scale-up of QCD was already ruled out by the FCNC as was discussed before, the real problem is whether or not the walking/conformal TC which solved the FCNC problem is also consistent with the S parameter constraint above. There have been many arguments[11,12] that the S parameter value could be reduced in the walking/conformal TC than in the simple scale-up of QCD. Recently such a reduction has also been argued[37,38] in a version of the holographic QCD[39] deformed to the walking/conformal TC by tuning a parameter to simulate the large anomalous dimension $\gamma_m \simeq 1$.

Here we present the most straightforward computation of the S parameter for the large N_f QCD, based on the SD equation and (inhomogeneous) Bethe-Salpeter (BS) equation in the ladder approximation.[12] The S parameter $S = (N_f/2)\hat{S}$ is defined by the slope of the the current correlators $\Pi_{JJ}(Q^2)$ at $Q^2 = 0$:

$$\hat{S} = -4\pi \frac{d}{dQ^2} \left[\Pi_{VV}(Q^2) - \Pi_{AA}(Q^2) \right] \Big|_{Q^2=0} , \tag{22}$$

where $\delta^{ab} \left(q_\mu q_\nu / q^2 - g_{\mu\nu} \right) \Pi_{JJ}(q^2) = \mathcal{F.T.}\ i\langle 0|T J_\mu^a(x) J_\nu^b(0)|0\rangle$, $(J_\mu^a(x) = V_\mu^a(x), A_\mu^a(x))$, with $F_\pi^2 = \Pi_{VV}(0) - \Pi_{AA}(0)$. The current correlators are obtained by closing the fermion legs of the BS amplitudes $\chi_\mu^{(J)}(p;q) \sim \mathcal{F.T.}$ $\langle 0|T\ \psi(r/2)\ \bar{\psi}(-r/2)\ J_\mu(x)\ |0\rangle$, which is determined by the ladder BS equation (Fig.3). Solving the BS equation with the fermion propagator given as the solution

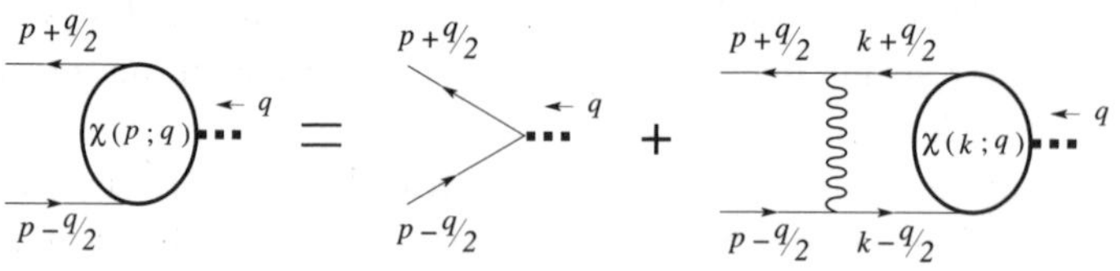

Fig. 3. Graphical expression of the BS equation in the ladder approximation.

of the ladder SD equation, we can evaluate the $\Pi_{VV}(Q^2) - \Pi_{AA}(Q^2)$ numerically. From this result we may read its slope at $Q^2 = 0$ to get $\hat{S}$.

The results show definitely smaller values of $\hat{S}$ than that in the ordinary QCD and moreover there is a tendency of the $\hat{S}$ getting reduced when approaching the conformal window $\alpha_* \searrow \alpha_{cr}$ ($N_f \nearrow N_f^{cr}$). However, due to technical limitation of the present computation getting very close to the conformal window, the reduction does not seem to be so dramatic as the walking TC being enough to be consistent with the experimental constraints. It is highly desirable to extend the computation further close to the conformal window.

Another approach to this problem is the deformation of the holographic QCD by the anomalous dimension. The reduction of S parameter in the SWC TC has been argued in a version of the hard-wall type bottom up holographic QCD[39] deformed to the SWCTC by tuning a parameter to simulate the large anomalous dimension $\gamma_m \simeq 1$.[37] We examined[38] such a possibility paying attention to the renormalization point dependence of the condensate. We explicitly calculated the S parameter in entire parameter space of the holographic SWC TC. We here take a set of F_π/M_ρ and γ_m. We find that $S > 0$ and it monotonically decreases to zero in accord with the previous results[37] . However, our result turned out fairly independent of the value of the anomalous dimension γ_m, yielding no particular suppression solely by tuning the anomalous dimension large, $\hat{S} \sim B(F_\pi/M_\rho)^2 \to 0$ as $F_\pi/M_\rho \to 0$, with $B \simeq 27(32)$ for $\gamma_m \simeq 1(0)$, in sharp contrast to the previous claim[37] . Although B contains full contributions from the infinite tower of the vector/axial-vector Kaluza-Klein modes (gauge bosons of hidden local symmetries)[40] of the 5-dimensional gauge bosons, the resultant value of B turned out close to $B \simeq 4\pi a \simeq 8\pi$ of the single ρ meson dominance, where $a \simeq 2$ is the parameter of the hidden local symmetry only for the ρ meson.[40] This implies that as far as the pure TC dynamics (without ETC dynamics, etc.) is concerned, an obvious way to dynamically reduce S parameter is to tune F_π/M_ρ very small, namely techni-ρ mass very large to several TeV region. (See, however, footnote below Eq.(30).)

We would need more dynamical information other than the holographic recipe, since the parameter corresponding to F_π/M_ρ as well as the scale parameter is a pure input in all the holographic models, whether bottom up or top down approach, in contrast to the underlying gauge theory which has only a single parameter, a scale parameter like Λ_{QCD}.

Curiously enough, when we calculate F_π/M_ρ from the SD and the homogeneous BS equations[16] and S from the SD and the inhomogeneous BS equation[12] both in the straightforward calculation in the ladder approximation, a set of the calculated values of $(F_\pi/M_\rho, S)$ lies on the line of the holographic result.[38]

5. Techni-Dilaton

Now we come to the discussions of Techni-dilaton (TD). Existence of two largely separated scales, $\Lambda_\chi \sim m$ and Λ_{TC} such that $\Lambda_\chi \ll \Lambda_{TC}$, is the most important feature of SWC-TC, in sharp contrast to the ordinary QCD with small number of flavors (in the chiral limit) where all the mass parameters like dynamical mass of quarks are of order of the single scale parameter of the theory Λ_{QCD}, $m \sim \Lambda_\chi \sim \Lambda_{QCD}$. See Fig. 1. The intrinsic scale Λ_{TC} is related with the scale anomaly corresponding to the *perturbative* running effects of the coupling, with the ordinary two-loop beta function $\beta(\alpha)$ in the Region I, in the same sense as in QCD.

$$\langle \partial^\mu D_\mu \rangle = \langle \theta^\mu_\mu \rangle = \frac{\beta(\alpha)}{4\alpha^2} \langle \alpha G^2_{\mu\nu} \rangle = \mathcal{O}(\Lambda^4_{\rm TC}), \tag{23}$$

which implies that all the techni-glue balls have mass of $\mathcal{O}(\Lambda_{\rm TC})$.

On the other hand, the scale Λ_χ is related with totally different scale anomaly due to the dynamical generation of $m\,(\sim \Lambda_\chi)$ which does exist even in the idealized case with non-running coupling $\alpha(\mu) \equiv \alpha (> \alpha_{\rm cr})$ such as the Maskawa-Nakajima solution,[1] as was discussed some time ago.[36] Such an idealized case well simulates the dynamics of Region II of Fig. 1 ,[8,9] with anomalous dimension $\gamma_m \simeq 1$ and $m \ll \Lambda_{\rm TC}$ in the numerical calculations,[16] with the *perturbative* coupling constant in Region II being almost constant slightly larger than $\alpha_{\rm cr}$, $\alpha_{\rm cr} < \alpha(\mu) (< \alpha_*)$, for a wide infrared region. The coupling $\alpha \equiv \alpha_*$ in the "idealized Region II" actually runs *non-perturbatively* according to the *essential-singularity* scaling (Miransky scaling[21]) of mass generation, Eq.(11), with the *non-perturbative* beta function $\beta_{\rm NP}(\alpha)$, Eq.(12), having a *multiple zero* at $\alpha = \alpha_{\rm cr}$. Then the *non-perturbative* scale anomaly reads[9,i]

$$\langle \partial^\mu D_\mu \rangle_{\rm NP} = \langle \theta^\mu_\mu \rangle_{\rm NP} = \frac{\beta_{\rm NP}(\alpha)}{4\alpha^2} \langle \alpha G^2_{\mu\nu} \rangle_{\rm NP} = -m^4 \cdot \frac{4 N_f N_{\rm TC}}{\pi^4} = -\mathcal{O}(\Lambda^4_\chi), \tag{24}$$

where $\langle \cdots \rangle_{\rm NP}$ is the quantity with the perturbative contributions subtracted:[36] $\langle \cdots \rangle_{\rm NP} \equiv \langle \cdots \rangle - \langle \cdots \rangle_{\rm Perturbative}$. Eq.(24) coincides with Eq.(20) and $\langle \partial^\mu D_\mu \rangle_{\rm NP} / \Lambda^4_{\rm TC}$ vanishes with $\langle \partial^\mu D_\mu \rangle_{\rm NP} / m^4 \to$ const. $\neq 0$, when we approach the conformal window from the broken phase $\alpha_* \searrow \alpha_{\rm cr}$ $(m/\Lambda_{\rm TC} \to 0)$. All the techni-fermion bound states have mass of order of m,[41] while there are no light bound states in the symmetric phase (conformal window) $\alpha_* < \alpha_{\rm cr}$, a characteristic feature of the conformal phase transition.[9] The TD is associated with the latter scale anomaly and should have mass on order of $m (\ll \Lambda_{\rm TC})$.

5.1. *Calculation from Gauged NJL Model in the Ladder SD Equation*[15]

More concretely, the mass of TD or scalar bound state in the SWC-TC was estimated in various methods: The first method [15] was based on the the ladder SD equation for the gauged NJL model which well simulates[8,9] the conformal phase transition in the large N_f QCD. The result was already given by Eq.(21):

$$M_{\rm TD} \simeq \sqrt{2} m\,. \tag{25}$$

[i] In terms of the gauged NJL model mentioned in Section 3 this is the expression of the scale anomaly for $\alpha \to \alpha_{\rm cr}$ with $g =$ const. $= g_*$, in contrast to Eq.(20) for $g \to g_*$ with $\alpha =$ const. $= \alpha_* = \alpha_{\rm cr}$. Both yield the same vacuum energy and hence the same scale anomaly.

5.2. *Straightforward Calculation from Ladder SD and BS Equations*

Also a straightforward calculation[16] of mass of TD, the scalar bound state was made in the vicinity of the CBZ-IR fixed point in the large N_f QCD, based on the coupled use of the ladder SD equation and (*homogeneous*) BS equation lacking the first term in Fig. 3: All the bound states masses are $M = \mathcal{O}(m)$ and $M/\Lambda_{\rm TC} \to 0$, when approaching the conformal window $\alpha_* \to \alpha_{\rm cr}$ ($N_f \to N_f^{\rm cr}$) such that $m/\Lambda_{\rm TC} \to 0$. Near the conformal window ($N_f \nearrow N_f^{\rm cr}$) the calculated values are $M_\rho/F_\pi \simeq 11, M_{a_1}/F_\pi \simeq 12$ (near degenerate !). On the other hand, the scalar mass sharply drops near the the conformal window, $M_{\rm TD}/F_\pi \searrow 4$, or

$$M_{\rm TD} \searrow 1.5m \simeq \sqrt{2}m \, (< M_\rho, M_{a_1})\,. \tag{26}$$

Note that in this calculation $M_{\rm TD}/F_\pi \to$ const. $\neq 0$ and hence there is no isolated massless scalar bound states even in the limit $N_f \to N_f^{\rm cr}$. The result is consistent with Eq.(25) and is contrasted to the ordinary QCD where the scalar mass is larger than those of the vector mesons ("higgsless") within the same framework of ladder SD/BS equation approach. The result would imply

$$m_{\rm TD} \simeq 500\,\text{GeV} \tag{27}$$

in the case of the one-family TC model with $F_\pi \simeq 125\,$GeV.

5.3. *Holographic Techni-Dilaton*[13]

Recently, we have calculated[13] mass of TD in an extension of the previous paper[38] on the hard-wall-type bottom-up holographic SWC-TC by including effects of (techni-) gluon condensation parameterized as

$$\Gamma \equiv \left(\frac{\left(\frac{1}{\pi}\langle \alpha G_{\mu\nu}^2\rangle / F_\pi^4\right)}{\left(\frac{1}{\pi}\langle \alpha G_{\mu\nu}^2\rangle / f_\pi^4\right)_{\rm QCD}} \right)^{1/4} \tag{28}$$

through the bulk flavor/chiral-singlet scalar field Φ_X, in addition to the conventional bulk scalar field Φ dual to the chiral condensate.

The five-dimensional action is given by

$$S_5 = \int d^4x \int_\epsilon^{z_m} dz\, \sqrt{-g}\, \frac{1}{g_5^2} e^{c g_5^2 \Phi_X(z)} \Big(-\frac{1}{4} \text{Tr}\, \left[L_{MN}L^{MN} + R_{MN}R^{MN} \right] \\ + \text{Tr}\, \left[D_M \Phi^\dagger D^M \Phi - m_\Phi^2 \Phi^\dagger \Phi \right] + \frac{1}{2} \partial_M \Phi_X \partial^M \Phi_X \Big)\,, \tag{29}$$

where the anti-de-Sitter space (AdS$_5$) with the curvature radius L of AdS$_5$ is described by the metric $ds^2 = g_{MN}dx^M dx^N = (L/z)^2 \left(\eta_{\mu\nu}dx^\mu dx^\nu - dz^2\right)$ with $\eta_{\mu\nu} = \text{diag}[1,-1,-1,-1], g = \det[g_{MN}] = -(L/z)^{10}$; g_5 denotes the gauge coupling in five-dimension and c is the dimensionless coupling constant, and $L_M(R_M) =$

$L_M^a(R_M^a)T^a$ with the generators of $SU(N_f)$ are normalized by $\mathrm{Tr}[T^aT^b] = \delta^{ab}$; $L(R)_{MN} = \partial_M L(R)_N - \partial_N L(R)_M - i[L(R)_M, L(R)_N]$. The covariant derivative acting on Φ is defined as $D_M\Phi = \partial_M\Phi + iL_M\Phi - i\Phi R_M$.

The TD, a flavor-singlet scalar bound state of techni-fermion and anti-techni-fermion, will be identified with the lowest KK mode coming from the bulk scalar field Φ, not Φ_X. Thanks to the additional explicit bulk scalar field Φ_X, we naturally improve the matching with the OPE of the underlying theory (QCD and SWC-TC) for current correlators so as to reproduce gluonic $1/Q^4$ term, which is clearly distinguished from the same $1/Q^4$ terms from chiral condensate in the case of SWC-TC with $\gamma_m \simeq 1$. Our model with $\gamma_m = 0$ and $N_f = 3$ well reproduces the real-life QCD.

It is rather straightforward[37,39] to compute masses of the techni-ρ meson (M_ρ), the techni-a_1 meson (M_{a_1}), while for that of the TD, flavor-singlet scalar meson ($M_{\rm TD}$), we would need additional IR potential with quartic coupling λ to stabilize the SχSB vacuum.[42] Such an IR potential might be regarded as generated by techni-fermion loop effects and we naturally expect $\lambda \sim N_{\rm TC}/(4\pi)^2$. The S parameter was also calculated through the current correlators by the standard way. We found general tendency of the dependence of the meson masses relative to F_π, (M_ρ/F_π, M_{a_1}/F_π, $M_{\rm TD}/F_\pi$) on γ_m, S, and Γ.

We find a characteristic feature of the techni-dilaton mass related to the conformality of SWC-TC: For fixed S and γ_m, absolute values of (M_ρ/F_π) and (M_{a_1}/F_π) are not sensitive to Γ, although they get *degenerate for large* Γ. On the contrary, $(M_{\rm TD}/F_\pi)$ *substantially decreases as* Γ *increases.* Actually, in the formal limit $\Gamma \to \infty$ we would have $(M_{\rm TD}/F_\pi) \to 0$ (This is contrast to the straightforward computation through ladder SD and BS equations mentioned before[16]). For fixed S and Γ, again (M_ρ/F_π) and (M_{a_1}/F_π) are not sensitive to γ_m, while $(M_{\rm TD}/F_\pi)$ substantially decreases as γ_m increases.

Particularly for the case of $\gamma_m = 1$, we study the dependence of the S parameter on (M_ρ/F_π) for typical values of Γ. It is shown that the techni-gluon contribution reduces the value of S about 10% in the region of $\hat{S} \lesssim 0.1$, although the general tendency is similar to the previous paper[38] without techni-gluon condensation: $\hat{S}$ decreases to zero monotonically with respect to (F_π/M_ρ). This implies (M_ρ/F_π) necessarily increases when $\hat{S}$ is required to be smaller.

To be more concrete, we consider a couple of typical models of SWC-TC with $\gamma_m \simeq 1$ and $N_{\rm TC} = 2, 3, 4$ based on the CBZ-IRFP in the large N_f QCD. Using the non-perturbative conformal anomaly Eq.(24) together with the non-perturbative beta function Eq.(12) and Eq.(11), we find a concrete relation between Γ and $(\Lambda_{\rm ETC}/F_\pi)$: In the case of $N_{\rm TC} = 3$ ($N_f = 4N_{\rm TC}$) and $S \simeq 0.1$, we have $\Gamma \simeq 7$ for $(\Lambda_{\rm ETC}/F_\pi) = 10^4$–$10^5$ (required by the FCNC constraint). Thanks to the large anomalous dimension $\gamma_m \simeq 1$ and large techni-gluon condensation $\Gamma \simeq 7$, we obtain a relatively light techni-dilaton with mass

$$M_{\rm TD} \simeq 600\,\mathrm{GeV} \tag{30}$$

compared with $M_\rho \simeq M_{a_1} \simeq 3.8\,\text{TeV}$ (almost degenerate). Eq. (30) is consistent with the perturbative unitarity of $W_L W_L$ scattering even for large M_ρ, M_{a_1}. Note that largeness of M_ρ and M_{a_1} is essentially determined by the requirement of $S = 0.1$ fairly independently of techni-gluon condensation. [j]

The essential reason for the large Γ is due to the existence of the wide conformal region $F_\pi(\sim m) < \mu < \Lambda_{\text{ETC}}$ with $\Lambda_{\text{ETC}}/F_\pi = 10^4$–$10^5$, which yields the smallness of the beta function (see Eq.(12) and Eq. (11)) and hence amplifies the techni-gluon condensation compared with the ordinary QCD with $\Gamma = 1$. Actually, in the idealized (phenomenologically uninteresting) limit $\Lambda_{\text{ETC}}/F_\pi \to \infty$, we would have $\Gamma \to \infty$, which in turn would imply $M_{\text{TD}}/F_\pi \to 0$ as mentioned above. [k]

To conclude, various methods predicted the mass of the techni-dilaton ("conformal Higgs") in the range of $M_{\text{TD}} \simeq 500 - 600$ GeV, which is within reach of LHC discovery.

Since the SWC TC models are strong coupling theories and the ladder approximation/holographic calculations would be no more than a qualitative hint, more reliable calculations are certainly needed, including the lattice simulations, before drawing a definite conclusion about the physics predictions. Besides the phase diagram including the TC-induced/ETC-driven four-fermion couplings on the lattice, more reliable calculations such as the spectra as well as anomalous dimensions, non-perturbative beta functions, S parameter, etc. are highly desired.

Acknowledgments

We thank K. Haba, M. Harada, M. Hashimoto, M. Kurachi, and S. Matsuzaki for collaborations on the relevant topics and useful discussions. This work was supported by the JSPS Grant-in-Aid for Scientific Research(S) # 22224003 and by the Nagoya University Foundation.

References

1. T. Maskawa and H. Nakajima, Prog. Theor. Phys. **52** (1974), 1326; **54** (1975), 860.
2. S. Weinberg, Phys. Rev. D **13** (1976), 974; D **19** (1979), 1277; L. Susskind, Phys. Rev. D **20** (1979), 2619: See for a review of earlier literature, E. Farhi and L. Susskind, Phys. Rep. **74** (1981), 277.

[j] The calculated S parameter here was from the TC dynamics alone and could be drastically changed by incorporating contributions from the generation mechanism of the mass of SM fermions such as the strong ETC dynamics. For instance, the fermion delocalization[43] in the Higgsless models as a possible analogue of certain ETC effects in fact can cancel large positive S arising from the 5-dimensional gauge sector which corresponds to the pure TC dynamics. If it is the case in the explicit ETC model, then the overall mass scale of techni-hadrons including TD would be much lower than the above estimate down to, say, 300 GeV.

[k] This does not mean existence of true (exactly massless) NG boson of the broken scale symmetry, since such a would-be NG boson gets decoupled in our case: The decay constant of techni-dilaton F_{TD} would diverge, $F_{\text{TD}}/F_\pi \to \infty$, in that limit through the PCDC relation $F_{\text{TD}}^2 = -4\langle\theta_\mu^\mu\rangle/M_{\text{TD}}^2 \sim m^4/M_{\text{TD}}^2 \sim F_\pi^4/M_{\text{TD}}^2$. The scale symmetry is broken explicitly as well as spontaneously.

3. M. E. Peskin and T. Takeuchi, Phys. Rev. Lett. **65** (1990), 964; Phys. Rev. D **46** (1992), 381; B. Holdom and J. Terning, Phys. Lett. B **247** (1990), 88; M. Golden and L. Randall, Nucl. Phys. B **361** (1991), 3.
4. K. Yamawaki, M. Bando and K. Matumoto, Phys. Rev. Lett. **56** (1986), 1335; M. Bando, K. Matumoto and K. Yamawaki, Phys. Lett. B **178** (1986), 308. See also M. Bando, T. Morozumi, H. So and K. Yamawaki, Phys. Rev. Lett. **59** (1987), 389 . The solution of the FCNC problem by the large anomalous dimension was first considered by B. Holdom, based on a pure assumption of the existence of UV fixed point without explicit dynamics and hence without definite prediction of the value of the anomalous dimension. See B. Holdom, Phys. Rev. D **24** (1981), 1441.
5. T. Akiba and T. Yanagida, Phys. Lett. B **169** (1986), 432; T. W. Appelquist, D. Karabali and L. C. R. Wijewardhana, Phys. Rev. Lett. **57** (1986), 957; T. Appelquist and L. C. R. Wijewardhana, Phys. Rev. D **36** (1987), 568. For an earlier work on this line based on pure numerical analysis see B. Holdom, Phys. Lett. B **150** (1985), 301.
6. C. T. Hill and E. H. Simmons, Phys. Rept. **381** (2003), 235 [Erratum-ibid. **390** (2004), 553]; K. Yamawaki, Lecture at 14th Symposium on Theoretical Physics, Cheju, Korea, July 1995, arXiv:hep-ph/9603293.
7. K. D. Lane and M. V. Ramana, Phys. Rev. D **44**, 2678 (1991).
8. T. Appelquist, J. Terning and L. C. Wijewardhana, Phys. Rev. Lett. **77** (1996), 1214; T. Appelquist, A. Ratnaweera, J. Terning and L. C. Wijewardhana, Phys. Rev. D **58** (1998), 105017.
9. V. A. Miransky and K. Yamawaki, Phys. Rev. D **55** (1997), 5051 [Errata; **56** (1997), 3768].
10. W. E. Caswell, Phys. Rev. Lett. **33**, 244 (1974); T. Banks and A. Zaks, Nucl. Phys. B **196** (1982), 189.
11. R. Sundrum and S. D. H. Hsu, Nucl. Phys. B **391** (1993), 127; T. Appelquist and G. Triantaphyllou, Phys. Lett. B **278** (1992), 345; T. Appelquist and F. Sannino, Phys. Rev. D **59** (1999), 067702.
12. M. Harada, M. Kurachi and K. Yamawaki, Prog. Theor. Phys. **115** (2006), 765 ; M. Kurachi and R. Shrock, Phys. Rev. D **74**, 056003 (2006); M. Kurachi, R. Shrock and K. Yamawaki, ibid D **76**, 035003 (2007).
13. K. Haba, S. Matsuzaki and K. Yamawaki, arXiv:1003.2841 [hep-ph]; arXiv:1006.2526 [hep-ph].
14. K. Yamawaki, Prog. Theor. Phys. Suppl. **180**, 1 (2010) [arXiv:0907.5277 [hep-ph]]; See also Prog. Theor. Phys. Suppl. **167**, 127 (2007).
15. S. Shuto, M. Tanabashi and K. Yamawaki, in *Proc. 1989 Workshop on Dynamical Symmetry Breaking*, Dec. 21-23, 1989, Nagoya, eds. T. Muta and K. Yamawaki (Nagoya Univ., Nagoya, 1990) 115-123; M. S. Carena and C. E. M. Wagner, Phys. Lett. B **285**, 277 (1992); M. Hashimoto, Phys. Lett. B **441**, 389 (1998).
16. M. Harada, M. Kurachi and K. Yamawaki, Phys. Rev. D **68** (2003), 076001; M. Kurachi and R. Shrock, JHEP **0612**, 034 (2006).
17. S. Dimopoulos and L. Susskind, Nucl. Phys. B **155** (1979), 237; E. Eichten and K. D. Lane, Phys. Lett. B **90** (1980), 125.
18. K. Yamawaki and T. Yokota, Phys. Lett. B **113** (1982), 293; Nucl. Phys. B **223** (1983), 144.
19. K. Johnson, M. Baker and R. Willey, Phys. Rev. **136** (1964), B1111; M. Baker and K. Johnson, Phys. Rev. **D3** (1971), 2516; R. Jackiw and K. Johnson Phys. Rev. **D8** (1973), 2386; J.M. Cornwall and R.E. Norton, ibid, 3338.
20. R. Fukuda and T. Kugo, Nucl. Phys. B **117** (1976), 250.
21. V. A. Miransky, Nuovo Cim. A **90** (1985), 149.

22. F. Sannino and K. Tuominen, Phys. Rev. D **71**, 051901 (2005); D. K. Hong, S. D. H. Hsu and F. Sannino, Phys. Lett. B **597**, 89 (2004); For a review F. Sannino, arXiv:0804.0182 [hep-ph].
23. Y. Iwasaki, K. Kanaya, S. Kaya, S. Sakai and T. Yoshie, Nucl. Phys. (Proc. Suppl.) **53** (1997), 449; Prog. Theor. Phys. Suppl. No. 131 (1998), 415.
24. T. Appelquist, G. T. Fleming and E. T. Neil, Phys.Rev.Lett.100:171607,2008, Erratum-ibid.102:149902,2009; Phys. Rev. D **79**, 076010 (2009); A. Deuzeman, M. P. Lombardo and E. Pallante, Phys. Lett. B **670**, 41 (2008); Talks presented at SCGT 09, http://www.eken.phys.nagoya-u.ac.jp/scgt09/ .
25. Talks at SCGT 09: http://www.eken.phys.nagoya-u.ac.jp/scgt09/ .
26. H. Georgi, Phys. Rev. Lett. **98**, 221601 (2007).
27. Y. Kikukawa and K. Yamawaki, Phys. Lett. B **234**, 497 (1990).
28. K. i. Kondo, M. Tanabashi and K. Yamawaki, Prog. Theor. Phys. **89**, 1249 (1993).
29. W. A. Bardeen, C. N. Leung and S. T. Love, Phys. Rev. Lett. **56**, 1230 (1986).
30. K. i. Kondo, H. Mino and K. Yamawaki, Phys. Rev. D **39**, 2430 (1989); K. Yamawaki, in *Proc. Johns Hopkins Workshop on Current Problems in Particle Theory 12, Baltimore, June 8-10, 1988*, edited by G. Domokos and S. Kovesi-Domokos (World Scientific Pub. Co., Singapore 1988); T. Appelquist, M. Soldate, T. Takeuchi, and L. C. R. Wijewardhana, *ibid.*
31. V. A. Miransky and K. Yamawaki, Mod. Phys. Lett. A **4** (1989), 129.
32. K. i. Kondo, S. Shuto and K. Yamawaki, Mod. Phys. Lett. A **6**, 3385 (1991).
33. K. I. Aoki, K. Morikawa, J. I. Sumi, H. Terao and M. Tomoyose, Prog. Theor. Phys. **102**, 1151 (1999).
34. D. B. Kaplan, J. W. Lee, D. T. Son and M. A. Stephanov, Phys. Rev. D **80**, 125005 (2009).
35. T. Nonoyama, T. B. Suzuki and K. Yamawaki, Prog. Theor. Phys. **81**, 1238 (1989).
36. V. A. Miransky and V. P. Gusynin, Prog. Theor. Phys. **81** (1989) 426.
37. D. K. Hong and H. U. Yee, Phys. Rev. D **74** (2006), 015011; M. Piai, arXiv:hep-ph/0608241; K. Agashe, C. Csaki, C. Grojean and M. Reece, JHEP **0712**, 003 (2007).
38. K. Haba, S. Matsuzaki and K. Yamawaki, Prog. Theor. Phys. **120**, 691 (2008).
39. J. Erlich, E. Katz, D. T. Son and M. A. Stephanov, Phys. Rev. Lett. **95** (2005), 261602; L. Da Rold and A. Pomarol, Nucl. Phys. B **721**, 79 (2005).
40. M. Bando, T. Kugo, S. Uehara, K. Yamawaki and T. Yanagida, Phys. Rev. Lett. **54** (1985), 1215; M. Bando, T. Kugo and K. Yamawaki, Nucl. Phys. B **259** (1985), 493; Phys. Rept. **164** (1988), 217; M. Harada and K. Yamawaki, Phys. Rept. **381** (2003), 1.
41. R. S. Chivukula, Phys. Rev. D **55** (1997), 5238.
42. L. Da Rold and A. Pomarol, JHEP **0601**, 157 (2006).
43. G. Cacciapaglia, C. Csaki, C. Grojean and J. Terning, Phys. Rev. D **71**, 035015 (2005); R. Foadi, S. Gopalakrishna and C. Schmidt, Phys. Lett. B **606**, 157 (2005). R. S. Chivukula, E. H. Simmons, H. J. He, M. Kurachi and M. Tanabashi, Phys. Rev. D **72**, 015008 (2005)

Phase Diagram of Strongly Interacting Theories

Francesco Sannino
CP^3-Origins, University of Southern Denmark, Odense, 5230 M, Denmark
E-mail: sannino@cp3.sdu.dk, sannino@cp3-origins.net
www.cp3-origins.dk

We summarize the phase diagrams of SU, SO and Sp gauge theories as function of the number of flavors, colors, and matter representation as well as the ones of phenomenologically relevant chiral gauge theories such as the Bars-Yankielowicz and the generalized Georgi-Glashow models. We finally report on the intriguing possibility of the existence of gauge-duals for nonsupersymmetric gauge theories and the impact on their conformal window.

1. Phases of Gauge Theories

Models of dynamical breaking of the electroweak symmetry are theoretically appealing and constitute one of the best motivated natural extensions of the standard model (SM). We have proposed several models[1–8] possessing interesting dynamics relevant for collider phenomenology[9–12] and cosmology.[7,13–31] The structure of one of these models, known as Minimal Walking Technicolor, has led to the construction of a new supersymmetric extension of the SM featuring the maximal amount of supersymmetry in four dimension with a clear connection to string theory, i.e. Minimal Super Conformal Technicolor.[32] These models are also being investigated via first principle lattice simulations[33–48] [a]. An up-to-date review is Ref. 59 while an excellent review updated till 2003 is Ref. 60. These are also among the most challenging models to work with since they require deep knowledge of gauge dynamics in a regime where perturbation theory fails. In particular, it is of utmost importance to gain information on the nonperturbative dynamics of non-abelian four dimensional gauge theories. The phase diagram of $SU(N)$ gauge theories as functions of number of flavors, colors and matter representation has been investigated in.[1,61–64] The analytical tools which will be used here for such an exploration are: i) The conjectured *physical* all orders beta function for nonsupersymmetric gauge theories with fermionic matter in arbitrary representations of the gauge group;[63] ii) The truncated Schwinger-Dyson equation (SD)[65–67] (referred also as the ladder approximation in the literature); The Appelquist-Cohen-Schmaltz (ACS) conjecture[68] which makes use of the counting of the thermal degrees of freedom at high and low temperature. These are the methods which we have used in our investigations. However several very interesting and competing analytic approaches[69–80] have been pro-

[a] Earlier interesting models[49–51] have contributed triggering the lattice investigations for the conformal window with theories featuring fermions in the fundamental representation[52–58]

posed in the literature. What is interesting is that despite the very different starting point the various methods agree qualitatively on the main features of the various conformal windows presented here.

1.1. *Physical all orders Beta Function - Conjecture*

Recently we have conjectured an all orders beta function which allows for a bound of the conformal window[63] of $SU(N)$ gauge theories for any matter representation. The predictions of the conformal window coming from the above beta function are nontrivially supported by all the recent lattice results.[33,34,81–85]

In[86] we further assumed the form of the beta function to hold for $SO(N)$ and $Sp(2N)$ gauge groups and further extended in[59] to chiral gauge theories. Consider a generic gauge group with $N_f(r_i)$ Dirac flavors belonging to the representation r_i, $i = 1, \ldots, k$ of the gauge group. The conjectured beta function reads:

$$\beta(g) = -\frac{g^3}{(4\pi)^2} \frac{\beta_0 - \frac{2}{3} \sum_{i=1}^{k} T(r_i)\, N_f(r_i)\, \gamma_i(g^2)}{1 - \frac{g^2}{8\pi^2} C_2(G) \left(1 + \frac{2\beta_0'}{\beta_0}\right)} , \tag{1}$$

with

$$\beta_0 = \frac{11}{3} C_2(G) - \frac{4}{3} \sum_{i=1}^{k} T(r_i) N_f(r_i) \quad \text{and} \quad \beta_0' = C_2(G) - \sum_{i=1}^{k} T(r_i) N_f(r_i) \,. \tag{2}$$

The generators T_r^a, $a = 1 \ldots N^2 - 1$ of the gauge group in the representation r are normalized according to $\text{Tr}\left[T_r^a T_r^b\right] = T(r)\delta^{ab}$ while the quadratic Casimir $C_2(r)$ is given by $T_r^a T_r^a = C_2(r) I$. The trace normalization factor $T(r)$ and the quadratic Casimir are connected via $C_2(r)d(r) = T(r)d(G)$ where $d(r)$ is the dimension of the representation r. The adjoint representation is denoted by G.

The beta function is given in terms of the anomalous dimension of the fermion mass $\gamma = -d\ln m/d\ln\mu$ where m is the renormalized mass, similar to the supersymmetric case.[87–89] The loss of asymptotic freedom is determined by the change of sign in the first coefficient β_0 of the beta function. This occurs when

$$\sum_{i=1}^{k} \frac{4}{11} T(r_i) N_f(r_i) = C_2(G) \,, \qquad \text{Loss of AF.} \tag{3}$$

At the zero of the beta function we have

$$\sum_{i=1}^{k} \frac{2}{11} T(r_i) N_f(r_i) \left(2 + \gamma_i\right) = C_2(G) \,, \tag{4}$$

Hence, specifying the value of the anomalous dimensions at the IRFP yields the last constraint needed to construct the conformal window. Having reached the zero of the beta function the theory is conformal in the infrared. For a theory to be conformal the dimension of the non-trivial spinless operators must be larger than one in order not to contain negative norm states.[90–92] Since the dimension of the chiral condensate is $3 - \gamma_i$ we see that $\gamma_i = 2$, for all representations r_i, yields the maximum possible bound

$$\sum_{i=1}^{k} \frac{8}{11} T(r_i) N_f(r_i) = C_2(G) , \qquad \gamma_i = 2 . \tag{5}$$

In the case of a single representation this constraint yields

$$N_f(r)^{\rm BF} \geq \frac{11}{8} \frac{C_2(G)}{T(r)} , \qquad \gamma = 2 . \tag{6}$$

The actual size of the conformal window can be smaller than the one determined by the bound above, Eq. (3) and (5). It may happen, in fact, that chiral symmetry breaking is triggered for a value of the anomalous dimension less than two. If this occurs the conformal window shrinks. Within the ladder approximation[65,66] one finds that chiral symmetry breaking occurs when the anomalous dimension is close to one. Picking $\gamma_i = 1$ we find:

$$\sum_{i=1}^{k} \frac{6}{11} T(r_i) N_f(r_i) = C_2(G) , \qquad \gamma = 1 . \tag{7}$$

In the case of a single representation this constraint yields

$$N_f(r)^{\rm BF} \geq \frac{11}{6} \frac{C_2(G)}{T(r)} , \qquad \gamma = 1 . \tag{8}$$

When considering two distinct representations the conformal window becomes a three dimensional volume, i.e. the conformal *house*.[62] Of course, we recover the results by Banks and Zaks[93] valid in the perturbative regime of the conformal window.

We note that the presence of a physical IRFP requires the vanishing of the beta function for a certain value of the coupling. The opposite however is not necessarily true; the vanishing of the beta function is not a sufficient condition to determine if the theory has a fixed point unless the beta function is *physical*. By *physical* we mean that the beta function allows to determine simultaneously other scheme-independent quantities at the fixed point such as the anomalous dimension of the mass of the fermions. This is exactly what our beta function does. In fact, in the case of a single representation, one finds that at the zero of the beta function one has:

$$\gamma = \frac{11 C_2(G) - 4 T(r) N_f}{2 T(r) N_f} . \tag{9}$$

1.2. *Schwinger-Dyson in the Rainbow Approximation*

For nonsupersymmetric theories another way to get quantitative estimates is to use the *rainbow* approximation to the Schwinger-Dyson equation.[94,95] After a series of approximations (see[59] for a review) one deduces for an $SU(N)$ gauge theory with N_f Dirac fermions transforming according to the representation r the critical number of flavors above which chiral symmetry maybe unbroken:

$$N_f^{\rm SD} = \frac{17 C_2(G) + 66 C_2(r)}{10 C_2(G) + 30 C_2(r)} \frac{C_2(G)}{T(r)} . \tag{10}$$

Comparing with the previous result obtained using the all orders beta function we see that it is the coefficient of $C_2(G)/T(r)$ which is different. We note that in[96] it has been advocated a coefficient similar to the one of the all-orders beta function.

1.3. *The SU, SO and Sp phase diagrams*

We consider here gauge theories with fermions in any representation of the $SU(N)$ gauge group[1,61–63,97] using the various analytic methods described above.

Here we plot in Fig. 1 the conformal windows for various representations predicted with the physical all orders beta function and the SD approaches.

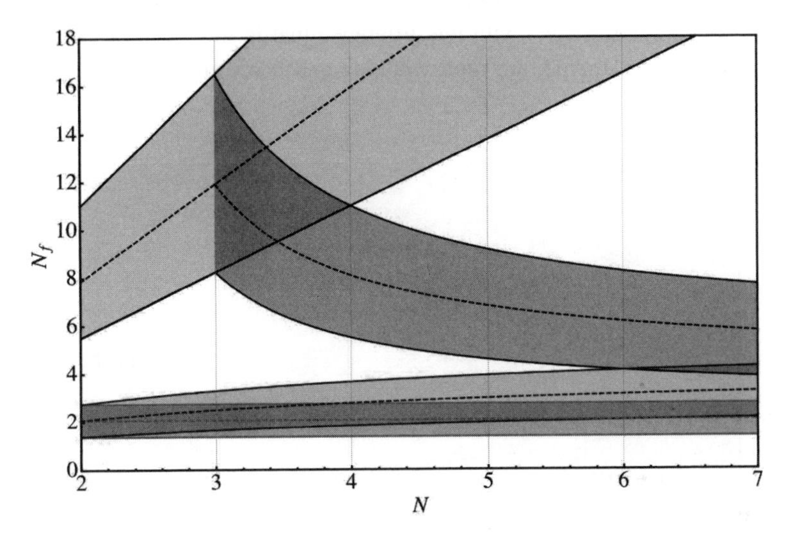

Fig. 1. Phase diagram for nonsupersymmetric theories with fermions in the: i) fundamental representation (black), ii) two-index antisymmetric representation (blue), iii) two-index symmetric representation (red), iv) adjoint representation (green) as a function of the number of flavors and the number of colors. The shaded areas depict the corresponding conformal windows. Above the upper solid curve the theories are no longer asymptotically free. In between the upper and the lower solid curves the theories are expected to develop an infrared fixed point according to the all orders beta function. The area between the upper solid curve and the dashed curve corresponds to the conformal window obtained in the ladder approximation.

The ladder result provides a size of the window, for every fermion representation, smaller than the maximum bound found earlier. This is a consequence of the value of the anomalous dimension at the lower bound of the window. The unitarity constraint corresponds to $\gamma = 2$ while the ladder result is closer to $\gamma \sim 1$. Indeed if we pick $\gamma = 1$ our conformal window approaches the ladder result. Incidentally, a value of γ larger than one, still allowed by unitarity, is a welcomed feature when using this window to construct walking technicolor theories. It may allow for the physical value of the mass of the top while avoiding a large violation of flavor changing neutral currents[98] which were investigated in[99] in the case of the ladder approximation for minimal walking models.

1.3.1. *The $Sp(2N)$ phase diagram*

$Sp(2N)$ is the subgroup of $SU(2N)$ which leaves the tensor $J^{c_1 c_2} = (\mathbf{1}_{N\times N} \otimes i\sigma_2)^{c_1 c_2}$ invariant. Irreducible tensors of $Sp(2N)$ must be traceless with respect to $J^{c_1 c_2}$. Here we consider $Sp(2N)$ gauge theories with fermions transforming according to a given irreducible representation. Since $\pi^4\left[Sp(2N)\right] = Z_2$ there is a Witten topological anomaly[100] whenever the sum of the Dynkin indices of the various matter fields is odd. The adjoint of $Sp(2N)$ is the two-index symmetric tensor.

In Figure 2 we summarize the relevant zero temperature and matter density phase diagram as function of the number of colors and Weyl flavors (N_{Wf}) for $Sp(2N)$ gauge theories. For the vector representation $N_{Wf} = 2N_f$ while for the two-index theories $N_{Wf} = N_f$. The shape of the various conformal windows are very similar to the ones for $SU(N)$ gauge theories[1,61,63] with the difference that in this case the two-index symmetric representation is the adjoint representation and hence there is one less conformal window.

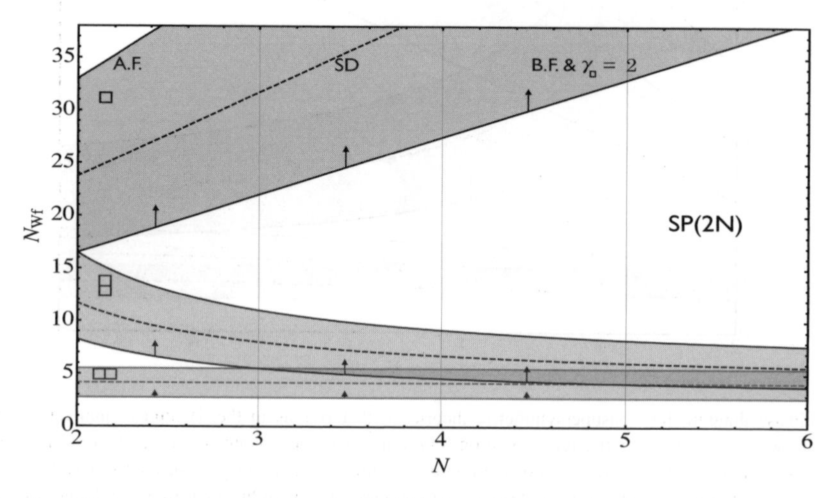

Fig. 2. Phase Diagram, from top to bottom, for $Sp(2N)$ Gauge Theories with $N_{Wf} = 2N_f$ Weyl fermions in the vector representation (light blue), $N_{Wf} = N_f$ in the two-index antisymmetric representation (light red) and finally in the two-index symmetric (adjoint) (light green). The arrows indicate that the conformal windows can be smaller and the associated solid curves correspond to the all orders beta function prediction for the maximum extension of the conformal windows.

1.3.2. *The $SO(N)$ phase diagram*

We shall consider $SO(N)$ theories (for $N > 5$) since they do not suffer of a Witten anomaly[100] and, besides, for $N < 7$ can always be reduced to either an SU or an Sp theory.

In Figure 3 we summarize the relevant zero temperature and matter density phase diagram as function of the number of colors and Weyl flavors (N_f) for $SO(N)$ gauge theories. The shape of the various conformal windows are very similar to the ones for $SU(N)$ and

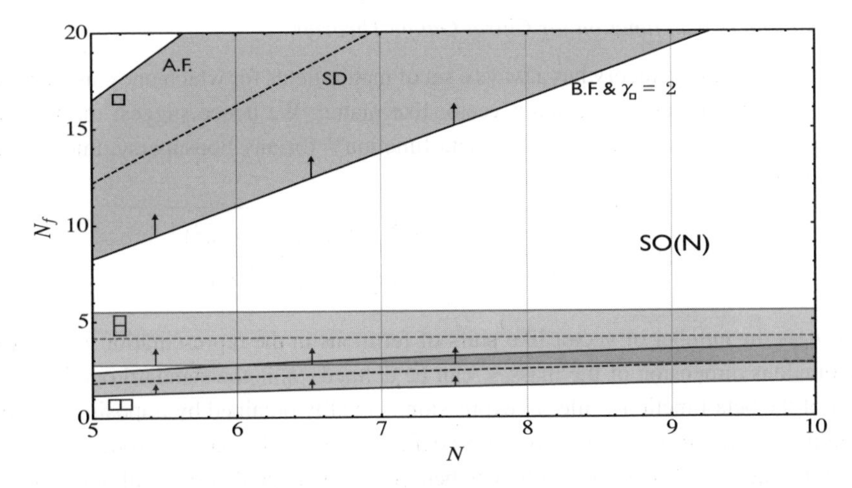

Fig. 3. Phase diagram of $SO(N)$ gauge theories with N_f Weyl fermions in the vector representation, in the two-index antisymmetric (adjoint) and finally in the two-index symmetric representation. The arrows indicate that the conformal windows can be smaller and the associated solid curves correspond to the all orders beta function prediction for the maximum extension of the conformal windows.

$Sp(2N)$ gauge with the difference that in this case the two-index antisymmetric representation is the adjoint representation. We have analyzed only the theories with $N \geq 6$ since the remaining smaller N theories can be deduced from Sp and SU using the fact that $SO(6) \sim SU(4)$, $SO(5) \sim Sp(4)$, $SO(4) \sim SU(2) \times SU(2)$, $SO(3) \sim SU(2)$, and $SO(2) \sim U(1)$.

The phenomenological relevance of orthogonal gauge groups for models of dynamical electroweak symmetry breaking has been shown in.[7]

1.4. *Phases of Chiral Gauge Theories*

Chiral gauge theories, in which at least part of the matter field content is in complex representations of the gauge group, play an important role in efforts to extend the SM. These include grand unified theories, dynamical breaking of symmetries, and theories of quark and lepton substructure. Chiral theories received much attention in the 1980's.[101,102]

Here we confront the results obtained in Ref.[103,104] using the thermal degree of count freedom with the generalization of the all orders beta function useful to constrain chiral gauge theories appeared in.[59] The two important class of theories we are going to investigate are the Bars-Yankielowicz (BY)[105] model involving fermions in the two-index symmetric tensor representation, and the other is a generalized Georgi-Glashow (GGG) model involving fermions in the two-index antisymmetric tensor representation. In each case, in addition to fermions in complex representations, a set of p anti fundamental-fundamental pairs are included and the allowed phases are considered as a function of p. An independent relevant study of the phase diagrams of chiral gauge theories appeared in.[75] Here the authors also compare their results with the ones presented below.

1.4.1. *All-orders beta function for Chiral Gauge Theories*

A generic chiral gauge theory has always a set of matter fields for which one cannot provide a mass term, but it can also contain vector-like matter. We hence suggest the following minimal modification of the all orders beta function[63] for any nonsupersymmetric chiral gauge theory:

$$\beta_{\chi}(g) = -\frac{g^3}{(4\pi)^2}\frac{\beta_0 - \frac{2}{3}\sum_{i=1}^{k} T(r_i)p(r_i)\gamma_i(g^2)}{1 - \frac{g^2}{8\pi^2}C_2(G)\left(1 + \frac{2\beta'_{\chi}}{\beta_0}\right)}\,, \tag{11}$$

where p_i is the number of vector like pairs of fermions in the representation r_i for which an anomalous dimension of the mass γ_i can be defined. β_0 is the standard one loop coefficient of the beta function while β'_{χ} expression is readily obtained by imposing that when expanding β_{χ} one recovers the two-loop coefficient correctly and its explicit expression is not relevant here. According to the new beta function gauge theories without vector-like matter but featuring several copies of purely chiral matter will be conformal when the number of copies is such that the first coefficient of the beta function vanishes identically. Using topological excitations an analysis of this case was performed in.[74]

1.4.2. *The Bars Yankielowicz (BY) Model*

This model is based on the single gauge group $SU(N \geq 3)$ and includes fermions transforming as a symmetric tensor representation, $S = \psi_L^{\{ab\}}$, $a, b = 1, \cdots, N$; $N + 4 + p$ conjugate fundamental representations: $\bar{F}_{a,i} = \psi^c_{a,iL}$, where $i = 1, \cdots, N + 4 + p$; and p fundamental representations, $F^{a,i} = \psi_L^{a,i}$, $i = 1, \cdots, p$. The $p = 0$ theory is the basic chiral theory, free of gauge anomalies by virtue of cancellation between the antisymmetric tensor and the $N + 4$ conjugate fundamentals. The additional p pairs of fundamentals and conjugate fundamentals, in a real representation of the gauge group, lead to no gauge anomalies.

The global symmetry group is

$$G_f = SU(N + 4 + p) \times SU(p) \times U_1(1) \times U_2(1)\,. \tag{12}$$

Two $U(1)$'s are the linear combination of the original $U(1)$'s generated by $S \to e^{i\theta_S}S$, $\bar{F} \to e^{i\theta_{\bar{F}}}\bar{F}$ and $F \to e^{i\theta_F}F$ that are left invariant by instantons, namely that for which $\sum_j N_{R_j}T(R_j)Q_{R_j} = 0$, where Q_{R_j} is the $U(1)$ charge of R_j and N_{R_j} denotes the number of copies of R_j.

Thus the fermionic content of the theory is where the first $SU(N)$ is the gauge group, indicated by the square brackets.

Table 1. The Bars Yankielowicz (BY) model.

Fields	$[SU(N)]$	$SU(N+4+p)$	$SU(p)$	$U_1(1)$	$U_2(1)$
S	$\square\square$	1	1	$N+4$	$2p$
$\bar{F}$	$\bar{\square}$	$\bar{\square}$	1	$-(N+2)$	$-p$
F	$\square$	1	$\square$	$N+2$	$-(N-p)$

From the numerator of the chiral beta function and the knowledge of the one-loop coefficient of the BY perturbative beta function the predicted conformal window is:

$$3\frac{(3N-2)}{2+\gamma^*} \leq p \leq \frac{3}{2}(3N-2)\,, \tag{13}$$

with γ^* the largest possible value of the anomalous dimension of the mass. The maximum value of the number of p flavors is obtained by setting $\gamma^* = 2$:

$$\frac{3}{4}(3N-2) \leq p \leq \frac{3}{2}(3N-2)\,, \qquad \gamma^* = 2\,, \tag{14}$$

while for $\gamma^* = 1$ one gets:

$$(3N-2) \leq p \leq \frac{3}{2}(3N-2)\,, \qquad \gamma^* = 1\,. \tag{15}$$

The chiral beta function predictions for the conformal window are compared with the thermal degree of freedom investigation as shown in the left panel of Fig. 4. In order to derive

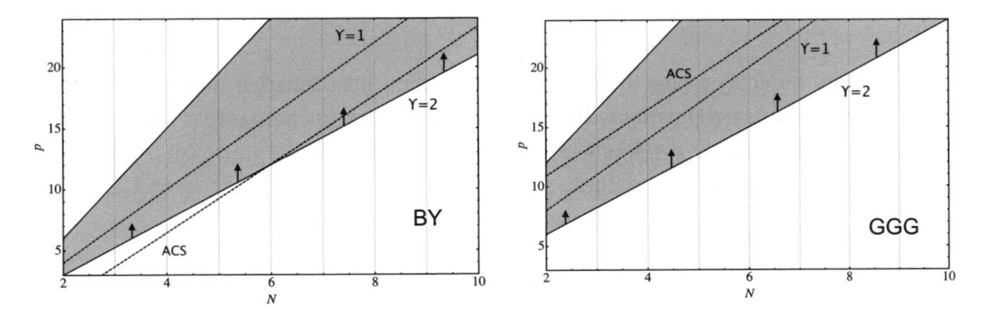

Fig. 4. *Left panel*: Phase diagram of the BY generalized model. The upper solid (blue) line corresponds to the loss of asymptotic freedom; the dashed (blue) curve corresponds to the chiral beta function prediction for the breaking/restoring of chiral symmetry. The dashed black line corresponds to the ACS bound stating that the conformal region should start above this line. We have augmented the ACS method with the Appelquist-Duan-Sannino[104] extra requirement that the phase with the lowest number of massless degrees of freedom wins among all the possible phases in the infrared a chiral gauge theory can have. We hence used $f_{IR}^{\mathrm{brk+sym}}$ and f_{UV} to determine this curve. According to the all orders beta function (B.F.) the conformal window cannot extend below the solid (blue) line, as indicated by the arrows. This line corresponds to the anomalous dimension of the mass reaching the maximum value of 2. *Right panel*: The same plot for the GGG model.

a prediction from the ACS method we augmented it with the Appelquist-Duan-Sannino[104] extra requirement that the phase with the lowest number of massless degrees of freedom wins among all the possible phases in the infrared a chiral gauge theory can have. The thermal critical number is:

$$p^{\mathrm{Therm}} = \frac{1}{4}\left[-16 + 3N + \sqrt{208 - 196N + 69N^2}\right]\,. \tag{16}$$

1.4.3. *The Generalized Georgi-Glashow (GGG) Model*

This model is similar to the BY model just considered. It is an $SU(N \geq 5)$ gauge theory, but with fermions in the anti-symmetric, rather than symmetric, tensor representation. The complete fermion content is $A = \psi_L^{[ab]}$, $a, b = 1, \cdots, N$; an additional $N-4+p$ fermions in the conjugate fundamental representations: $\bar{F}_{a,i} = \psi^c_{a,iL}$, $i = 1, \cdots, N-4+p$; and p fermions in the fundamental representations, $F^{a,i} = \psi_L^{a,i}$, $i = 1, \cdots, p$.

The global symmetry is

$$G_f = SU(N-4+p) \times SU(p) \times U_1(1) \times U_2(1) \,. \tag{17}$$

where the two $U(1)$'s are anomaly free. With respect to this symmetry, the fermion content is

Table 2. The Generalized Georgi-Glashow (GGG) model.

Fields	$[SU(N)]$	$SU(N-4+p)$	$SU(p)$	$U_1(1)$	$U_2(1)$
A	$\begin{array}{c}\Box\\\Box\end{array}$	1	1	$N-4$	$2p$
$\bar{F}$	$\Box$	$\bar{\Box}$	1	$-(N-2)$	$-p$
F	$\Box$	1	$\Box$	$N-2$	$-(N-p)$

Following the analysis for the BY model the chiral beta function predictions for the conformal window are compared with the thermal degree of freedom investigation and the result is shown in the right panel of Fig. 4.

1.5. *Conformal Chiral Dynamics*

Our starting point is a nonsupersymmetric non-abelian gauge theory with sufficient massless fermionic matter to develop a nontrivial IRFP. The cartoon of the running of the coupling constant is represented in Fig. 5. In the plot Λ_U is the dynamical scale below which the IRFP is essentially reached. It can be defined as the scale for which α is $2/3$ of the fixed point value in a given renormalization scheme. If the theory possesses an IRFP the chiral condensate must vanish at large distances. Here we want to study the behavior of the condensate when a flavor singlet mass term is added to the underlying Lagrangian $\Delta L = -m\widetilde{\psi}\psi + \text{h.c.}$ with m the fermion mass and ψ^f_c as well as $\widetilde{\psi}^c_f$ left transforming two component spinors, c and f represent color and flavor indices. The omitted color and

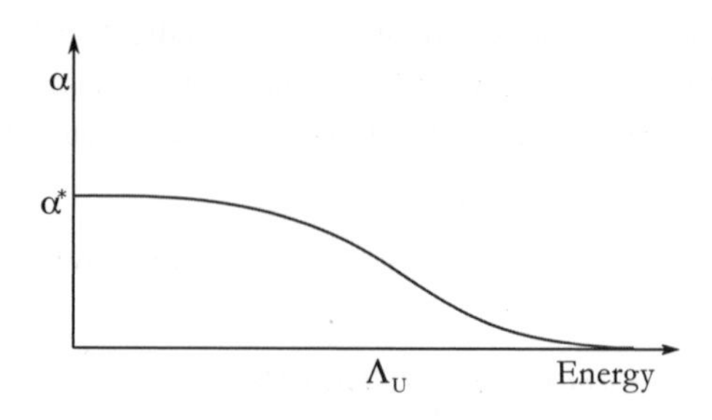

Fig. 5. Running of the coupling constant in an asymptotically free gauge theory developing an infrared fixed point for a value $\alpha = \alpha^*$.

flavor indices, in the Lagrangian term, are contracted. We consider the case of fermionic matter in the fundamental representation of the $SU(N)$ gauge group. The effect of such a term is to break the conformal symmetry together with some of the global symmetries of the underlying gauge theory. The composite operator $\mathcal{O}_{\widetilde{\psi}\psi}{}^{f'}_{f} = \widetilde{\psi}^{f'}\psi_f$ has mass dimension $d_{\widetilde{\psi}\psi} = 3 - \gamma$ with γ the anomalous dimension of the mass term. At the fixed point γ is a positive number smaller than two.[90] We assume $m \ll \Lambda_U$. Dimensional analysis demands $\Delta L \to -m\,\Lambda_U^{\gamma}\,\mathrm{Tr}[\mathcal{O}_{\widetilde{\psi}\psi}] + \mathrm{h.c.}$. The mass term is a relevant perturbation around the IRFP driving the theory away from the fixed point. It will induce a nonzero vacuum expectation value for $\mathcal{O}_{\widetilde{\psi}\psi}$ itself proportional to $\delta_f^{f'}$. It is convenient to define $\mathrm{Tr}[\mathcal{O}_{\widetilde{\psi}\psi}] = N_f\mathcal{O}$ with $\mathcal{O}$ a flavor singlet operator. The relevant low energy Lagrangian term is then $-m\,\Lambda_U^{\gamma}\,N_f\mathcal{O} + \mathrm{h.c.}$. To determine the vacuum expectation value of $\mathcal{O}$ we follow.[106,107]

The induced physical mass gap is a natural infrared cutoff. We, hence, identify Λ_{IR} with the physical value of the condensate. We find:

$$\langle \widetilde{\psi}_c^f \psi_f^c \rangle \propto -m\Lambda_U^2 \,, \qquad 0 < \gamma < 1 \,, \tag{18}$$

$$\langle \widetilde{\psi}_c^f \psi_f^c \rangle \propto -m\Lambda_U^2 \log \frac{\Lambda_U^2}{|\langle \mathcal{O} \rangle|} \,, \quad \gamma \to 1 \,, \tag{19}$$

$$\langle \widetilde{\psi}_c^f \psi_f^c \rangle \propto -m^{\frac{3-\gamma}{1+\gamma}} \Lambda_U^{\frac{4\gamma}{1+\gamma}} \,, \quad 1 < \gamma \leq 2 \,. \tag{20}$$

We used $\langle \widetilde{\psi}\psi \rangle \sim \Lambda_U^{\gamma}\langle \mathcal{O} \rangle$ to relate the expectation value of $\mathcal{O}$ to the one of the fermion condensate. Via an allowed axial rotation m is now real and positive. The effects of the Instantons on the conformal dynamics has been investigated in.[108] Here it was shown that the effects of the instantons can be sizable only for a very small number of flavors given that, otherwise, the instanton induced operators are highly irrelevant.

1.6. *Gauge Duals and Conformal Window*

One of the most fascinating possibilities is that generic asymptotically free gauge theories have magnetic duals. In fact, in the late nineties, in a series of ground breaking papers Seiberg[109,110] provided strong support for the existence of a consistent picture of such a duality within a supersymmetric framework. Arguably the existence of a possible dual of a generic nonsupersymmetric asymptotically free gauge theory able to reproduce its infrared dynamics must match the 't Hooft anomaly conditions.

We have exhibited several solutions of these conditions for QCD in[111] and for certain gauge theories with higher dimensional representations in.[112] An earlier exploration already appeared in the literature.[113] The novelty with respect to these earlier results are: i) The request that the gauge singlet operators associated to the magnetic baryons should be interpreted as bound states of ordinary baryons;[111] ii) The fact that the asymptotically free condition for the dual theory matches the lower bound on the conformal window obtained using the all orders beta function.[63] These extra constraints help restricting further the number of possible gauge duals without diminishing the exactness of the associate solutions with respect to the 't Hooft anomaly conditions.

We will briefly summarize here the novel solutions to the 't Hooft anomaly conditions for QCD. The resulting *magnetic* dual allows to predict the critical number of flavors above which the asymptotically free theory, in the electric variables, enters the conformal regime as predicted using the all orders conjectured beta function.[63]

1.6.1. *QCD Duals*

The underlying gauge group is $SU(3)$ while the quantum flavor group is

$$SU_L(N_f) \times SU_R(N_f) \times U_V(1) , \tag{21}$$

and the classical $U_A(1)$ symmetry is destroyed at the quantum level by the Adler-Bell-Jackiw anomaly. We indicate with $Q^i_{\alpha;c}$ the two component left spinor where $\alpha = 1, 2$ is the spin index, $c = 1, ..., 3$ is the color index while $i = 1, ..., N_f$ represents the flavor. $\widetilde{Q}^{\alpha;c}_i$ is the two component conjugated right spinor.

The global anomalies are associated to the triangle diagrams featuring at the vertices three $SU(N_f)$ generators (either all right or all left), or two $SU(N_f)$ generators (all right or all left) and one $U_V(1)$ charge. We indicate these anomalies for short with:

$$SU_{L/R}(N_f)^3 , \qquad SU_{L/R}(N_f)^2 \; U_V(1) . \tag{22}$$

For a vector like theory there are no further global anomalies. The cubic anomaly factor, for fermions in fundamental representations, is 1 for Q and -1 for $\tilde{Q}$ while the quadratic anomaly factor is 1 for both leading to

$$SU_{L/R}(N_f)^3 \propto \pm 3 , \quad SU_{L/R}(N_f)^2 U_V(1) \propto \pm 3 . \tag{23}$$

If a magnetic dual of QCD does exist one expects it to be weakly coupled near the critical number of flavors below which one breaks large distance conformality in the electric variables. This idea is depicted in Fig 6.

Determining a possible unique dual theory for QCD is, however, not simple given the few mathematical constraints at our disposal. The saturation of the global anomalies is an important tool but is not able to select out a unique solution. We shall see, however, that one of the solutions, when interpreted as the QCD dual, leads to a prediction of a critical number of flavors corresponding exactly to the one obtained via the conjectured all orders beta function.

We seek solutions of the anomaly matching conditions for a gauge theory $SU(X)$ with global symmetry group $SU_L(N_f) \times SU_R(N_f) \times U_V(1)$ featuring *magnetic* quarks q and $\widetilde{q}$ together with $SU(X)$ gauge singlet states identifiable as baryons built out of the *electric* quarks Q. Since mesons do not affect directly global anomaly matching conditions we could add them to the spectrum of the dual theory. We study the case in which X is a linear combination of number of flavors and colors of the type $\alpha N_f + 3\beta$ with α and β integer numbers.

We add to the *magnetic* quarks gauge singlet Weyl fermions which can be identified with the baryons of QCD but massless.

Having defined the possible massless matter content of the gauge theory dual to QCD one computes the $SU_L(N_f)^3$ and $SU_L(N_f)^2 \; U_V(1)$ global anomalies in terms of the new

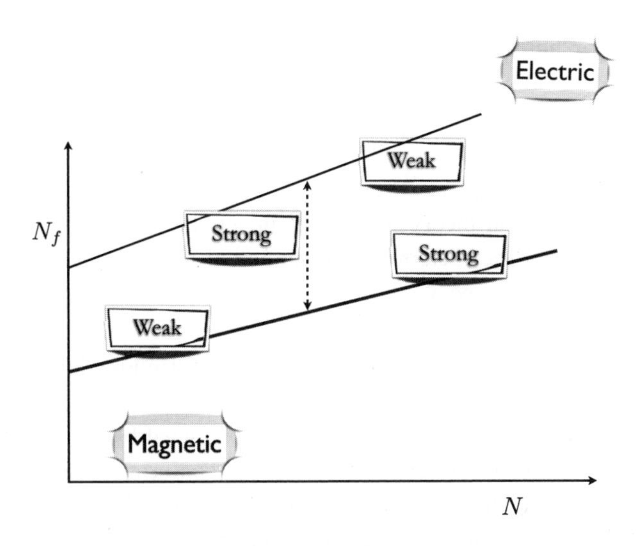

Fig. 6. Schematic representation of the phase diagram as function of number of flavors and colors. For a given number of colors by increasing the number flavors within the conformal window we move from the lowest line (violet) to the upper (black) one. The upper black line corresponds to the one where one looses asymptotic freedom in the electric variables and the lower line where chiral symmetry breaks and long distance conformality is lost. In the *magnetic* variables the situation is reverted and the perturbative line, i.e. the one where one looses asymptotic freedom in the magnetic variables, correspond to the one where chiral symmetry breaks in the electric ones.

fields. We have found several solutions to the anomaly matching conditions presented above. Some were found previously in.[113] Here we display a new solution in which the gauge group is $SU(2N_f - 5N)$ with the number of colors N equal to 3. It is, however, convenient to keep the dependence on N explicit. X must assume a value strictly larger

Table 3. Massless spectrum of *magnetic* quarks and baryons and their transformation properties under the global symmetry group. The last column represents the multiplicity of each state and each state is a Weyl fermion.

Fields	$[SU(2N_f - 5N)]$	$SU_L(N_f)$	$SU_R(N_f)$	$U_V(1)$	# of copies
q	$\Box$	$\Box$	1	$\frac{N(2N_f-5)}{2N_f-5N}$	1
$\widetilde{q}$	$\overline{\Box}$	1	$\overline{\Box}$	$-\frac{N(2N_f-5)}{2N_f-5N}$	1
A	1	$\begin{smallmatrix}\Box\\\Box\\\Box\end{smallmatrix}$	1	3	2
B_A	1	$\begin{smallmatrix}\Box\\\Box\end{smallmatrix}$	$\Box$	3	-2
D_A	1	$\Box$	$\begin{smallmatrix}\Box\\\Box\end{smallmatrix}$	3	2
$\widetilde{A}$	1	1	$\begin{smallmatrix}\Box\\\Box\\\Box\end{smallmatrix}$	-3	2

than one otherwise it is an abelian gauge theory. This provides the first nontrivial bound on the number of flavors:

$$N_f > \frac{5N+1}{2}, \tag{24}$$

which for $N = 3$ requires $N_f > 8$. Asymptotic freedom of the newly found theory is dictated by the coefficient of the one-loop beta function :

$$\beta_0 = \frac{11}{3}(2N_f - 5N) - \frac{2}{3}N_f \, . \tag{25}$$

To this order in perturbation theory the gauge singlet states do not affect the magnetic quark sector and we can hence determine the number of flavors obtained by requiring the dual theory to be asymptotic free. i.e.:

$$N_f \geq \frac{11}{4}N \qquad\qquad \text{Dual Asymptotic Freedom} \, . \tag{26}$$

Quite remarkably this value *coincides* with the one predicted by means of the all orders conjectured beta function for the lowest bound of the conformal window, in the *electric* variables, when taking the anomalous dimension of the mass to be $\gamma = 2$. We recall that for any number of colors N the all orders beta function requires the critical number of flavors to be larger than:

$$N_f^{BF}|_{\gamma=2} = \frac{11}{4}N \, . \tag{27}$$

For N=3 the two expressions yield 8.25 [b]. We consider this a nontrivial and interesting result lending further support to the all orders beta function conjecture and simultaneously suggesting that this theory might, indeed, be the QCD magnetic dual. The actual size of the conformal window matching this possible dual corresponds to setting $\gamma = 2$. We note that although for $N_f = 9$ and $N = 3$ the magnetic gauge group is $SU(3)$ the theory is not trivially QCD given that it features new massless fermions and their interactions with massless mesonic type fields.

Recent suggestions to analyze the conformal window of nonsupersymmetric gauge theories based on different model assumptions[74] are in qualitative agreement with the precise results of the all orders beta function conjecture. It is worth noting that the combination $2N_f - 5N$ appears in the computation of the mass gap for gauge fluctuations presented in.[74,114] It would be interesting to explore a possible link between these different approaches in the future.

We have also find solutions for which the lower bound of the conformal window is saturated for $\gamma = 1$. The predictions from the gauge duals are, however, entirely and surprisingly consistent with the maximum extension of the conformal window obtained using the all orders beta function.[63] Our main conclusion is that the 't Hooft anomaly conditions alone do not exclude the possibility that the maximum extension of the QCD conformal window is the one obtained for a large anomalous dimension of the quark mass.

By computing the same gauge singlet correlators in QCD and its suggested dual, one can directly validate or confute this proposal via lattice simulations.

---

[b] Actually given that X must be at least 2 we must have $N_f \geq 8.5$ rather than 8.25

1.7. *Conclusions*

We investigated the conformal windows of chiral and non-chiral nonsupersymmetric gauge theories with fermions in any representation of the underlying gauge group using four independent analytic methods. For vector-like gauge theories one observes a universal value, i.e. independent of the representation, of the ratio of the area of the maximum extension of the conformal window, predicted using the all orders beta function, to the asymptotically free one, as defined in.[62] It is easy to check from the results presented that this ratio is not only independent on the representation but also on the particular gauge group chosen.

The four methods we used to unveil the conformal windows are the all orders beta function (BF), the SD truncated equation, the thermal degrees of freedom method and possible gauge - duality. They have vastly different starting points and there was no, a priori, reason to agree with each other, even at the qualitative level.

Several questions remain open such as what happens on the right hand side of the infrared fixed point as we increase further the coupling. Does a generic strongly coupled theory develop a new UV fixed point as we increase the coupling beyond the first IR value[115]? If this were the case our beta function would still be a valid description of the running of the coupling of the constant in the region between the trivial UV fixed point and the neighborhood of the first IR fixed point. One might also consider extending our beta function to take into account of this possibility as done in.[76] It is also possible that no non-trivial UV fixed point forms at higher values of the coupling constant for any value of the number of flavors within the conformal window. Gauge-duals seem to be in agreement with the simplest form of the beta function. The extension of the all orders beta function to take into account fermion masses has appeared in.[116]

Our analysis substantially increases the number of asymptotically free gauge theories which can be used to construct SM extensions making use of (near) conformal dynamics. Current Lattice simulations can test our predictions and lend further support or even disprove the emergence of a universal picture possibly relating the phase diagrams of gauge theories of fundamental interactions.

References

1. F. Sannino and K. Tuominen, Phys. Rev. **D71**, 051901 (2005), hep-ph/0405209.
2. D. D. Dietrich, F. Sannino, and K. Tuominen, Phys. Rev. **D73**, 037701 (2006), hep-ph/0510217.
3. D. D. Dietrich, F. Sannino, and K. Tuominen, Phys. Rev. **D72**, 055001 (2005), hep-ph/0505059.
4. S. B. Gudnason, T. A. Ryttov, and F. Sannino, Phys. Rev. **D76**, 015005 (2007), hep-ph/0612230.
5. T. A. Ryttov and F. Sannino, Phys. Rev. **D78**, 115010 (2008), 0809.0713.
6. M. T. Frandsen, I. Masina, and F. Sannino, (2009), 0905.1331.
7. M. T. Frandsen and F. Sannino, (2009), 0911.1570.
8. O. Antipin, M. Heikinheimo, and K. Tuominen, JHEP **10**, 018 (2009), 0905.0622.
9. R. Foadi, M. T. Frandsen, T. A. Ryttov, and F. Sannino, Phys. Rev. **D76**, 055005 (2007), 0706.1696.
10. A. Belyaev *et al.*, Phys. Rev. **D79**, 035006 (2009), 0809.0793.
11. M. Antola, M. Heikinheimo, F. Sannino, and K. Tuominen, (2009), 0910.3681.
12. O. Antipin, M. Heikinheimo, and K. Tuominen, (2010), 1002.1872.
13. S. Nussinov, Phys. Lett. **B165**, 55 (1985).

14. S. M. Barr, R. S. Chivukula, and E. Farhi, Phys. Lett. **B241**, 387 (1990).
15. J. Bagnasco, M. Dine, and S. D. Thomas, Phys. Lett. **B320**, 99 (1994), hep-ph/9310290.
16. S. B. Gudnason, C. Kouvaris, and F. Sannino, Phys. Rev. **D73**, 115003 (2006), hep-ph/0603014.
17. S. B. Gudnason, C. Kouvaris, and F. Sannino, Phys. Rev. **D74**, 095008 (2006), hep-ph/0608055.
18. K. Kainulainen, K. Tuominen, and J. Virkajarvi, Phys. Rev. **D75**, 085003 (2007), hep-ph/0612247.
19. C. Kouvaris, Phys. Rev. **D76**, 015011 (2007), hep-ph/0703266.
20. C. Kouvaris, Phys. Rev. **D77**, 023006 (2008), 0708.2362.
21. M. Y. Khlopov and C. Kouvaris, Phys. Rev. **D77**, 065002 (2008), 0710.2189.
22. M. Y. Khlopov and C. Kouvaris, Phys. Rev. **D78**, 065040 (2008), 0806.1191.
23. C. Kouvaris, Phys. Rev. **D78**, 075024 (2008), 0807.3124.
24. K. Belotsky, M. Khlopov, and C. Kouvaris, Phys. Rev. **D79**, 083520 (2009), 0810.2022.
25. J. M. Cline, M. Jarvinen, and F. Sannino, Phys. Rev. **D78**, 075027 (2008), 0808.1512.
26. E. Nardi, F. Sannino, and A. Strumia, JCAP **0901**, 043 (2009), 0811.4153.
27. R. Foadi, M. T. Frandsen, and F. Sannino, Phys. Rev. **D80**, 037702 (2009), 0812.3406.
28. M. Jarvinen, T. A. Ryttov, and F. Sannino, Phys. Lett. **B680**, 251 (2009), 0901.0496.
29. M. Jarvinen, C. Kouvaris, and F. Sannino, (2009), 0911.4096.
30. K. Kainulainen, K. Tuominen, and J. Virkajarvi, JCAP **1002**, 029 (2010), 0912.2295.
31. K. Kainulainen, K. Tuominen, and J. Virkajarvi, (2010), 1001.4936.
32. M. Antola, S. Di Chiara, F. Sannino, and K. Tuominen, (2010), 1001.2040.
33. S. Catterall and F. Sannino, Phys. Rev. **D76**, 034504 (2007), 0705.1664.
34. S. Catterall, J. Giedt, F. Sannino, and J. Schneible, JHEP **11**, 009 (2008), 0807.0792.
35. L. Del Debbio, A. Patella, and C. Pica, (2008), 0805.2058.
36. A. Hietanen, J. Rantaharju, K. Rummukainen, and K. Tuominen, PoS **LATTICE2008**, 065 (2008), 0810.3722.
37. A. J. Hietanen, K. Rummukainen, and K. Tuominen, Phys. Rev. **D80**, 094504 (2009), 0904.0864.
38. C. Pica, L. Del Debbio, B. Lucini, A. Patella, and A. Rago, (2009), 0909.3178.
39. S. Catterall, J. Giedt, F. Sannino, and J. Schneible, (2009), 0910.4387.
40. B. Lucini, (2009), 0911.0020.
41. F. Bursa, L. Del Debbio, L. Keegan, C. Pica, and T. Pickup, Phys. Rev. **D81**, 014505 (2010), 0910.4535.
42. T. DeGrand, Phys. Rev. **D80**, 114507 (2009), 0910.3072.
43. T. DeGrand, Y. Shamir, and B. Svetitsky, Phys. Rev. **D79**, 034501 (2009), 0812.1427.
44. T. DeGrand and A. Hasenfratz, Phys. Rev. **D80**, 034506 (2009), 0906.1976.
45. Z. Fodor, K. Holland, J. Kuti, D. Nogradi, and C. Schroeder, PoS **LATTICE2008**, 058 (2008), 0809.4888.
46. Z. Fodor, K. Holland, J. Kuti, D. Nogradi, and C. Schroeder, JHEP **11**, 103 (2009), 0908.2466.
47. Z. Fodor, K. Holland, J. Kuti, D. Nogradi, and C. Schroeder, JHEP **08**, 084 (2009), 0905.3586.
48. J. B. Kogut and D. K. Sinclair, (2010), 1002.2988.
49. T. Appelquist and R. Shrock, Phys. Lett. **B548**, 204 (2002), hep-ph/0204141.
50. T. Appelquist and R. Shrock, Phys. Rev. Lett. **90**, 201801 (2003), hep-ph/0301108.
51. T. Appelquist, M. Piai, and R. Shrock, Phys. Rev. **D69**, 015002 (2004), hep-ph/0308061.
52. T. Appelquist, G. T. Fleming, and E. T. Neil, Phys. Rev. **D79**, 076010 (2009), 0901.3766.
53. T. Appelquist *et al.*, (2009), 0910.2224.
54. Z. Fodor, K. Holland, J. Kuti, D. Nogradi, and C. Schroeder, Phys. Lett. **B681**, 353 (2009), 0907.4562.
55. Z. Fodor, K. Holland, J. Kuti, D. Nogradi, and C. Schroeder, PoS **LATTICE2008**, 066 (2008), 0809.4890.
56. A. Deuzeman, M. P. Lombardo, and E. Pallante, (2009), 0904.4662.

57. Z. Fodor, K. Holland, J. Kuti, D. Nogradi, and C. Schroeder, (2009), 0911.2934.
58. Z. Fodor, K. Holland, J. Kuti, D. Nogradi, and C. Schroeder, (2009), 0911.2463.
59. F. Sannino, (2009), 0911.0931.
60. C. T. Hill and E. H. Simmons, Phys. Rept. **381**, 235 (2003), hep-ph/0203079.
61. D. D. Dietrich and F. Sannino, Phys. Rev. **D75**, 085018 (2007), hep-ph/0611341.
62. T. A. Ryttov and F. Sannino, Phys. Rev. **D76**, 105004 (2007), 0707.3166.
63. T. A. Ryttov and F. Sannino, Phys. Rev. **D78**, 065001 (2008), 0711.3745.
64. F. Sannino, (2008), 0804.0182.
65. T. Appelquist, K. D. Lane, and U. Mahanta, Phys. Rev. Lett. **61**, 1553 (1988).
66. A. G. Cohen and H. Georgi, Nucl. Phys. **B314**, 7 (1989).
67. V. A. Miransky and K. Yamawaki, Phys. Rev. **D55**, 5051 (1997), hep-th/9611142.
68. T. Appelquist, A. G. Cohen, and M. Schmaltz, Phys. Rev. **D60**, 045003 (1999), hep-th/9901109.
69. G. Grunberg, Phys. Rev. **D65**, 021701 (2002), hep-ph/0009272.
70. E. Gardi and G. Grunberg, JHEP **03**, 024 (1999), hep-th/9810192.
71. G. Grunberg, (1996), hep-ph/9608375.
72. H. Gies and J. Jaeckel, Eur. Phys. J. **C46**, 433 (2006), hep-ph/0507171.
73. J. Braun and H. Gies, Phys. Lett. **B645**, 53 (2007), hep-ph/0512085.
74. E. Poppitz and M. Unsal, JHEP **09**, 050 (2009), 0906.5156.
75. E. Poppitz and M. Unsal, JHEP **12**, 011 (2009), 0910.1245.
76. O. Antipin and K. Tuominen, (2009), 0909.4879.
77. O. Antipin and K. Tuominen, (2009), 0912.0674.
78. M. Jarvinen and F. Sannino, (2009), 0911.2462.
79. J. Braun and H. Gies, (2009), 0912.4168.
80. J. Alanen and K. Kajantie, (2009), 0912.4128.
81. L. Del Debbio, M. T. Frandsen, H. Panagopoulos, and F. Sannino, JHEP **06**, 007 (2008), 0802.0891.
82. T. Appelquist, G. T. Fleming, and E. T. Neil, Phys. Rev. Lett. **100**, 171607 (2008), 0712.0609.
83. Y. Shamir, B. Svetitsky, and T. DeGrand, Phys. Rev. **D78**, 031502 (2008), 0803.1707.
84. A. Deuzeman, M. P. Lombardo, and E. Pallante, Phys. Lett. **B670**, 41 (2008), 0804.2905.
85. B. Lucini and G. Moraitis, PoS **LAT2007**, 058 (2007), 0710.1533.
86. F. Sannino, Phys. Rev. **D79**, 096007 (2009), 0902.3494.
87. V. A. Novikov, M. A. Shifman, A. I. Vainshtein, and V. I. Zakharov, Nucl. Phys. **B229**, 381 (1983).
88. M. A. Shifman and A. I. Vainshtein, Nucl. Phys. **B277**, 456 (1986).
89. D. R. T. Jones, Phys. Lett. **B123**, 45 (1983).
90. G. Mack, Commun. Math. Phys. **55**, 1 (1977).
91. M. Flato and C. Fronsdal, Lett. Math. Phys. **8**, 159 (1984).
92. V. K. Dobrev and V. B. Petkova, Phys. Lett. **B162**, 127 (1985).
93. T. Banks and A. Zaks, Nucl. Phys. **B196**, 189 (1982).
94. T. Maskawa and H. Nakajima, Prog. Theor. Phys. **52**, 1326 (1974).
95. R. Fukuda and T. Kugo, Nucl. Phys. **B117**, 250 (1976).
96. A. Armoni, Nucl. Phys. **B826**, 328 (2010), 0907.4091.
97. T. A. Ryttov and F. Sannino, (2009), 0906.0307.
98. M. A. Luty and T. Okui, JHEP **09**, 070 (2006), hep-ph/0409274.
99. N. Evans and F. Sannino, (2005), hep-ph/0512080.
100. E. Witten, Phys. Lett. **B117**, 324 (1982).
101. R. D. Ball, Phys. Rept. **182**, 1 (1989).
102. S. Raby, S. Dimopoulos, and L. Susskind, Nucl. Phys. **B169**, 373 (1980).
103. T. Appelquist, A. G. Cohen, M. Schmaltz, and R. Shrock, Phys. Lett. **B459**, 235 (1999), hep-th/9904172.

104. T. Appelquist, Z.-y. Duan, and F. Sannino, Phys. Rev. **D61**, 125009 (2000), hep-ph/0001043.
105. I. Bars and S. Yankielowicz, Phys. Lett. **B101**, 159 (1981).
106. M. A. Stephanov, Phys. Rev. **D76**, 035008 (2007), 0705.3049.
107. F. Sannino and R. Zwicky, Phys. Rev. **D79**, 015016 (2009), 0810.2686.
108. F. Sannino, Phys. Rev. **D80**, 017901 (2009), 0811.0616.
109. N. Seiberg, Phys. Rev. **D49**, 6857 (1994), hep-th/9402044.
110. N. Seiberg, Nucl. Phys. **B435**, 129 (1995), hep-th/9411149.
111. F. Sannino, Phys. Rev. **D80**, 065011 (2009), 0907.1364.
112. F. Sannino, Nucl. Phys. **B830**, 179 (2010), 0909.4584.
113. J. Terning, Phys. Rev. lett. **80**, 2517 (1998), hep-th/9706074.
114. E. Poppitz and M. Unsal, JHEP. **0903**, 027 (2009), 0812.2085.
115. D. B. Kaplan, J.-W. Lee, D. T. Son, and M. A. Stephanov, Phys. Rev. **D80**, 125005 (2009), 0905.4752.
116. D. D. Dietrich, Phys. Rev. **D80**, 065032 (2009), 0908.1364.

Resizing Conformal Windows

O. Antipin

Department of Physics, P.O.Box 35
FI-00014 University of Jyväskylä, Finland
and
Helsinki Institute of Physics, P.O. Box 64
FI-00014, University of Helsinki, Finland
E-mail: oleg.a.antipin@jyu.fi

K. Tuominen

CP3-Origins, Campusvej 55
DK-5230, Odense N, Denmark
and
Helsinki Institute of Physics, P.O. Box 64
FI-00014 University of Helsinki, Finland
E-mail: kimmo.tuominen@jyu.fi

Different mechanisms for the loss of conformality and analytic ansätze for the beta-function of a generic gauge theory are reviewed and the implications on the conformal windows considered.

Keywords: Conformal window, gauge theories.

1. Introduction

Recent progress on the phase diagrams of strongly coupled gauge theory as a function of number of colors and flavors as well as matter representations[1,2] shows that infrared conformal theories with simple particle content exist. The key feature in this analysis is the use of higher representations with respect to the new strong dynamics. Recently these models have been investigated also on the lattice.[3,5,7–9] These theories are applied in beyond the Standard Model physics as phenomenologically viable candidates of walking Technicolor theories.[10,11]

If the β-function for a generic SU(N) gauge theory with matter was known beyond perturbation theory, the determination of the conformal window would be a simple matter. For $\mathcal{N} = 1$ super-QCD (SQCD) the exact β-function is known,[12] and for $3N_c/2 < N_f < 3N_c$ SQCD has an infrared stable fixed point.[13,14] In[15] a β-function ansatz for non-supersymmetric gauge theories was introduced. The parallel between SQCD and QCD was pursued further in[16] where a proposal for the electromagnetic dual of QCD was constructed. In[17] it was argued that there are three generic mechanisms how conformality is lost in quantum theories. Namely,

1) the fixed point can go to zero coupling, 2) it can go to infinite coupling or 3) the Infrared fixed point (IRFP) may annihilate with an ultraviolet fixed point (UVFP) whence they both disappear into the complex plane.

Mechanism 3) seems not to be allowed either with the β-function of SQCD or with the β-function ansatz of[15] for non-supersymmetric SU(N_c) gauge theory. On the other hand, it was shown in[17] that mechanism 3) is realized in a wide class of non-supersymmetric theories. Quantitative indications for this phenomena in QCD with many flavors were first presented in.[18]

In this talk we will discuss how the ideas of[15] and[17] could be combined.[19] We propose a concrete implementation of the mechanism 3) in an all order β-function for a generic Yang–Mills (YM) theory and show how this determines conformal windows; see also.[19]

2. An all order β-function ansatz

We begin by a brief review of the basic results of[15] to define some notations. The rescaled 't Hooft coupling is defined as $a \equiv g^2 N_c/(4\pi)^2$. In the perturbative β-function for a generic non-supersymmetric YM theory with N_f Dirac fermions in a given representation R of the gauge group, the two lowest order coefficients are

$$\overline{\beta}_0(x) = 11\frac{C_2(G)}{N_c} - 4T(R)x \tag{1}$$

$$\overline{\beta}_1(x) = 34\frac{C_2^2(G)}{N_c^2} - 20\frac{C_2(G)}{N_c}T(R)x - 12\frac{C_2(R)}{N_c}T(R)x \; . \tag{2}$$

These two coefficients are universal, and the two-loop β-function is independent of the renormalization scheme, $\beta(a) = -\frac{2}{3}\overline{\beta}_0(x)a^2 - \frac{2}{3}\overline{\beta}_1(x)a^3$. The conventions for the group theory factors for the representations we will consider can be found explicitly in.[15]

The proposal of[15] was to write all-order β-function as

$$\beta(a) = -\frac{2}{3}a^2 \frac{\overline{\beta}_0(X) - X\,\gamma(a)}{1 - \frac{2aC_2(G)}{N_c}\left(1 + \frac{6\overline{\beta}_0'(X)}{\overline{\beta}_0(X)}\right)} \,, \tag{3}$$

with $\overline{\beta}_0(X) = 11\frac{C_2(G)}{N_c} - 2X$, $\overline{\beta}_0'(X) = C_2(G)/N_c - X/2$ and γ denotes the anomalous dimension of the quark mass operator. For the group SU(N_c), $C_2(G) = N_c$ which simplifies the above equations. Perturbative two-loop β-function is obtained upon expansion to $\mathcal{O}(a^3)$. Since only the two-loop β-function has universal coefficients, one assumes that there is a scheme in which the proposed β-function is complete. The ansatz in (3) determines the conformal window by assuming the value of γ at the lower boundary of the conformal window; by the unitarity bound $\gamma \leq 2$.

As discussed in the introduction, we will next modify Eq. (3) to be able to include conformality loss via mechanism 3).

3. Fixed point merger in an all order β-function ansatz

The simplest way to allow for the additional non-trivial UVFP in β-function of Eq. (3), and thus anticipate for the 'fixed point merger', is to include a term $\sim \gamma^2$ in the numerator of the β-function suggesting a modified form

$$\beta(a) = -\frac{2}{3}a^2 \cdot \frac{\overline{\beta}_0(X) - X\,\gamma(a) + r(X)\gamma^2(a)}{1 - \frac{2aC_2(G)}{N_c}\left(1 + \frac{6\overline{\beta}_0'(X)}{\overline{\beta}_0(X)}\right)}, \tag{4}$$

where $\overline{\beta}_0(X)$ and $\overline{\beta}_0'(X)$ are as defined earlier and we introduced unknown function $r(X)$. We will determine this function first it in the limit $N_c \to \infty$; possible $\mathcal{O}(1/N_c)$ corrections will be discussed briefly in a later subsection.

3.1. *Large N_c limit*

In the large N_c limit we will apply the holographic expectation[17,20] that operator dimensions of the quark mass operators at the two fixed points satisfy $\Delta_+ + \Delta_- = d = 4$, which translates for the anomalous dimensions into equation

$$\gamma_1 + \gamma_2 = 2. \tag{5}$$

Note, that this implies that the fixed point merger, a point where conformality is lost, will occur at $\gamma_1 = \gamma_2 = 1$.

Setting the numerator of the β-function (4) to zero and solving the resulting quadratic equation, with the constraint Eq. (5) for the two roots, we find that $r(X) = X/2$. The numerator of Eq. (4) hence becomes

$$\overline{\beta}_0(X) + \frac{X}{2}\gamma(\gamma - 2). \tag{6}$$

Note the following features: First, the modified β-function, in the large N_c limit, predicts the lower end of the conformal window. For a given theory, we have to solve for X by setting the numerator of modified beta-function, Eq. (6), equal to zero under the condition $\gamma = 1$ on the basis of the constraint (5) from holography. Therefore, by construction, the maximum value of the anomalous dimension at the IRFP is $\gamma^{*\max} = 1$. This explicitly shows that the β-function of Eq. (4) is different from the one proposed in[15] since scheme independent quantities, like the location of the conformal window and the values of the anomalous dimension, are different. Second, the upper end of the conformal window corresponds to $\gamma = 0$ and $\beta_0 = 0$. Notice also that the second value of the anomalous dimension satisfying $\beta_0 = 0$ in Eq. (6) is $\gamma = 2$ which coincides with the maximum value generally allowed in conformal field theory.

We can then predict the size of the conformal window with the new β-function ansatz. Setting the expression in Eq. (6) to zero and using $\gamma = 1$ leads to the prediction that, in the large N_c limit, the lower end of the conformal window is at

$$X_{\min} = \frac{22C_2(G)}{5N_c} = \frac{22}{5}. \tag{7}$$

where we used $G_2(G) = N_c$. For the SU(N_c) gauge theory with the fundamental quarks this predicts that $N_f^{\rm cr}/N_c \equiv x_{\rm min} = 22/5$ in the large N_c limit. Note that for fundamental fermions this implies also that we have $N_f \to \infty$, i.e. the Veneziano limit. The extrapolation of this result for the SU(3) theory with fundamental matter is shown in the left panel of Fig. 2. There, above $x = 11/2 = 5.5$ asymptotic freedom is lost (corresponding to the upper end of the conformal window), while below $x = 22/5 = 4.4$ the two zeros for γ become complex. Hence, for the SU(3), our extrapolation from large N_c result gives the critical number of flavors $N_f^{cr} = 13.2$ below which the conformality is lost. This disagrees with the lattice result that the lower end of the conformal window occurs for $8 \leq N_f \leq 12$[9] while agrees with e.g.[21] Clearly more refined lattice studies are needed.

As another example we consider calculating the lower end of the conformal window in terms of critical number of Dirac fermions N_f^{cr} from (7) for fermions in the two-index symmetric (2S) representation of the SU(N_c) gauge group. The results are presented in Fig.1 together with results corresponding to the β-function ansatz of[15] and with recent results[22] obtained using deformation theory to establish the role of topological excitations in generating a mass gap in a gauge theory. Also the result from the ladder approximation is shown.

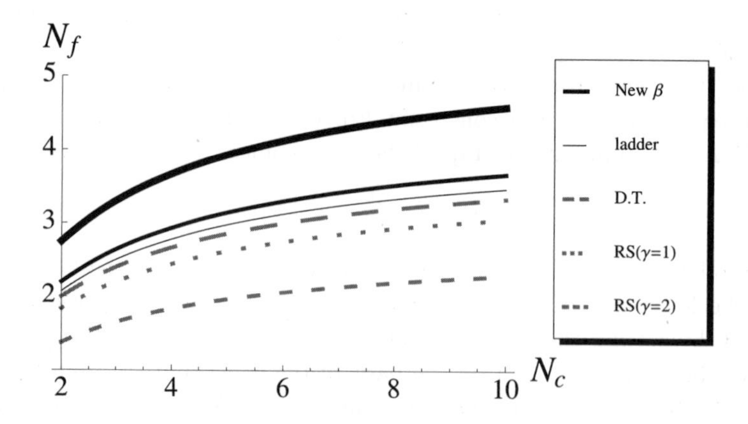

Fig. 1. Size of the conformal window determined using different methods discussed in the text with N_f fermions in 2-index symmetric representation of SU(N_c). The topmost thick solid curve denotes the upper boundary of the conformal window, i.e. loss of asymptotic freedom. For the lower boundary the lower solid curve is the prediction from our β-function proposal, Eq.(7), the long dashed curve is the prediction from[22] and the curve with shorter dashes correspond to the prediction from[15] for $\gamma = 2$ (the dotted curve shows the corresponding result for $\gamma = 1$). The thin line between the solid and long-dashed curves corresponds to the ladder approximation.

3.2. *Away from the large N_c limit*

However, in principle there could be important $\mathcal{O}(1/N_c)$ corrections to (5) which, in turn, would modify the $r(X)$ coefficient of the γ^2 term in (4). To get an idea of how such corrections would affect the results, we estimate these effects by means of a phenomenological modifying parameter ϵ as $\gamma_1 + \gamma_2 = 2 + \epsilon$. Repeating then the exercise of Sec. 3.1, we arrive at the following coefficient of the γ^2 term

$$r(X) = \frac{X}{2+\epsilon}, \tag{8}$$

and, thus, zero of the numerator of the new beta-function will occur at:

$$\overline{\beta}_0(X) + X\,\gamma(\frac{\gamma}{2+\epsilon} - 1) = 0. \tag{9}$$

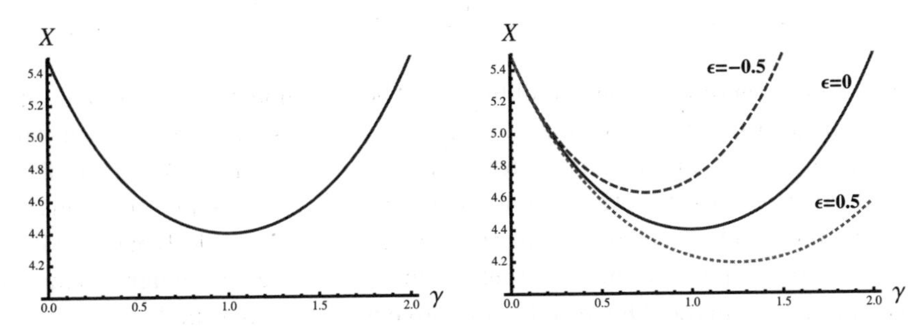

Fig. 2. Left: The line $X(\gamma)$ along which $\beta(a) = 0$. Above $X = 11/2 = 5.5$ asymptotic freedom is lost while below $X = 22/5 = 4.4$ no real solutions exist. Right: The effects of $1/N_c$ corrections. Solid curve is the same as in the left panel. Dashed curve corresponds to $\epsilon = -0.5$ and dotted to $\epsilon = 0.5$

In the right panel of Fig. 2 we plot the solution $X(\gamma)$ of this equation for the SU(3) fundamental with the illustrative values $\epsilon = \pm 0.5$ and $\epsilon = 0$. The latter coincides with the curve in the left panel of Fig. 2. The upper end of the conformal window at X=11/2=5.5 corresponding to $\gamma_1 = 0$ remains unchanged, but the size of the conformal window increases (decreases) with positive (negative) ϵ. For positive ϵ the anomalous dimension at the UVFP will exceed the unitarity limit $\gamma \leq 2$ before the IRFP merges with the free UVFP, i.e. before the other solution reaches $\gamma = 0$ value. In other words, as we decrease γ at the IRFP (by increasing ϵ with fixed X or vice versa) there is a point where UVFP disappears due to violation of the unitarity bound. Thus, past this point our picture becomes qualitatively similar to the one predicted by Ryttov-Sannino β-function. The value of γ at the IRFP below which the UVFP ceases to exist is given by $\gamma = \epsilon$.

Above, we have treated ϵ as a phenomenological parameter whose value is expected to be $\mathcal{O}(1)$. Considering positive ϵ and solving Eq. (9) at the point where the UV and IR fixed points merge (i.e. where $\gamma_1 = \gamma_2 = 1 + \epsilon/2$) we obtain

$$X_{\min} = \frac{22C_2(G)}{N_c(5+\epsilon/2)} = \frac{22}{5+\epsilon/2}, \tag{10}$$

where the second equality again applies for SU(N_c) gauge theory. For illustration we set $\epsilon = 1$ which leads to $X_{\min} = 4$.

Let us apply this result first for SU(3) with fundamental fermions. Using $x_{\min} = X_{\min} = 4$ this translates to $N_f^{\min} = 12$. Then let us consider various values of N_c

Table 1. Lower end of the conformal window using procedure outlined in text for fundamental (F), 2-index (anti)symmetric (2AS) 2S and adjoint (A) representations.

N_c	$N_{f,\min}$ Fund.	$N_{f,\min}$ 2AS	$N_{f,\min}$ 2S	$N_{f,\min}$ A.
2	8	-	2	2
3	12	12	2.4	2
4	16	8	2.67	2
5	20	6.67	2.86	2
6	24	6	3	2
10	40	5	3.33	2
$\to\infty$	$4N_c$	4	4	2

and also higher representations in addition to the fundamental one. We collect the results into the Table 1. Interestingly, our results are consistent with present lattice results. Remarkably, our results match exactly on the corresponding predictions from deformation theory both for fundamental[23] and higher representations.[22] Of course any relation between our β-function ansatz and some underlying microscopic dynamics is pure speculation; nevertheless this coincidence, although unexpected, is temptingly systematic. Our results also agree with the ones in.[24]

4. Conclusions

In this paper we exploited recently proposed mechanism of IR and UV fixed point annihilation as a key difference between supersymmetric and non-supersymmetric YM gauge theories. We incorporated this mechanism of fixed point merging into a proposal for a nonperturbative β-function whose functional form is similar to the Ryttov-Sannino β-function.

We also compared our results with results from Ryttov-Sannino β-function[15] as well as with results of.[22] As a speculative note, the agreements and disagreements between these three approaches suggest that there might be important differences in how conformality is lost in supersymmetric versus non-supersymmetric gauge theories. Combining the insights from the analytic approaches with the numerical studies, performed recently and currently in progress, is expected to lead to significant improvement in our understanding of non-perturbative dynamics of (quasi)conformal gauge theories.

References

1. F. Sannino and K. Tuominen, Phys. Rev. D **71**, 051901 (2005) [arXiv:hep-ph/0405209].
2. D. D. Dietrich, F. Sannino and K. Tuominen, Phys. Rev. D **72**, 055001 (2005) [arXiv:hep-ph/0505059]; D. D. Dietrich, F. Sannino and K. Tuominen, Phys. Rev. D **73**, 037701 (2006) [arXiv:hep-ph/0510217].
3. A. J. Hietanen, J. Rantaharju, K. Rummukainen and K. Tuominen, JHEP **0905**, 025 (2009) [arXiv:0812.1467 [hep-lat]].
4. A. J. Hietanen, K. Rummukainen and K. Tuominen, Phys. Rev. D **80**, 094504 (2009) [arXiv:0904.0864 [hep-lat]].

5. S. Catterall and F. Sannino, Phys. Rev. D **76**, 034504 (2007) [arXiv:0705.1664 [hep-lat]]; L. Del Debbio, A. Patella and C. Pica, arXiv:0805.2058 [hep-lat];
6. S. Catterall, J. Giedt, F. Sannino and J. Schneible, JHEP **0811**, 009 (2008) [arXiv:0807.0792 [hep-lat]].
7. Y. Shamir, B. Svetitsky and T. DeGrand, Phys. Rev. D **78**, 031502 (2008) [arXiv:0803.1707 [hep-lat]].
8. T. DeGrand, Y. Shamir and B. Svetitsky, Phys. Rev. D **79**, 034501 (2009) [arXiv:0812.1427 [hep-lat]].
9. T. Appelquist, G. T. Fleming and E. T. Neil, Phys. Rev. Lett. **100** (2008) 171607 [Erratum-ibid. **102** (2009) 149902] [arXiv:0712.0609 [hep-ph]]; T. Appelquist, G. T. Fleming and E. T. Neil, Phys. Rev. D **79**, 076010 (2009) [arXiv:0901.3766 [hep-ph]].
10. O. Antipin, M. Heikinheimo and K. Tuominen, JHEP **0910**, 018 (2009) [arXiv:0905.0622 [hep-ph]]. M. T. Frandsen, I. Masina and F. Sannino, arXiv:0905.1331 [hep-ph]; R. Foadi, M. T. Frandsen and F. Sannino, Phys. Rev. D **77**, 097702 (2008) [arXiv:0712.1948 [hep-ph]]; O. Antipin and K. Tuominen, Phys. Rev. D **79**, 075011 (2009) [arXiv:0901.4243 [hep-ph]].
11. A. Belyaev, R. Foadi, M. T. Frandsen, M. Jarvinen, F. Sannino and A. Pukhov, Phys. Rev. D **79**, 035006 (2009) [arXiv:0809.0793 [hep-ph]].
12. V. A. Novikov, M. A. Shifman, A. I. Vainshtein and V. I. Zakharov, Nucl. Phys. B **229** (1983) 381; M. A. Shifman and A. I. Vainshtein, Nucl. Phys. B **277** (1986) 456 [Sov. Phys. JETP **64** (1986 ZETFA,91,723-744.1986) 428].
13. N. Seiberg, Nucl. Phys. B **435** (1995) 129 [arXiv:hep-th/9411149].
14. K. A. Intriligator and N. Seiberg, Nucl. Phys. Proc. Suppl. **45BC**, 1 (1996) [arXiv:hep-th/9509066].
15. T. A. Ryttov and F. Sannino, Phys. Rev. D **78** (2008) 065001 [arXiv:0711.3745 [hep-th]].
16. F. Sannino, Phys. Rev. D **80**, 065011 (2009) [arXiv:0907.1364 [hep-th]].
17. D. B. Kaplan, J. W. Lee, D. T. Son and M. A. Stephanov, Phys. Rev. D **80**, 125005 (2009) [arXiv:0905.4752 [hep-th]].
18. H. Gies and J. Jaeckel, Eur. Phys. J. C **46**, 433 (2006) [arXiv:hep-ph/0507171].
19. O. Antipin and K. Tuominen, arXiv:0909.4879 [hep-ph].
20. I. R. Klebanov and E. Witten, Nucl. Phys. B **556**, 89 (1999) [arXiv:hep-th/9905104].
21. Z. Fodor, K. Holland, J. Kuti, D. Nogradi and C. Schroeder, arXiv:0911.2463 [hep-lat].
22. E. Poppitz and M. Unsal, JHEP **0909**, 050 (2009) [arXiv:0906.5156 [hep-th]].
23. E. Poppitz and M. Unsal, arXiv:0910.1245 [hep-th].
24. A. Armoni, Nucl. Phys. B **826**, 328 (2010) [arXiv:0907.4091 [hep-ph]].

Nearly Conformal Gauge Theories on the Lattice

Zoltán Fodor
Department of Physics, University of Wuppertal, Gauss Strasse 20, D-42119, Germany
and
Institute for Theoretical Physics, Eotvos University Budapest, Pazmany P. 1, H-1117, Hungary
E-mail: fodor@bodri.elte.hu

Kieran Holland
Department of Physics, University of the Pacific
3601 Pacific Ave, Stockton CA 95211, USA
E-mail: kholland@pacific.edu

Julius Kuti*
Department of Physics 0319, University of California, San Diego
9500 Gilman Drive, La Jolla CA 92093, USA
E-mail: jkuti@ucsd.edu

Dániel Nógrádi
Institute for Theoretical Physics, Eotvos University Budapest, Pazmany P. 1, H-1117, Hungary
E-mail: nogradi@bodri.elte.hu

Chris Schroeder
Department of Physics, University of Wuppertal, Gauss Strasse 20, D-42119, Germany
E-mail: crs@physics.ucsd.edu

We present selected new results on chiral symmetry breaking in nearly conformal gauge theories with fermions in the fundamental representation of the SU(3) color gauge group. We found chiral symmetry breaking (χSB) for all flavors between $N_f = 4$ and $N_f = 12$ with most of the results discussed here for $N_f = 4, 8, 12$ as we approach the conformal window. To identify χSB we apply several methods which include, within the framework of chiral perturbation theory, the analysis of the Goldstone spectrum in the p-regime and the spectrum of the fermion Dirac operator with eigenvalue distributions of random matrix theory in the ϵ-regime. Chiral condensate enhancement is observed with increasing N_f when the electroweak symmetry breaking scale F is held fixed in technicolor language. Important finite-volume consistency checks from the theoretical understanding of the $SU(N_f)$ rotator spectrum of the δ-regime are discussed. We also consider these gauge theories at $N_f = 16$ inside the conformal window. Our work on the running coupling is presented separately.[1]

Keywords: Beyond Standard Model, lattice, near-conformal, running coupling.

*Speaker at the conference.

1. Introduction

It is an intriguing possibility that new physics beyond the Standard Model might take the form of some new strongly-interacting gauge theory building on the original technicolor idea.[2–4] This approach has lately been revived by new explorations of the multi-dimensional theory space of nearly conformal gauge theories.[5–8] Model building of a strongly interacting electroweak sector requires the knowledge of the phase diagram of nearly conformal gauge theories as the number of colors N_c, number of fermion flavors N_f, and the fermion representation R of the technicolor group are varied in theory space. For fixed N_c and R the theory is in the chirally broken phase for low N_f, and asymptotic freedom is maintained with a negative β function. On the other hand, if N_f is large enough, the β function is positive for all couplings, and the theory is trivial. There is some range of N_f for which the β function might have a non-trivial zero, an infrared fixed point, where the theory is in fact conformal.[11,12] This method has been refined by estimating the critical value of N_f, above which spontaneous chiral symmetry breaking no longer occurs.[13–15]

Interesting models require the theory to be very close to, but below, the conformal window, with a running coupling which is almost constant over a large energy range .[16–21] The nonperturbative knowledge of the critical N_f^{crit} separating the two phases is essential and this has generated much interest and many new lattice studies.[22–55] To provide theoretical framework for the analysis of simulation results, we review first a series of tests expected to hold in the setting of χPT in finite volume and in the infinite volume limit.

2. Chiral symmetry breaking below the conformal window

We will identify in lattice simulations the chirally broken phases with $N_f = 4, 8, 12$ flavors of staggered fermions in the fundamental SU(3) color representation using finite volume analysis. We deploy staggered fermions with exponential (stout) smearing[56] in the lattice action to reduce well-known cutoff effects with taste breaking in the Goldstone spectrum.[57] The presence of taste breaking requires careful analysis of staggered χPT following the important work of Lee, Sharpe, Aubin and Bernard.[58–60]

2.1. *Finite volume analysis in the p-regime*

Three different regimes can be selected in simulations to identify the chirally broken phase from finite volume spectra and correlators. For a lattice size $L_s^3 \times L_t$ in euclidean space and in the limit $L_t \gg L_s$, the conditions $F_\pi L_s > 1$ and $M_\pi L_s > 1$ select the the p-regime, in analogy with low momentum counting.[61,62]

For arbitrary N_f, in the continuum and in infinite volume, the one-loop chiral corrections to M_π and F_π of the degenerate Goldstone pions are given by

$$M_\pi^2 = M^2\left[1 - \frac{M^2}{8\pi^2 N_f F^2} ln\left(\frac{\Lambda_3}{M}\right)\right], \tag{1}$$

$$F_\pi = F\left[1 + \frac{N_f M^2}{16\pi^2 F^2} ln\left(\frac{\Lambda_4}{M}\right)\right], \tag{2}$$

where $M^2 = 2B \cdot m_q$ and $F, B, \Lambda_3, \Lambda_4$ are four fundamental parameters of the chiral Lagrangian, and the small quark mass m_q explicitly breaks the symmetry.[63] The chiral parameters F, B appear in the leading part of the chiral Lagrangian, while Λ_3, Λ_4 enter in next order. There is the well-known GMOR relation $\Sigma_{cond} = BF^2$ in the $m_q \to 0$ limit for the chiral condensate per unit flavor.[64] It is important to note that the one-loop correction to the pion coupling constant F_π is enhanced by a factor N_f^2 compared to M_π^2. The chiral expansion for large N_f will break down for F_π much faster for a given M_π/F_π ratio. The NNLO terms have been recently calculated[65] showing potentially dangerous N_f^2 corrections to Eqs. (1,2).

The finite volume corrections to M_π and F_π are given in the p-regime by

$$M_\pi(L_s, \eta) = M_\pi\left[1 + \frac{1}{2N_f}\frac{M^2}{16\pi^2 F^2} \cdot \widetilde{g}_1(\lambda, \eta)\right], \tag{3}$$

$$F_\pi(L_s, \eta) = F_\pi\left[1 - \frac{N_f}{2}\frac{M^2}{16\pi^2 F^2} \cdot \widetilde{g}_1(\lambda, \eta)\right], \tag{4}$$

where $\widetilde{g}_1(\lambda, \eta)$ describes the finite volume corrections with $\lambda = M \cdot L_s$ and aspect ratio $\eta = L_t/L_s$. The form of $\widetilde{g}_1(\lambda, \eta)$ is a complicated infinite sum which contains Bessel functions and requires numerical evaluation.[62] Eqs. (1-4) provide the foundation of the p-regime fits in simulations.

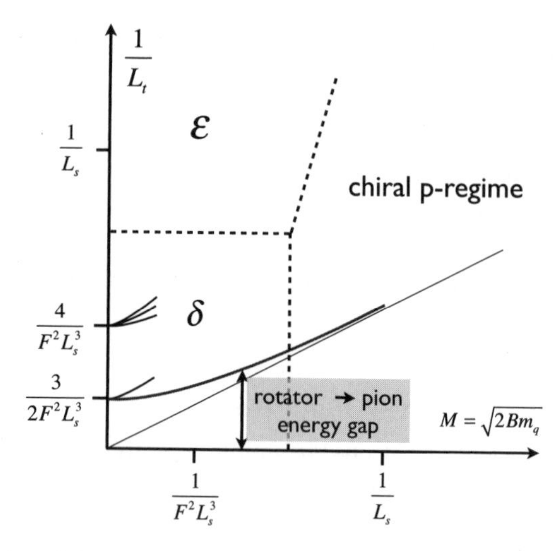

Fig. 1. Schematic plot of the regions in which the three low energy chiral expansions are valid. The vertical axis shows the finite temperature scale (euclidean time in the path integral) which probes the rotator dynamics of the δ-regime and the ϵ-regime. The first two low lying rotator levels are also shown on the vertical axis for the simple case of $N_f = 2$. The fourfold degenerate lowest rotator excitation at $m_q = 0$ will split into an isotriplet state (lowest energy level), which evolves into the p-regime pion as m_q increases, and into an isosinglet state representing a multi-pion state in the p-regime. Higher rotator excitations have similar interpretations.

2.2. δ-regime and ϵ-regime

At fixed L_s and in cylindrical geometry $L_t/L_s \gg 1$, a crossover occurs from the p-regime to the δ-regime when $m_q \to 0$, as shown in Fig. 1. The dynamics is dominated by the rotator states of the chiral condensate in this limit[66] which is characterized by the conditions $FL_s > 1$ and $ML_s \ll 1$. The densely spaced rotator spectrum scales with gaps of the order $\sim 1/F^2L_s^3$, and at $m_q = 0$ the chiral symmetry is apparently restored. However, the rotator spectrum, even at $m_q = 0$ in the finite volume, will signal that the infinite system is in the chirally broken phase for the particular parameter set of the Lagrangian. This is often misunderstood in the interpretation of lattice simulations. Measuring finite energy levels with pion quantum numbers at fixed L_s in the $m_q \to 0$ limit is not a signal for chiral symmetry restoration of the infinite system.[40]

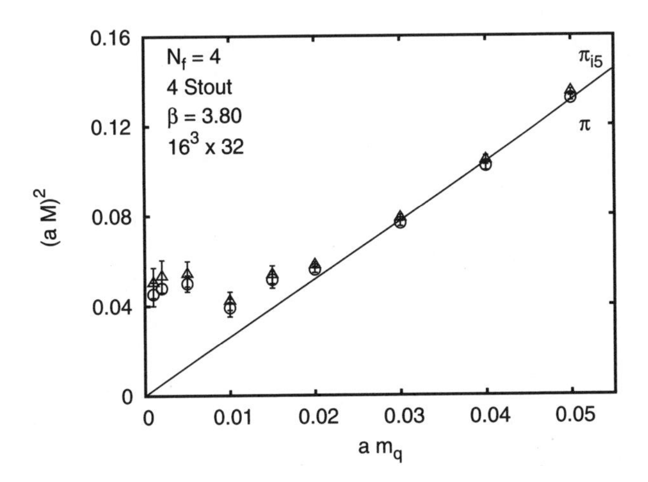

Fig. 2. The crossover from the p-regime to the δ-regime is shown for the π and π_{i5} states at $N_f = 4$.

If $L_t \sim L_s$ under the conditions $FL_s > 1$ and $ML_s \ll 1$, the system will be driven into the ϵ-regime which can be viewed as the high temperature limit of the δ-regime quantum rotator. Although the δ-regime and ϵ-regime have an overlapping region, there is an important difference in their dynamics. In the δ-regime of the quantum rotator, the mode of the pion field $U(x)$ with zero spatial momentum dominates with time-dependent quantum dynamics. The ϵ-regime is dominated by the four-dimensional zero momentum mode of the chiral Lagrangian.

We report simulation results of all three regimes in the chirally broken phase of the technicolor models we investigate. The analysis of the three regimes complement each other and provide cross-checks for the correct identification of the phases. First, we will probe Eqs. (1-4) in the p-regime, and follow with the study of Dirac spectra and RMT eigenvalue distributions in the ϵ-regime. The spectrum in the δ-regime is used as a signal to monitor p-regime spectra as m_q decreases. Fig. 2 is an illustrative example of this crossover in our simulations. It is important to note that the energy levels in the chiral limit do not always match the rotator spectrum at the small $F \cdot L_s$ values of the simulations. This squeezing of insufficiently large enough $F \cdot L_s$ for undistorted, finite volume chiral behavior in the p-regime, ϵ-regime, and δ-regime is a serious limitation of all recent simulations.

3. Goldstone spectrum and χSB at $N_f = 12$

We find the chiral symmetry breaking pattern for the controversial $N_f = 12$ case similar to the $N_f = 8, 9$ cases. The Goldstone spectrum remains separated from the technicolor scale of the ρ-meson. The true Goldstone pion and two additional split pseudo-Goldstone states are shown again in Fig. 3 with different slopes as $a \cdot m_q$ increases. The trends and the underlying explanation are similar to the $N_f = 8, 9$ cases. The chiral fit to M_π^2/m_q shown at the top right side of Fig. 3 is based on Eq. (1) only since the F_π data points are outside the convergence range of the chiral expansion. At $\beta = 2.2$ the fitted value of B is $a \cdot B = 2.7(2)$ in lattice units with $a \cdot F = 0.0120(1)$ and $a \cdot \Lambda_3 = 0.50(3)$ also fitted. The fitted ρ-mass in the chiral limit is $a \cdot M_\rho = 0.115(15)$ from $a \cdot m_q = 0.025 - 0.045$ with $M_\rho/F = 10(1)$. The fitted value of $B/F = 223(17)$ is not very reliable but consistent with the enhancement of the chiral condensate found at $N_f = 8, 9$ without including renormalization scale effects. Again, at fixed lattice spacing, the small chiral condensate $\langle\overline{\psi}\psi\rangle$ summed over all flavors is dominated by the linear term in m_q from UV contributions. The linear fit gives $\langle\overline{\psi}\psi\rangle = 0.0033(13)$ in the chiral limit which came out unexpectedly close the GMOR relation of $\langle\overline{\psi}\psi\rangle = 12F^2B$ with $12F^2B = 0.0046(4)$ fitted. Issues and concerns in the systematics are similar to the $N_f = 8, 9$ cases but finite volume limitations and the convergence of the chiral expansion are more problematic. Currently we are investigating the important $N_f = 12$ model on larger lattices to probe the possible influence of unwanted squeezing effects on the spectra. This should also clarify the mass splitting pattern of the ρ and A_1 states we are seeing in the chiral limit as N_f is varied.

Our findings at $N_f = 12$ are not consistent with some earlier work[31,32] where the runing coupling was studied. Lessons from the Dirac spectra and RMT to complement p-regime tests are discussed in the next section.

4. Epsilon regime, Dirac spectrum and RMT

If the bare parameters of a gauge theory are tuned to the ϵ-regime in the chirally broken phase, the low-lying Dirac spectrum follows the predictions of random matrix theory. The corresponding random matrix model is only sensitive to the pattern of chiral symmetry breaking, the topological charge and the rescaled fermion mass once the eigenvalues are also rescaled by the same factor $\Sigma_{cond}V$. This idea has been confirmed in various settings both in quenched and fully dynamical simulations. The same method is applied here to nearly conformal gauge models.

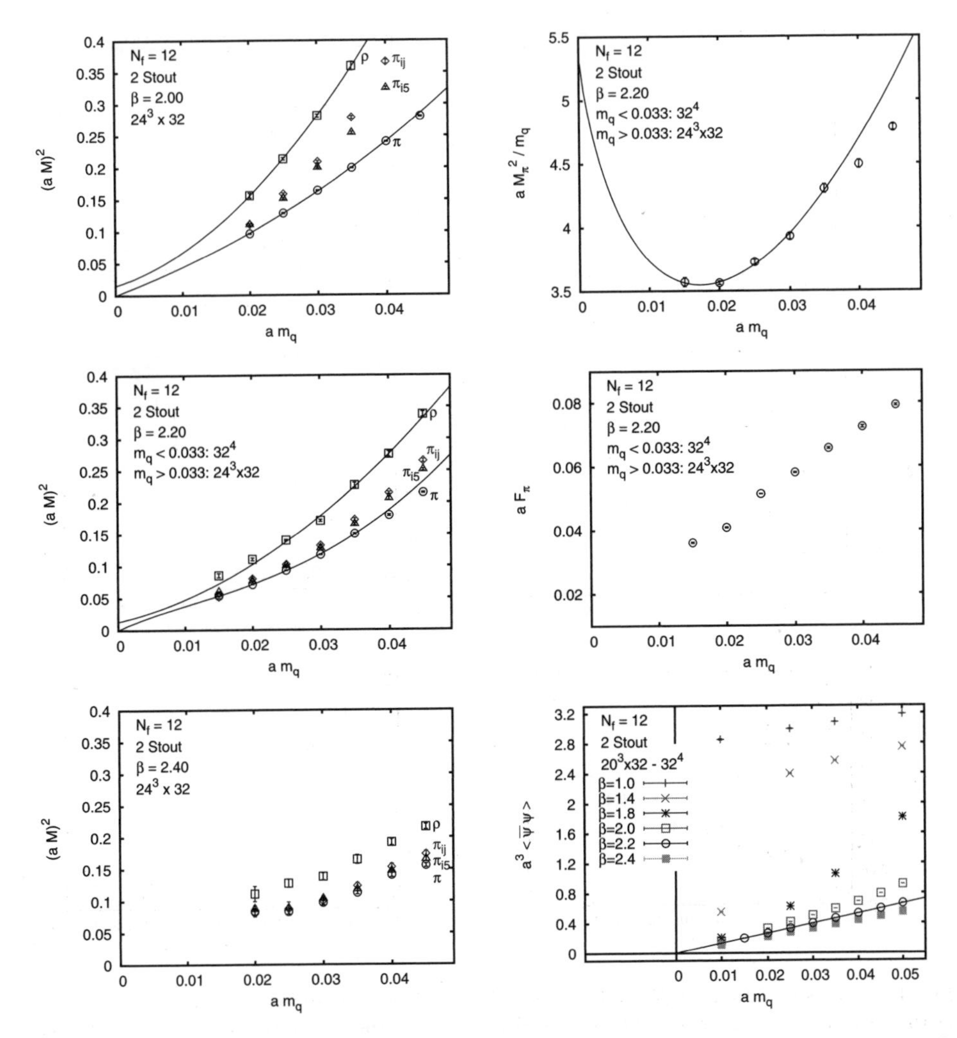

Fig. 3. The pseudo-Goldstone spectrum and chiral fits are shown for $N_f = 12$ simulations with lattice size $24^3 \times 32$ and 32^4. The left column shows the pseudo-Goldstone spectrum with decreasing taste breaking as the gauge coupling is varied from $\beta = 2.0$ to $\beta = 2.4$. Although the bottom figure on the left at $\beta = 2.4$ illustrates the continued restoration of taste symmetry, the volume is too small for the Goldstone spectrum. The middle value at $\beta = 2.2$ was chosen in the top right figure with fitting range $a \cdot m_q = 0.015 - 0.035$ of the NLO chiral fit to M_π^2/m_q which approaches $2B$ in the chiral limit. The middle figure on the right shows the F_π data with no NLO fit far away from the chiral limit. The bottom right figure, with its additional features discussed in the text, is the linear fit to the chiral condensate with fitting range $a \cdot m_q = 0.02 - 0.04$. The physical fit parameters B, F, Λ_3 are discussed in the text. Two stout steps were used in all $N_f = 12$ simulations.

The connection between the eigenvalues λ of the Dirac operator and chiral symmetry breaking is given in the Banks-Casher relation,[70]

$$\Sigma_{cond} = -\langle \overline{\Psi}\Psi \rangle = \lim_{\lambda \to 0} \lim_{m \to 0} \lim_{V \to \infty} \frac{\pi \rho(\lambda)}{V},$$

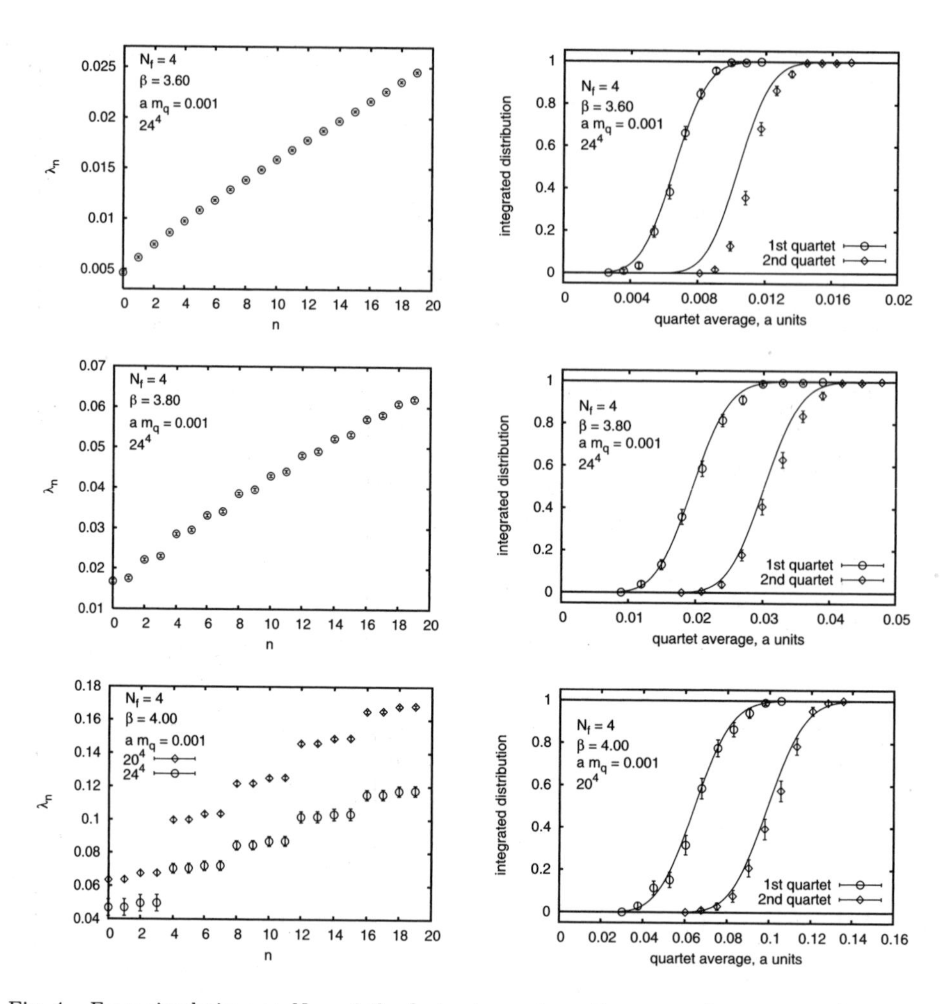

Fig. 4. From simulations at $N_f = 4$ the first column shows the approach to quartet degeneracy of the spectrum as β increases. The second column shows the integrated distribution of the two lowest quartets averaged. The solid line compares this procedure to RMT with $N_f = 4$.

where Σ_{cond} designates the quark condensate normalized to a single flavor. To generate a non-zero density $\rho(0)$, the smallest eigenvalues must become densely packed as the volume increases, with an eigenvalue spacing $\Delta\lambda \approx 1/\rho(0) = \pi/(\Sigma_{cond}V)$. This allows a crude estimate of the quark condensate Σ_{cond}. One can do better by exploring the ϵ-regime: If chiral symmetry is spontaneously broken, tune the volume and quark mass such that $\frac{1}{F_\pi} \ll L \ll \frac{1}{M_\pi}$, so that the Goldstone pion is much lighter than the physical value, and finite volume effects are dominant as we discussed in Section 2. The chiral Lagrangian is dominated by the zero-momentum mode from the mass term and all kinetic terms are suppressed. In this limit, the distributions of the lowest eigenvalues are identical to those of random matrix theory, a theory of large matrices obeying certain symmetries.[71–73] To connect with RMT, the eigenvalues and quark mass are rescaled as $z = \lambda\Sigma_{cond}V$ and $\mu = m_q\Sigma_{cond}V$, and the eigenvalue distributions also depend on the topological charge ν and the number of quark flavors N_f. RMT is a very useful tool to calculate analytically all of the eigenvalue distributions.[74] The eigenvalue distributions in various topological sectors are measured via lattice simulations, and via comparison with RMT, the value of the condensate Σ_{cond} can be extracted.

After we generate large thermalized ensembles, we calculate the lowest twenty eigenvalues of the Dirac operator using the PRIMME package.[75] In the continuum limit, the staggered eigenvalues form degenerate quartets, with restored taste symmetry. The first column of Fig. 4 shows the change in the eigenvalue structure for $N_f = 4$ as the coupling constant is varied. At $\beta = 3.6$ grouping into quartets is not seen, the Goldstone pions are somewhat still split, and staggered perturbation theory is just beginning to kick in. At $\beta = 3.8$ doublet pairing appears and at $\beta = 4.0$ the quartets are nearly degenerate. The Dirac spectrum is collapsed as required by the Banks-Casher relation. In the second column we show the integrated distributions of the two lowest eigenvalue quartet averages,

$$\int_0^\lambda p_k(\lambda')d\lambda', \qquad k = 1, 2 \tag{5}$$

which is only justified close to quartet degeneracy. All low eigenvalues are selected with zero topology. To compare with RMT, we vary $\mu = m_q\Sigma_{cond}V$ until we satisfy

$$\frac{\langle\lambda_1\rangle_{\rm sim}}{m} = \frac{\langle z_1\rangle_{\rm RMT}}{\mu}, \tag{6}$$

where $\langle\lambda_1\rangle_{\rm sim}$ is the lowest quartet average from simulations and the RMT average $\langle z\rangle_{\rm RMT}$ depends implicitly on μ and N_f. With this optimal value of μ, we can predict the shapes of $p_k(\lambda)$ and their integrated distributions, and compare to the simulations. The agreement with the two lowest integrated RMT eigenvalue shapes is excellent for the larger β values.

The main qualitative features of the RMT spectrum are very similar in our $N_f = 8$ simulations as shown in Fig. 5. One marked quantitative difference is a noticeable slowdown in response to change in the coupling constant. As β grows the recovery of the quartet degeneracy is considerably delayed in comparison with the onset of p-regime Goldstone dynamics. Overall, for the $N_f = 4, 8$ models we find consistency between the p-regime analysis and the RMT tests. Earlier, using Asqtad fermions at a particular β value, we found agreement with RMT even at $N_f = 12$ which indicated a chirally broken phase.[23] Strong taste breaking with Asqtad fermions leaves the quartet averaging in question and the bulk pronounced crossover of the Asqtad action as β grows is also an issue. Currently we are investigating the RMT picture for $N_f = 9, 10, 11, 12$ with our much improved action with stout smearing. This action shows no artifact transitions and handles taste breaking much more effectively. Firm conclusions on the $N_f = 12$ model to support our findings of χSB in the p-regime will require continued investigations.

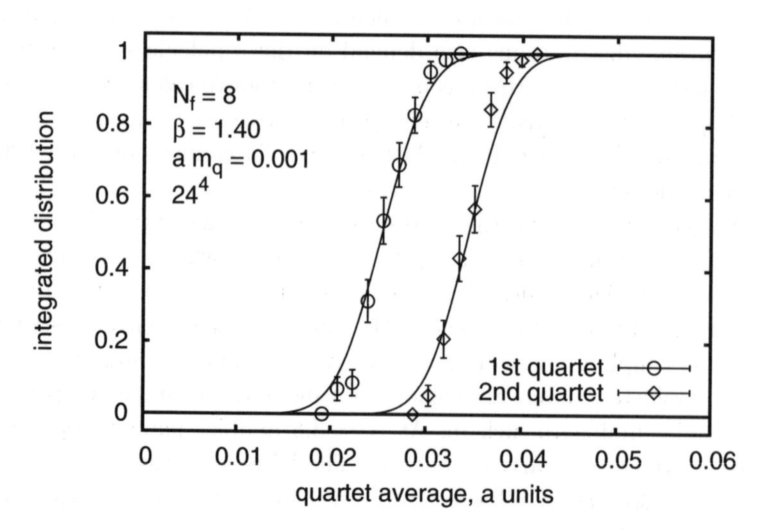

Fig. 5. The solid lines compare the integrated distribution of the two lowest quartet averages to RMT predictions with $N_f = 8$.

5. Inside the conformal window

A distinguished feature of the $N_f = 16$ conformal model is how the renormalized coupling $g^2(L)$ runs with L, the linear size of the spatial volume in a Hamiltonian or Transfer Matrix description. On very small scales the running coupling $g^2(L)$ grows with L as in any other asymptotically free theory. However, $g^2(L)$ will not grow large, and in the $L \to \infty$ limit it will converge to the fixed point g^{*2} which is rather weak,[76] within the reach of perturbation theory. There is non-trivial, small-volume dynamics which is illustrated first in the pure gauge sector.

At small g^2, without fermions, the zero-momentum components of the gauge field are known to dominate the dynamics.[77–79] With $SU(3)$ gauge group, there are twenty-seven degenerate vacuum states, separated by energy barriers which are generated by the integrated effects of the non-zero momentum components of the gauge field in the Born-Oppenheimer approximation. The lowest-energy excitations of the gauge field Hamiltonian scale as $\sim g^{2/3}(L)/L$ evolving into glueball states and becoming independent of the volume as the coupling constant grows with L. Non-trivial dynamics evolves through three stages as L grows. In the first regime, in very small boxes, tunneling is suppressed between vacua which remain isolated. In the second regime, for larger L, tunneling sets in and electric flux states will not be exponentially suppressed. Both regimes represent small worlds with zero-momentum spectra separated from higher momentum modes of the theory with energies on the scale of $2\pi/L$. At large enough L the gauge dynamics overcomes the energy barrier, and wave functions spread over the vacuum valley. This third regime is the crossover to confinement where the electric fluxes collapse into thin string states wrapping around the box.

It is likely that a conformal theory with a weak coupling fixed point at $N_f = 16$ will have only the first two regimes which are common with QCD. Now the calculations have to include fermion loops.[80,81] The vacuum structure in small enough volumes, for which the wave functional is sufficiently localized around the vacuum configuration, remains calculable by adding in one-loop order the quantum effects of the fermion field fluctuations. The spatially constant abelian gauge fields parametrizing the vacuum valley are given by $A_i(\mathbf{x}) = T^a C_i^a / L$ where T_a are the (N-1) generators for the Cartan subalgebra of $SU(N)$. For $SU(3)$, $T_1 = \lambda_3/2$ and $T_2 = \lambda_8/2$. With N_f flavors of massless fermion fields the effective potential of the constant mode is given by

$$V_{\rm eff}^{\mathbf{k}}(\mathbf{C}^b) = \sum_{i>j} V(\mathbf{C}^b[\mu_b^{(i)} - \mu_b^{(j)}]) - N_f \sum_i V(\mathbf{C}^b \mu_b^{(i)} + \pi \mathbf{k}), \tag{7}$$

with $\mathbf{k} = \mathbf{0}$ for periodic, or $\mathbf{k} = (1,1,1)$, for antiperiodic boundary conditions on the fermion fields. The function $V(\mathbf{C})$ is the one-loop effective potential for $N_f = 0$ and the weight vectors $\mu^{(i)}$ are determined by the eigenvalues of the abelian generators. For SU(3) $\mu^{(1)} = (1,1,-2)/\sqrt{12}$ and $\mu^{(2)} = \frac{1}{2}(1,-1,0)$. The correct quantum vacuum is found at the minimum of this effective potential which is dramatically changed by the fermion loop contributions. The Polyakov loop observables remain center elements at the new vacuum configurations with complex values; for $SU(N)$

$$P_j = \frac{1}{N} \,\mathrm{tr}\, \left(\exp(iC_j^b T_b)\right) = \frac{1}{N} \sum_n \exp(i\mu_b^{(n)} C_j^b) = \exp(2\pi i l_j / N). \tag{8}$$

This implies $\mu_b^{(n)} \mathbf{C}^b = 2\pi \mathbf{l}/N \pmod{2\pi}$, independent of n, and $V_{\rm eff}^{\mathbf{k}} = -N_f N V(2\pi \mathbf{l}/N + \pi \mathbf{k})$. In the case of antiperiodic boundary conditions, $\mathbf{k} = (1,1,1)$, this is minimal only when $\mathbf{l} = \mathbf{0} \pmod{2\pi}$. The quantum vacuum in this case is the naive one, $A = 0$ $(P_j = 1)$. In the case of periodic boundary conditions, $\mathbf{k} = \mathbf{0}$, the vacua have $\mathbf{l} \neq \mathbf{0}$, so that P_j correspond to non-trivial center elements. For

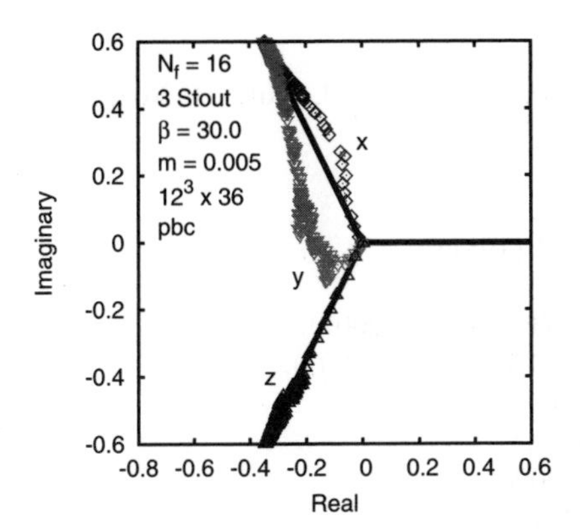

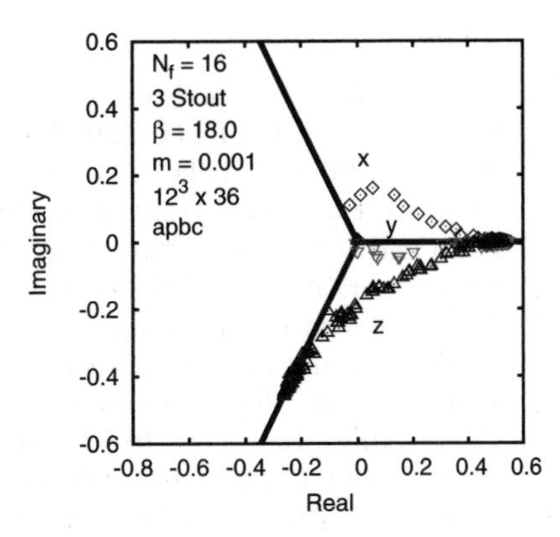

Fig. 6. The time evolution of complex Polyakov loop distributions are shown from our $N_f = 16$ simulations with $12^3 \times 36$ lattice volume. Tree-level Symanzik-improved gauge action is used in the simulations and staggered fermions with three stout steps and very small fermion masses.

SU(3), there are now 8 degenerate vacua characterized by eight different Polyakov loops, $P_j = \exp(\pm 2\pi i/3)$. Since they are related by coordinate reflections, in a small volume parity (P) and charge conjugation (C) are spontaneously broken, although CP is still a good symmetry.[80]

Our simulations of the $N_f = 16$ model below the conformal fixed point g^{*2} confirm the theoretical vacuum structure. Fig. 6 shows the time evolution of Polyakov loop distributions monitored along the three separate spatial directions. On the left side, with periodic spatial boundary conditions, the time evolution is shown starting from randomized gauge configuration with the Polyakov loop at the origin. The system evolves into one of the eight degenerate vacua selected by the positive imaginary part of the complex Polyakov loop along the x and y direction and negative imaginary part along the z direction. On the right, with antiperiodic spatial boundary conditions, the vacuum is unique and trivial with real Polyakov loop in all three directions. The time evolution is particularly interesting in the z direction with a swing first from the randomized gauge configuration to a complex metastable minimum first, and eventually tunneling back to the trivial vacuum and staying there, as expected. The measured fermion-antifermion spectra and the spectrum of the Dirac operator further confirm this vacuum structure.

Acknowledgments

Julius Kuti is greatful to the organizers for a most enjoyable meeting. We thank Sandor Katz and Kalman Szabo for the Wuppertal RHMC code. For some calculations, we used the publicly available MILC code. We performed simulations on the Wuppertal GPU cluster, Fermilab clusters under the auspices of USQCD and SciDAC, and the Ranger cluster of the Teragrid organization. This research is supported by the NSF under grant 0704171, by the DOE under grants DOE-FG03-97ER40546, DOE-FG-02-97ER25308, by the DFG under grant FO 502/1 and by SFB-TR/55.

References

1. K. Holland, *Talk in this Proceedings*; Z. Fodor, K. Holland, J. Kuti, D. Nogradi and C. Schroeder, PoS **LATTICE2009**, 058 (2009).
2. S. Weinberg, Phys. Rev. D **19**, 1277 (1979).
3. L. Susskind, Phys. Rev. D **20**, 2619 (1979).
4. E. Farhi and L. Susskind, Phys. Rept. **74**, 277 (1981).
5. F. Sannino, arXiv:0902.3494 [hep-ph].
6. T. A. Ryttov and F. Sannino, Phys. Rev. D **76**, 105004 (2007).
7. D. D. Dietrich and F. Sannino, Phys. Rev. D **75**, 085018 (2007).
8. D. K. Hong et al., Phys. Lett. B **597**, 89 (2004).
9. H. Georgi, Phys. Rev. Lett. **98**, 221601 (2007).
10. M. A. Luty and T. Okui, JHEP **0609**, 070 (2006).
11. W. E. Caswell, Phys. Rev. Lett. **33**, 244 (1974).
12. T. Banks and A. Zaks, Nucl. Phys. B **196**, 189 (1982).
13. T. Appelquist et al., Phys. Rev. Lett. **61**, 1553 (1988).
14. A. G. Cohen and H. Georgi, Nucl. Phys. B **314**, 7 (1989).

15. T. Appelquist et al., Phys. Rev. Lett. **77**, 1214 (1996).
16. B. Holdom, Phys. Rev. D **24**, 1441 (1981).
17. K. Yamawaki et al., Phys. Rev. Lett. **56**, 1335 (1986).
18. T. W. Appelquist et al., Phys. Rev. Lett. **57**, 957 (1986).
19. V. A. Miransky and K. Yamawaki, Phys. Rev. D **55**, 5051 (1997).
20. M. Kurachi and R. Shrock, JHEP **0612**, 034 (2006).
21. E. Eichten and K. D. Lane, Phys. Lett. B **90**, 125 (1980).
22. Z. Fodor, K. Holland, J. Kuti, D. Nogradi and C. Schroeder, Phys. Lett. B **681**, 353 (2009).
23. Z. Fodor et al., PoS **LATTICE2008**, 066 (2008).
24. Z. Fodor et al., PoS **LATTICE2008**, 058 (2008).
25. Z. Fodor, K. Holland, J. Kuti, D. Nogradi and C. Schroeder, arXiv:0908.2466 [hep-lat].
26. Z. Fodor, K. Holland, J. Kuti, D. Nogradi and C. Schroeder, JHEP **0908**, 084 (2009) [arXiv:0905.3586 [hep-lat]].
27. T. DeGrand et al., Phys. Rev. D **79**, 034501 (2009).
28. T. DeGrand et al., arXiv:0809.2953 [hep-lat].
29. B. Svetitsky et al., arXiv:0809.2885 [hep-lat].
30. Y. Shamir et al., Phys. Rev. D **78**, 031502 (2008).
31. T. Appelquist et al., Phys. Rev. Lett. **100**, 171607 (2008).
32. T. Appelquist et al., arXiv:0901.3766 [hep-ph].
33. G. T. Fleming, PoS **LATTICE2008**, 021 (2008).
34. L. Del Debbio et al., arXiv:0812.0570 [hep-lat].
35. L. Del Debbio et al., arXiv:0805.2058 [hep-lat].
36. L. Del Debbio et al., JHEP **0806**, 007 (2008).
37. A. J. Hietanen et al., arXiv:0904.0864 [hep-lat].
38. A. J. Hietanen et al., arXiv:0812.1467 [hep-lat].
39. A. Hietanen et al., PoS **LATTICE2008**, 065 (2008).
40. A. Deuzeman et al., arXiv:0904.4662 [hep-ph].
41. A. Deuzeman et al., arXiv:0810.3117 [hep-lat].
42. A. Deuzeman et al., PoS **LATTICE2008**, 060 (2008).
43. A. Deuzeman et al., Phys. Lett. B **670**, 41 (2008).
44. S. Catterall and F. Sannino, Phys. Rev. D **76**, 034504 (2007).
45. S. Catterall et al., JHEP **0811**, 009 (2008).
46. X. Y. Jin and R. D. Mawhinney, PoS **LATTICE2008**, 059 (2008).
47. A. Hasenfratz, arXiv:0907.0919 [hep-lat].
48. T. DeGrand and A. Hasenfratz, arXiv:0906.1976 [hep-lat].
49. T. DeGrand, arXiv:0906.4543 [hep-lat].
50. L. Del Debbio et al., arXiv:0907.3896 [hep-lat].
51. T. DeGrand, arXiv:0910.3072 [hep-lat].
52. T. Appelquist *et al.*, arXiv:0910.2224 [hep-ph].
53. X. Y. Jin and R. D. Mawhinney, PoS **LAT2009**, 049 (2009).
54. A. Hasenfratz, arXiv:0911.0646 [hep-lat].
55. D. K. Sinclair and J. B. Kogut, arXiv:0909.2019 [hep-lat].
56. C. Morningstar and M. J. Peardon, Phys. Rev. D **69**, 054501 (2004).
57. Y. Aoki *et al.*, JHEP **0601**, 089 (2006); Phys. Lett. B **643**, 46 (2006).
58. W. J. Lee and S. R. Sharpe, Phys. Rev. D **60**, 114503 (1999).
59. C. Aubin and C. Bernard, Phys. Rev. D **68**, 034014 (2003).
60. C. Aubin and C. Bernard, Phys. Rev. D **68**, 074011 (2003).
61. J. Gasser and H. Leutwyler, Nucl. Phys. B **307**, 763 (1988).
62. F. C. Hansen and H. Leutwyler, Nucl. Phys. B **350**, 201 (1991).

63. J. Gasser and H. Leutwyler, Annals Phys. **158**, 142 (1984).
64. M. Gell-Mann et al., Phys. Rev. **175**, 2195 (1968).
65. J. Bijnens and J. Lu, arXiv:0910.5424 [hep-ph].
66. H. Leutwyler, Phys. Lett. B **189**, 197 (1987).
67. J. Noaki *et al.* [JLQCD and TWQCD Collaborations], Phys. Rev. Lett. **101**, 202004 (2008).
68. P. Hasenfratz and F. Niedermayer, Z. Phys. B bf 92, 91 (1993), arXiv:hep-lat/9212022.
69. P. Hasenfratz, arXiv:0909.3419 [hep-th].
70. T. Banks and A. Casher, Nucl. Phys. B **169**, 103 (1980).
71. E. V. Shuryak and J. J. M. Verbaarschot, Nucl. Phys. A **560**, 306 (1993).
72. P. H. Damgaard, Nucl. Phys. Proc. Suppl. **128**, 47 (2004).
73. J. J. M. Verbaarschot and T. Wettig, Ann. Rev. Nucl. Part. Sci. **50**, 343 (2000).
74. P. H. Damgaard and S. M. Nishigaki, Phys. Rev. D **63**, 045012 (2001).
75. A. Stathopoulos and J. R. McCombs, SIAM J. Sci. Comput., Vol. **29, No. 5**, 2162 (2007).
76. U. M. Heller, Nucl. Phys. Proc. Suppl. **63**, 248 (1998).
77. G. 't Hooft, Nucl. Phys. B **153**, 141 (1979).
78. M. Luscher, Nucl. Phys. B **219**, 233 (1983).
79. P. van Baal and J. Koller, Annals Phys. **174**, 299 (1987).
80. P. van Baal, Nucl. Phys. B **307**, 274 (1988).
81. J. Kripfganz and C. Michael, Nucl. Phys. B **314**, 25 (1989).

Going Beyond QCD in Lattice Gauge Theory

G. T. Fleming
Sloane Physics Laboratory, Yale University
New Haven, CT 06520-8120, USA
E-mail: George.Fleming@Yale.edu

Strongly coupled gauge theories (SCGT's) have been studied theoretically for many decades using numerous techniques. The obvious motivation for these efforts stemmed from a desire to understand the source of the strong nuclear force: Quantum Chromodynamics (QCD). Guided by experimental results, theorists generally consider QCD to be a well-understood SCGT. Unfortunately, it is not clear how to extend the lessons learned from QCD to other SCGT's. Particularly urgent motivators for new studies of other SCGT's are the ongoing searches for physics beyond the standard model (BSM) at the Large Hadron Collider (LHC) and the Tevatron. Lattice gauge theory (LGT) is a technique for systematically-improvable calculations in many SCGT's. It has become the standard for non-perturbative calculations in QCD and it is widely believed that it may be useful for study of other SCGT's in the realm of BSM physics. We will discuss the prospects and potential pitfalls for these LGT studies, focusing primarily on the flavor dependence of SU(3) gauge theory.

Keywords: Beyond the standard model, dynamical electroweak symmetry breaking, lattice gauge theory, technicolor.

1. Introduction

The Tevatron is now scheduled to run through the end of 2011 with the expectation of providing three-sigma evidence against the existence of a standard model Higgs boson if, in fact, such a boson does not exist in nature. Given the current commissioning status, it is possible that the Large Hadron Collider (LHC) may be able to meaningfully contribute to the Higgs search in the same time frame. If the Higgs boson is ruled out, the case for new strong interactions to be discovered at higher energies in LHC detectors will be very much enhanced.

If a strongly coupled gauge theory (SCGT) is responsible for the dynamical breaking of electroweak symmetry (DEWSB) in Nature, then the Tevatron and LHC may not observe a Standard Model Higgs boson. Such experimental *non-observation* of the Higgs boson will accelerate already growing interest in DEWSB models. It is fortuitous that calculational capabilities of lattice gauge theory (LGT) are now able to make a significant impact just when new strong physics has the best chance of being discovered at the TeV scale. As a result, many lattice groups have now started calculations in theories which describe possible new TeV-scale strong interactions. See[2,3] for reviews and additional references therein.

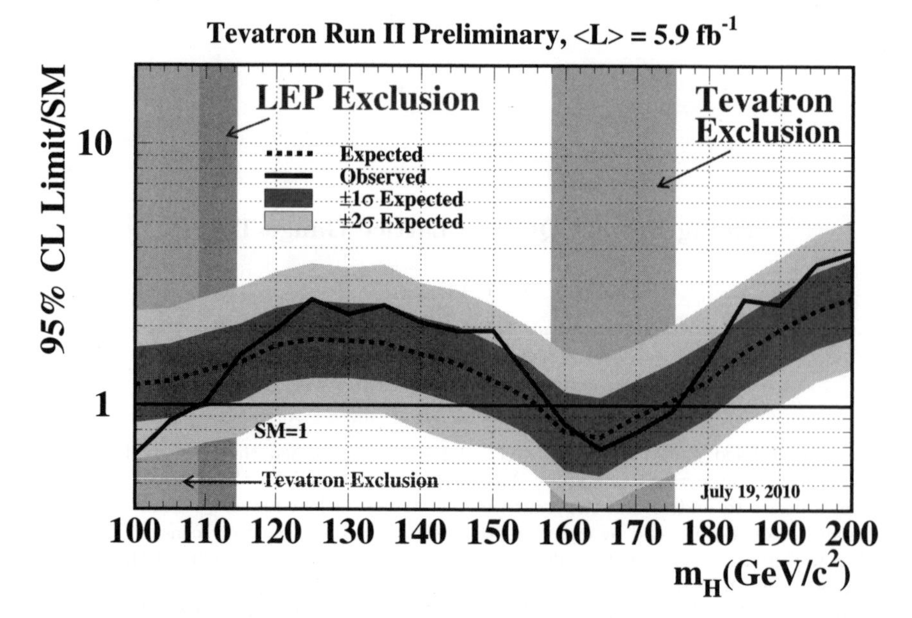

Fig. 1. Observed and expected (median, for the background-only hypothesis) 95% C.L. upper limits on the ratios to the SM cross section, as functions of the Higgs boson mass for the combined CDF and D0 analyses. The limits are expressed as a multiple of the SM prediction for test masses (every 5 GeV/c^2) for which both experiments have performed dedicated searches in different channels. The points are joined by straight lines for better readability. The bands indicate the 68% and 95% probability regions where the limits can fluctuate, in the absence of signal. The limits displayed in this figure are obtained with the Bayesian calculation.[1]

2. Conformal Window

With a small number of massless fermions, a vector-like gauge field theory such as QCD exhibits confinement and dynamical chiral symmetry breaking. If the number of massless fermions in the fundamental representation, N_f, is larger than some easily computed number $N_f^{\rm af} = (11/2)N_c$, then asymptotic freedom is lost and the theory is not even well defined because the renormalized coupling diverges at short distances. But if N_f is just below value $N_f^{\rm af}$, the theory is conformal in the infrared, governed by a weak infrared fixed point (IRFP) which appears already in the two-loop beta function.[4,5] There is no confinement, and chiral symmetry is unbroken. This IRFP persists down to some critical value $N_f^{\rm c}$, where the coupling grows sufficiently strong at long distances that a transition to the confined, chirally broken phase takes place. The range $N_f^{\rm c} < N_f < N_f^{\rm af}$ is the *"conformal window"*, where the theory is in the "non-Abelian Coulomb phase". The initial indications, based on the non-perturbative LGT calculations of the running coupling in the Schrödinger functional scheme[6–8] as shown in Fig. 2, lead us to the tentative conclusion that $8 < N_f^{\rm c} < 12$.

However, the most recent LGT studies[14–17,19–34] for SU(3) gauge theory with N_f flavors in the fundamental representation use a wide array of calculational tech-

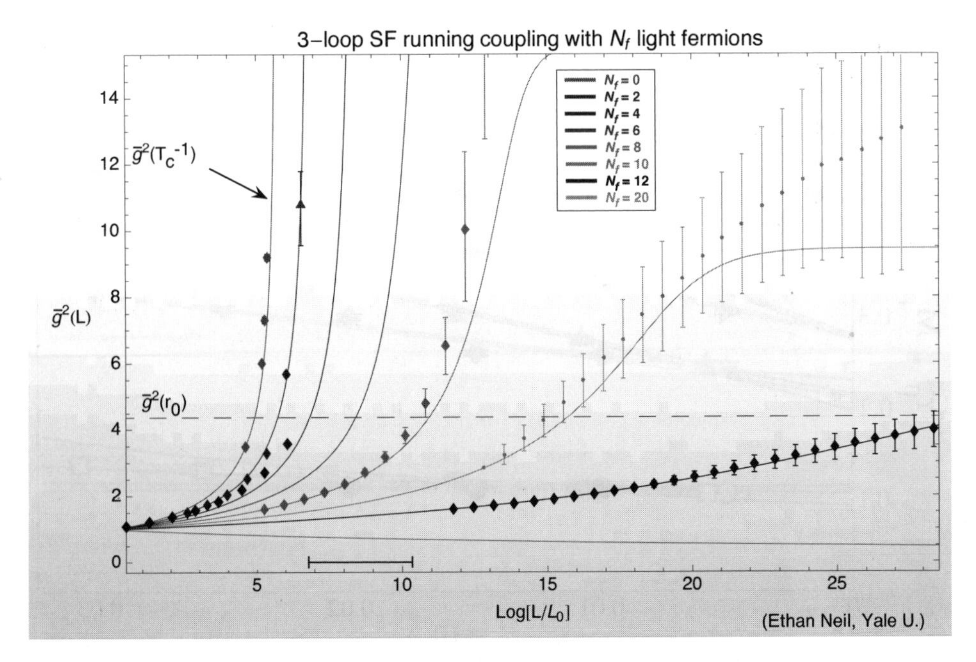

Fig. 2. Non-perturbative LGT calculations of the SU(3) Yang-Mills running coupling in the Schrödinger functional scheme for N_f=0,2,4,8,10,12 flavors in the fundamental representation.[9–18] The **preliminary** N_f=10 points were presented by N. Yamada at the second *Workshop on Lattice Gauge Theory for LHC Physics*, 6–7 Nov. 2009, Boston, MA. The horizontal bar at the bottom of the figure indicates the range of scales covered by a lattice calculation on a 32^3 spatial volume.

niques to address the question of whether N_f=12 is inside or outside the conformal window and the situation is very unclear. In particular, LGT methods commonly used to study QCD seem poorly suited for the study of theories with IRFP's. Thus, it will likely be some time before these methods can definitively confirm (or rule out) the existence of an IRFP for $N_f = 12$ massless flavors.

Just below this critical number of flavors, it is plausible that Yang-Mills exhibits behavior at intermediate scales that is more like the physics of the non-Abelian Coulomb phase than of QCD. Of course, at very long distances, the theory must eventually confine otherwise it would be inside the conformal window. Such behavior is called "walking" because the renormalized coupling is expected to run much slower or "walk" over intermediate distance scales when compared to QCD.[35–39]

This "walking" phenomenon can play an important phenomenological role in a technicolor theory of electroweak symmetry breaking. If there are $N_f/2$ electroweak doublets, then $F = F_{EW}/\sqrt{N_f/2} \simeq 250\ \text{GeV}/\sqrt{N_f/2}$. Flavor-changing neutral currents (FCNC's), which are present when the technicolor theory is extended to provide for the generation of quark masses, can be too large unless the associated scale Λ_{ETC} is high enough. But then the first- and second-generation quark masses are typically much too small. They are proportional to the quantity $\langle\overline{\psi}\psi\rangle/\Lambda^2_{ETC}$, where ψ is a technifermion field and $\langle\overline{\psi}\psi\rangle$ is the bilinear fermion condensate defined

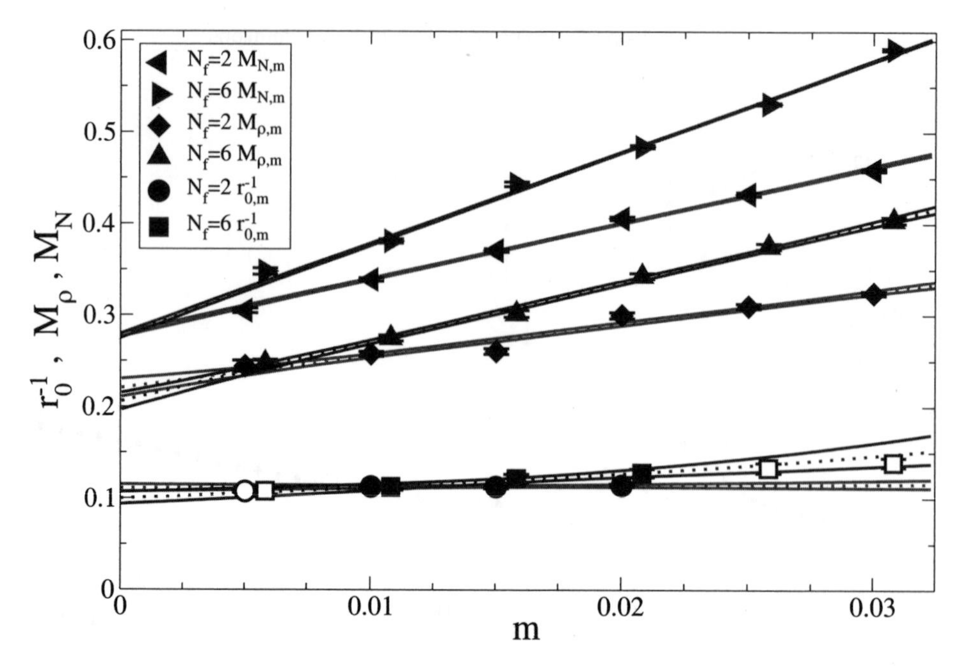

Fig. 3. Chiral extrapolations of the nucleon mass M_N, the vector meson mass M_ρ and the Sommer scale $r_{0,m}$, in lattice units. All three quantities show good agreement between $N_f = 2$ and $N_f = 6$ in the chiral limit, suggesting that the lattice spacings in the two calculations are matched within 10% accuracy.

(cut off) at Λ_{ETC}. Walking can lift the quark masses by enhancing the condensate $\langle\overline{\psi}\psi\rangle$ significantly above its value ($O(4\pi F^3)$) in a QCD-like theory while keeping Λ^2_{ETC} large enough to suppress FCNC's.

One main difficulty of studying a walking theory using LGT methods is the emergence of two dynamically-generated, and presumably widely-separated, scales: approximate conformal invariance at intermediate distances and confinement at long distances. It has been quite challenging to produce lattices large enough to simultaneously exhibit both phenomena without significant distortions due to finite volume or finite lattice spacing effects. An examination in Fig. 2 of the range of scales covered by a typical LGT calculation on a 32^3 spatial volume indicates the challenge of simultaneously observing both weakly-coupled behavior on short distance scales and strongly-coupled behavior at long distance scales in a single ensemble as the running coupling slows with the increase in N_f. Thus, it is prudent to perform studies away from the conformal window and closer to QCD, where finite volume effects are less severe.

3. Lattice Strong Dynamics

Recently, the Lattice Strong Dynamics (LSD) collaboration has formed to compute the relevant observables in SU(3) Yang-Mills with N_f = 2, 6, 10 light flavors and compare these results with N_f=2+1 calculations of the RBC collaboration.[40–42] In Fig. 3 we show the chiral extrapolation (at fixed lattice spacing) of three physical quantities whose energy scales are characteristically associated with confinement: the nucleon mass, the vector meson mass, and the Sommer parameter. Our inverse lattice spacings for the $N_f = 2$ and 6 flavor calculations are matched at a scale of $a^{-1} \approx 5M_\rho$, about twice as large as the a^{-1} used by the RBC collaboration. We intentionally chose the finer lattice spacing in order to study the sensitivity of the chiral condensate to the short distance cutoff.

The focus of our first analysis[43] has been to calculate the dimensionless ratio of the chiral condensate to the pseudoscalar decay constant $\langle\overline{\psi}\psi\rangle/F^3$ by looking at three different dimensionless ratios

$$R_{XY,m} = \overbrace{\frac{\langle\overline{\psi}\psi\rangle}{F_\pi^3}}^{\text{CF}} = \overbrace{\frac{M_\pi^3}{\sqrt{(2m)^3\langle\overline{\psi}\psi\rangle}}}^{\text{CM}} = \overbrace{\frac{M_\pi^2}{2mF_\pi}}^{\text{FM}} \quad \text{as} \quad m \to 0 \tag{1}$$

where the labels "X" and "Y" should be replaced with appropriate combinations of "C", "F" and "M" to reflect the pair of observables used to construct the ratio. The three ratios are equivalent in the chiral limit by the Gell-Mann-Oakes-Renner (GMOR) relation[44]

$$2m\langle\overline{\psi}\psi\rangle = M_\pi^2 F_\pi^2 \quad \text{as} \quad m \to 0\,. \tag{2}$$

The ratio of ratios for two SU(N_c) theories with different number of flavors, *e.g.* $\mathcal{R}_{FM,\widetilde{m}} \equiv [M_\pi^2/2mF_\pi]_{N_f}/[M_\pi^2/2mF_\pi]_{N_f=2}$, can be used to compare the relative amount of condensate enhancement provided we have computed all ratios using the same UV cutoff (lattice spacing) in physical units. In terms of the geometric mean of the quark masses, an expansion of the ratio of ratios around the chiral limit produces a simple expression to next-to-leading order (NLO)

$$\mathcal{R}_{XY,\widetilde{m}} = \frac{R^{(N_f)}}{R^{(2)}}\left[1 + \widetilde{m}\left(\alpha_{XY10} + \alpha_{11}\log\widetilde{m}\right)\right]\,, \quad \widetilde{m} = \sqrt{m^{(N_f)}m^{(2)}} \tag{3}$$

with the remarkable feature that the coefficient of the log, α_{11}, is identical for all three ratios. The ratios and their chiral extrapolation is shown in Fig. 4. Since this enhancement arises dominantly from physics at the UV cutoff, where even the $N_f = 6$ coupling is relatively weak, we compare it with an estimate based on the perturbative computation of the anomalous dimension γ of $\langle\overline{\psi}\psi\rangle$,[45] finding a perturbative enhancement in the range $5 - 10\%$ in the $\overline{\text{MS}}$ scheme. Converting to the lattice regularization scheme could increase the enhancement to $20-30\%$, quite a bit smaller than the factor of two computed non-perturbatively. The much larger value suggests that $\gamma(\mu)$ for $N_f = 6$ is considerably underestimated by perturbation theory over scales ranging from $\mu = M_\rho - 5M_\rho$.

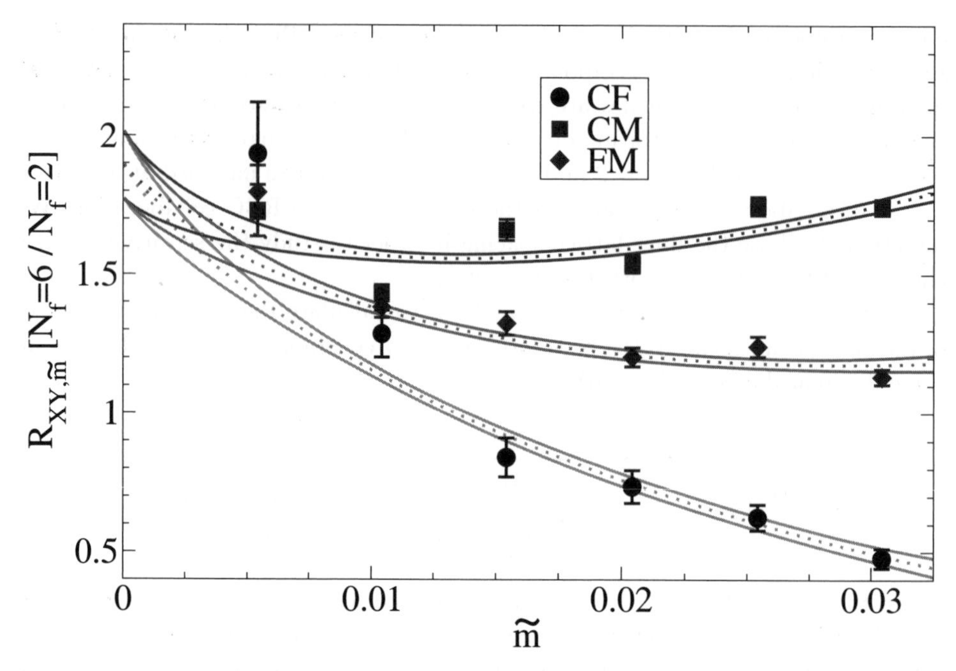

Fig. 4. The ratio of ratios $\mathcal{R}_{XY,\tilde{m}}$ implies a surprisingly large amount of enhancement in the chiral limit for $\langle\overline{\psi}\psi\rangle/F_\pi^3$ between N_f=2 and N_f=6 when $a^{-1} \approx 5M_\rho$.

4. Conclusions

The theoretical exploration of SCGT's for the last several decades has been largely dominated by the study of QCD, for the obvious reason that a large number of experimental observations are available to guide our understanding. Recently, LGT has advanced to become the method of choice for calculating most non-perturbative quantities in QCD. The use of SCGT's in building models of BSM physics has not been nearly as successful, presumably due to the lack of experimental observations. LGT offers the possibility of first-principles calculations of SCGT's for BSM physics. But, current phenomenology favors SCGT's that exhibit "walking", *i.e.* nearly conformal behavior over a range of intermediate scales. LGT, as currently optimized for the study of QCD, may not yet have developed the optimal set of tools for studying walking SCGT's. We emphasized the conservative approach of the LSD collaboration: to slowly approach the walking regime while varying the number of flavors looking for trends which indicate the onset of anticipated walking behavior while keeping an eye on studying the reliability of the methods at hand. Current indications are that the methods are still reliable up to N_f=6.

Acknowledgments

We would like to thank the current members of the Lattice Strong Dynamics (LSD) collaboration without whom this work would not have been possible: T. Appelquist,

R. Babich, R. Brower, M. Buchoff, M. Cheng, M. Clark, S. Cohen, J. Kiskis, M. Lin, E. Neil, J. Osborn, C. Rebbi, D. Schaich, G. Voronov, P. Vranas, J. Wasem. This material is based upon work supported by the National Science Foundation under Grant No. PHY-0801068.

References

1. The Tevatron New-Phenomena and Higgs Working Group (2010), arXiv:1007.4587 [hep-ex].
2. G. T. Fleming, *PoS* **LATTICE2008**, p. 021 (2008).
3. E. Pallante, *PoS* **LAT2009**, p. 015 (2009).
4. W. E. Caswell, *Phys. Rev. Lett.* **33**, p. 244 (1974).
5. T. Banks and A. Zaks, *Nucl. Phys.* **B196**, p. 189 (1982).
6. M. Lüscher, R. Narayanan, P. Weisz and U. Wolff, *Nucl. Phys.* **B384**, 168 (1992).
7. S. Sint, *Nucl. Phys.* **B421**, 135 (1994).
8. U. M. Heller, *Nucl. Phys.* **B504**, 435 (1997).
9. M. Lüscher, R. Sommer, P. Weisz and U. Wolff, *Nucl. Phys.* **B413**, 481 (1994).
10. U. M. Heller, *Nucl. Phys. Proc. Suppl.* **63**, 248 (1998).
11. S. Capitani, M. Lüscher, R. Sommer and H. Wittig, *Nucl. Phys.* **B544**, 669 (1999).
12. J. Heitger, H. Simma, R. Sommer and U. Wolff, *Nucl. Phys. Proc. Suppl.* **106**, 859 (2002).
13. M. Della Morte *et al.*, *Nucl. Phys.* **B713**, 378 (2005).
14. T. Appelquist, G. T. Fleming and E. T. Neil, *Phys. Rev. Lett.* **100**, p. 171607 (2008).
15. T. Appelquist, G. T. Fleming and E. T. Neil, *Phys. Rev.* **D79**, p. 076010 (2009).
16. N. Yamada *et al.*, *PoS* **LAT2009**, p. 066 (2009).
17. N. Yamada *et al.* (2010), arXiv:1003.3288 [hep-lat] and these proceedings.
18. F. Tekin, R. Sommer and U. Wolff (2010), arXiv:1006.0672 [hep-lat].
19. A. Deuzeman, M. P. Lombardo and E. Pallante, *Phys. Lett.* **B670**, 41 (2008).
20. Z. Fodor, K. Holland, J. Kuti, D. Nogradi and C. Schroeder, *PoS* **LATTICE2008**, p. 066 (2008).
21. A. Deuzeman, M. P. Lombardo and E. Pallante, *PoS* **LATTICE2008**, p. 060 (2008).
22. A. Deuzeman, E. Pallante and M. P. Lombardo, *PoS* **LATTICE2008**, p. 056 (2008).
23. X.-Y. Jin and R. D. Mawhinney, *PoS* **LATTICE2008**, p. 059 (2008).
24. E. T. Neil, T. Appelquist and G. T. Fleming, *PoS* **LATTICE2008**, p. 057 (2008).
25. A. Deuzeman, M. P. Lombardo and E. Pallante (2009), arXiv:0904.4662 [hep-ph].
26. A. Hasenfratz, *Phys. Rev.* **D80**, p. 034505 (2009).
27. Z. Fodor, K. Holland, J. Kuti, D. Nogradi and C. Schroeder, *Phys. Lett.* **B681**, 353 (2009).
28. K.-i. Nagai, G. Carrillo-Ruiz, G. Koleva and R. Lewis, *Phys. Rev.* **D80**, p. 074508 (2009).
29. X.-Y. Jin and R. D. Mawhinney, *PoS* **LAT2009**, p. 049 (2009).
30. A. Hasenfratz, *PoS* **LAT2009**, p. 052 (2009).
31. A. Deuzeman, E. Pallante and M. P. Lombardo, *PoS* **LAT2009**, p. 044 (2009).
32. Z. Fodor, K. Holland, J. Kuti, D. Nogradi and C. Schroeder, *PoS* **LAT2009**, p. 055 (2009).
33. Z. Fodor, K. Holland, J. Kuti, D. Nogradi and C. Schroeder, *PoS* **LAT2009**, p. 058 (2009).
34. A. Hasenfratz, *Phys. Rev.* **D82**, p. 014506 (2010).
35. B. Holdom, *Phys. Rev.* **D24**, p. 1441 (1981).
36. B. Holdom, *Phys. Lett.* **B150**, p. 301 (1985).

37. K. Yamawaki, M. Bando and K.-i. Matumoto, *Phys. Rev. Lett.* **56**, p. 1335 (1986).
38. T. Appelquist and L. C. R. Wijewardhana, *Phys. Rev.* **D35**, p. 774 (1987).
39. T. Appelquist and L. C. R. Wijewardhana, *Phys. Rev.* **D36**, p. 568 (1987).
40. D. J. Antonio *et al.*, *Phys. Rev.* **D75**, p. 114501 (2007).
41. C. Allton *et al.*, *Phys. Rev.* **D76**, p. 014504 (2007).
42. C. Allton *et al.*, *Phys. Rev.* **D78**, p. 114509 (2008).
43. T. Appelquist *et al.*, *Phys. Rev. Lett.* **104**, p. 071601 (2010).
44. M. Gell-Mann, R. J. Oakes and B. Renner, *Phys. Rev.* **175**, 2195 (1968).
45. J. A. M. Vermaseren, S. A. Larin and T. van Ritbergen, *Phys. Lett.* **B405**, 327 (1997).

Phases of QCD from Small to Large N_f: (Some) Lattice Results

A. Deuzeman and E. Pallante
Centre for Theoretical Physics, University of Groningen
9747 AG, Netherlands
E-mail: a.deuzeman@rug.nl, e.pallante@rug.nl

M. P. Lombardo
INFN-Laboratori Nazionali di Frascati
I-00044, Frascati (RM), Italy
E-mail: mariapaola.lombardo@lnf.infn.it

We analyze the phases of strong interactions in the space of the flavor number N_f, bare coupling g_L, and temperature T by lattice MonteCarlo simulations for two and three unrooted staggered flavors, corresponding to eight and twelve continuum flavors, respectively. We observe a Coulomb-like phase at intermediate lattice couplings which we interpret as the avatar of a continuum conformal theory for $N_f = 12$. We comment on the possible occurrence of an UVFP associated with the bulk phase transition between the strong coupling lattice phase and the Coulomb-like phase.

1. Introduction

The subject of this investigation is the phase diagram of continuum QCD,[1] in the chiral limit, and as a function of the number of flavors, that we sketch in Fig 1. The motivations for this study has been beautifully introduced in early talks in this meeting, and we do not need to reiterate them here. We just emphasize that the chiral dynamics which is at the very hearth of the QCD phase diagram is eminently non-perturbative. Lattice simulations of QCD are then the method of choice, and indeed the properties of gauge theories with a potential conformal or near-conformal dynamics have been the subject of several lattice studies.[2–14]

Once QCD is discretized on a lattice, another direction automatically appears on the phase diagram, the lattice bare coupling. Lattice simulations cannot study directly the continuum limit: they work at a non-zero bare coupling, or, equivalently, at non-zero lattice spacing. Understanding the phase structure in the lattice parameters space is a mandatory step towards continuum physics.

Let us then consider the lattice phase diagram for N_f massless flavors proposed by V. Miransky and K. Yamawaki,[15] which we sketch in Fig. 2. The continuum physics - Fig.1 – appears at $g_L = 0$, the leftmost axes. Let us mention right at the onset of the talk that we could as well discuss this phase diagram as a function of one

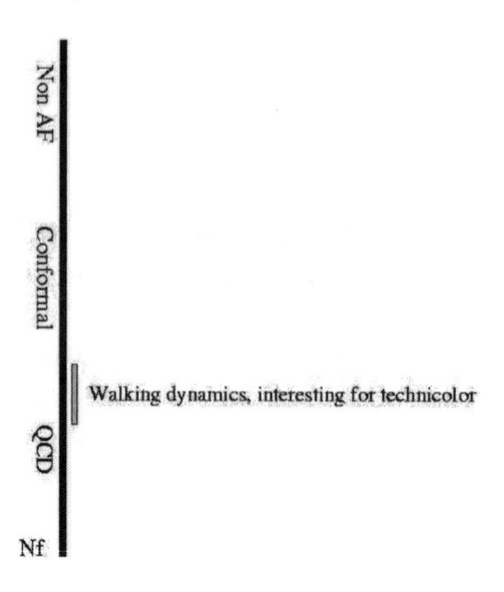

Fig. 1. The phase diagram of QCD as a function of the number of flavor.

component staggered flavors: nothing would change, since the continuum number of flavors does not play any role in this analysis (that would contrast, for instance, with a study of the thermal equation of state where the continuum number of flavor enters as a parameter). The line stemming from $N_f = 16.5$, where asymptotic freedom is lost, is the line of Banks-Zaks fixed points:[16] its precise location and shape depends on the choice of the lattice regularization. Since the first two coefficients of the beta function are universal, the shape of the line close to $g_L = 0$ depends very little on the regularization, and different choices of regularization are likely to deform that line only smoothly. The line of the phase diagram which separates the Coulomb phase from the strong coupling phase, instead, depends non trivially on the lattice action: for instance, positivity-violating operators which are often used by improvement schemes, and which are irrelevant in the continuum, might well originate a rich and peculiar phase structure at strong coupling. There is a robust feature, though: chiral symmetry must be broken in the strong coupling limit regardless the number of flavors. Hence, if a continuum symmetric phase exists, a bulk phase transition separating such a symmetric phase from a broken phase should exist as well.[16] Strictly speaking, the reverse is not true: in principle, one could conceive a so called re-entrant behavior, and we will comment on this possibility later on.

The goal of our investigation[4,5] is to map out quantitatively the phase diagram of Fig. 2. We will first describe our results from $N_f = 12$. We will base our analysis on the chiral condensate and the spectrum properties, and we will show how the mass spectrum, besides being a probe of the symmetry of the system, can be used to build lines of constant physics and to infer the sign of the beta function, much in the same spirit as in early work for $N_f = 16$.[17] Next, we will turn to our $N_f = 8$ results. It will be useful to consider one further parameter, the temperature. A bona fide transition to Quark Gluon Plasma persisting till the continuum limit will be

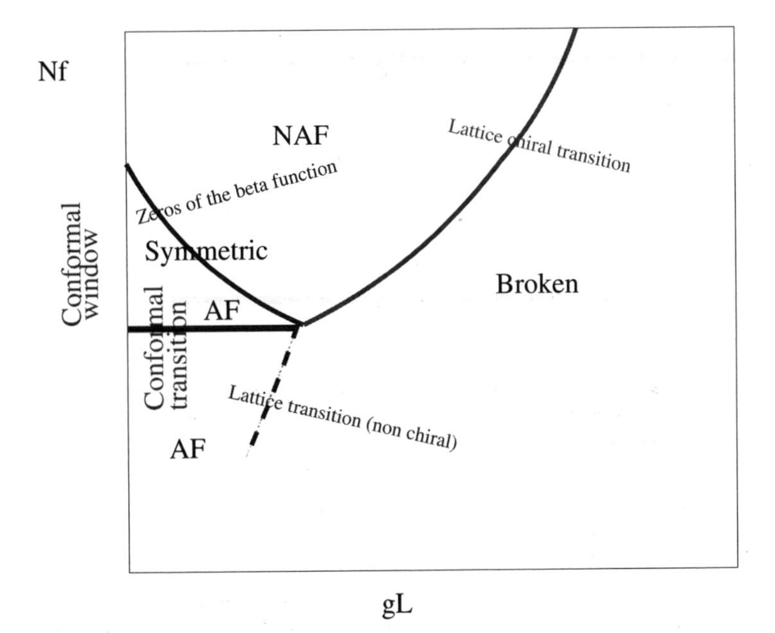

Fig. 2. The phase diagram in the space of the bare lattice coupling g_L and the number of continuum flavor N_f.[15] The continuum physics – Fig.1– is the $g_L = 0$ axes. See text for discussions.

clearly exposed there. This will confirm the hadronic nature of the theory with eight continuum flavors, and allow a natural interpretation of the conformal phase as the zero temperature limit of the Quark Gluon Plasma phase.

2. Zero temperature: $N_f = 12$

The results for the chiral condensate show a clear crossover, or transition, at zero temperature, and call for a careful statistic and systematic analysis in order to distinguish between a real transition from a broken to a symmetric phase from a rapid crossover.

To interpret quantitatively the results and establish the pattern of chiral symmetry, it might be useful to consider lattice QED as a paradigm for the conformal phase of QCD. Let us consider Fig. 3: in either cases a symmetric theory in the continuum has a bulk phase transition to a broken phase. In QCD the conformal symmetry, in the proximity of the bulk transition, is perturbatively broken by Coulomb forces: this suggests the functional form $\langle\bar{\psi}\psi\rangle = am^d$, $d = \frac{3-\gamma}{1+\gamma}$ for the chiral condensate. Note that d was actually measured in QED.[18] It was found $d \simeq 0.95$, increasing towards $d = 1.$, as expected, by weakening the coupling towards the free limit. Motivated by this discussion, we performed fits to $\langle\bar{\psi}\psi\rangle = am^d + \langle\bar{\psi}\psi\rangle_0$, either with all parameters open, or constraining $d = 1$ or $\langle\bar{\psi}\psi\rangle_0 = 0$ To make a long story short, data favor chiral symmetry restoration. For instance, at $\beta = 3.9$ and on a 24^4 lattice a fit to $Y = bX^d$ gives a $(\chi^2/ndf) = 0.4$, b $= 2.697(2)$ d $= 0.9642(3)$, which, taking into account $d = \frac{3-\gamma}{1+\gamma}$ gives $\gamma \simeq 1.05 > 1$. A linear fit with the same

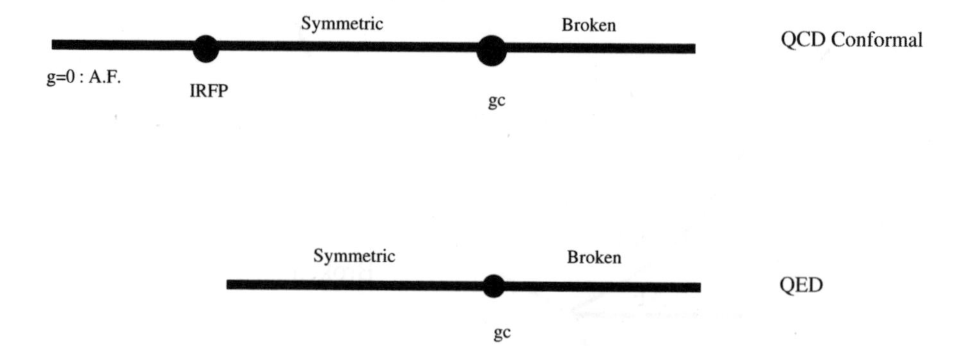

Fig. 3. A scenario for conformal QCD on the lattice vis-a-vis the known structure of lattice QED: in either cases a symmetric theory in the continuum has a bulk phase transition to a broken phase. We will note several similarities between the two bulk transitions. At a variance with QED, of course, QCD in the conformal phase still enjoys asymptotic freedom.

number of free parameters $Y = bX + K$ has instead a $(\chi^2/ndf) = 3.9$, so chiral symmetry breaking seems excluded within 2 σ. Once more the interested reader is referred to our work[5] for all the details. To avoid confusion, here we just note that we do not know yet how the anomalous dimension γ we have measured on the lattice at intermediate coupling would evolve towards the continuum limit: the anomalous dimension in the conformal phase of continuum QCD at the moment is only estimated by approximate analytic studies.[1]

Obviously one could consider more complicated fits motivated by different considerations: certainly there are many good reasons to go beyond the leading scaling form! The overall outcome is that chiral symmetry restoration remains favored. Most physically motivated corrections would enhance the bending of the chiral condensate at lower quark masses, so the practical effect of subleading corrections, if detectable at this level of accuracy, is to push chiral symmetry restoration towards stronger couplings.

Spectrum dynamics carries important information on the symmetry of a phase. For instance, in the conformal phase $m_\pi \simeq m_\sigma \simeq 2m_q$, and this limit has been searched for on the lattice.[11] It is interesting, by carrying over our analogy with QED, to see how this limit can be approached. Consider the ratio (Fig.4) m_π/m_σ as a function of the bare mass in the symmetric and the broken phase, for a fixed value of the lattice coupling.[19] At $m = 0$ the ratio is fixed by symmetry to be zero in the strong coupling and one in the weak coupling side. As mentioned above, close to the transition conformality is broken and the dynamics is dominated by weak Coulomb forces, so one observes deviations from the conformal value, $m_\pi/m_\sigma = 1$. By further decreasing the coupling, towards the continuum, conformal limit, indeed the ratio approaches one. One further comment concerns the heavy quark limit, where the ratio is one as well. Note that in the same limit the chiral condensate becomes a decreasing function of the quark mass: the only way to have degeneracy between chiral partners is a zero chiral condensate, which in the heavy quark limit

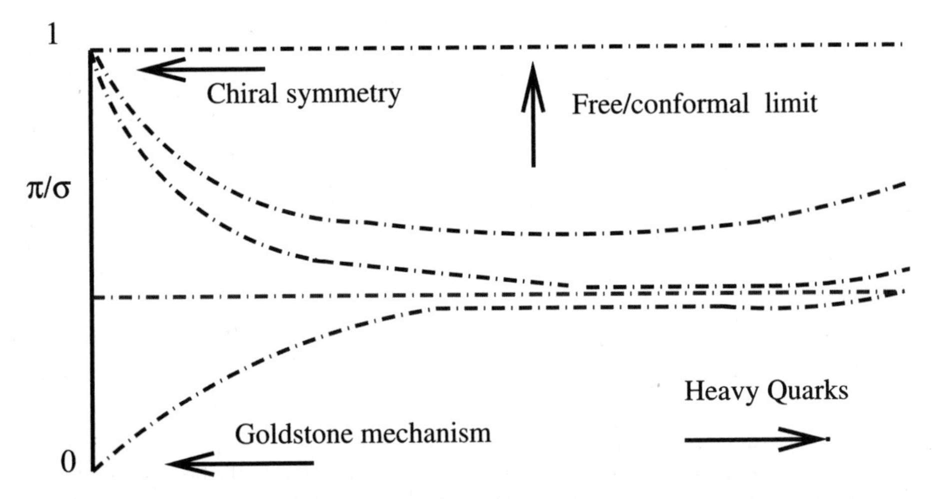

Fig. 4. A sketchy view of the behavior of the π to σ mass ratio as a function of the bare mass. Each line corresponds to one lattice coupling, and the coupling decreases from bottom to top. In the chiral limit the ratio is constrained by chiral symmetry: zero in the symmetric phase (bottom), one in the broken phase. In the conformal phase the ratio approaches one for all masses, which is the same as the heavy quark limit, and the residual Coulomb forces of the symmetric phase are responsible for a mass ratio which is below one. Note that the behavior of the chiral condensate can be used to discriminate between the heavy quark limit and the conformal limit.

is achieved since $< \bar{\psi}\psi > \propto 1/m$. So the trend of the chiral condensate as a function of the mass can be used to discriminate between a trivial heavy quark limit and physical conformal limit when the ratio is close to one. In Fig. 5(left) we show the ratios of measured pseudoscalar and vector masses from our simulations, for a fixed coupling and as a function of the bare quark mass. We have superimposed the ratios of the best fits to the raw mass data, confirming the good quality of the interpolations. The analogy with the predicted trend in the symmetric phase, Fig.4, including the ratio at fixed mass approaching one when decreasing the coupling, is very well exposed by our data. Combining this observation with the decreasing of the chiral condensate with the bare mass, we conclude again in favor of chiral symmetry in this phase, with the slow approach to one a precursor effect of the conformal phase (it would be desirable to have a direct measurement of the σ mass, which turns out to be noisy, but anyway very close to the ρ mass).

Further, we used the spectrum results to determine the lines of constant physics in the two dimensional parameter space g_L and am, the bare quark mass of degenerate fermions: we found that the lattice spacing increases for weaker couplings, signature of the Coulomb phase(see Fig.5,right). The same trend is observed in $N_f = 16$ QCD,[17] and indeed our strategy has been inspired by this study. Let us look again at Fig.2: our Coulomb phase looks pretty much like the one of that figure. So it seems that an emerging conformal phase in the continuum limit is a physical scenario consistent with our data. Can we challenge Fig.2 - and hence Ref.3? Of course we can - but to do so we should imagine a rather unusual scenario in which chiral

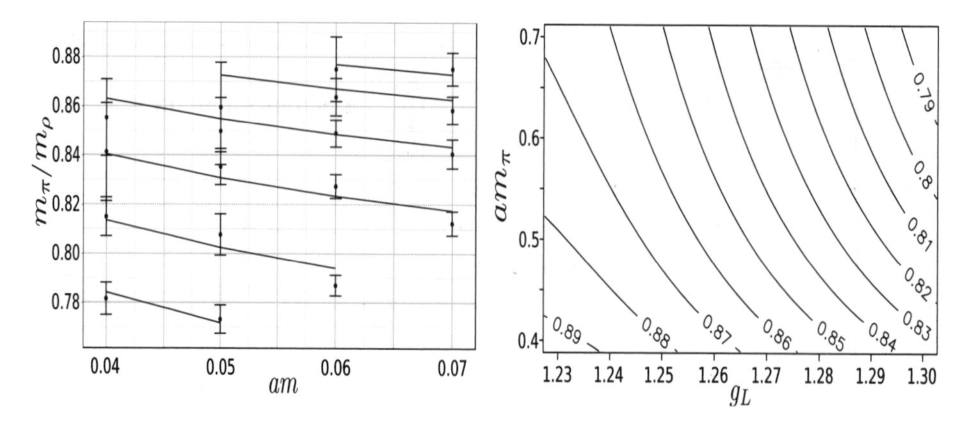

Fig. 5. The measured π to ρ mass ratio as a function of the bare mass and decreasing coupling g_L, bottom to top $6/g_L^2 = 3.5$ to 4. The superimposed lines are ratios of the best fits as a function of the bare quark mass (left); Pseudoscalar mass along lines of constant physics. The physical pion mass is identical along lines of constant physics, such that a decreasing pion mass in lattice units for increasing g_L, implies an increase of the lattice spacing for weaker couplings, signature of the Coulomb phase. Labels give the value of the ratio m_π/mp along the isolines (right).

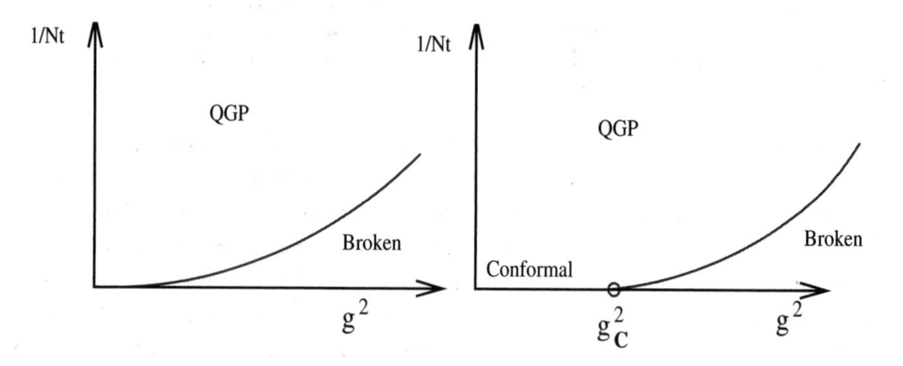

Fig. 6. The thermal behavior of QCD in the hadron (left) and conformal phase (right).

symmetry is spontaneously broken again by decreasing the coupling. A somewhat analogous behavior has been observed in condensed matter - for instance one can have a reversible liquid-to-crystal transition, with passage from a low-temperature liquid to a high-temperature crystal. Such re-entrant behavior usually emerges as a result of a complicated interplay among competing mechanisms. Indeed we have scanned the coupling space from the values discussed here towards the continuum, finding no trace of any further chiral restoring phase transition.

3. $N_f = 8$, and Quark Gluon Plasma

We now extend further our phase space and include the temperature axes.

We note that in ordinary QCD, as well as in the strong coupling phase of conformal QCD, there will be a high temperature chiral restoration transition at $g = g_T$.

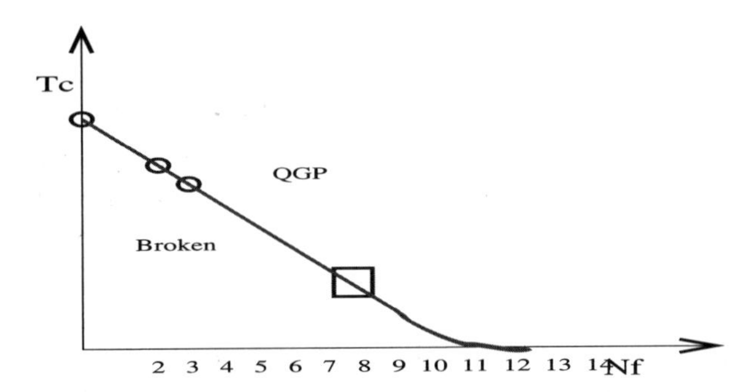

Fig. 7. The phase diagram in the T, N_f plane.[20] The circles indicate earlier lattice results, which fit very nicely the RG prediction. Our results for $N_f = 8$ indicate the existence of a thermal transition (marked as a square), whose temperature in physical units has yet to be estimated.

In the conformal phases $\lim_{N_t \to \infty} g_T^2 = g_c^2$, with g_c the bulk phase transition discussed above, while in ordinary QCD $\lim_{N_t \to \infty} g_T^2 = 0$ as $1/N_t = T_{phys} a(g^2)$. In a three dimensional picture a sheet of phase transitions in the N_t, g, N_f plane continues smoothly till $g_L = 0$ in ordinary QCD, while it is bounded at $g = g_c$ in the conformal phase. We can see the conformal phase as the T=0 limit of the ordinary, albeit cold, Quark Gluon Plasma phase, either on the lattice and in the continuum. We have indeed checked that for $N_f = 8$ the critical temperature computed on lattice of various sizes $T_c = \frac{1}{a(\beta_c) N_t}$ satisfies asymptotic scaling within 2 %

The results can be placed on the phase diagram in the T, N_f plane,[20] see Fig.7: results for a small number of flavors are available in the literature, and our results for $N_f = 8$ nicely extend these studies. Remarkably, the lattice results fare quite well on the near-linear behavior found in RG studies. The $N_f = 8$ critical temperature in physical units remains to be computed, and checked against the Braun-Gies[20] prediction.

4. Summary and discussion

The most plausible scenario accommodating our data is a continuum conformal phase for and SU(3) Yang–Mills gauge theory with $N_f = 12$, in agreement with calculations of the renormalized coupling in this model.[6,7] At the intermediate couplings which we have studied the physics is that of a symmetric phase of three unrooted staggered flavors, where conformal symmetry is broken by weak Coulombic forces. Upon further decreasing the coupling Coulombic forces progressively weaken, flavor symmetry will be restored and the physics of the continuum, conformal window should emerge. We underscore that this happens without any change of chiral symmetry. In this scenario, the Coulomb phase at intermediate coupling is just an avatar of the continuum conformal phase. Clearly, it would be extremely desirable to go beyond the avatar and expose directly the conformal physics of the continuum limit.

One final comment concerns the search for an UVFP in a strongly interacting four dimensional gauge theory. This search has a long history, which was also recorded by past editions of the Nagoya SCGT Meetings. Most attempts in the past focused on the strong coupling transition of QED, which, as we have discussed above, has a natural analogy with the bulk transition in the conformal window. Indeed, since the very first lattice studies of this model an UVFP was searched along the strong coupling line.[21,22] Clearly if such theory existed in large N_f QCD, the non–abelian nature of the gauge fields should play a major role, given that the QED transition appears to be trivial. More recently, Kaplan, Lee, Son and Stephanov have further elaborated on this fascinating possibility,[23] providing more motivation for such a search. It is interesting to note that even if the bulk phase transition for $N_f > N_f^c$ appeared to be of first order, its endpoint at $N_f = N_f^c$, where "conformality is lost"[23] should indeed be of a second order, following general statistical mechanics arguments, and might be a natural candidate for the existence of a new theory.

Acknowledgments

MPL wishes to thank the organisers of this enjoyable and interesting workshop for their most kind hospitality, and the participants for many lively and stimulating discussions.

References

1. F. Sannino, arXiv:0911.0931 [hep-ph].
2. G. T. Fleming, PoS **LATTICE2008** (2008) 021.
3. E. Pallante, arXiv:0912.5188 [hep-lat].
4. A. Deuzeman, M. P. Lombardo and E. Pallante, *Phys. Lett. B* **670**, 41.
5. A. Deuzeman, M. P. Lombardo and E. Pallante arXiv:0904.4662 [hep-ph].
6. T. Appelquist, G. T. Fleming and E. T. Neil, *Phys. Rev. Lett.* **100**, 171607 .
7. T. Appelquist, G. T. Fleming and E. T. Neil, *Phys. Rev. D* **79**, 076010 .
8. Z. Fodor, K. Holland, J. Kuti, D. Nogradi and C. Schroeder, arXiv:0911.2463 [hep-lat].
9. S. Catterall and F. Sannino, *Phys. Rev. D* **76**,p. 34504 (2007).
10. S. Catterall, J. Giedt, F. Sannino and J. Schneible, *J. High Energy Phys.* **2008**,p. 9.
11. L. Del Debbio, B. Lucini, A. Patella, C. Pica and A. Rago, *Phys. Rev. D* **80**, 074507.
12. T. DeGrand and A. Hasenfratz, *Phys. Rev. D* **80**, 034506.
13. Y. Shamir, B. Svetitsky and T. DeGrand, *Phys. Rev. D* **78**, 031502
14. J. B. Kogut and D. K. Sinclair, arXiv:1002.2988 [hep-lat].
15. V. A. Miransky and K. Yamawaki, *Phys. Rev. D* **55**, p. 5051.
16. T. Banks and A. Zaks, *Nuc. Phys. B* **196**, 189.
17. P. Damgaard, U. Heller, A. Krasnitz and P. Olesen, *Phys. Lett. B* **400**, 169.
18. A. Kocic, S. Hands, J. B. Kogut and E. Dagotto, *Nucl. Phys.* **B347**, 217 .
19. A. Kocic, J. B. Kogut and M.P. Lombardo, *Nucl. Phys.* **B398**, 376.
20. J. Braun and H. Gies, *J. High Energy Phys.* **2006**, p. 024 .
21. J. B. Kogut, J. Polonyi, H. W. Wyld and D. K. Sinclair, *Phys. Rev. Lett.* **54**, 1475.
22. J. B. Kogut and D. K. Sinclair, *Nucl. Phys.* **B295**, 465 .
23. D. B. Kaplan, J.-W. Lee, D. T. Son and M. A. Stephanov, *Phys. Rev. D* **80**, 125005.

Lattice Gauge Theory and (Quasi)-Conformal Technicolor

D. K. Sinclair

HEP Division, Argonne National Laboratory, 9700 South Cass Avenue
Argonne, Illinois 60439, USA
E-mail: dks@hep.anl.gov

J. B. Kogut

Department of Energy, Division of High Energy Physics
Washington, DC 20585, USA
and
Department of Physics – TQHN, University of Maryland, 82 Regents Drive
College Park, Maryland 20742, USA
E-mail: jbkogut@umd.edu

QCD with 2 flavours of massless colour-sextet quarks is studied as a theory which might exhibit a range of scales over which the running coupling constant evolves very slowly (walks). We simulate lattice QCD with 2 flavours of sextet staggered quarks to determine whether walks, or if it has an infrared fixed point, making it a conformal field theory. Our initial simulations are performed at finite temperatures $T = 1/N_t a$ ($N_t = 4$ and $N_t = 6$), which allows us to identify the scales of confinement and chiral-symmetry breaking from the deconfinement and chiral-symmetry restoring transitions. Unlike QCD with fundamental quarks, these two transitions appear to be well-separated. The change in coupling constants at these transitions between the two different temporal extents N_t, is consistent with these being finite temperature transitions for an asymptotically free theory, which favours walking behaviour. In the deconfined phase, the Wilson Line shows a 3-state signal. Between the confinement and chiral transitions, there is an additional transition where the states with Wilson Lines oriented in the directions of the complex cube roots of unity disorder into a state with a negative Wilson Line.

Keywords: Lattice gauge theory, Walking Technicolor.

1. Introduction

Technicolor theories are QCD-like gauge theories with massless fermions, whose pion-like excitations play the role of the Higgs field in giving masses to the W and Z.[1,2] We search for Yang-Mills gauge theories whose fermion content is such that the running coupling constant evolves very slowly – walks. Such theories can avoid the phenomenological problems which plague other (extended-)Technicolor theories.[3–6]

While many studies have used fermions in the fundamental representation, with large numbers of flavours,[7–22] we are concentrating on higher representations of the colour group (in particular the symmetric tensor), where conformality/walking can be achieved at much lower N_f. There have been some studies with $SU(2)$ colour with two adjoint (symmetric tensor) fermions.[23–30] We are considering QCD ($SU(3)$ colour) with colour-sextet (symmetric tensor) quarks.

The 2-loop β-function for QCD with N_f massless flavours of colour-sextet quarks, suggests that for $1\frac{28}{125} \leq N_f < 3\frac{3}{10}$, either this theory will have an infrared-stable fixed point, or a chiral condensate will form and this fixed point will be avoided. In the first case the theory will be conformal; in the second case it will walk. For $N_f = 3$ conformal behaviour is expected. $N_f = 2$ could, a priori, exhibit either behaviour. Because the quadratic Casimir operator for sextet quarks is $2\frac{1}{2}$ times that for fundamental quarks, it is easier for them to form a chiral condensate.

Lattice QCD gives us a direct method to determine which option the $N_f = 2$ theory chooses. We are studying the $N_f = 2$ theory using staggered fermions. We are currently performing simulations at finite temperature (T). Finite temperature enables us to study the scales of confinement and chiral symmetry breaking, and yields information on the running of the coupling constant.

Simulations using Wilson fermions by DeGrand, Shamir and Svetitsky suggest that this theory is conformal.[31,32] Our simulations suggest that it walks. The scales of confinement and chiral symmetry breaking appear to be very different. (This also contrasts with what was reported by DeGrand, Shamir and Svetitsky for simulations using Wilson quarks.[33]) Hence the phenomenology is expected to be different from that of QCD with fundamental quarks and N_f in the walking window, where these two scales appear to be the same. Preliminary studies of this theory using domain-wall quarks have been reported in.[34]

In the deconfined phase we observe states where the phase of the Wilson Line is $\pm 2\pi/3$, and at weaker couplings, π in addition to the expected states with positive Wilson Lines. These have since been predicted and observed by Machtey and Svetitsky using Wilson quarks.[35]

2. Simulations and Results

We use the standard Wilson (triplet) plaquette action for the gauge fields and an unimproved staggered-fermion action for the quarks. The only new feature is that the quark fields are six-vectors in colour space and the gauge fields on the links of the quark action are in the sextet representation of colour. The RHMC algorithm is used to tune the number of flavours, N_f, to 2.

We run on $8^3 \times 4$, $12^3 \times 4$ and $12^3 \times 6$ lattices at quark masses $m = 0.005$, $m = 0.01$ and $m = 0.02$ in lattice units to allow extrapolation to the chiral ($m = 0$) limit. $\beta = 6/g^2$ is varied over a range of values large enough to include the deconfinement and chiral transitions. Run lengths of 10,000–200,000 trajectories per (m, β) are used.

More details of the results presented here are given in ref.[36]

Since the results from the 2 $N_t = 4$ lattices are consistent, we present only results from our $12^3 \times 4$ simulations. Figure 1 shows the colour-triplet Wilson Line(Polyakov Loop) and the chiral condensate($\langle\bar{\psi}\psi\rangle$) as functions of $\beta = 6/g^2$, for each of the 3

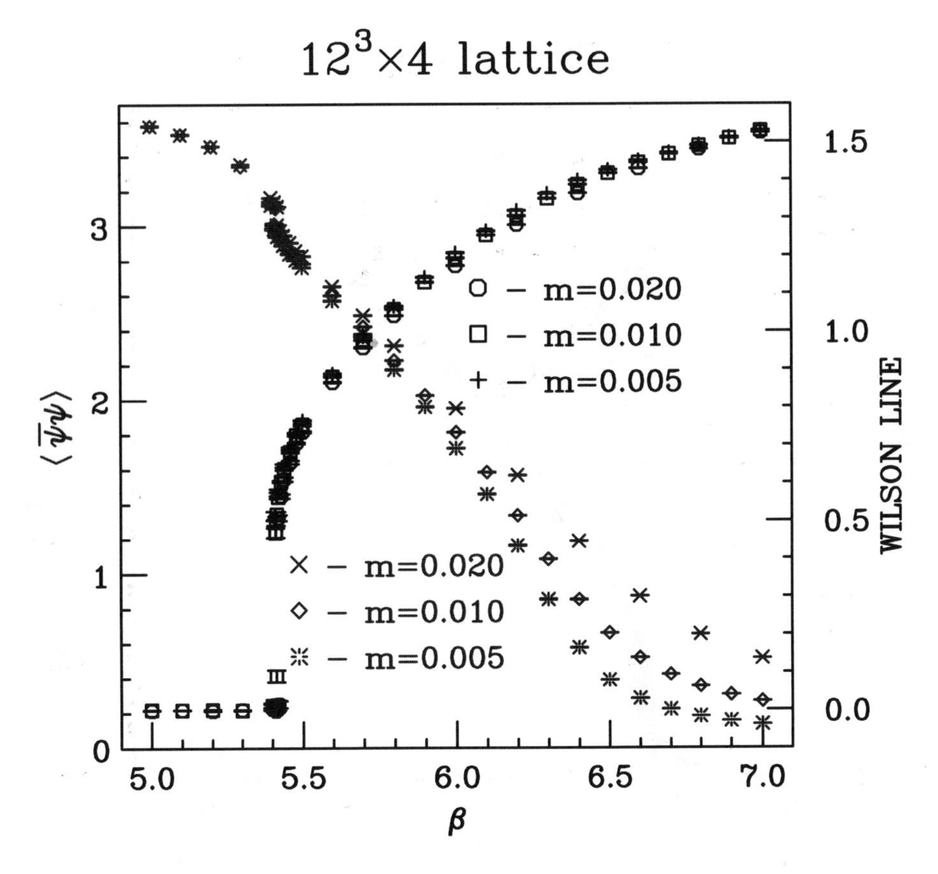

Fig. 1. Wilson line and $\langle\bar{\psi}\psi\rangle$ as functions of β on a $12^3 \times 4$ lattice.

quark masses on a $12^3 \times 4$ lattice. The deconfinement transition is marked by an abrupt increase in the value of the Wilson Line. Chiral symmetry restoration occurs where the chiral condensate vanishes in the chiral limit.

In contrast to what was found by DeGrand, Shamir and Svetitsky, we find well separated deconfinement and chiral-symmetry restoration transitions. The deconfinement transition occurs at $\beta = \beta_d$ where $\beta_d(m = 0.005) = 5.405(5)$, $\beta_d(m = 0.01) = 5.4115(5)$ and $\beta_d(m = 0.02) = 5.420(5)$. The chiral transition, estimated from the peaks in the chiral susceptibility curves, occurs at $\beta_\chi = 6.3(1)$.

Figure 1 only accounts for the state with a real positive Wilson Line in the deconfined regime. However, from the deconfinement transition up to $\beta \approx 5.9$ there exist long-lived states with the Wilson Line oriented in the directions of the other 2 cube roots of unity. However, these states are metastable, eventually decaying into the state with a positive Wilson Line. Above $\beta \approx 5.9$, these complex Wilson Line states disorder into a state with a negative Wilson Line.

Let us now consider our $12^3 \times 6$ simulations. Again, the deconfinement and chiral-symmetry restoring transitions are well-separated. Above the deconfinement

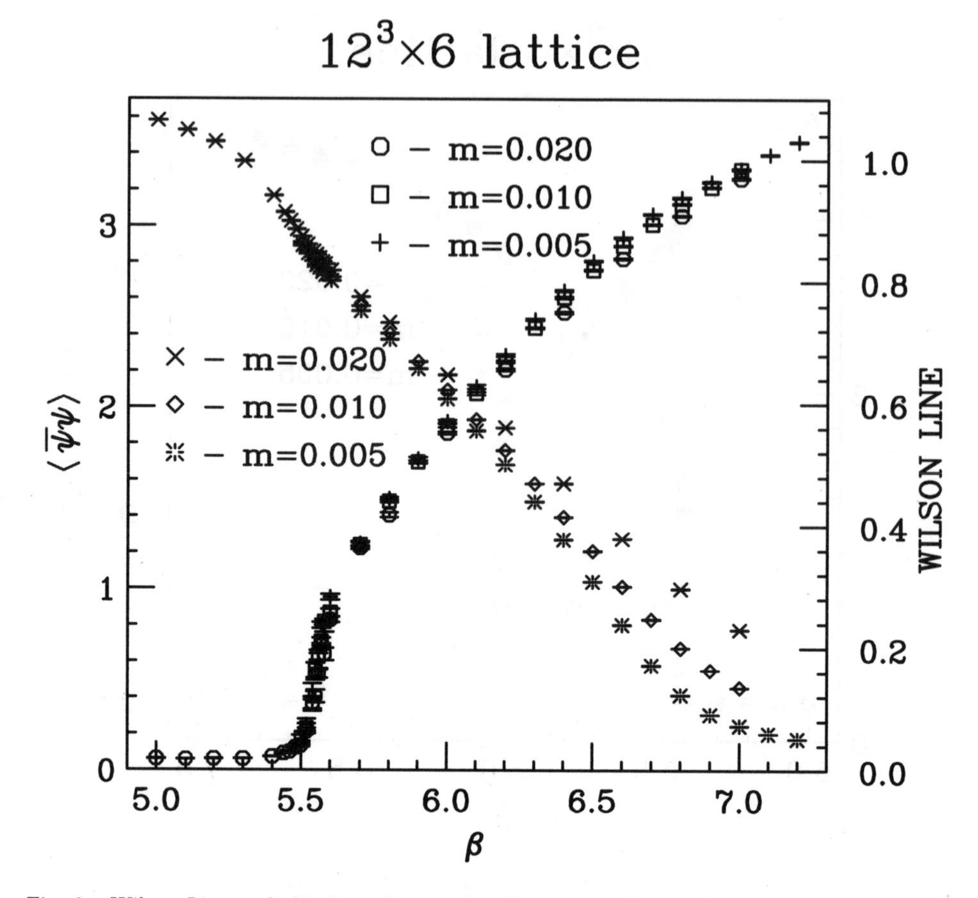

Fig. 2. Wilson Line and chiral condensate for the state with a real positive Wilson Line as functions of β for each of the 3 masses on a $12^3 \times 6$ lattice.

transition, we again find a clear 3-state signal. This time, however, all 3 states appear equally stable. The system tunnels between these 3 states for the duration of the run until we are so far above the transition that the relaxation time for tunneling exceeds the lengths of our runs. We therefore artificially bin our 'data' according to the phase of the Wilson Line, into bins $(-\pi, -\pi/3)$, $(-\pi/3, \pi/3)$, $(\pi/3, \pi)$.

Figure 2 shows the Wilson Lines and chiral condensates $\langle\bar{\psi}\psi\rangle$ for the central 'positive' Wilson Line bin. Figure 3 shows the Wilson Lines and chiral condensates for the first and last 'complex' and 'negative' Wilson Line bins. The deconfinement transitions occur at $\beta_d(m = 0.005) = 5.545(5)$, $\beta_d(m = 0.01) = 5.550(5)$ and $\beta_d(m = 0.02) = 5.560(5)$. Chiral-symmetry restoration occurs at $\beta_\chi = 6.6(1)$.

As for $N_t = 4$, there is further transition between the deconfinement and chiral transitions where the states with complex Wilson Lines disorder to produce a state with a negative Wilson Line. This transition occurs at $\beta \approx 6.4$ for $m = 0.01$ and $\beta \approx 6.5$ for $m = 0.02$. This transition can be seen in fig. 3.

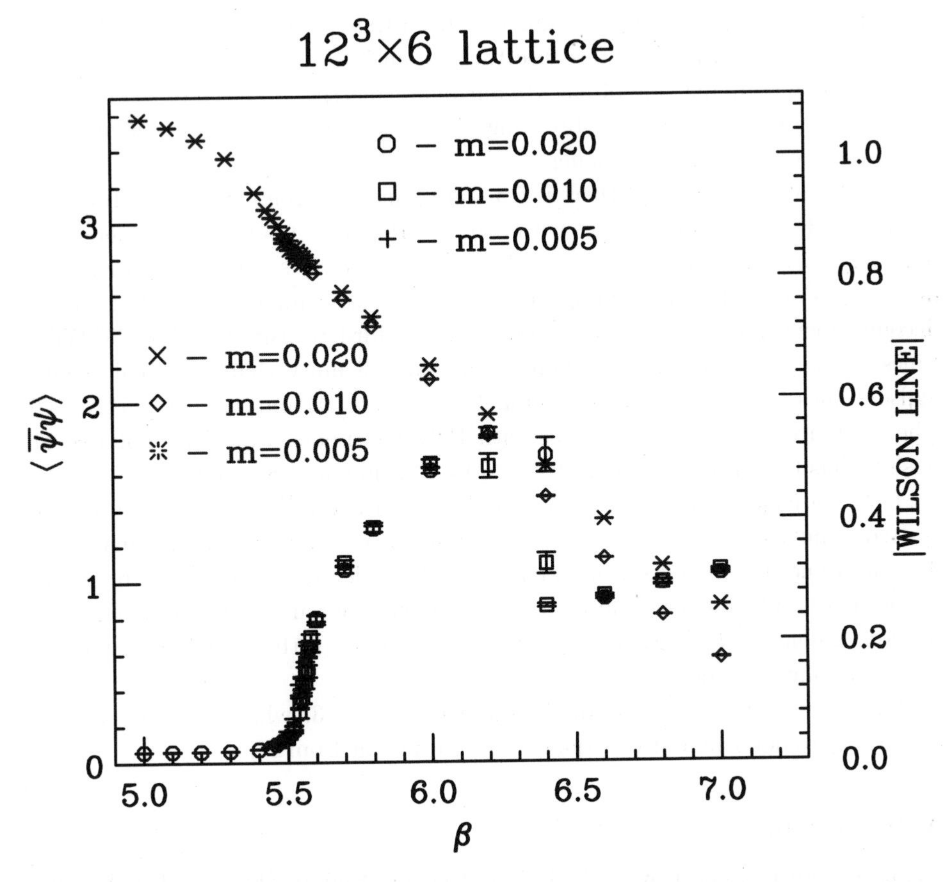

Fig. 3. Magnitude of the Wilson Line and chiral condensate for the state with a complex or negative Wilson Line as functions of β for each of the 3 masses on a $12^3 \times 6$ lattice.

3. Discussion and conclusions

We are studying the thermodynamics of Lattice QCD with 2 flavours of staggered colour-sextet quarks. We find well separated deconfinement and chiral-symmetry restoration transitions. This contrasts with the case of fundamental quarks, where these 2 transitions are coincident, but is similar to the case of adjoint quarks where again these 2 transitions are separate.[37,38]

We denote the value of $\beta = 6/g^2$ at the deconfinement transition by β_d and that at the chiral transition by β_χ. In the chiral limit $\beta_d \approx 5.40$ and $\beta_\chi = 6.3(1)$ at $N_t = 4$. At $N_t = 6$ these values become $\beta_d \approx 5.54$ and $\beta_\chi = 6.6(1)$. The increase in the βs for both transitions from $N_t = 4$ to $N_t = 6$ is consistent with their being finite temperature transitions for an asymptotically free theory (rather than bulk transitions). If there is an IR fixed point, we have yet to observe it. Our results suggest a Walking rather than a conformal behaviour.

Why is this phase diagram so different from that for Wilson quarks (DeGrand, Shamir and Svetitsky)? Is it because there is an infrared fixed point, and we are on the strong-coupling side of it? Are our quark masses too large to see the chiral limit? Is it because the flavour breaking of staggered quarks does not allow a true chiral limit at fixed lattice spacing?

For the deconfined phase there is a 3-state signal, the remnant of now-broken Z_3 symmetry. For $N_t = 4$ the states with complex Polyakov Loops appear metastable. For $N_t = 6$ all 3 states appear stable. Breaking of Z_3 symmetry is seen in the magnitudes of the Polyakov Loops for the real versus complex states. Between the deconfinement and chiral transitions, we find a third transition where the Wilson Lines in the directions of the 2 non-trivial roots of unity change to real negative Wilson Lines. This transition occurs for $\beta \approx 5.9$ ($N_t = 4$) and $\beta \approx 6.4$–6.5 ($N_t = 6$). The existence of these extra states with Polyakov Loops which are not real and positive has been predicted and observed by Machtey and Svetitsky.

Drawing conclusions from $N_t = 4$ and $N_t = 6$ is dangerous. We have recently started simulations with $N_t = 8$. We should also use smaller quark masses. At $N_t = 6$, we need a second spatial lattice size. To understand this theory more fully, we need to study its zero temperature behaviour, measuring its spectrum, string tension, potential, f_π.... Measurement of the running of the coupling constant for weak coupling is needed.

We have recently started simulations with $N_f = 3$, which is expected to be conformal, to determine if it is qualitatively different from $N_f = 2$.

Acknowledgments

DKS is supported in part by the U.S. Department of Energy, Division of High Energy Physics, Contract DE-AC02-06CH11357, and in part by the Argonne/University of Chicago Joint Theory Institute. JBK is supported in part by NSF grant NSF PHY03-04252. These simulations were performed on the Cray XT4, Franklin at NERSC under an ERCAP allocation, and on the Cray XT5, Kraken at NICS under an LRAC/TRAC allocation.

DKS thanks J. Kuti, D. Nogradi and F. Sannino for helpful discussions.

References

1. S. Weinberg, Phys. Rev. D **19**, 1277 (1979).
2. L. Susskind, Phys. Rev. D **20**, 2619 (1979).
3. B. Holdom, Phys. Rev. D **24**, 1441 (1981).
4. K. Yamawaki, M. Bando and K. i. Matumoto, Phys. Rev. Lett. **56**, 1335 (1986).
5. T. Akiba and T. Yanagida, Phys. Lett. B **169**, 432 (1986).
6. T. W. Appelquist, D. Karabali and L. C. R. Wijewardhana, Phys. Rev. Lett. **57**, 957 (1986).
7. J. B. Kogut, J. Polonyi, H. W. Wyld and D. K. Sinclair, Phys. Rev. Lett. **54**, 1475 (1985).
8. M. Fukugita, S. Ohta and A. Ukawa, Phys. Rev. Lett. **60**, 178 (1988).

9. S. Ohta and S. Kim, Phys. Rev. D **44**, 504 (1991).
10. S. y. Kim and S. Ohta, Phys. Rev. D **46**, 3607 (1992).
11. F. R. Brown, H. Chen, N. H. Christ, Z. Dong, R. D. Mawhinney, W. Schaffer and A. Vaccarino, Phys. Rev. D **46**, 5655 (1992) [arXiv:hep-lat/9206001].
12. Y. Iwasaki, K. Kanaya, S. Sakai and T. Yoshie, Phys. Rev. Lett. **69**, 21 (1992).
13. Y. Iwasaki, K. Kanaya, S. Kaya, S. Sakai and T. Yoshie, Phys. Rev. D **69**, 014507 (2004) [arXiv:hep-lat/0309159].
14. A. Deuzeman, M. P. Lombardo and E. Pallante, Phys. Lett. B **670**, 41 (2008) [arXiv:0804.2905 [hep-lat]].
15. A. Deuzeman, M. P. Lombardo and E. Pallante, arXiv:0904.4662 [hep-ph].
16. T. Appelquist, G. T. Fleming and E. T. Neil, Phys. Rev. D **79**, 076010 (2009) [arXiv:0901.3766 [hep-ph]].
17. T. Appelquist, G. T. Fleming and E. T. Neil, Phys. Rev. Lett. **100**, 171607 (2008) [Erratum-ibid. **102**, 149902 (2009)] [arXiv:0712.0609 [hep-ph]].
18. X. Y. Jin and R. D. Mawhinney, PoS **LATTICE2008**, 059 (2008) [arXiv:0812.0413 [hep-lat]].
19. X. Y. Jin and R. D. Mawhinney, PoS **LAT2009**, 049 (2009) [arXiv:0910.3216 [hep-lat]].
20. Z. Fodor, K. Holland, J. Kuti, D. Nogradi and C. Schroeder, arXiv:0907.4562 [hep-lat].
21. Z. Fodor, K. Holland, J. Kuti, D. Nogradi and C. Schroeder, arXiv:0911.2463 [hep-lat].
22. N. Yamada, M. Hayakawa, K. I. Ishikawa, Y. Osaki, S. Takeda and S. Uno, arXiv:0910.4218 [hep-lat].
23. S. Catterall and F. Sannino, Phys. Rev. D **76**, 034504 (2007) [arXiv:0705.1664 [hep-lat]].
24. S. Catterall, J. Giedt, F. Sannino and J. Schneible, JHEP **0811**, 009 (2008) [arXiv:0807.0792 [hep-lat]].
25. S. Catterall, J. Giedt, F. Sannino and J. Schneible, arXiv:0910.4387 [hep-lat].
26. L. Del Debbio, A. Patella and C. Pica, arXiv:0805.2058 [hep-lat].
27. L. Del Debbio, B. Lucini, A. Patella, C. Pica and A. Rago, arXiv:0907.3896 [hep-lat].
28. F. Bursa, L. Del Debbio, L. Keegan, C. Pica and T. Pickup, arXiv:0910.4535 [hep-ph].
29. A. J. Hietanen, J. Rantaharju, K. Rummukainen and K. Tuominen, JHEP **0905**, 025 (2009) [arXiv:0812.1467 [hep-lat]].
30. A. J. Hietanen, K. Rummukainen and K. Tuominen, arXiv:0904.0864 [hep-lat].
31. Y. Shamir, B. Svetitsky and T. DeGrand, Phys. Rev. D **78**, 031502 (2008) [arXiv:0803.1707 [hep-lat]].
32. T. DeGrand, Phys. Rev. D **80**, 114507 (2009) [arXiv:0910.3072 [hep-lat]].
33. T. DeGrand, Y. Shamir and B. Svetitsky, Phys. Rev. D **79**, 034501 (2009) [arXiv:0812.1427 [hep-lat]].
34. Z. Fodor, K. Holland, J. Kuti, D. Nogradi and C. Schroeder, arXiv:0809.4888 [hep-lat].
35. O. Machtey and B. Svetitsky, Phys. Rev. D **81**, 014501 (2010) [arXiv:0911.0886 [hep-lat]].
36. J. B. Kogut and D. K. Sinclair, arXiv:1002.2988 [hep-lat].
37. F. Karsch and M. Lutgemeier, Nucl. Phys. B **550**, 449 (1999) [arXiv:hep-lat/9812023].
38. J. Engels, S. Holtmann and T. Schulze, Nucl. Phys. B **724**, 357 (2005) [arXiv:hep-lat/0505008].

Study of the Running Coupling Constant in 10-Flavor QCD with the Schrödinger Functional Method

N. Yamada[a,b], M. Hayakawa[c], K.-I. Ishikawa[d], Y. Osaki[d], S. Takeda[e] and S. Uno[c]

[a]*KEK Theory Center, Institute of Particle and Nuclear Studies*
High Energy Accelerator Research Organization (KEK), Tsukuba 305-0801, Japan
[b]*School of High Energy Accelerator Science*
The Graduate University for Advanced Studies (Sokendai), Tsukuba 305-0801, Japan
[c]*Department of Physics, Nagoya University, Nagoya 464-8602, Japan*
[d]*Department of Physics, Hiroshima University, Higashi-Hiroshima 739-8526, Japan*
[e]*School of Mathematics and Physics, College of Science and Engineering*
Kanazawa University, Kakuma-machi, Kanazawa, Ishikawa 920-1192, Japan

The electroweak gauge symmetry is allowed to be spontaneously broken by the strongly interacting vector-like gauge dynamics. When the gauge coupling of a theory runs slowly in a wide range of energy scale, the theory is a candidate for walking technicolor. This may open up the possibility that the origin of all masses may be traced back to the gauge theory. We use the Schrödinger functional method to see whether the gauge coupling of 10-flavor QCD "walks" or not. Preliminary result is reported.

Keywords: Lattice gauge theory, LHC.

1. Introduction

The main goal of Large Hadron Collider (LHC) is to confirm the Higgs mechanism and to find particle contents and the physics law above the electroweak scale. So far many new physics models beyond the standard model have been proposed. Among them, Technicolor (TC)[1] is one of the most attractive candidates[2] as it does not require elementary scalar particles which cause, so-called, the fine-tuning problem. This model is basically a QCD-like, strongly interacting vector-like gauge theory. Therefore, lattice gauge theory provides the best way to study this class of model,[3] and the predictions can be as precise as those for QCD, in principle.

The simple, QCD-like TC model, *i.e.* an SU(3) gauge theory with two or three flavors of techniquarks, has been already ruled out by, for instance, the S-parameter[4] and the FCNC constraints. However, it has been argued that, if the gauge coupling runs very slowly ("walks") in a wide range of energy scale before spontaneous chiral symmetry breaking occurs, at least, the FCNC problem may disappear.[5] Such TC models are called walking technicolor (WTC) and several explicit candidates are discussed in semi-quantitative manner in Ref.[6] Since the dynamics in WTC might be completely different from that in QCD and hence the use of the naive scaling in N_c or N_f to estimate various quantities may not work, the S-parameter must be evaluated from the first principles.[7] Although really important quantity is the anomalous dimension of $\bar{\psi}\psi$ operator, looking for theories showing the walking behavior is a good starting point. Recently many groups started quantitative studies using lattice technique to answer the question what gauge theory shows walking behavior. In Ref.,[8] the running couplings of 8- and 12-flavor QCD are studied on the lattice using the Schrödinger functional (SF) scheme.[9] Their conclusion is that while 8-flavor QCD does not show walking behavior 12-flavor QCD reaches an infrared fixed point (IRFP) at $g^2_{\rm IR} \sim 5$. In spite of the scheme-dependence of running and its value of IRFP, the speculation inferred from Schwinger-Dyson equation[10] suggests that $g^2_{\rm IR} \sim 5$ is not large enough to trigger spontaneous chiral symmetry breaking. Although 12-flavor QCD is still an attractive candidate and is open to debate,[11] we explore other N_f. In the following, we report the preliminary results on the running coupling in 10-flavor QCD. Since the conference, statistics and analysis method are changed. The following analysis is based on the increased statistics and a slightly different analysis method.

2. Perturbative analysis

Before going into the simulation details, let us discuss some results from perturbative analysis. In this work, we adopt the β function defined by

$$\beta(g^2(L)) = L\,\frac{\partial\, g^2(L)}{\partial L} = b_1\, g^4(L) + b_2\, g^6(L) + b_3\, g^8(L) + b_4\, g^{10}(L) + \cdots\,, \tag{1}$$

where L denotes a length scale. The first two coefficients are scheme-independent, and given by

$$b_1 = \frac{2}{(4\pi)^2}\left[11 - \frac{2}{3}N_f\right], \qquad b_2 = \frac{2}{(4\pi)^4}\left[102 - \frac{38}{3}N_f\right]. \tag{2}$$

Other higher order coefficients are scheme-dependent and are known only in the limited schemes and orders. In this section, we analyze the perturbative running in the four different schemes/approximations: i) two-loop (universal), ii) three-loop in the $\overline{\rm MS}$ scheme, iii) four-loop in the $\overline{\rm MS}$ scheme, iv) three-loop in the Schrödinger functional scheme. The perturbative coefficients relevant to the following analysis are

$$b_3^{\overline{\rm MS}} = \frac{2}{(4\pi)^6}\left[\frac{2857}{2} - \frac{5033}{18}N_f + \frac{325}{54}N_f^2\right], \tag{3}$$

$$b_4^{\overline{\rm MS}} = \frac{2}{(4\pi)^8}\left[29243.0 - 6946.30\,N_f + 405.089\,N_f^2 + 1.49931\,N_f^3\,\right], \tag{4}$$

$$b_3^{\rm SF} = b_3^{\overline{\rm MS}} + \frac{b_2\, c_2^{\theta}}{2\pi} - \frac{b_1\,(c_3^{\theta} - c_2^{\theta\,2})}{8\pi^2}, \tag{5}$$

N_f	4	6	8	10	12	14	16
2-loop universal				27.74	9.47	3.49	0.52
3-loop SF	43.36	23.75	15.52	9.45	5.18	2.43	0.47
3-loop $\overline{\text{MS}}$		159.92	18.40	9.60	5.46	2.70	0.50
4-loop $\overline{\text{MS}}$			19.47	10.24	5.91	2.81	0.50

Tab. 1 The IRFP from perturbative analysis.

where the coefficients c_2^θ and c_3^θ depend on the spatial boundary condition of the SF used in calculations, *i.e* θ. Those for $\theta = \pi/5$ and c_2^θ for $\theta = 0$ are known as

$$c_2^{\theta=\pi/5} = 1.25563 + 0.039863 \times N_f, \tag{6}$$

$$c_3^{\theta=\pi/5} = (c_2^{\theta=\pi/5})^2 + 1.197(10) + 0.140(6) \times N_f - 0.0330(2) \times N_f^2, \tag{7}$$

$$c_2^{\theta=0} = 1.25563 + 0.022504 \times N_f, \tag{8}$$

but $c_3^{\theta=0}$ is not. Therefore, the case iv) is studied with θ=$\pi/5$. Notice that in our numerical simulation θ=0 and thus, rigorously speaking, the example iv) is not applied to our numerical result.

The perturbative infrared fixed point (IRFP) are numerically solved and summarized in Tab. 1. As seen from Tab. 1, with $N_f \geq 8$ the fixed point value is, to some extent, stable against the change of schemes/approximations. It is interesting that the IRFP of 12-flavor QCD in the SF scheme is consistent with the non-perturbative calculation by Ref.[8] Now looking at the perturbative IRFP at $N_f = 10$, it appears to be stable at $g^2 \sim 10$. Furthermore, according to the analysis based on Schwinger-Dyson equation, there is an argument that chiral symmetry breaking occurs at around $g^2 \sim 4\pi^2/(3\,C_2(R)) = \pi^2$.[10] In summary, the perturbative analysis suggests that 10-flavor QCD is the most attractive candidate for WTC.

3. Simulation parameters and setup

We employ the Schrödinger functional method[9] to calculate the running coupling constant. Unimproved Wilson fermion action and the standard plaquette gauge action without any boundary counter terms are used. The parameter of the spatial boundary condition for fermions, θ, is set to zero. The bare gauge coupling $\beta = 6/g_0^2$ is explored in the range of 4.4–24.0. In this analysis, we report the results obtained from $(L/a)^4 = 6^4$, 8^4 and 12^4 lattices. The calculation on 16^4 lattice is in progress. The numerical simulation is carried out on several architectures including GPGPU and PC cluster. The standard HMC algorithm is used with some improvements in the solver part like the mixed precision algorithm. So far, we have accumulated 5,000 to 200,000 trajectories depending on $(\beta,\ L/a)$.

Since the Wilson type fermion explicitly violates chiral symmetry, the critical value of κ has to be tuned to the massless limit. We performed this tuning for every pair of $(\beta,\ L/a)$. At around $\beta \sim 4.4$, for massless fermions we encounter a (probably first order) phase transition independently of L/a, where the plaquette

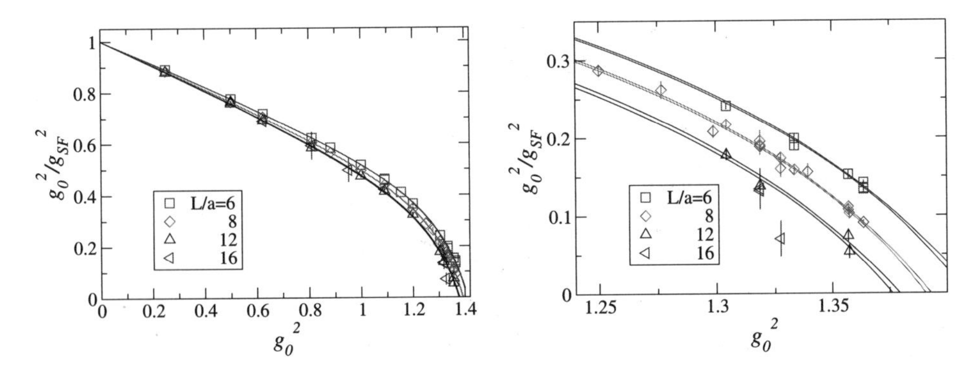

Fig. 1. g_0^2-dependence of $g_0^2/g_{\rm SF}^2(L)$ at $L/a = 6,\ 8,\ 12,\ 16$. The right panel is just an enlargement of the left.

value suddenly jumps to a smaller value. Since this bulk phase transition is inferred to be lattice artifact, whenever this happens we discard the configurations. Thus the position of the critical β (~ 4.4) sets the lower limit on β at which simulations make sense.

4. Preliminary results

Figure 1 shows the Schrödinger functional coupling calculated on the lattices, where $g_0^2/g_{\rm SF}^2$ is plotted as a function of the bare coupling. The solid curve is the result (and statistical error) of the fit to

$$\frac{g_0^2}{{g_{\rm SF}^{\rm lat}}^2(g_0^2, L/a)} = \frac{1 - a_{L/a,1}\, g_0^4}{1 + p_{1,L/a} \times g_0^2 + \sum_{n=2}^{N} a_{L/a,n} \times g_0^{2\,n}}, \tag{9}$$

where $p_{1,L/a}$ is the L/a-dependent coefficient and is found, by perturbative calculation, to be

$$p_{1,L/a} = \begin{cases} 0.4477107831 & \text{for } L/a = 6 \\ 0.4624813408 & \text{for } L/a = 8 \\ 0.4756888260 & \text{for } L/a = 12 \end{cases} . \tag{10}$$

We optimize the degree of polynomial N in the denominator of (9) by monitoring χ^2/dof, and take $N = 5$ for $L/a = 6$ and $N = 4$ for $L/a = 8$, 12.

Since we do not implement any $O(a)$ improvements, large scaling violation is expected to exist. One promising prescription to improve discretization errors has been proposed in Ref.[12] Let us parameterize the lattice artifact in the step scaling by

$$\delta(u, s, L/a) = \frac{\dfrac{\Sigma_0(u, s, L/a)}{1 + \delta^{(1)}(s, L/a)\, u} - \sigma(u, s)}{\sigma(u, s)} = \delta^{(2)}(s, L/a) u^2 + \cdots, \tag{11}$$

where $u = g_{\rm SF}^2(L)$, $\sigma(u, s) = g_{\rm SF}^2(sL)$ and $\Sigma_0(u, s, L/a)$ is ${g_{\rm SF}^{\rm lat}}^2(g_0^2, sL/a)$ at g_0^2 satisfying ${g_{\rm SF}^{\rm lat}}^2(g_0^2, L/a) = u$.

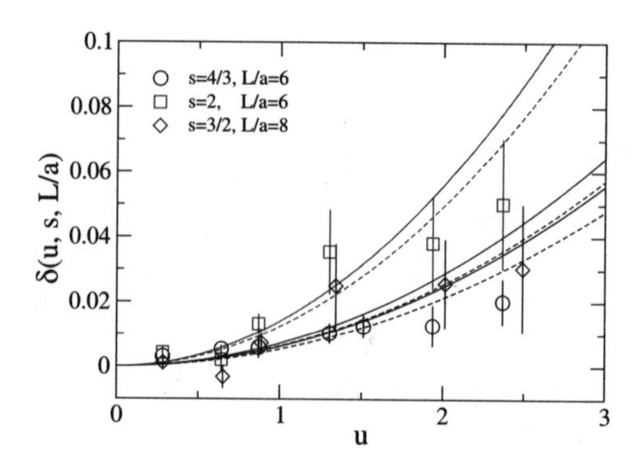

Fig. 2. δ as a function of u. $\delta^{(2)}(u,s,L/a)$ is determined as a coefficient of the quadratic term in u. (solid and dashed curves are obtained from different fit ranges.)

The coefficient $\delta^{(1)}(s,L/a)$ in eq. (11) is given by

$$\delta^{(1)}(s,L/a) = \Big(p_{1,sL/a} - b_1 \ln(sL/a)\Big) - \Big(p_{1,L/a} - b_1 \ln(L/a)\Big). \qquad (12)$$

Dividing the lattice data $\Sigma_0(u,s,L/a)$ by the factor $(1+\delta^{(1)}(s,L/a))$ improves the $O(u)$ discretization error and hence $\delta(u,s,L/a)$ starts from $O(u^2)$ as already indicated in eq. (11).

$\sigma(u,s)$ has perturbative expansion,

$$\sigma(u,s) = u + s_0 u^2 + s_1 u^3 + s_2 u^4 + s_3 u^5 + \cdots, \qquad (13)$$

$$s_0 = b_1 \ln(s), \quad s_1 = \ln(s)\left({b_1}^2 \ln(s) + b_2\right), \qquad (14)$$

$$s_2 = \ln(s)\left({b_1}^3 \ln^2(s) + \frac{5}{2} b_1 b_2 \ln(s) + b_3\right), \qquad (15)$$

$$s_3 = \ln(s)\left\{{b_1}^4 \ln^3(s) + \frac{13}{3}{b_1}^2 b_2 \ln^2(s) + \ln(s)\left(3b_1 b_3 + \frac{3}{2}{b_2}^2\right) + b_4\right\}, \qquad (16)$$

where b_i's are the coefficients of the β-function introduced in sec. 2. Since we know the first two coefficients b_1 and b_2 and hence $\sigma(u,s)$ can be numerically determined to $O(u^3)$, with such σ the $O(u^2)$ term in $\delta(u,s,L/a)$ is attributed to discretization error. The coefficient of u^2 term, $\delta^{(2)}(s,L/a)$, is then obtained by fitting $\delta(u,s,L/a)$ to a quadratic function of u. This fit must be done in the small-coupling region where the perturbative series is reliable. Since at this moment we have only a limited number of data points in such a region, the fit range is forced to extend to $u \sim 2.5$.

Figure 2 shows the u dependence of $\delta(u,s,L/a)$ and the fit results. The obtained coefficients are

$$\delta^{(2)}(s,L/a) = \begin{cases} 0.0062(9) & \text{for } s=4/3,\ L/a=6 \\ 0.0138(26) & \text{for } s=2,\ L/a=6 \\ 0.0070(24) & \text{for } s=3/2,\ L/a=8 \end{cases}, \qquad (17)$$

$$\delta^{(2)}(s,L/a) = \begin{cases} 0.0053(8) & \text{for } s=4/3,\ L/a=6 \\ 0.0122(21) & \text{for } s=2,\ L/a=6 \\ 0.0062(19) & \text{for } s=3/2,\ L/a=8 \end{cases}. \qquad (18)$$

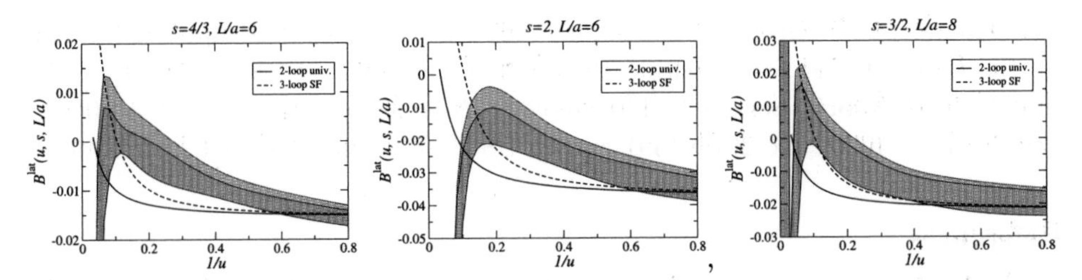

Fig. 3. Discrete beta functions for $(s,\ L/a)$=(4/3, 6), (2, 6) and (3/2, 8) from left to right. Two colored bands are the results with statistical error, obtained by the two-loop improvement with eq. (17) (red) and eq. (18) (blue), respectively. The black solid (dashed) curve is the corresponding discrete beta function obtained by integrating perturbative two-loop universal (three-loop SF scheme) beta function.

In the fit, two different fit ranges, $u \in [0,\ 2.02]$ and $[0,\ 2.50]$, are applied to examine the fit range dependence of the result. Eqs. (17) and (18) correspond to the former and the latter fit range, respectively. Using $\delta^{(2)}(s, L/a)$ thus extracted, we define the improved lattice data by

$$\Sigma^{\text{imp}}(u, s, L/a) = \frac{\Sigma_0(u, s, L/a)}{1 + \delta^{(1)}(s, L/a)\, u + \delta^{(2)}(s, L/a)\, u^2}. \tag{19}$$

To see the running in detail, we introduce the discrete beta function,[13]

$$B^{\text{lat}}(u, s, L/a) = \frac{1}{\Sigma^{\text{imp}}(u, s, L/a)} - \frac{1}{u}. \tag{20}$$

Figure 3 shows the $1/u$ dependence of $B^{\text{lat}}(u, s, L/a)$. As usual β function, large negative value of B^{lat} means rapid increase with length scale of the coupling, and B^{lat} flipping the sign indicate the existence of IRFP. As seen from the figure, in either pairs of $(s,\ L/a)$ the discrete beta function approaches to zero from below when $1/u$ decreases from 0.8 to 0.2, and this happens independently of the choice of $\delta^{(2)}$. This means that at around u=1.25 the running starts to slow down, *i.e.* "walk". In order to confirm that this behavior remains even in the continuum limit, simulations on a larger lattice is necessary.

5. Summary

The running coupling constant of 10-flavor QCD is studied. The perturbative analysis suggests that this theory is extremely interesting. The preliminary result obtained without continuum limit seems to indicate the walking behavior. In order to draw definite conclusions, we clearly need larger lattices to take the continuum limit. Such calculations are in progress.

A part of numerical simulations is performed on Hitachi SR11000 and the IBM System Blue Gene Solution at High Energy Accelerator Research Organization (KEK) under a support of its Large Scale Simulation Program (No. 09-05), on GCOE (Quest for Fundamental Principles in the Universe) cluster system at Nagoya University and on the INSAM (Institute for Numerical Simulations and Applied

Mathematics) GPU cluster at Hiroshima University. This work is supported in part by the Grant-in-Aid for Scientific Research of the Japanese Ministry of Education, Culture, Sports, Science and Technology (Nos. 20105001, 20105002, 20105005, 21684013, 20540261 and 20740139), and by US DOE grant #DE-FG02-92ER40699.

References

1. S. Weinberg, Phys. Rev. D **13**, 974 (1976); L. Susskind, Phys. Rev. D **20**, 2619 (1979);
2. For a recent review, see, for example, C. T. Hill and E. H. Simmons, Phys. Rept. **381**, 235 (2003) [Erratum-ibid. **390**, 553 (2004)]; F. Sannino, arXiv:0804.0182 [hep-ph].
3. For recent review, see, for example, G. T. Fleming, PoS **LATTICE2008**, 021 (2008) [arXiv:0812.2035 [hep-lat]]; E. Pallante, in these proceedings.
4. M. E. Peskin and T. Takeuchi, Phys. Rev. Lett. **65**, 964 (1990); Phys. Rev. D **46**, 381 (1992).
5. B. Holdom, Phys. Rev. D **24**, 1441 (1981); K. Yamawaki, M. Bando and K. i. Matumoto, Phys. Rev. Lett. **56**, 1335 (1986); T. W. Appelquist, D. Karabali and L. C. R. Wijewardhana, Phys. Rev. Lett. **57**, 957 (1986); T. Akiba and T. Yanagida, Phys. Lett. B **169**, 432 (1986); M. Bando, T. Morozumi, H. So and K. Yamawaki, Phys. Rev. Lett. **59**, 389 (1987).
6. D. D. Dietrich and F. Sannino, Phys. Rev. D **75**, 085018 (2007) [arXiv:hep-ph/0611341].
7. E. Shintani *et al.* [JLQCD Collaboration], Phys. Rev. Lett. **101**, 242001 (2008); P. A. Boyle *et al.* [RBC and UKQCD collaborations], arXiv:0909.4931 [hep-lat].
8. T. Appelquist, G. T. Fleming and E. T. Neil, Phys. Rev. Lett. **100**, 171607 (2008); Phys. Rev. D **79**, 076010 (2009).
9. M. Luscher, R. Narayanan, P. Weisz and U. Wolff, Nucl. Phys. B **384** (1992) 168; M. Luscher, R. Sommer, P. Weisz and U. Wolff, Nucl. Phys. B **413**, 481 (1994); S. Sint and R. Sommer, Nucl. Phys. B **465**, 71 (1996); M. Luscher and P. Weisz, Nucl. Phys. B **479**, 429 (1996); A. Bode, P. Weisz and U. Wolff [ALPHA collaboration], Nucl. Phys. B **576**, 517 (2000) [Erratum-ibid. B **600**, 453 (2001 ERRAT,B608,481.2001)].
10. T. Appelquist, K. D. Lane and U. Mahanta, Phys. Rev. Lett. **61**, 1553 (1988); A. G. Cohen and H. Georgi, Nucl. Phys. B **314**, 7 (1989).
11. For the works suggesting the other possibility, see, for example, A. Hasenfratz, Phys. Rev. D **80**, 034505 (2009) [arXiv:0907.0919 [hep-lat]]; X. Y. Jin and R. D. Mawhinney, arXiv:0910.3216 [hep-lat].
12. S. Aoki *et al.* [PACS-CS Collaboration], JHEP **0910**, 053 (2009) [arXiv:0906.3906 [hep-lat]].
13. Y. Shamir, B. Svetitsky and T. DeGrand, Phys. Rev. D **78**, 031502 (2008) [arXiv:0803.1707 [hep-lat]].

Study of the Running Coupling in Twisted Polyakov Scheme

T. Aoyama[1], H. Ikeda[2], E. Itou[3], M. Kurachi[4], C.-J. D. Lin[5], H. Matsufuru[6], H. Ohki[7], T. Onogi[8], E. Shintani[8,*] and T. Yamazaki[9]

[1] *Graduate School of Science, Nagoya University, Aichi 464-8602, Japan*
[2] *School of High Energy Accelerator Science*
The Graduate University for Advanced Studies (Sokendai), Ibaraki 305-8081, Japan
[3] *Academic Support Center, Kogakuin University, Nakanomachi Hachioji, 192-0015, Japan*
[4] *Department of Physics, Tohoku University, Sendai, 980-8578, Japan*
[5] *National Chiao-Tung University, and National Center for Theoretical Sciences Hsinchu 300, Taiwan*
[6] *KEK Computing Center, High Energy Accelerator Research Organization (KEK) Ibaraki 305-8081, Japan*
[7] *Department of Physics, Kyoto University, Kyoto 606-8501, Japan*
[8] *Department of Physics, Osaka University, Toyonaka, Osaka 560-0043, Japan*
[9] *Center for Computational Sciences, University of Tsukuba, Tsukuba, Ibaraki 305-8577, Japan*

We present a non-perturbative study of the running coupling constant in the Twisted Polyakov Loop (TPL) scheme. We investigate how the systematic and statistical errors can be controlled *via* a feasibility study in SU(3) pure Yang-Mills theory. We show that our method reproduces the perturbative determination of the running coupling in the UV and gives consistent results with the theoretical prediction in the IR. We also present our preliminary results for $N_f = 12$ QCD.

1. Introduction

The search for a non-trivial infrared fixed point (IRFP) in gauge theories is an interesting topic. In addition to its intrinsic field-theoretic interests, it is relevant to the phenomenology of dynamical electroweak symmetry breaking models such as the technicolor (TC) models [1]. The existence of the (approximate) IRFP provides a slowly running, or "walking," coupling, which is required to suppress the flavor-changing neutral current (FCNC) by a large anomalous dimension of the bilinear techni-fermion operator [2, 3]. The walking behavior of the coupling might also help to reduce the technicolor contribution to the S parameter. Non-trivial IRFP is also interesting to understand phase structure, e.g. whether the chiral broken phase and confinement phase are different or not.

SU(N) gauge theory with N_f massless fermions (also known as large flavor QCD) has been extensively investigated as a possible candidate for the theory with walking dynamics [4]. An analysis using the two-loop beta function and Schwinger-Dyson

*Speaker, E-mail: shintani@het.phys.sci.osaka-u.ac.jp

equation suggests the existence of the IRFP for $N_f^{\rm cr} < N_f \leq 16$ with $N_f^{\rm cr} \simeq 12$ in the case of $N_c = 3$. However to search for the conformal window the full non-perturbative analysis is necessary.

Recently non-perturbative analysis of the running coupling constant using lattice gauge theory with the Schrödinger functional (SF) scheme has been done by Appelquist *et al.* [5]. While they discovered IRFP-like behavior in the case of $N_f = 12$ QCD, there is no such evidence in $N_f = 8$ QCD. (Two groups have also studied the phase structure of the $N_f = 12$ theory [6, 7].)

In this work we focus on the confirmation of IRFP in $N_f = 12$ QCD with different scheme from previous IRFP studies. We here employ the Twisted Polyakov Loop (TPL) scheme [8, 9]. This scheme is not only automatically evaded from $O(a/L)$ discretization errors, but also predicts the behavior of running coupling toward infrared limit. In addition compared with Wilson loop scheme [10] TPL scheme done not need relatively large lattice volume and link smearing. Therefore we can precisely search non-trivial IRFP with the present machine resources.

We first give a brief review of TPL scheme in section 2, and in section 3 we present a validity study of this scheme by calculating in SU(3) pure Yang-Mills theory. Our preliminary results for $N_f = 12$ SU(3) gauge theory is reported in section 4.

2. Twisted Polyakov Loop scheme

In this section, we present the definition of TPL scheme in SU(3) gauge theory. Here we consider twisted boundary condition for the link variables in x and y directions on the lattice:

$$U_\mu(x + \hat{\nu}L/a) = \Omega_\nu U_\mu(x)\Omega_\nu^\dagger. \quad (\nu = 1, 2) \tag{1}$$

Ω_ν are the twist matrices which satisfy

$$\Omega_1\Omega_2 = e^{i2\pi/3}\Omega_2\Omega_1, \quad \Omega_\mu\Omega_\mu^\dagger = 1, \quad (\Omega_\mu)^3 = 1, \quad \mathrm{Tr}[\Omega_\mu] = 0. \tag{2}$$

In order to keep the gauge invariance and translation invariance, the Polyakov loops in the twisted directions are defined as

$$P_1(y, z, t) = \mathrm{Tr}\Big(\prod_j U_1(x = j, y, z, t)\Omega_1 e^{i2\pi y/3L}\Big), \tag{3}$$

for x direction, and for y direction we obtain by $1 \to 2$ and $x \leftrightarrow y$. In the above definition we add Ω_ν for the invariance under gauge transformation, and phase $e^{i2\pi y/3L}$ for the translational invariance.

The renormalized coupling is then given by

$$g_{TP}^2 = \frac{1}{k}\frac{\langle \sum_{y,z} P_1(y, z, L/2a)P_1(0, 0, 0)^\dagger\rangle}{\langle \sum_{x,y} P_3(x, y, L/2a)P_3(0, 0, 0)^\dagger\rangle}. \tag{4}$$

At the tree level, the ratio of twisted and untwisted Polyakov-loop correlator is proportional to the bare coupling. The coefficient k is obtained by analytical calculation of the one-gluon-exchange diagram:

$$k = \frac{1}{24\pi^2} \sum \frac{(-1)^n}{n^2 + (1/3)^2} = 0.03184\cdots. \tag{5}$$

in the case of SU(3) gauge group. The twist matrices are chosen in the same form as those in Ref.[11],

$$\Omega_1 = \begin{pmatrix} 0\,1\,0 \\ 0\,0\,1 \\ 1\,0\,0 \end{pmatrix}, \quad \Omega_2 = \begin{pmatrix} e^{-i2\pi/3} & 0 & 0 \\ 0 & e^{i2\pi/3} & 0 \\ 0 & 0 & 1 \end{pmatrix}. \tag{6}$$

In the infrared limit ($L \to \infty$), twisted and untwisted Polyakov-loop correlators are not different since the correlation between Polyakov loops vanishes. Therefore the ratio of those is unit, namely g^2_{TP} consists of $1/k \sim 32$. This fact is also applicable to check our simulation.

The naive twisted boundary condition for lattice fermions is inconsistent when we change the order of translations, i.e. $\psi(x + \hat{\nu}L/a + \hat{\rho}L/a) = \Omega_\rho \Omega_\nu \psi(x) \neq \Omega_\nu \Omega_\rho \psi(x)$. To avoid this problem, we introduce "smell" degree of freedom [12], which is a copy of color degree of freedom. We identify the fermion field as a $N_c \times N_s$ matrix $(\psi^a_\alpha(x))$. Then we impose the twisted boundary condition for fermion fields to be

$$\psi^a_\alpha(x + \hat{\nu}L/a) = e^{i\pi/3} \Omega^{ab}_\nu \psi^b_\beta (\Omega_\nu)^\dagger_{\beta\alpha} \tag{7}$$

for $\nu = 1, 2$ directions. Here, the smell index can be interpreted as a "flavor" index, then the number of flavors should be a multiple of $N_s (= N_c = 3)$. We use staggered fermion in our simulation. Since this describes four tastes of fermions, we can perform simulations in $N_f = 12$ QCD, which is same number of flavor as that in Ref. [5], with twisted boundary condition.

3. Quenched QCD case

Before carrying out the simulation for $N_f = 12$, we first measure the TPL running coupling in quenched QCD to confirm that the scheme provides the expected behavior of running coupling constant. The gauge configurations are generated by the pseudo-heatbath algorithm and the overrelaxation algorithm mixed in the ratio 1:5. To reduce large statistical fluctuation of the TPL coupling, we measure Polyakov loops at every Monte Calro sweep and perform a jackknife analysis with large bin size, typically of $O(10^3)$. The simulations are carried out with several lattice sizes ($L/a = 4, 6, 8, 10, 12, 14, 16$) in the range $6.2 \leq \beta \leq 16$. We generate 200,000–400,000 configurations for each data point.

Figure 1 shows the β dependence of the renormalized coupling in our scheme at various lattice sizes. We fit the data at each fixed lattice size by an interpolating function:

$$g^2_{\rm TP}(\beta) = \sum_{i=1}^{n} \frac{A_i}{(\beta - B)^i}, \tag{8}$$

where A_i and B are the fit parameters, $4 \leq B \leq 5$, and $n = 3, 4$ are employed. We can see that the interpolating function in Eq.(8) well describes all lattice data as shown in Figure 1. In the step scaling we use $L/a = 4, 6, 8, 10$ for starting points. The step-scaling parameter is $s = 1.5$, and we estimate the coupling constant for $L/a = 9, 15$ from interpolations.

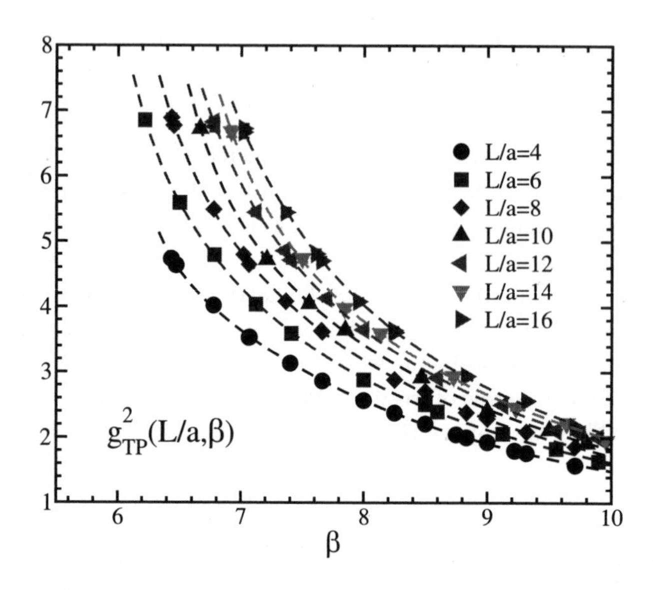

Fig. 1. TPL renormalized coupling in the each β and L/a in quenched QCD.

We take the continuum limit using a linear function in $(a/L)^2$ in each step of step scaling procedure. TPL scheme involves tiny $O((a/L)^2)$ discretization effect, and no $O(a/L)$ error below $g^2_{TP} \simeq 5$, as shown in Figure 2.

Figure 3 shows the g^2_{TP} normalized by $1/k$ in Eq.(5) in the infrared limit. As expected g^2_{TP} goes to $1/k$ in the $L \to \infty$.

g^2_{TP} in quenched QCD as a function of energy scale is shown in Figure 4 together with one- and two-loop perturbative results. The step-scaling function starts from $g^2(L_0/L) = 0.65$. The running coupling constant obtained in TPL scheme is consistent with perturbation theory above $L_0/L = 0.01$. On the other hand, in $L_0/L < 0.01$, the running is slower than one-loop and two-loop perturbation because of higher order effects. From this quenched test, we can see that TPL scheme provides reasonable behavior for running coupling constant.

4. $N_f = 12$ case

In this section, we present a preliminary result in $N_f = 12$. The simulation parameters are $4.0 \leq \beta \leq 25.0$ with lattice sizes $L/a = 4, 6, 8, 10, 12, 16$. Figure 5 shows the β dependence of g^2_{TP} normalized by $1/k$ for each lattice size. Note that in low-β region the growth of g^2_{TP} toward coarse lattice saturates, especially for small lattice.

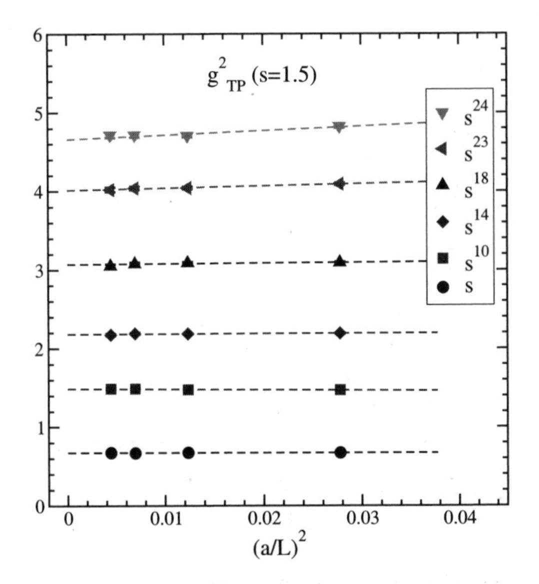

Fig. 2. The continuum extrapolation of g^2_{TP} with $s = 1.5$. The fit function is a linear function of $(a/L)^2$. The statistical error bars are of the same size of the symbols.

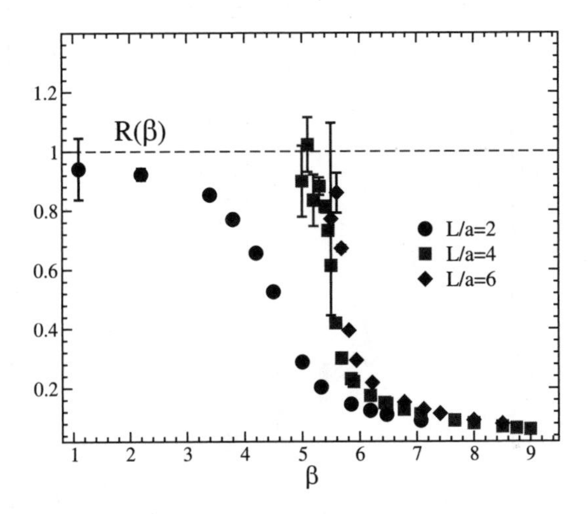

Fig. 3. $g^2_{TP}k$ as a function of β, which corresponds to the inverse lattice spacing, in $L/a = 2$, 4, 6.

It might be due to the lattice artifacts e.g. the effect of taste breaking for staggered fermions, but it is necessary to do further detailed analysis. To perform the global fit, we consider the following interpolation function $kg^2_{TP}(\beta, a/L) = f(\beta, a/L)$:

$$f(\beta, a/L) = \frac{6}{\beta + c_0(L/a)} + c_1(L/a)\Big(\frac{6}{\beta + c_0(L/a)}\Big)^2 + c_2(L/a)\Big(\frac{6}{\beta + c_0(L/a)}\Big)^3, \quad (9)$$

with $c_i(L/a) = x_i + y_i \ln(L/a)$, $i = 1, 2$. x_i, y_i denote the fit parameters. The solid lines in Figure 5 denote the fit function with fit range $6 \leq \beta \leq 20$. In the step-

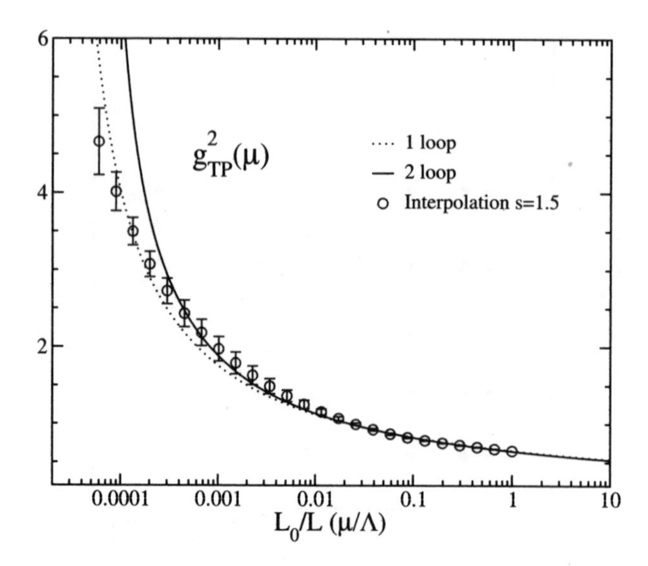

Fig. 4. The running coupling constants in TPL scheme, one-loop and two-loop.

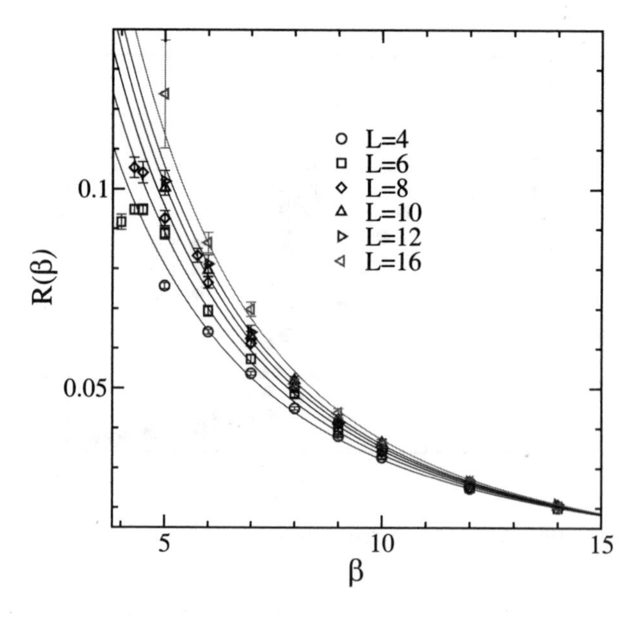

Fig. 5. The ratio of Polyakov loop in each β and L/a.

scaling function, we use three points, $L = 4, 6, 8$ and $sL = 8, 12, 16$. Figure 6 shows the continuum extrapolation in each steps with $s = 2$. In $N_f = 12$ case the $O(a/L)$ error is also absent. Accumulating more statistics and data points of low-β region, we will check the consistency with perturbation theory in high-β and study the running behavior in strong coupling region. These works are under way.

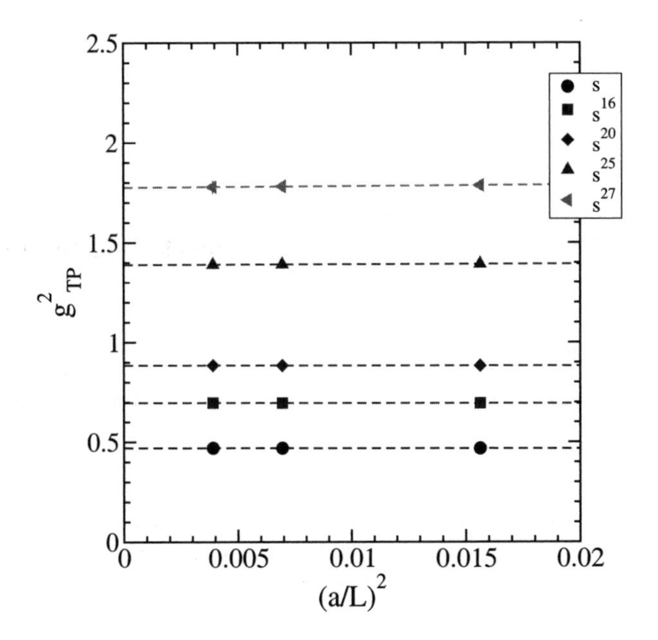

Fig. 6. The continuum extrapolation of g^2_{TP} at several step-scaling points.

References

1. S. Weinberg, Phys. Rev. D **13**, 974 (1976); Phys. Rev. D **19**, 1277 (1979); L. Susskind, Phys. Rev. D **20**, 2619 (1979).
2. B. Holdom, Phys. Rev. D **24**, 1441 (1981).
3. K. Yamawaki, M. Bando and K. i. Matumoto, Phys. Rev. Lett. **56**, 1335 (1986).
4. T. Banks and A. Zaks, Nucl. Phys. B **196** (1982) 189; T. Appelquist, J. Terning and L. C. R. Wijewardhana, Phys. Rev. Lett. **77**, 1214 (1996) [arXiv:hep-ph/9602385].
5. T. Appelquist, G. T. Fleming and E. T. Neil, Phys. Rev. Lett. **100** (2008) 171607, [Erratum-ibid. **102** (2009) 149902], Phys. Rev. D **79** (2009) 076010
6. A. Deuzeman, M. P. Lombardo and E. Pallante, arXiv:0904.4662 [hep-ph].
7. Z. Fodor, K. Holland, J. Kuti, D. Nogradi and C. Schroeder, arXiv:0809.4890 [hep-lat].
8. G. M. de Divitiis, R. Frezzotti, M. Guagnelli and R. Petronzio, Nucl. Phys. B **422** (1994) 382
9. G. M. de Divitiis, R. Frezzotti, M. Guagnelli and R. Petronzio, Nucl. Phys. B **433** (1995) 390
10. E. Bilgici, *et al.*, Phys. Rev. D **80** (2009) 034507
11. H. D. Trottier, N. H. Shakespeare, G. P. Lepage and P. B. Mackenzie, Phys. Rev. D **65** (2002) 094502
12. G. Parisi,Published in Cargese Summer Inst. 1983:0531
13. G. M. de Divitiis *et al.*, (Alpha Collaboration), Nucl. Phys. B **437** (1995) 447

Running Coupling in Strong Gauge Theories via the Lattice

Zoltán Fodor

Department of Physics, University of Wuppertal, Gauss Strasse 20, D-42119, Germany
and
Institute for Theoretical Physics, Eotvos University Budapest, Pazmany P. 1, H-1117, Hungary
E-mail: fodor@bodri.elte.hu

Kieran Holland

Department of Physics, University of the Pacific
3601 Pacific Ave, Stockton CA 95211, USA
E-mail: kholland@pacific.edu

Julius Kuti

Department of Physics 0319, University of California, San Diego
9500 Gilman Drive, La Jolla CA 92093, USA
E-mail: jkuti@ucsd.edu

Dániel Nógrádi

Institute for Theoretical Physics, Eotvos University Budapest, Pazmany P. 1, H-1117, Hungary
E-mail: nogradi@bodri.elte.hu

Chris Schroeder

Department of Physics, University of Wuppertal, Gauss Strasse 20, D-42119, Germany
E-mail: crs@physics.ucsd.edu

New strongly-interacting gauge theories are possible Beyond Standard Model candidates. It is essential to determine if a given non-abelian gauge theory is QCD-like or conformal. One way is to calculate the running of the renormalized gauge coupling. We describe a method to do this using Wilson loop ratios measured in lattice simulations. We demonstrate this in $SU(3)$ pure gauge theory and show initial results for dynamical fermions in the fundamental representation.

Keywords: Beyond Standard Model, lattice, near-conformal, running coupling.

1. Running coupling

There is great interest in strongly-coupled gauge theories as new physics candidates,[1–21] one example being technicolor. As a starting point, it is crucial to distinguish conformal gauge theories from those with a mass gap like QCD. There are many recent lattice studies of various theories, attempting to answer this question.[22–54] One signal of conformal behavior is if the running gauge coupling flows to an infrared fixed point. In a QCD-like theory, no such fixed point exists. For a given theory, this should be consistent with other signals, such as the mass spectrum, the Dirac operator eigenvalues, the existence of finite-temperature transitions, or the renormalization group flow in bare parameter space.

We define the renormalized coupling using Wilson loops $W(R,T,L)$, R and T being the space- and time-like loop extents, and L the finite volume extent. One definition uses the quark-antiquark potential and its derivative

$$V(R) = \lim_{T\to\infty} \left(-\frac{\partial}{\partial T} \ln\langle W(R,T,L)\rangle\right), \qquad F = \frac{dV}{dR} = C_F \frac{g^2(R)}{4\pi R^2}, \tag{1}$$

where $C_F = 4/3$, and L is large enough that finite-volume effects are absent. The infinite-volume coupling $g^2(R)$ runs with the quark separation R. In lattice simulations, the separation of scales L, T, R and lattice spacing a is challenging. An alternate definition is[55]

$$g^2(R,L) = -\frac{R^2}{k} \frac{\partial^2}{\partial R \partial T} \ln\langle W(R,T,L)\rangle \,|_{T=R}\,, \tag{2}$$

for convenience restricted to square Wilson loops. The numerical factor k depends on R/L and is calculated perturbatively, such that the leading term in $g^2(R,L)$ is the tree-level gauge coupling. In the infinite-volume limit, the coupling $g^2(R)$ again runs with the loop size. Alternatively, one can hold R/L fixed in which case $g^2(L)$ runs with the finite volume. This may be advantageous in simulations. The fixed value of R/L defines a renormalization scheme. The role of lattice simulations is to measure expectation values $\langle ... \rangle$ non-perturbatively, with derivatives replaced by finite differences, giving

$$\begin{aligned} g^2(R,L) &= \frac{1}{k}(R+1/2)^2 \chi(R+1/2,L)\,, \\ \chi(R+1/2,L) &= -\ln\left[\frac{\langle W(R+1,T+1,L)\rangle\langle W(R,T,L)\rangle}{\langle W(R+1,T,L)\rangle\langle W(R,T+1,L)\rangle}\right] |_{T=R}\,, \end{aligned} \tag{3}$$

χ being the Creutz ratio[56] and the renormalization scheme is $r = (R+1/2)/L$ fixed.

The continuum limit corresponds to $L \to \infty$, where the physical length scale $L_{\rm phys}$ is held fixed while the lattice spacing $a \to 0$. One starts the RG flow from some reference physical point $L_{\rm phys,0}$ which is implicitly set by the initial choice e.g. $g^2(r, L_{\rm phys,0}) = 0.8$. In a QCD-like theory, g^2 increases with increasing $L_{\rm phys}$, flowing in the infrared direction. However in a conformal theory, g^2 flows towards some non-trivial infrared fixed point g^{*2} as $L_{\rm phys}$ increases.

One way to take the continuum limit of the RG flow is via step-scaling. The bare lattice gauge coupling is defined in the usual way $\beta = 6/g_0^2$ in the Wilson lattice gauge action. On a sequence of lattice sizes $L_1, L_2, ..., L_n$, the bare coupling is tuned on each lattice so that identical values $g^2(r, L_i, \beta_i) = g^2(r, L_{\rm phys})$ are measured in simulations. Next a new set of simulations is performed, on a sequence of lattice sizes $2L_1, 2L_2, ..., 2L_n$, using the corresponding tuned couplings $\beta_1, \beta_2, ..., \beta_n$. From the simulations, one measures $g^2(r, 2L_i, \beta_i)$, which vary with the bare coupling i.e. the lattice spacing. These data are extrapolated to the continuum as a function of $1/L_i^2$.

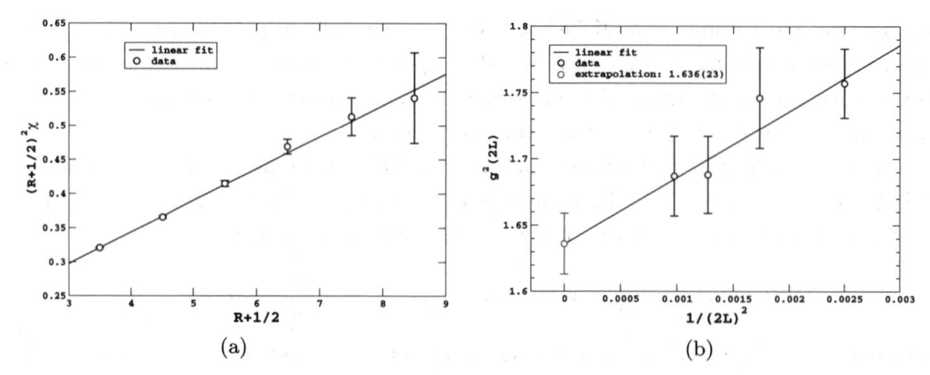

(a) (b)

Fig. 1. (a) The rescaled Creutz ratio on a 28^4 lattice at $\beta = 6.99$. We interpolate the data linearly to $r = (R + 1/2)/L = 0.25$, giving a chi squared per degree of freedom 2.0/4. (b) The measured coupling $g^2(2L_i, \beta_i)$ for $2L_i = 20, 24, 28$ and 32, where β_i is tuned such that $g^2(L_i, \beta_i) = 1.44$. A linear continuum extrapolation gives $g^2(2L_{\rm phys}) = 1.636(23)$, with $\chi^2/{\rm dof} = 0.57/2$.

This gives one blocking step $g^2(r, L_{\rm phys}) \to g^2(r, 2L_{\rm phys})$ in the continuum RG flow. The whole procedure is then iterated. The chain of measurements gives the flow $g^2(r, L_{\rm phys}) \to g^2(r, 2L_{\rm phys}) \to g^2(r, 4L_{\rm phys}) \to g^2(r, 8L_{\rm phys}) \to ...$, as far as is feasible. One is free to choose a different blocking factor, say $L_{\rm phys} \to (3/2)L_{\rm phys}$, in which case more blocking steps are required to cover the same energy range.

2. $SU(3)$ pure gauge theory

As a first test, we study $SU(3)$ pure gauge theory in four dimensions, calculating the renormalized coupling using square Wilson loops. We simulate using the standard Wilson lattice gauge action, with a mixture of five over-relaxation updates for every heatbath update. Our renormalization scheme is $r = 0.25$, for brevity we omit the label r. In this R/L range, the Creutz ratio can be accurately measured, and the factor k converges quickly to its continuum-limit value. In the pure gauge theory test, we use the finite-L values of k, as this may remove some of the cutoff dependence of the renormalized coupling. For the RG flow we choose the blocking step $L \to 2L$. We simulate on small lattices of size $L = 10, 12, 14, 16, 18, 20$ and 22, and the corresponding doubled lattices $2L = 20, 24, 28, 32, 36, 40$ and 44. Wilson loop measurements are separated by 1000 sweeps, which we find is sufficient to generate statistically independent configurations.

To tune β_i for each small lattice size L_i, we typically run separate simulations at $5 - 10$ different β values in the relevant range of renormalized coupling g^2. Each of these runs contains $300 - 500$ measurements i.e. up to 5×10^5 sweeps. Simulating on the doubled lattices at the tuned β_i values, we generate between 200 and 1000 measurements each, typically more than 500. The signal of the Creutz ratio disappears into the noise as the size of the Wilson loop increases. One way to suppress the noise in measurements is to gauge fix the configurations to Coulomb gauge, and replace the thin-link Wilson loop with the correlator of the products of the time-like gauge

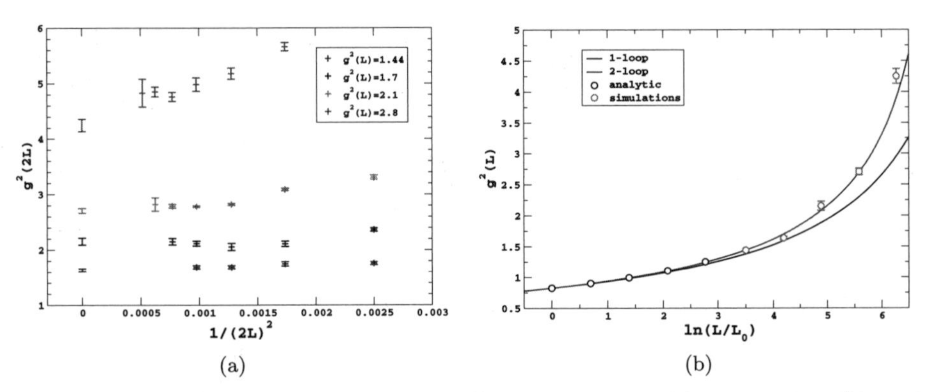

Fig. 2. (a) The continuum extrapolations of four discrete RG steps. (b) The RG flow $g^2(L_{\text{phys}})$, combining analytic lattice perturbation theory and the simulation results. The running starts at $g^2(L_{\text{phys},0}) = 0.825$. There is excellent agreement with continuum 2-loop running, at the strongest coupling, the simulation results begin to break away from perturbation theory.

links.[57,58] Note that gauge fixing is not implemented in the actual Monte Carlo updating algorithm. An alternative method to suppress noise is to smear the gauge links and measure the fat-link Wilson loop operator. In the pure gauge theory test, we use the gauge-fixing method, in the dynamical fermion simulations we describe later, we use the smearing method. These improvement methods do not correspond to calculating the original thin-link Wilson loop operator.

We show in Fig. 1(a) a typical result for the rescaled Creutz ratio $(R+1/2)^2\chi$. The doubled lattice is 28^4 and the bare coupling $\beta = 6.99$ is tuned from simulations on 14^4 volumes. Errorbars are calculated using the jackknife method. We interpolate the data linearly to the point $r = (R+1/2)/L = 0.25$, obtaining $\chi^2/\text{dof} = 2.0/4$. The data at different R are highly correlated, being measured on the same gauge configurations. For error estimation, we bin the gauge configurations, and analyze and interpolate separately each bin. An example of the step-scaling method is shown in Fig. 1(b). The bare couplings are tuned such that $g^2(L_i, \beta_i) = 1.44$ for $L_i = 10, 12, 14$ and 16, the figure shows the data $g^2(2L_i, \beta_i)$ for $2L_i = 20, 24, 28$ and 32. The leading lattice artifacts in the Creutz ratio are expected to be of order $\mathcal{O}(a^2)$, corresponding to $\mathcal{O}(1/L^2)$, since the physical lattice size is La. Extrapolating linearly in $1/(2L)^2$ gives $g^2(2L_{\text{phys}}) = 1.636(23)$ and $\chi^2/\text{dof} = 0.57/2$. For a systematic error, we omit one data point at a time and repeat the extrapolation. Combined in quadrature with the statistical error, our continuum result is $g^2(2L_{\text{phys}}) = 1.636(25)$.

We iterate the procedure, giving four discrete RG steps as shown in Fig. 2(a). At stronger coupling, we need lattices up to $2L_i = 44$ for the continuum extrapolation. The use of the finite-L values of k does not appear to reduce the cutoff effects. The continuum RG flow is shown in Fig. 2(b). At weak coupling, we use analytic/numeric lattice perturbation theory to calculate the Wilson loop ratios in finite volumes.[59,60] The Wilson loops are calculated to 1-loop in the bare coupling in finite volume. The series is reexpanded in the boosted coupling constant at the relevant scale of the

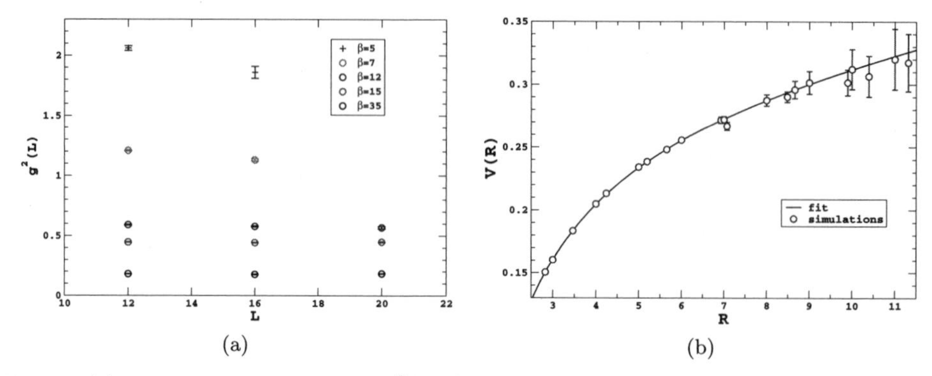

Fig. 3. (a) The renormalized coupling $g^2(L,\beta)$ at fixed bare coupling for $N_f = 16$ fundamental flavors. For $g^2(L,\beta) > 0.5$ the coupling decreases with increasing L, for $g^2(L,\beta) < 0.5$ the coupling is independent of L within the errors. This is consistent with the existence of an infrared fixed point. (b) The quark-antiquark potential $V(R)$ for $N_f = 12$ fundamental flavors measured in a $32^3 \times 64$ lattice at $\beta = 2.2$ and $m = 0.015$. The data indicate QCD-like behavior.

Creutz ratio. Step-scaling of the finite volume ratios can be used in exactly the same way as for the simulations results, to determine the RG flow in the continuum. The analytic RG flow starts at the reference point $g^2(L_{\mathrm{phys},0}) = 0.825$. At weak coupling there is complete agreement with 2-loop perturbation theory. We connect lattice perturbation theory to the simulation results by matching the flows at $g^2(L_{\mathrm{phys}}) = 1.44$, where the simulation RG flow begins. There is continued agreement with 2-loop perturbation theory at even stronger coupling, only at the strongest coupling do we see deviation from the perturbative flow.

3. Fundamental fermions

We have begun studies of $SU(3)$ gauge theory with fermions in the fundamental representation. For $N_f = 16$ flavors, 2-loop perturbation theory predicts the theory is conformal with an infrared fixed point $g^{*2} \approx 0.5$. At such weak coupling, this perturbative result is quite trustworthy. Because of the computational expense of step-scaling with dynamical fermions, we have not yet performed a continuum extrapolation. Hence the running coupling is contaminated with cutoff effects. If the linear lattice size L is large enough, the trend from the volume dependence of $g^2(L,\beta)$ should indicate the location of the fixed point. For $g^2(L,\beta) > g^{*2}$ we expect decrease in the running coupling as L grows, although the cutoff of the flow cannot be removed above the fixed point. Below the fixed point with $g^2(L,\beta) < g^{*2}$ we expect the running coupling to grow as L increases and the continuum limit of the flow could be determined. The first results are shown in Fig. 3(a). We use stout-smeared[61] staggered fermions[62,63] and the RHMC algorithm, simulating at quark mass $m_q = 0.01$, with some runs at $m_q = 0.001$ to test that the mass dependence is negligible. For the Wilson loop ratios, we smear the gauge fields and measure the fat-link Wilson operator. Our experience in the pure gauge theory test is cutoff effects are not reduced using finite-L values of k, hence we use the infinite-volume k

value to convert the Wilson loop ratios to a renormalized coupling. The results are consistent with the above picture. For $g^2(L,\beta) > 0.5$, the cutoff-dependent renormalized coupling decreases with L. For $g^2(L,\beta) < 0.5$, the renormalized coupling is L-independent within errors. The theory appears conformal but precise determination of the infrared fixed point requires further study.

We also show results for $N_f = 12$ flavors in the fundamental representation. Perturbatively this theory is predicted to be conformal, at 2-loop $g^{*2} \approx 9$, with significant corrections coming from 3-loop for various schemes, for example $g^{*2} \approx 4.3$ in the Wilson-loop scheme. This theory is controversial, some recent lattice studies find conformal behavior, others indicate it is QCD-like. In Fig. 3(b) we show the quark-antiquark potential $V(R)$ in a $32^3 \times 64$ lattice, using the same improved action as for $N_f = 16$. The data suggest the theory is in fact QCD-like, $V(R)$ does not display the purely $1/R$ dependence of a conformal theory. This is consistent with our other findings for $N_f = 12$, that the pion mass spectrum and eigenvalues of the Dirac operator indicate spontaneously broken chiral symmetry. The running coupling study of fundamental fermions is continuing.

Acknowledgments

We wish to thank Urs Heller for the use of his code to calculate Wilson loops in lattice perturbation theory, and Paul Mackenzie for related discussions. We are grateful to Sandor Katz and Kalman Szabo for helping us in using the Wuppertal RHMC code. In some calculations we use the publicly available MILC code, and the simulations were performed on computing clusters at Fermilab, under the auspices of USQCD and SciDAC, and on the Wuppertal GPU cluster. This research is supported by the NSF under grant 0704171, by the DOE under grants DOE-FG03-97ER40546, DOE-FG-02-97ER25308, by the DFG under grant FO 502/1 and by SFB-TR/55.

References

1. S. Weinberg, Phys. Rev. D **19**, 1277 (1979).
2. L. Susskind, Phys. Rev. D **20**, 2619 (1979).
3. E. Farhi and L. Susskind, Phys. Rept. **74**, 277 (1981).
4. F. Sannino and K. Tuominen, Phys. Rev. D **71**, 051901 (2005)
5. D. D. Dietrich, F. Sannino and K. Tuominen, Phys. Rev. D **72**, 055001 (2005)
6. D. D. Dietrich and F. Sannino, Phys. Rev. D **75**, 085018 (2007).
7. T. A. Ryttov and F. Sannino, Phys. Rev. D **76**, 105004 (2007).
8. D. K. Hong, S. D. H. Hsu and F. Sannino, Phys. Lett. B **597**, 89 (2004).
9. H. Georgi, Phys. Rev. Lett. **98**, 221601 (2007).
10. M. A. Luty and T. Okui, JHEP **0609**, 070 (2006).
11. W. E. Caswell, Phys. Rev. Lett. **33**, 244 (1974).
12. T. Banks and A. Zaks, Nucl. Phys. B **196**, 189 (1982).
13. T. Appelquist, K. D. Lane and U. Mahanta, Phys. Rev. Lett. **61**, 1553 (1988).
14. A. G. Cohen and H. Georgi, Nucl. Phys. B **314**, 7 (1989).
15. T. Appelquist et al., Phys. Rev. Lett. **77**, 1214 (1996).

16. B. Holdom, Phys. Rev. D **24**, 1441 (1981).
17. K. Yamawaki, M. Bando and K. i. Matumoto, Phys. Rev. Lett. **56**, 1335 (1986).
18. T. W. Appelquist et al., Phys. Rev. Lett. **57**, 957 (1986).
19. V. A. Miransky and K. Yamawaki, Phys. Rev. D **55**, 5051 (1997).
20. E. Eichten and K. D. Lane, Phys. Lett. B **90**, 125 (1980).
21. T. Appelquist, M. Piai and R. Shrock, Phys. Rev. D **69**, 015002 (2004)
22. T. Appelquist, G. T. Fleming and E. T. Neil, Phys. Rev. Lett. **100**, 171607 (2008)
23. T. Appelquist, G. T. Fleming and E. T. Neil, Phys. Rev. D **79**, 076010 (2009)
24. T. Appelquist *et al.*, arXiv:0910.2224 [hep-ph].
25. S. Catterall and F. Sannino, Phys. Rev. D **76**, 034504 (2007)
26. S. Catterall, J. Giedt, F. Sannino and J. Schneible, JHEP **0811**, 009 (2008)
27. S. Catterall, J. Giedt, F. Sannino and J. Schneible, arXiv:0910.4387 [hep-lat].
28. Z. Fodor et al., PoS **LATTICE2008**, 058 (2008)
29. Z. Fodor et al., PoS **LATTICE2008**, 066 (2008)
30. Z. Fodor et al., JHEP **0908**, 084 (2009)
31. Z. Fodor et al., Phys. Lett. B **681**, 353 (2009)
32. Z. Fodor et al., JHEP **0911**, 103 (2009)
33. Z. Fodor et al., PoS **LATTICE2009**, 055 (2009)
34. Z. Fodor et al., PoS **LATTICE2009**, 058 (2009)
35. A. J. Hietanen et al., JHEP **0905**, 025 (2009)
36. A. J. Hietanen, K. Rummukainen and K. Tuominen, Phys. Rev. D **80**, 094504 (2009)
37. A. Deuzeman, M. P. Lombardo and E. Pallante, Phys. Lett. B **670**, 41 (2008)
38. A. Deuzeman, M. P. Lombardo and E. Pallante, arXiv:0904.4662 [hep-ph].
39. A. Deuzeman, M. P. Lombardo and E. Pallante, arXiv:0911.2207 [hep-lat].
40. L. Del Debbio, A. Patella and C. Pica, arXiv:0805.2058 [hep-lat].
41. L. Del Debbio, B. Lucini, A. Patella, C. Pica and A. Rago, arXiv:0907.3896 [hep-lat].
42. F. Bursa et al., Phys. Rev. D **81**, 014505 (2010)
43. Y. Shamir, B. Svetitsky and T. DeGrand, Phys. Rev. D **78**, 031502 (2008)
44. T. DeGrand, Y. Shamir and B. Svetitsky, Phys. Rev. D **79**, 034501 (2009)
45. O. Machtey and B. Svetitsky, Phys. Rev. D **81**, 014501 (2010)
46. T. DeGrand and A. Hasenfratz, Phys. Rev. D **80**, 034506 (2009)
47. T. DeGrand, Phys. Rev. D **80**, 114507 (2009)
48. A. Hasenfratz, Phys. Rev. D **80**, 034505 (2009)
49. X. Y. Jin and R. D. Mawhinney, PoS **LATTICE2008**, 059 (2008)
50. X. Y. Jin and R. D. Mawhinney, PoS **LAT2009**, 049 (2009)
51. D. K. Sinclair and J. B. Kogut, arXiv:0909.2019 [hep-lat].
52. J. B. Kogut and D. K. Sinclair, arXiv:1002.2988 [hep-lat].
53. E. Bilgici *et al.*, Phys. Rev. D **80**, 034507 (2009)
54. E. Bilgici *et al.*, arXiv:0910.4196 [hep-lat].
55. M. Campostrini, P. Rossi and E. Vicari, Phys. Lett. B **349**, 499 (1995)
56. M. Creutz, Phys. Rev. D **21**, 2308 (1980).
57. A. Bazavov *et al.*, arXiv:0903.3598 [hep-lat].
58. C. W. Bernard *et al.*, Phys. Rev. D **62**, 034503 (2000)
59. U. M. Heller and F. Karsch, Nucl. Phys. B **251**, 254 (1985).
60. G. P. Lepage and P. B. Mackenzie, Phys. Rev. D **48**, 2250 (1993)
61. C. Morningstar and M. J. Peardon, Phys. Rev. D **69**, 054501 (2004)
62. Y. Aoki, Z. Fodor, S. D. Katz and K. K. Szabo, JHEP **0601**, 089 (2006)
63. Y. Aoki, Z. Fodor, S. D. Katz and K. K. Szabo, Phys. Lett. B **643**, 46 (2006)

Higgsinoless Supersymmetry and Hidden Gravity*

Michael L. Graesser
Theoretical Division T-2, Los Alamos National Laboratory, Los Alamos, NM 87545, USA

Ryuichiro Kitano and Masafumi Kurachi
Department of Physics, Tohoku University, Sendai 980-8578, Japan

We present a simple formulation of non-linear supersymmetry where superfields and partnerless fields can coexist. Using this formalism, we propose a supersymmetric Standard Model without the Higgsino as an effective model for the TeV-scale supersymmetry breaking scenario. We also consider an application of the Hidden Local Symmetry in non-linear supersymmetry, where we can naturally incorporate a spin-two resonance into the theory in a manifestly supersymmetric way. Possible signatures at the LHC experiments are discussed.

1. Introduction

Technicolor is an attractive idea in which electroweak symmetry is dynamically broken by a strong dynamics operating around the TeV energy scale.[2,3] The big hierarchy, $m_W \ll M_{\rm Pl}$, is elegantly explained by the very same reason as $\Lambda_{\rm QCD} \ll M_{\rm Pl}$. However, it is well-known that there are two phenomenological difficulties in this idea. One is that the electroweak precision measurements seem to prefer scenarios with a weakly coupled light Higgs boson.[4] Another is the difficulty in writing down the Yukawa interactions to generate fermion masses in the Standard Model.

After the LEP-I experiments, supersymmetry (SUSY) has become very popular as a natural scenario for the light weakly coupled Higgs boson. However, with the experimental bound on the lightest Higgs boson mass from the LEP-II experiments, parameters in the minimal SUSY standard model (MSSM) are required to be more and more fine-tuned, at least in the conventional scenarios.[5–7]

In this situation, it may be interesting to (re)consider a hybrid of technicolor and SUSY along the similar spirit of the early attempts of SUSY model building.[8,9] We assume that strong dynamics breaks SUSY at the multi–TeV energy scale (which we call the scale Λ), with electroweak symmetry breaking triggered by the dynamics through direct couplings between the Higgs field and the dynamical sector. This scenario has several virtues: (1) the Yukawa interactions can be written down by

*Based on the work in Ref. 1. The talk was given by R. Kitano.

assuming an existence of elementary Higgs fields in the UV theory, which mix with (or remain as) the Higgs field to break electroweak symmetry at low energy;[10–12] (2) the hierarchy problem, $\Lambda \ll M_{\rm Pl}$, is explained by dynamical SUSY breaking;[13] (3) one can hope that the little hierarchy, $m_W \sim m_h \ll \Lambda$, is explained by either SUSY or some other mechanisms such as the Higgs boson as the pseudo-Nambu-Goldstone particle in the strong dynamics;[14] (4) the cosmological gravitino problem is absent;[15] (5) one can expect additional contributions to the Higgs boson mass from the SUSY breaking sector, with which the mass bound from the LEP-II experiments can be evaded;[16] and (6), there is an interesting possibility that the LHC experiments can probe the SUSY breaking dynamics directly. SUSY is phenomenologically motivated from the point (1) (and also (6)) in this framework in addition to the connection to string theory.

Although the TeV-scale SUSY breaking scenario is an interesting possibility, an explicit model realizing this scenario will not be attempted here. In this paper, we take a less ambitious approach and construct an effective Lagrangian for the scenario without specifying (while hoping for the existence of) a UV theory responsible for SUSY breaking and its mediation.

In constructing the effective Lagrangian, we take the following as organizing principles: (1) the Lagrangian possesses non-linearly realized supersymmetry; (2) the quarks/leptons and gauge fields are only weakly coupled to the SUSY breaking sector, so that the typical mass splitting between bosons and fermions are $O(100)$ GeV (in other words, the matter and gauge fields are introduced as superfields which transform linearly under SUSY); and (3) the Higgs boson is introduced as a non-linearly transforming field because it is assumed to be directly coupled to the SUSY breaking sector. The Higgsino field is absent in the minimal model.

In this Higgsinoless model, the Higgs potential receives quadratic divergences from loop diagrams with the gauge interactions and the Higgs quartic interaction although the top-quark loops can be cancelled by the loops of the scalar top quarks as usual. The rough estimate of the correction to the Higgs boson mass is of the order of $(\alpha/4\pi)\Lambda^2$ and $(k/16\pi^2)\Lambda^2$ with k being the coupling constant of the Higgs quartic interaction. By comparing with the quadratic term needed for electroweak symmetry breaking, $m_H^2 = k\langle H\rangle^2/2$, naturalness suggests $\Lambda \lesssim 4\pi\langle H\rangle \sim$ (a few) $\times$ TeV. Precision electroweak constraints, on the other hand, obtained from the LEP-II and SLC experiments do not generically allow such a low scale without fine-tuning.[17] The dynamical scale may therefore have to be larger, $\Lambda \simeq O(6-10\text{ TeV})$. To obtain a light Higgs boson at this larger scale either requires fine-tuning, or some new weakly coupled new physics below Λ. (Or simply the Higgsino appears around a few TeV.) It is also true that the direct coupling to the dynamical sector generically gives the Higgs boson mass to be $O(\Lambda)$. We may therefore need to assume that the Higgs boson is somewhat special in the dynamics, e.g., a pseudo-Goldstone boson. In this paper we simply ignore the issue because its resolution depends on the UV completion, and here we only concerned with the effective theory below the TeV scale.

The stop potential also receives quadratic divergences, in this case from a loop diagram involving the Higgs boson (and proportional to λ_t^2). This divergence is not cancelled, simply because the Higgsinos are not present in the low energy theory. One therefore expects the stops to have a mass no smaller than a loop factor below the scale Λ.

The Lagrangian we construct needs to contain interaction terms among superfields and also partnerless fields such as the Higgs boson. Since these two kinds of fields are defined on different spaces – one superspace and the other the usual Minkowski space – one needs to convert the partnerless fields into superfields or vice-versa. One approach is to utilize established formulations for constructing superfields out of partnerless fields[18–20] where the Goldstino field is also promoted to a superfield. In this paper, we present a simple manifestly supersymmetric formulation where we do not try to convert partnerless fields into superfields, although it is totally equivalent to the known formalisms. The essence is to prepare two kinds of spaces: the superspace and the Minkowski space, on which superfields and partnerless fields are defined. By embedding the Minkowski space into the superspace by using a SUSY invariant map, one can define a Lagrangian density on a single spacetime. By using the formalism, one can write down a SUSY invariant Lagrangian, in particular the Yukawa interactions, only with a single Higgs field. We also find that the coupling constant of the Higgs quartic interaction can be a free parameter, unrelated to the gauge coupling constant. Therefore, the Higgs boson mass can be treated as a free parameter in this model.

As a related topic, a model in which the MSSM is only partly supersymmetric has been proposed in Ref.[21] There SUSY is broken explicitly at the Planck scale, and only the Higgs sector is remained to be supersymmetric which is made possible by a warped extra-dimension (or a conformal dynamics). Our philosophy is opposite to that and is, relatively speaking, closer to Ref.[22] by the same authors, where SUSY is broken on the IR brane (or equivalently by some strong dynamics at the $O(\text{TeV})$ scale).

As a possible signature of the TeV-scale dynamics, we construct a model "Hidden Gravity," which is an analogy of the Hidden Local Symmetry[23–27] in the chiral Lagrangian. The Hidden Local Symmetry is a manifestly chiral symmetric model to describe the vector resonance (the ρ meson) as the gauge boson of the hidden vectorial SU(2) symmetry (the unbroken symmetry of the chiral Lagrangian). When we apply this technique to SUSY, we obtain a supersymmetric Lagrangian for a massive spin-two field which is introduced as a graviton associated with a hidden general covariance because the unbroken symmetry is the Poincaré symmetry.

One can consistently incorporate the resonance as a non-strongly coupled field for a range of parameters and small range of energy. Indeed, we show that there is a sensible parameter region where we can perform a perturbative calculation of the resonant single-graviton production cross section. At energies not far above the graviton mass the effective theory becomes strongly coupled and incalculable. If the graviton is much lighter than the cut-off scale, new physics is required to complete the theory up to Λ, another direction not pursued here. We discuss signatures of this graviton scenario at the LHC.

2. Non-linear SUSY and invariant Lagrangian

In this section we present a method to construct a Lagrangian invariant under the non-linearly realized global supersymmetry. We will introduce the Higgs boson as a non-linearly transforming field (which we call a non-linear field) and also matter and gauge fields as superfields. We therefore need a formulation to write down a supersymmetric Lagrangian where both kinds of fields are interacting. Roček,[18] Ivanov and Kapustnikov,[19] and Samuel and Wess[20] have established a superfield formalism of non-linear SUSY by upgrading the Goldstino fermion and other non-linear fields to constrained superfields. (See[28] for a recent work.) Although the formalism is somewhat complicated, using superfields is motivated there as a first step towards embedding the theory into supergravity. As we are not interested in supergravity in this paper, we will use a simpler formalism where the Goldstino field remains as a non-linearly transforming field. We will also use results from earlier work by Ivanov and Kapustnikov[29] that establishes the correspondence between superfields and non-linear fields [a].

2.1. *Linear and non-linear SUSY*

Linearly realized SUSY is most simply formulated in the superspace. Under a group element,

$$g = e^{ic^a P_a + i\eta Q + i\bar{\eta}\bar{Q}}, \tag{1}$$

the superspace coordinate $(x^a, \theta_\alpha, \bar{\theta}_{\dot{\alpha}})$ transforms as[31]

$$x^a \to x^{a\prime} = x^a + c^a + \Delta^a(\eta, \theta), \quad \theta_\alpha \to \theta'_\alpha = \theta_\alpha + \eta_\alpha, \quad \bar{\theta}_{\dot{\alpha}} \to \bar{\theta}'_{\dot{\alpha}} = \bar{\theta}_{\dot{\alpha}} + \bar{\eta}_{\dot{\alpha}}, \tag{2}$$

where the Δ^a factor is defined by

$$\Delta^a(\eta, \xi) \equiv i\eta\sigma^a\bar{\xi} - i\xi\sigma^a\bar{\eta}. \tag{3}$$

On the other hand, the non-linear SUSY transformation under g in Eq. (1) is defined by Volkov and Akulov in Ref.[32] It is

$$\tilde{x}^\mu \to \tilde{x}^{\mu\prime} = \tilde{x}^\mu + c^\mu + \Delta^\mu(\eta, \lambda(\tilde{x})), \tag{4}$$

$$\lambda_\alpha(\tilde{x}) \to \lambda'_\alpha(\tilde{x}') = \lambda_\alpha(\tilde{x}) + \eta_\alpha, \tag{5}$$

$$\bar{\lambda}_{\dot{\alpha}}(\tilde{x}) \to \bar{\lambda}'_{\dot{\alpha}}(\tilde{x}') = \bar{\lambda}_{\dot{\alpha}}(\tilde{x}) + \bar{\eta}_{\dot{\alpha}}. \tag{6}$$

The fields λ and $\bar{\lambda}$ are the Goldstino fermion and its complex conjugate, respectively. (See[33] for a formalism for constructing the non-linear Lagrangian.)

---

[a]The generalization of this relationship to local supersymmetry can be found in Ref.[30]

We are interested in constructing a Lagrangian where non-linear fields (such as the Goldstino) and superfields coexist and they are interacting. (See[18,20,29,34–37] for constructing superfields out of non-linear fields.) However, as one can see two kinds of fields are living in different spaces x and $\tilde{x}$ which we cannot identify as the same space at this stage since their SUSY transformations are different. This situation can be cured by introducing a SUSY invariant delta function:

$$1 = \int d^4\tilde{x} \det X \delta^4(x^\mu - \tilde{x}^\mu - \Delta^\mu(\lambda(\tilde{x}), \theta)), \tag{7}$$

with

$$X_\mu{}^a = \eta_\mu{}^a - i\theta\sigma^a\partial_\mu\bar{\lambda} + i\partial_\mu\lambda\sigma^a\bar{\theta}. \tag{8}$$

The invariant action can now be written down as

$$\begin{aligned} S = &\int d^4x d^4\theta d^4\tilde{x} \det X \ \delta^4(x^\mu - \tilde{x}^\mu - \Delta^\mu(\lambda(\tilde{x}), \theta)) \\ &\times \mathcal{K}\left[\Psi(x, \theta, \bar{\theta}), \phi(\tilde{x}), \theta - \lambda, \bar{\theta} - \bar{\lambda}, \nabla_a\lambda, \nabla_a\bar{\lambda}, A, X, \cdots\right], \end{aligned} \tag{9}$$

where Ψ and ϕ represents arbitrary superfields and non-linear fields, respectively. The function $\mathcal{K}$ must be real and scalar under the general coordinate transformation about the $\tilde{x}$ coordinate. As an example of the invariant action, we can take $\mathcal{K} = \Psi(x, \theta, \bar{\theta})$ which gives the supersymmetric action,

$$S = \int d^4x d^4\theta \Psi(x, \theta, \bar{\theta}). \tag{10}$$

If we take $\mathcal{K} = \delta^4(\theta - \lambda) \cdot (-f^4/2)$, we obtain the Volkov-Akulov action[32]

$$S = -\frac{f^4}{2} \int d^4\tilde{x} \det A, \tag{11}$$

where the matrix A is

$$A_\mu{}^a = \eta_\mu{}^a - i\lambda\sigma^a\partial_\mu\bar{\lambda} + i\partial_\mu\lambda\sigma^a\bar{\lambda}. \tag{12}$$

The action contains the kinetic term for the Goldstino. The parameter f is the decay constant which represents the size of the SUSY breaking.

3. Higgsinoless SUSY

We are now ready to construct a Lagrangian. For quarks/leptons and gauge superfields, one can simply write down the MSSM Lagrangian in the superspace. Soft SUSY breaking terms can be written down by using delta functions $\delta^2(\theta - \lambda)$ and $\delta^4(\theta - \lambda)$:

$$\mathcal{K}_{\text{soft}} = -\delta^4(\theta - \lambda) \cdot m_\Psi^2 \Psi^\dagger \Psi \tag{13}$$

and

$$\mathcal{W}_{\text{soft}} = -\delta^2(\theta - \lambda) \cdot \frac{m_{1/2}}{2} W^\alpha W_\alpha. \tag{14}$$

These are the same as the spurion method for the soft SUSY breaking terms. The appearance of the Goldstino interactions makes these terms manifestly supersymmetric. One can also add hard breaking terms by using covariant derivatives. We assume that such soft and hard breaking terms are somewhat suppressed because the quarks/leptons and gauge fields are not participating the SUSY breaking dynamics.

We introduce the Higgs field as a non-linear field on the $\tilde{x}$ space, $h(\tilde{x})$, motivated by an assumption that the SUSY breaking dynamics at the cut-off scale Λ has something to do with the origin of electroweak symmetry breaking. The way to construct interaction terms has been discussed already in the previous subsection. The Yukawa interactions for up-type quarks are

$$\mathcal{K}_{\text{up}} = \delta^4(\theta - \lambda) \left[y_u^{ij} h(\tilde{x}) \cdot \left(\frac{1}{2} D^2_{(\text{cov})} U_j^c Q_i \right) \right]. \tag{15}$$

For down-type quarks and leptons,

$$\begin{aligned} \mathcal{K}_{\text{down}} &= \delta^4(\theta - \lambda) \\ &\times \left[y_d^{ij} h(\tilde{x})^\dagger e^{-2gV} \left(\frac{1}{2} D^2_{(\text{cov})} D_j^c Q_i \right) + y_e^{ij} h(\tilde{x})^\dagger e^{-2gV} \left(\frac{1}{2} D^2_{(\text{cov})} E_j^c L_i \right) \right]. \end{aligned} \tag{16}$$

Here we have used the covariant derivative:

$$D^2_{(\text{cov})} \equiv e^{2gV} D^2 e^{-2gV}. \tag{17}$$

It is not necessary to introduce two kinds of Higgs fields for the Yukawa interactions. The A-terms can also be written down by taking

$$\mathcal{W}_{\text{A}} = \delta^2(\theta - \lambda) \left[A_u^{ij} h(\tilde{x}) \cdot (U_j^c Q_i) \right], \tag{18}$$

and

$$\mathcal{K}_{\text{A}} = \delta^4(\theta - \lambda) \left[A_d^{ij} h(\tilde{x})^\dagger e^{-2gV} (D_j^c Q_i) + A_e^{ij} h(\tilde{x})^\dagger e^{-2gV} (E_j^c L_i) \right]. \tag{19}$$

The kinetic term and potential for the Higgs boson can be written as

$$\mathcal{K}_{\text{kin.}} = \delta^4(\theta - \lambda) \left[(D_a \phi(\tilde{x}))^\dagger e^{-2gV} D^a \phi(\tilde{x}) \right], \tag{20}$$

and

$$\mathcal{K}_{\text{pot.}} = \delta^4(\theta - \lambda) \left[-m^2 \phi^\dagger(\tilde{x}) e^{-2gV} \phi(\tilde{x}) - \frac{k}{4} \left(\phi^\dagger(\tilde{x}) e^{-2gV} \phi(\tilde{x}) \right)^2 \right]. \tag{21}$$

The derivative in the kinetic term is

$$D_a \equiv \nabla_a - ig\mathcal{A}_a + g(\nabla_a \lambda^\alpha)\mathcal{A}_\alpha, \tag{22}$$

where

$$g\mathcal{A}_a(x, \theta, \bar{\theta}) \equiv \frac{1}{4} \bar{D} e^{2gV} \bar{\sigma}_a D e^{-2gV}, \tag{23}$$

and

$$gA_\alpha(x,\theta,\bar{\theta}) \equiv e^{2gV} D_\alpha e^{-2gV}. \tag{24}$$

Since the quartic coupling of the Higgs boson in Eq. (21) is a free parameter, the Higgs boson mass is not related to the Z-boson mass. It is not a very obvious result that we could write down a Lagrangian with a single Higgs boson with the enlarged gauge invariance. For example, in Ref.[35] it has been necessary to introduce an extra Higgs boson, and that is claimed to be a general requirement for constructing a realistic model with non-linear SUSY.

4. Hidden gravity

A SUSY transformation in the $\tilde{x}$ space is realized as a local coordinate transformation in Eq. (4). This local translation allows us to introduce a metric in the $\tilde{x}$ space having a local transformation law under the global SUSY. This provides a description of a composite spin-two field[b] in the SUSY breaking dynamics analogous to the ρ meson in QCD. We further elaborate on this comparison towards the end of this section.

Specifically, we introduce a second "metric" whose transformation under g is

$$g_{\mu\nu}(\tilde{x}) \to g'_{\mu\nu}(\tilde{x}') = \frac{\partial \tilde{x}^\rho}{\partial \tilde{x}'^\mu}\frac{\partial \tilde{x}^\sigma}{\partial \tilde{x}'^\nu} g_{\rho\sigma}(\tilde{x}), \tag{25}$$

where $\tilde{x}'$ is given in Eq. (4). Note that this is a *global* SUSY transformation, and one should not be confused with the actual general coordinate transformation on the x-space. The space-time is always flat. The deviation of $g_{\mu\nu}$ from the Minkowski metric describes the spin-two field.

The invariant action having the Fierz-Pauli form[42] is

$$S = \int d^4\tilde{x}\left[-\frac{f^4}{2}\det A - \frac{m_{\rm P}^2}{2}\sqrt{g}R(g) - \frac{m_{\rm P}^2 m^2}{8}\sqrt{g}g^{\mu\nu}g^{\alpha\beta}\left(H_{\mu\alpha}H_{\nu\beta} - H_{\mu\nu}H_{\alpha\beta}\right)\right] \tag{26}$$

where

$$H_{\mu\nu} = g_{\mu\nu} - G_{\mu\nu} \tag{27}$$

and

$$G_{\mu\nu} = A_\mu{}^a A_\nu{}^b \eta_{ab}. \tag{28}$$

is a covariant tensor, defined previously. The $H_{\mu\nu}$ field is therefore a SUSY covariant tensor. The scale $m_{\rm P}$ is a mass parameter of $O(\text{TeV})$, unrelated to the four-dimensional Planck mass $G_N^{-1/2}$ of Einstein gravity. With

$$g_{\mu\nu} = \eta_{\mu\nu} + \frac{2}{m_{\rm P}} h_{\mu\nu}\ , \tag{29}$$

[b] An attempt to describe a spin-two resonance in QCD as a massive graviton can be found in Ref.,[38] whose supergravity extension is discussed in Ref.[39] A more ambitious attempt to formulate Einstein gravity as a composite of the Goldstino fermions can be found in Ref.[40] The appearance of a massive bound-state graviton in open string field theory can be found in Ref.[41]

one has

$$H_{\mu\nu} = \frac{2}{m_{\rm P}} h_{\mu\nu} + \left(i\lambda\sigma_\mu\partial_\nu\bar{\lambda} - i\partial_\nu\lambda\sigma_\mu\bar{\lambda} + (\mu \leftrightarrow \nu)\right) + (\text{four-fermion terms}) \quad (30)$$

Note the relative coefficient (of -1) between the two terms appearing in the definition of $H_{\mu\nu}$ is fixed by requiring that the Fierz-Pauli mass term not introduce a tadpole for the graviton. The last term in the action gives a mass m to the spin-two field in a global SUSY invariant way.

In the chiral Lagrangian of QCD one can construct SU(2) vector- and axial-type 1-forms $j_{V,A}$ out of the pion fields. A chirally invariant Lagrangian can be constructed only out of j_A, since j_V transforms inhomogeneously under the chiral SU(2). By introducing an $\mathrm{SU}(2)_V$ vector boson V_μ, the term $\mathrm{Tr}\tilde{j}_V\tilde{j}_V$, with $\tilde{j}_V = j_V - V$, is made chirally invariant and can be added to the action. This gives a mass to the vector boson, but no kinetic term; it can be trivially integrated out. The key assumption of[23–27] is that this vector boson is dynamical (and describes the ρ vector meson). The action obtained in this way coincides with the spontaneously broken $\mathrm{SU}(2)_V$ gauge theory in the unitary gauge, which obviously can have a sensible description up to some high energy scale.

The analogy of the massive spin-two field as formulated here to the Hidden Local Symmetry (HLS) of QCD can now be drawn more closely, though imprecisely. Here the Maurer-Cartan 1-forms $A_\mu{}^a$ are analogous to the j_V in QCD. By introducing a spin-two field having a local and inhomogeneous transformation under the global symmetry, it is then possible to introduce the 1-forms into the action, in the form of the Fierz-Pauli mass term. Further assuming a Ricci scalar term in the action for the spin-two field is equivalent to the physical assumption in HLS that the gauge boson is dynamical.

4.1. *Perturbative unitarity*

A question to be addressed here is whether this new spin-two resonance can be consistently introduced in a weakly coupled regime, in which perturbative calculations make sense at energies of $O(m)$. We first check this by looking at the elastic scattering of two Goldstinos with the same helicity. Then we require that the spin-two field is not strongly coupled at threshold.

The s-wave amplitude $\mathcal{M}_{\lambda\lambda}$ for $\lambda\lambda \to \lambda\lambda$ receives contributions from both the Volkov-Akulov action and the action for the spin-two field. Specifically, one obtains

$$\mathcal{M}_{\lambda\lambda}^{\ell=0} = \frac{1}{16\pi}\frac{s^2}{f^4} - \frac{5}{32\pi}\frac{m_{\rm P}^2 m^2 s^2}{f^8} - \frac{1}{16\pi}\frac{m_{\rm P}^2 m^4 s}{f^8}\left[3 - \left(4 + \frac{3m^2}{s}\right)\ln\left(1 + \frac{s}{m^2}\right)\right]. \quad (31)$$

The first term in the RHS comes from the Goldstino self-interactions, while the remaining terms come from graviton exchanges. There is no parameter to control the relative sign of the two contributions. One can see that two $O(s^2)$ terms in

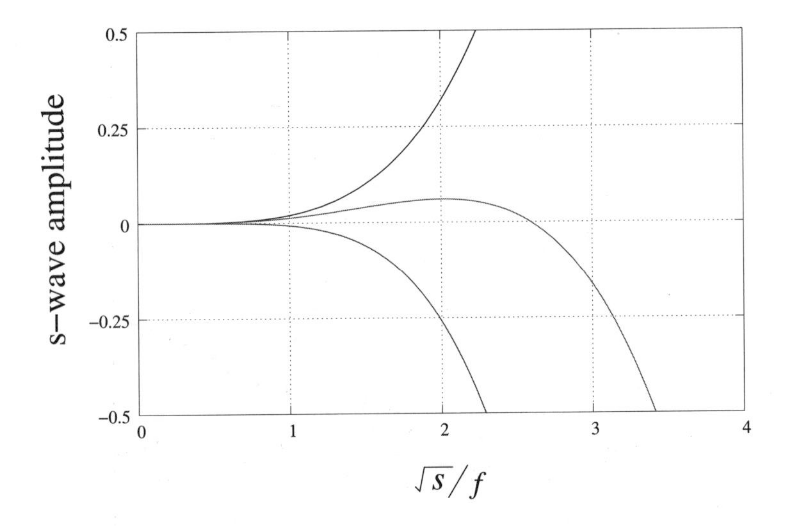

Fig. 1. The magnitude of the s-wave amplitude as a function of $\sqrt{s}/f$. Upper and lower curves represent contributions from $(-f^4/2)\det A$ and $(H_{\mu\nu}H^{\mu\nu} - H^\mu_\mu H^\nu_\nu)$ terms, and the middle curve represents the total amplitude. Here, $m_\mathrm{P} = 0.7f$ and $m = 1.2f$ are used as an example.

Eq. (31) always have an opposite sign, and thus the graviton contribution partially cancels the growth of the amplitude. The magnitude of the s-wave amplitude is plotted in Fig. 1 as a function of $\sqrt{s}/f$. The upper and lower curves represent contributions from the first and second row of the RHS of Eq. (31), respectively. The middle curve represents the total amplitude. The parameters $m_\mathrm{P} = 0.7f$ and $m = 1.2f$ are chosen for illustration.

We define the perturbative-unitarity-violation scale E_*, by the energy where the tree level s-wave amplitude of $\lambda\lambda$ scattering reaches the value 0.5. In the case of the example depicted in Fig. 1, the pure Goldstino amplitude (i.e., the upper curve) gives $E_* \sim 2.2f$. The contribution of the spin-two particle to this amplitude has the opposite sign (lower curve). This contribution partially cancels the pure Goldstino amplitude, delaying the onset of the strong coupling regime. For the parameter values used in Fig. 1, one finds $E_* \sim 3.4f$.

Combined with the cut-off scale of the massive spin-two theory, we show in Fig. 2 the parameter region in which the spin-two resonance can be consistently incorporated in the effective theory.

4.2. *Phenomenological signatures*

We estimate the cross section for the spin-two resonance at the LHC experiments. The cross section for $pp \to h_{\mu\nu} \to hh$ are shown in Fig. 3 for a choice of parameters. The spin-two couplings to the Higgs boson and the gluon are chosen to be $O(1)$. We note that the production cross-section is sensitive to the spin-two coupling δ_g to gluons (which is taken to be 0.1 in the figure), so that larger rates are possible.

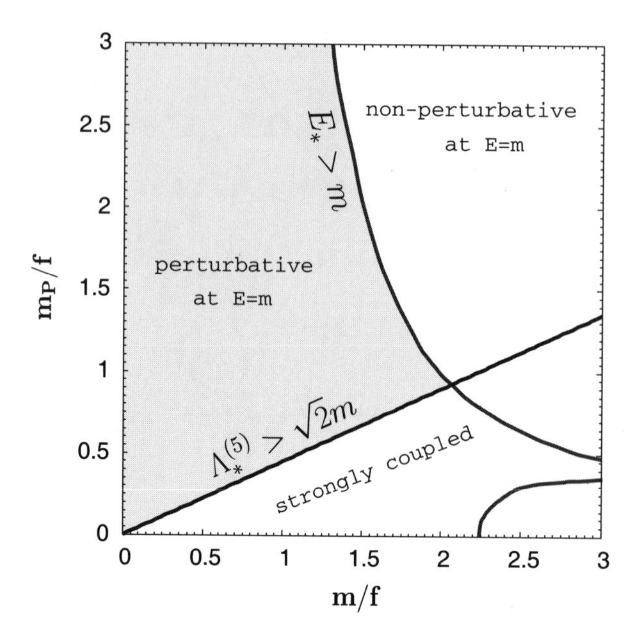

Fig. 2. The contour of $E_* = m$ and $\Lambda_*^{(5)} = \sqrt{2}m$ in the parameter space of m/f and m_P/f. In the region to the left of the thick lines, both $E_* > m$ and $\Lambda_*^{(5)} > \sqrt{2}m$ are satisfied.

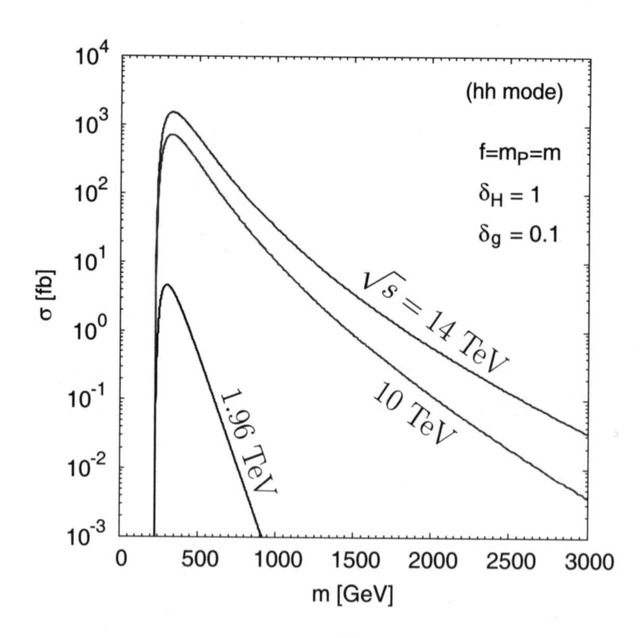

Fig. 3. The cross section (hh mode) as functions of the graviton mass for 1.96, 10 and 14 TeV center-of-mass energies. Using CTEQ6M parton distribution functions[43] and setting $m_h = 114$ GeV. The relation $f = m_P = m$ is kept fixed as m is varied.

5. Summary

If SUSY is broken near the TeV scale by strong dynamics then there may be composites that can be accessed at the LHC. It is desirable to have a formalism for writing the effective theory describing the interactions between the matter or gauge fields and the composites, especially in the situation where the matter and gauge fields are not participants of these dynamics. The challenge then in this case is that the matter and gauge fields appear in linearly realized multiplets, whereas the composites do not. We have presented a formulation of non-linearly realized global SUSY in which this can be done.

As an application, we consider two scenarios. In both, the Higgs boson is such a composite. No Higgsinos are present in the low-energy theory. We show that it is possible to write down SUSY invariant Yukawa couplings and A terms, despite the presence of only one Higgs boson.

Next, we further suppose that the composites include a light spin-two field in addition to the Higgs boson. The construction of the SUSY invariant action is in analogy to the Hidden Local Symmetry of chiral dynamics, where the ρ meson is a massive vector boson of a hidden local $SU(2)_V$ symmetry. Here though the hidden symmetry is local Poincaré. Thus we find that a massive graviton can naturally be incorporated in a theory with an enlarged spacetime symmetry, which in this case is global SUSY.

Some phenomenological signatures are discussed. Unlike a generic Kaluza-Klein graviton, the spin-two particle does not couple to the stress-energy tensor. Instead its interactions with matter and gauge fields are constrained only by gauge, Lorentz and non-linear SUSY invariance. Its dominant decay mode is to Goldstinos (invisible), electroweak gauge bosons and the Higgs boson. Search strategies to find boosted Higgs bosons are particularly interesting for this scenario. Vector boson fusion producing the spin-two particle occurs and may be of experimental interest. Rare decays to di-photons also occur but the rate is more model-dependent. Search strategies to find the Standard Model Higgs boson are therefore simultaneously sensitive to finding the spin-two particle.

This scenario has the usual low-energy SUSY experimental signatures (without Higgsinos), while in addition possessing signatures of both large[44] and warped extra dimensions[45] - monojets (ADD) and a spin-two resonance (RS) - even though there is no extra dimension. The discovery of SUSY particles and a single spin-two resonance is not sufficient to claim discovery of an extra (supersymmetric) dimension. It may just be due to four-dimensional strong SUSY breaking dynamics.

Acknowledgments

RK thanks the organizers of the SCGT 2009 workshop. He especially thanks Prof. Koichi Yamawaki for his hospitality in Nagoya. The work of MG is supported by the U.S. Department of Energy at Los Alamos National Laboratory under Contract No. DE-AC52-06NA25396. The work of RK is supported in part by the Grant-in-Aid for Scientific Research 21840006 of JSPS.

References

1. M. L. Graesser, R. Kitano and M. Kurachi, JHEP **0910**, 077 (2009) [arXiv:0907.2988 [hep-ph]].
2. S. Weinberg, Phys. Rev. D **13**, 974 (1976); Phys. Rev. D **19**, 1277 (1979); L. Susskind, Phys. Rev. D **20**, 2619 (1979).
3. See for reviews, e.g., E. Farhi and L. Susskind, Phys. Rept. **74**, 277 (1981); C. T. Hill and E. H. Simmons, Phys. Rept. **381**, 235 (2003) [Erratum-ibid. **390**, 553 (2004)] and references therein.
4. M. E. Peskin and T. Takeuchi, Phys. Rev. Lett. **65**, 964 (1990); Phys. Rev. D **46**, 381 (1992); G. Altarelli and R. Barbieri, Phys. Lett. B **253**, 161 (1991); G. Altarelli, R. Barbieri and S. Jadach, Nucl. Phys. B **369**, 3 (1992) [Erratum-ibid. B **376**, 444 (1992)].
5. R. Barbieri and A. Strumia, arXiv:hep-ph/0007265.
6. R. Kitano and Y. Nomura, Phys. Rev. D **73**, 095004 (2006) [arXiv:hep-ph/0602096].
7. G. F. Giudice and R. Rattazzi, Nucl. Phys. B **757**, 19 (2006) [arXiv:hep-ph/0606105].
8. M. Dine, W. Fischler and M. Srednicki, Nucl. Phys. B **189**, 575 (1981).
9. S. Dimopoulos and S. Raby, Nucl. Phys. B **192**, 353 (1981).
10. M. A. Luty, J. Terning and A. K. Grant, Phys. Rev. D **63**, 075001 (2001) [arXiv:hep-ph/0006224].
11. H. Murayama, arXiv:hep-ph/0307293.
12. R. Harnik, G. D. Kribs, D. T. Larson and H. Murayama, Phys. Rev. D **70**, 015002 (2004) [arXiv:hep-ph/0311349].
13. E. Witten, Nucl. Phys. B **188**, 513 (1981).
14. G. F. Giudice, C. Grojean, A. Pomarol and R. Rattazzi, JHEP **0706**, 045 (2007) [arXiv:hep-ph/0703164].
15. M. Viel, J. Lesgourgues, M. G. Haehnelt, S. Matarrese and A. Riotto, Phys. Rev. D **71**, 063534 (2005) [arXiv:astro-ph/0501562]; A. Boyarsky, J. Lesgourgues, O. Ruchayskiy and M. Viel, JCAP **0905**, 012 (2009) [arXiv:0812.0010 [astro-ph]].
16. A. Brignole, J. A. Casas, J. R. Espinosa and I. Navarro, Nucl. Phys. B **666**, 105 (2003) [arXiv:hep-ph/0301121]; M. Dine, N. Seiberg and S. Thomas, Phys. Rev. D **76**, 095004 (2007) [arXiv:0707.0005 [hep-ph]].
17. R. Barbieri and A. Strumia, Phys. Lett. B **462**, 144 (1999) [arXiv:hep-ph/9905281]; R. Barbieri, A. Pomarol, R. Rattazzi and A. Strumia, Nucl. Phys. B **703**, 127 (2004) [arXiv:hep-ph/0405040].
18. M. Roček, Phys. Rev. Lett. **41**, 451 (1978).
19. E. A. Ivanov and A. A. Kapustnikov, J. Phys. G **8** (1982) 167.
20. S. Samuel and J. Wess, Nucl. Phys. B **221**, 153 (1983).
21. T. Gherghetta and A. Pomarol, Phys. Rev. D **67**, 085018 (2003) [arXiv:hep-ph/0302001].
22. T. Gherghetta and A. Pomarol, Nucl. Phys. B **586**, 141 (2000) [arXiv:hep-ph/0003129].
23. M. Bando, T. Kugo, S. Uehara, K. Yamawaki and T. Yanagida, Phys. Rev. Lett. **54**, 1215 (1985).
24. M. Bando, T. Kugo and K. Yamawaki, Nucl. Phys. B **259**, 493 (1985).
25. M. Bando, T. Fujiwara and K. Yamawaki, Prog. Theor. Phys. **79**, 1140 (1988).
26. M. Bando, T. Kugo and K. Yamawaki, Phys. Rept. **164**, 217 (1988).
27. M. Harada and K. Yamawaki, Phys. Rept. **381**, 1 (2003) [arXiv:hep-ph/0302103].
28. Z. Komargodski and N. Seiberg, arXiv:0907.2441 [hep-th].
29. E. A. Ivanov and A. A. Kapustnikov, JINR-E2-10765, (1977); E. A. Ivanov and A. A. Kapustnikov, J. Phys. A **11**, 2375 (1978).

30. E. A. Ivanov and A. A. Kapustnikov, Phys. Lett. B **143**, 379 (1984); E. A. Ivanov and A. A. Kapustnikov, Nucl. Phys. B **333**, 439 (1990).
31. A. Salam and J. A. Strathdee, Fortsch. Phys. **26**, 57 (1978).
32. D. V. Volkov and V. P. Akulov, JETP Lett. **16**, 438 (1972) [Pisma Zh. Eksp. Teor. Fiz. **16**, 621 (1972)].
33. T. E. Clark and S. T. Love, Phys. Rev. D **70**, 105011 (2004) [arXiv:hep-th/0404162].
34. T. Uematsu and C. K. Zachos, Nucl. Phys. B **201**, 250 (1982).
35. S. Samuel and J. Wess, Nucl. Phys. B **233**, 488 (1984).
36. M. A. Luty and E. Ponton, Phys. Rev. D **57**, 4167 (1998) [arXiv:hep-ph/9706268].
37. I. Antoniadis and M. Tuckmantel, Nucl. Phys. B **697**, 3 (2004) [arXiv:hep-th/0406010].
38. C. J. Isham, A. Salam and J. A. Strathdee, Phys. Rev. D **3**, 867 (1971).
39. A. H. Chamseddine, A. Salam and J. A. Strathdee, Nucl. Phys. B **136** (1978) 248.
40. J. Lukierski, Phys. Lett. B **121**, 135 (1983).
41. W. Siegel, Phys. Rev. D **49**, 4144 (1994) [arXiv:hep-th/9312117].
42. M. Fierz and W. Pauli, Proc. Roy. Soc. Lond. A **173**, 211 (1939).
43. S. Kretzer, H. L. Lai, F. I. Olness and W. K. Tung, Phys. Rev. D **69**, 114005 (2004) [arXiv:hep-ph/0307022].
44. N. Arkani-Hamed, S. Dimopoulos and G. R. Dvali, Phys. Lett. B **429**, 263 (1998) [arXiv:hep-ph/9803315]; I. Antoniadis, N. Arkani-Hamed, S. Dimopoulos and G. R. Dvali, Phys. Lett. B **436**, 257 (1998) [arXiv:hep-ph/9804398]; N. Arkani-Hamed, S. Dimopoulos and G. R. Dvali, Phys. Rev. D **59**, 086004 (1999) [arXiv:hep-ph/9807344]; G. F. Giudice, R. Rattazzi and J. D. Wells, Nucl. Phys. B **544**, 3 (1999) [arXiv:hep-ph/9811291].
45. L. Randall and R. Sundrum, Phys. Rev. Lett. **83**, 3370 (1999) [arXiv:hep-ph/9905221].

The Latest Status of LHC and the EWSB Physics

S. Asai
Physics Department, University of Tokyo, Hongo Bunkyo-ku Tokyo 133-033 Japan
E-mail: Shoji.Asai@cern.ch

The latest status of LHC and the performances of ATLAS and CMS detectors are summarized in the first part. Physics potential to solve the origin of the ElectroWeak Symmetry Breaking is summarized in the 2nd part, focusing especially on two major scenarios, (1) the light Higgs boson plus SUSY and (2) the Strong Coupling Gauge Theory. Both ATLAS and CMS detectors have the excellent potential to discover them, and we can perform crucial test on the ElectroWeak Symmetry Breaking.

Keywords: LHC, Higgs, SUSY, technicolor.

1. Introduction

The most urgent and important topics of the particle physics is to understand the electroweak symmetry breaking and the origin of "Mass". These are the main purpose of the Large Hadron Collider (LHC). The LHC accelerator is composed with 1232 units of the 8.4 T superconducting dipole magnet (length is 14.2 m each) and 392 units of the quadruple magnet. They have already be arranged in the 26.6 km circumference tunnel. Protons will be accelerated upto 7 TeV and collide each other. Thus the center-of-mass energy($\sqrt{s}$) of pp-system is 14 TeV. The first physics collisions have been observed in 2009 with the lower $\sqrt{s}$'s of 900 GeV and 2.36 TeV. The design luminosity is $10^{34} cm^{-2} s^{-1}$, which corresponds to 100 fb^{-1} per year, will be achieved at 2015.

The production cross-sections are expected to be large at LHC for the various high p_T and high mass elementary processes as listed in Table 1, since gluon inside partons can contribute remarkably. Furthermore, because the LHC provides the high luminosity of 10–100 fb^{-1} per year, large numbers of the interesting events will be observed. LHC has an excellent potential to produce high mass particles, for example, the top quark, the Higgs boson, SUSY particles.

2. Status and future plan of LHC

2.1. *Incident on 19th September 2008*

The LHC has started on 10th September 2008. Protons were injected with the energy of 450GeV and were RF-captured successfully to store in the LHC ring. But there were the bad connections of the splicing in the copper bar between magnet

Table 1. Production cross-section and event numbers for major high p_T and high mass processes with an integrated luminosity of 10 fb^{-1}.

	σ (pb at 14TeV)	Event number (L=10 fb^{-1})
$W^{\pm} \to \ell^{\pm}\nu$	1.7×10^5	$\sim 10^9$
$Z^0 \to \ell^+\ell^-$	2.5×10^4	$\sim 10^8$
$t\bar{t}$	830	$\sim 10^7$
jj $p_T>$ 200GeV	10^5	$\sim 10^9$
SM Higgs (M=115GeV)	35	$\sim 10^5$
$\tilde{g}\tilde{g}$ (M=500GeV)	~ 100	$\sim 10^6$
(M=1TeV)	~ 1	$\sim 10^4$

units. The copper bar becomes the path of intense current (~10000 A), when the superconducting in the magnet is quenched. Since the bad connections have the resistance of ~50$\mu\Omega$, the temperature goes up quickly and the liquid He boils up. Total 6 ton of the liquid He leaks from the pipe and 53 units of the magnets have been destroyed.

To repair these magnets and to add the various safety systems for the quench (improving the quench detectors, increasing the safety valves, and making the magnet support stronger) it has taken more than one year.

2.2. *Restart at November 2009*

On 20th November 2009, the proton beam makes a turn successfully in the LHC rings and are stored in the beam pipe with the RF power. This test have been performed for both proton beams separately, and the LHC is back to the status before the incident.

On 23rd November, the first collision at $\sqrt{s}$ =900 GeV has been observed in the all detectors and the events observed at the ATLAS and CMS detectors are shown in Fig.1. These events are the soft proton-proton collision, called as "minimum bias". Since the cross-section of minimum bias at 900 GeV is very huge, $\sigma \sim 58$ mb, large number of the events were observed even with the lower luminosity of 10^{24}cm^{2}s^{-1}. Many soft π's whose p_T less than 2 GeV are emitted in the Minimum bias events, and these are useful to check the performance of the detectors as mentioned in Sec.3.

2.3. *Acceleration and collision at $\sqrt{s}$=2.36TeV*

On 30th November, both beams are accelerated upto 1.18 TeV. Since the beam just after injected has large emittance, the first step of the acceleration is important to control beam. The beam emittance decrease after acceleration, so this success shows that LHC clears the first critical point.

Fig. 1. The collision events at $\sqrt{s}$=900 GeV observed at ATLAS and CMS

On 8th December, we have the first collision at $\sqrt{s}$=2.36TeV, which is the highest collision energy in the world. The events observed at the ATLAS and CMS detectors are shown in Fig.2, and these events are multi-jet QCD. The production cross-section and jet-properties measured at both 900 GeV and 2.36 TeV are consistent with the Monte-Calro predictions, still preliminary.

Fig. 2. The collision events at $\sqrt{s}$=2.36 TeV observed at ATLAS and CMS

2.4. *Future plan*

The plan of LHC after 2010 is summarized here. There is the high risk when the LHC is operated at $\sqrt{s}$=10 TeV without the repair of the bad connection of the splicing mentioned in Sec.2.1. Magnet units should be warm in order to repair the connections, and it takes much time more of one year. Thus the LHC schedule is determined as listed in Table 2. LHC will be operated at $\sqrt{s}$=7 TeV at both 2010 and 2011, and the expected integrated luminosity (L) is about 1 fb^{-1} in the forthcoming two years. Physics potentials with these $\sqrt{s}$ and L are summarized in Sec.4.

Table 2. LHC schedule.

Year	ECM	expected Luminosity
2010	7 TeV	200 pb^{-1}
2011	7 TeV	$\sim 1fb^{-1}$
2012	Long shut down	
2013	13-14 TeV	$\sim afewfb^{-1}$
2014	14 TeV	$\sim 10fb^{-1}$

3. Status and performance of ATLAS/CMS detector

Two general-purpose experiments exist, ATLAS and CMS, at LHC. The ATLAS (A Toroidal LHC Apparatus) detector measures 22 m high, 44 m long, and weight 7,000 tons. The characteristics of the ATLAS detector are summarized as follows

- Precision inner tracking system is constituted with pixel, strip of silicon and TRT with 2 T solenoidal magnet. Good performance is expected on the B-tagging and the γ-conversion tagging.
- Liquid Argon electromagnetic calorimeter has fine granularity for space resolution and longitudinal segmentation for fine angular resolution and particle identifications. It has also good energy resolution of about 1.5% for e/γ with energy of 100 GeV.
- Large muon spectrometer with air core toroidal magnet will provide a precise measurement on muon momenta(about 2% for 100 GeV-μ) even in the forward region.

The CMS (Compact Muon Solenoid) detector measures 15 m high, 21 m long, and weight 12,500 tons, with the following features

- Precise measurement on high p_T track is performed with the strong 4 T solenoidal magnet.
- $PbW0_4$ crystal electromagnetic calorimeter is dedicated for $\mathrm{H}^0_{SM} \to \gamma\gamma$.
- High purity identification and precise measurement are expected on μ tracks using the compact muon system.

Both detectors are ready to observe the collision events well and the various performance (for example, tracking efficiency, resolution of p_T, resolution of the energy deposited on the calorimeters) are checked with the collision data at 900 GeV. Some performance plots are shown in Fig.3; Fig.3(a) shows the reconstructed pair of tracks with the displaced vertex for $K_S \to \pi^+\pi^-$, and the distribution of the invariant mass of the reconstructed tracks are shown in Fig.3(b). The clear peak is observed at K_s mass. These are obtained at the ATLAS detector. $\pi^0 \to \gamma\gamma$ is good process to check the EM calorimeters, and the invariant mass distribution of two γ's is shown in Fig.3(c). Clear peak is observed at π^0 mass and the good resolution of σ=10 MeV is obtained at the CMS detector, since the $PbW0_4$ crystal scintillators are used in CMS detector. The $\not{E}_T$ is the vector opposite to the sum of the all energy deposited on the calorimeters and muon tracks, and is crucial variable to detect neutrino or to discovery SUSY. The $\not\!\!E_T$ resolution is related to the energy sum deposited on the calorimeters, and it is also checked as shown in Fig.3(d). The resolution is proportional to $0.5 \times \sqrt{\Sigma E_T}$ and the data is consistent with the MC simulation.

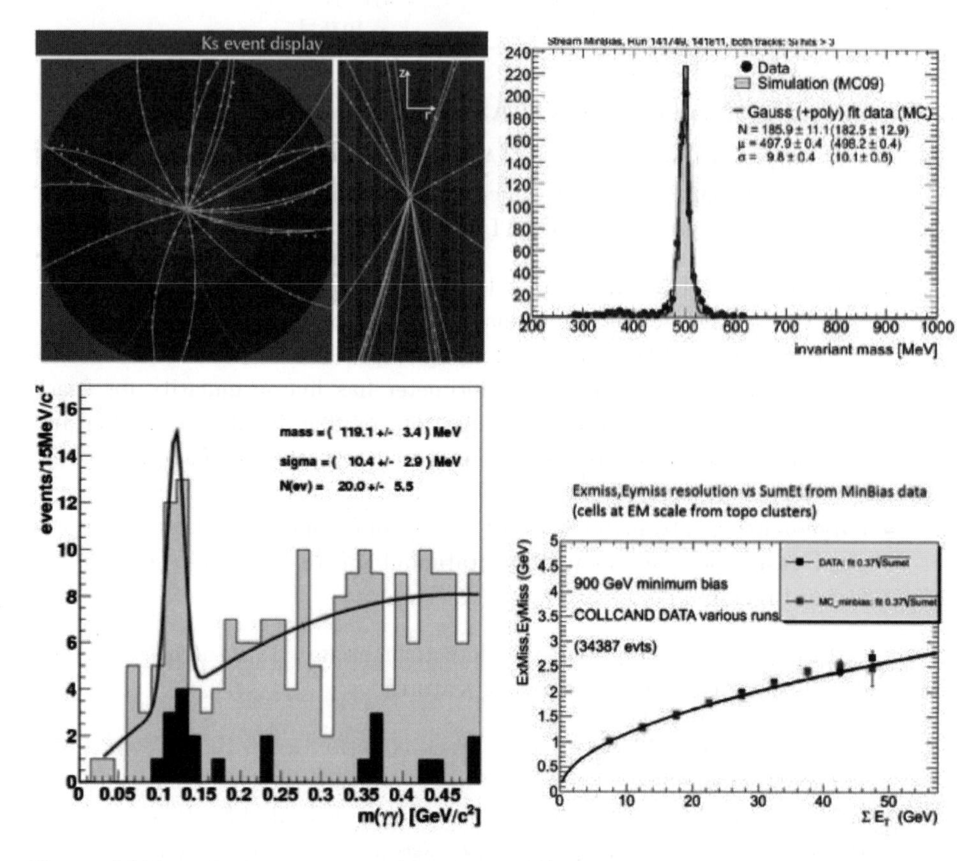

Fig. 3. (a) Event display observed at ATLAS with $\sqrt{s}$=900GeV. The displaced vertex has clearly been reconstructed for $K_S \rightarrow \pi^+\pi^-$. (b) The invariant mass distribution of $\pi^+\pi^-$ pair (ATLAS) (c) The invariant mass distribution of two photons $\pi^0 \rightarrow \gamma\gamma$ (CMS) (d) The resolution of $\not\!\!E_T$ as a function of ΣE_T

4. EWSB scenario 1: Light Higgs and SUSY

4.1. $\sqrt{s}$ *dependence of the production cross-section* σ

The studies shown in this note are based on $\sqrt{s}$=10 or 14 TeV. In order to extrapolate these results into $\sqrt{s}$=7 TeV, $\sqrt{s}$-dependences of the production cross-section σ are shown in Figs.4 (a) Higgs (b) SUSY. The production cross-sections at 7 TeV is smaller by factor 3 than σ at 10 TeV, and by factor 5–8 than σ at 14 TeV. The difference becomes larger for the heavier particles.

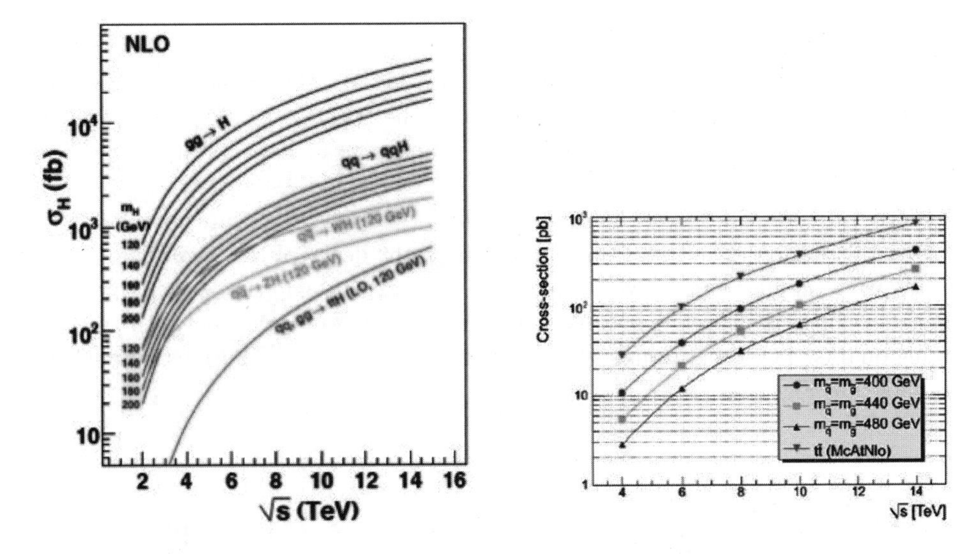

Fig. 4. The production cross-section as a function $\sqrt{s}$: (a) The light SM Higgs whose mass 120 to 200 GeV (b) The SUSY particles (gluino, squark masses 400-480 GeV)

4.2. *The Light SM Higgs boson*

A discovery of one or several Higgs bosons will give a definite experimental proof of the breaking mechanism of the Electroweak gauge symmetry, and detail studies of the Yukawa couplings between the Higgs boson and various fermions will give insights on the origin of lepton and quark masses. The mass of the SM Higgs boson itself is not theoretically predicted, but it's upper limit is considered to be about 160 GeV(95%C.L.) from the Electroweak precision measurements and the direct search at the Tevatron. Thee lower limit of the Higgs boson mass is set at 114 GeV(95%C.L.) by direct searches at LEP. The Higgs boson should exist in the narrow mass range of 114–160 GeV, and lighter than 130 GeV if the Supersymmetry exists.

The SM Higgs boson, H^0_{SM}, is produced at the LHC predominantly via gluon-gluon fusion(GF) and the second dominant process is vector boson fusion(VBF) H^0_{SM} decays mainly into $b\bar{b}$ and $\tau^+\tau^-$ for the lighter case ($\lesssim$130 GeV). Although its decay into $\gamma\gamma$, via the one-loop process including top quark or W boson, is suppressed ($\sim 2 \times 10^{-3}$), this decay mode is very important at the LHC for this light case. On the other hand, it decays into W^+W^- and ZZ with a large branching fraction for the heavier case ($\gtrsim$140 GeV).

4.2.1. $H^0_{SM} \to \gamma\gamma$ *in the GF and VBF*

This channel is promising for the light Higgs boson, whose mass is lighter than 140 GeV, and this mode indicates the spin of the Higgs boson candidate. Although the branching fraction of this decay mode is small and there is a large background processes via $q\bar{q} \to \gamma\gamma$, the distinctive features of the signal, high p_T isolated two photons with a mass peak, allows us to separate the signal from the large irreducible background. The mass resolution of the $H^0_{SM} \to \gamma\gamma$ process is expected to be 1.3 GeV(ATLAS) and 0.9GeV(CMS). Sharp peak appears at Higgs boson mass over the smooth distribution of background events as shown in Fig.5.

VBF provides additional signatures in which two high p_T jets are observed in the forward regions, and the only two photons from the decay of H^0_{SM} will be observed in the wide rapidity gap between these jets. The rapidity gap (no jet activity in the central region) is expected because there is no color-connection between two outgoing quarks. These signatures suppress the background contributions significantly as shown in Fig.5(b).

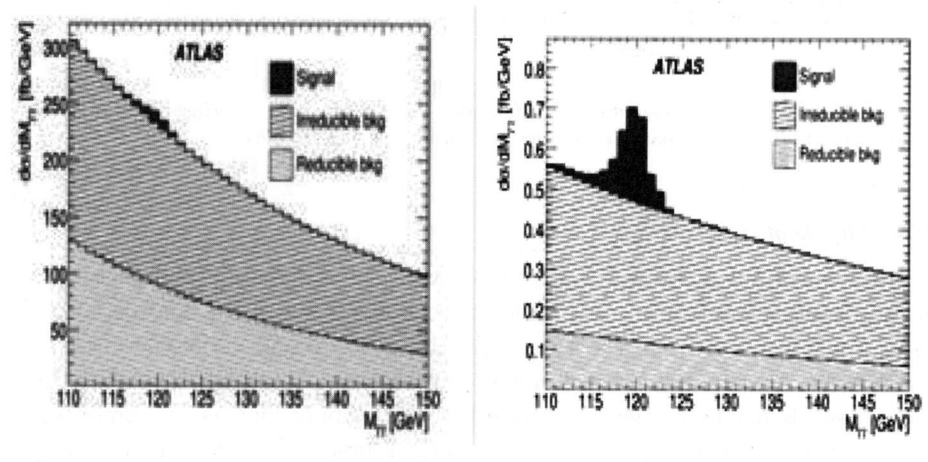

Fig. 5. The invariant mass distribution of $\gamma\gamma$ (ATLAS). H^0_{SM} mass is assumed to be 120 GeV. (a) No jet (mainly for GF process) (b) plus 2 jets (for VBF process) In both figures, red histogram shows the signal. Blue and green show the background contributions for real photo and for fake photon, respectively.

4.2.2. H^0_{SM} ($\to W^+W^- \to \ell^\pm\nu\ell^\pm\nu$) *in GF and VBF*

The mass range of 130-200 GeV is well covered by the analysis of $H^0_{SM} \to WW \to \ell\nu\ell\nu$, both in GF and VBF process. The transverse mass, $M_{\rm T}$, is defined as $\sqrt{2\not{E}_{\rm T} P_{\rm T}(\ell\ell)(1-\cos\phi)}$, in which ϕ is the azimuthal angle between $\not{E}_T$ and $P_{\rm T}(\ell\ell)$. Figure 6 shows the $M_{\rm T}$ distribution and a clear Jacobian peak is observed above smooth background distributions. The main background process is W^+W^- for the topology without-jet (GF) $t\bar{t}$ for that with two forward jets (VBF). These can be suppressed by using the azimuthal angle correlation between the dileptons because of the following reason. Since the Higgs boson is a spin zero particle, the helicities of the emitted W bosons are opposite. The leptons are then emitted preferably in the same direction due to the 100% parity violation in W decays.

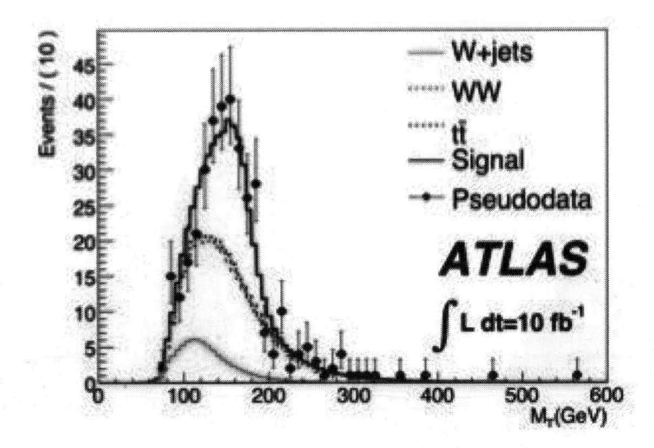

Fig. 6. The transverse mass distribution of $\not{E}_T$ and $\ell\ell$, in which $\mathrm{M}(\mathrm{H}^0_{SM})$=170 GeV is assumed. Open blue histogram shows the Higgs signal and the red(W^+W^-) and green($t\bar{t}$) histogram show the background contribution.

4.2.3. $\mathrm{H}^0_{SM} \to \tau^+\tau^-$ *in VBF*

$\mathrm{H}^0_{SM} \to \tau^+\tau^-$ provides high p_T $\ell^{\pm}$, when τ decays leptonically, and it can makes a clear trigger. Momenta carried by ν's emitted from τ decays can be solved approximately by using the $\not{E}_T$ information, and the Higgs mass can be reconstructed (collinear approximation). Figures 7 show the mass distributions of the reconstructed tau-pair for leptonic plus hadronic decays and both leptonic decays. The mass resolutions of about 10 GeV can be obtained, and the signal can be separated from the Z-boson production background. The performance of the $\not{E}_T$ is crucial in this analysis as shown in Fig.3(d). This process provides a direct information on the coupling between the Higgs boson and a fermion, the tau lepton.

4.2.4. $\mathrm{H}^0_{SM}\,(\to ZZ \to \ell^+\ell^-\ell^+\ell^-)$ *in GF*

The four-lepton channel ($\mathrm{H}^0_{SM} \to ZZ \to \ell^+\ell^-\ell^+\ell^-$) is very clean and a sharp mass peak is expected as shown in Figs. 8. Mass resolution of the four lepton system is typically 1%. As shown in Figures, the main background process is ZZ production which gives continuous distribution above 200 GeV. This channel is important for H^0_{SM} whose mass is heavier than 130 GeV.

4.2.5. *Discovery potential* H^0_{SM}

5σ-discovery potential and 98%C.L. exclusion of H^0_{SM} are summarized in Fig.9 as a function of the Higgs mass. Vertical axis shows the integrated luminosity for discovery and exclusion. The results at ATLAS and CMS are combined. Higgs whose mass is lighter than 130 GeV can be discovered or excluded in 2013 or 2011, respectively. Higgs, heavier 130 GeV, becomes more easy. The Crucial test for light Higgs boson scenario can be performed around 2011-2013.

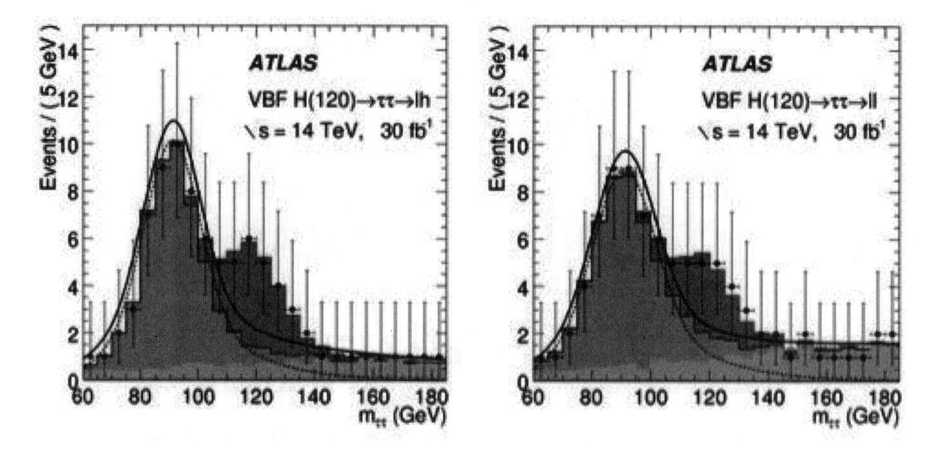

Fig. 7. The invariant mass distributions of $\tau^+\tau^-$ (L=30 fb^{-1} at ATLAS). (a) One tau decays leptonically and another decays hadronically. (b) both tau decays leptonically. In both figures, the red histogram shows the signal with $\mathrm{M(H}^0_{SM})$=120 GeV, and blue($\mathrm{Z}^0 \to \tau^+\tau^-$) and green($\mathrm{t\bar{t}}$) shows background contributions.

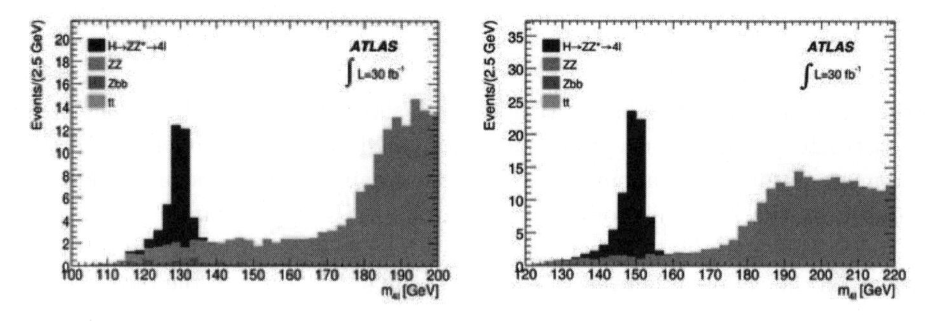

Fig. 8. The invariant mass distribution of $\ell\ell\ell\ell$ with luminosities of 30 fb^{-1}. $\mathrm{M(H}^0_{SM})$=130(a) and 150(b) GeV are assumed.

$\mathrm{H}^0_{SM} \to \gamma\gamma$ and $\tau^+\tau^-$ channels have good potential in the mass region lighter than 130 GeV. For the heavy mass case, ($\geq$ 130 GeV), decay to $ZZ(\to \ell^+\ell^-\ell^+\ell^-)$ and $\mathrm{W}^+\mathrm{W}^-$ have an excellent performance. After the discovery of the Higgs boson, we can perform to measure the mass and couplings between Higgs and particles with accuracies of about 0.1% and 10%, respectively.

4.3. *Supersymmetry*

If the Higgs boson is light, some mechanism to protect Higgs mass from the radiative correction is necessary. Supersymmetric (SUSY) standard models are most promising extensions of the SM, because the SUSY can naturally explain the weak boson mass scale. Furthermore, the SUSY models provide a natural candidate of

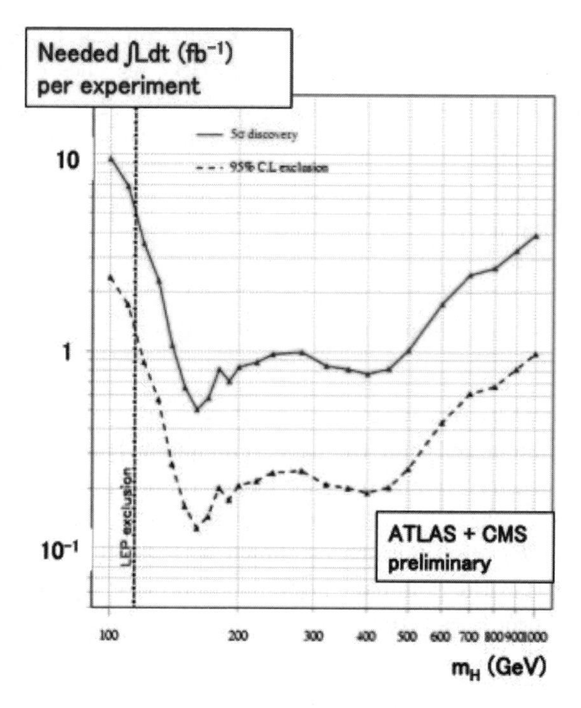

Fig. 9. The needed integrated luminosity as a function of Higgs mass. Red shows 5σ discover and blue shows 98%C.L. exclusion.

the cold dark matter and they have given a hint of the Grand Unification in which three gauge couplings of the SM are unified at around 2×10^{16} GeV. In SUSY, each elementary particle has a superpartner whose spin differs by 1/2 from that of the particle. Discovery of the SUSY particles should open a new epoch of the fundamental physics, which is another important purpose of the LHC project.

4.3.1. *Event topologies of SUSY events*

$\tilde{g}$ and/or $\tilde{q}$ are copiously produced at the LHC. High p_T jets are emitted from the decays of $\tilde{g}$ and $\tilde{q}$. If the R-parity is conserved, each event contains two $\tilde{\chi}_1^0$'s in the final state, which is stable, neutral and weakly interacting and escapes from the detection. This $\tilde{\chi}_1^0$ is an excellent candidate of the cold dark matter. The missing transverse energy($\not{E}_T$), which is carried away by the two $\tilde{\chi}_1^0$'s, plus multiple high p_T jets is the leading experimental signature of the SUSY at the LHC. Additional leptons, h($\rightarrow$ b$\bar{\text{b}}$) and τ, coming from the decays of $\tilde{\chi}_2^0$ and $\tilde{\chi}_1^{\pm}$, can also be detected in the some part of event.

4.3.2. *Inclusive searches*

Inclusive searches will be performed with the large $\not{E}_T$ and high p_T multi-jet topology (no-lepton mode). One additional hard lepton (one-lepton mode) or two same-sign (SS-dilepton) and opposite-sign leptons(OS-dilepton) can be added to the selections. These four are the promising modes of the SUSY searches, especially no-lepton and one-lepton modes are important for the discovery.

The following four SM processes can potentially have $\not{E}_T$ event topology with jets:

- $W^{\pm}(\to \ell\nu)$ + jets,
- $Z^0(\to \nu\bar{\nu}, \tau^+\tau^-)$ + jets,
- $t\bar{t}$ + jets,
- QCD multi-jets with mis-measurement and semi-leptonic decays of $b\bar{b}$ and $c\bar{c}$ with jets

We require at least three jets with $p_T \geq 50$ GeV, and p_T of the leading jets and $\not{E}_T$ are required to be larger than 100 GeV. The azimuthal angles between $\not{E}_T$ and the leading three jets are required to be larger than 0.2 for the no-lepton mode to reduce the QCD background. The excess coming from the SUSY signals can be clearly seen in the $\not{E}_T$ distributions as shown in Figs.10 for both no-lepton and one-lepton modes. $t\bar{t}$ is the dominant background process for the one-lepton mode, and all of the four processes listed above contribute to no-lepton mode.

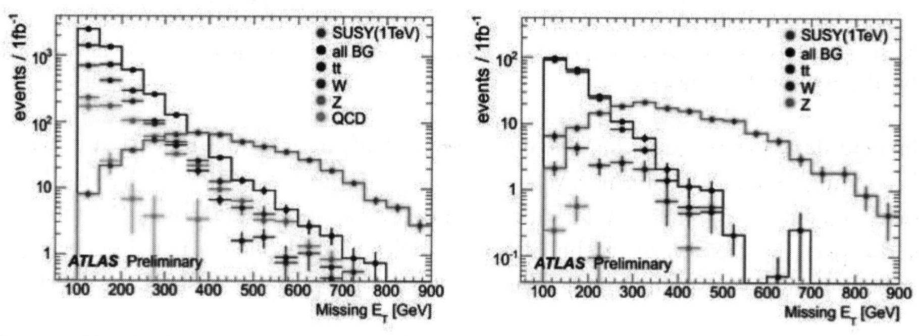

Fig. 10. The $\not{E}_T$ distributions of the SUSY signal and background processes with a luminosity of 1 fb^{-1} and $\sqrt{s}$=14 TeV. (a) No lepton (b) one lepton mode. In both figures, red open histogram show the SUSY signal with (m_0=100GeV,$m_{1/2}$=400GeV and $\tan\beta$=10) The black show the sum of the all SM backgrounds

4.3.3. *Discovery potential*

Figures 12 show 5σ-discovery potential in m_0-$m_{1/2}$ plane for (a) $\sqrt{s}$=10TeV L=200 pb^{-1} and (b) $\sqrt{s}$=14TeV L=30 fb^{-1}. Figure 12(a) is corresponding data in 2011, and $\tilde{g}$ and $\tilde{q}$ can be discovered upto $\sim$ 800 GeV. No lepton mode has slightly good sensitivity than one-lepton mode. With the naive SUGRA assumption, Chargino and the lightest Neutralino (dark matter candidate) can be discovered upto $\sim$ 250 GeV and 130 GeV, respectively. The interesting region in which Ω_{DM} is consistent with the observation is covered in 2011 run. Figure 12(b) is corresponding data in around 2014, and $\tilde{g}$ and $\tilde{q}$ can be discovered upto $\sim$ 2 TeV.

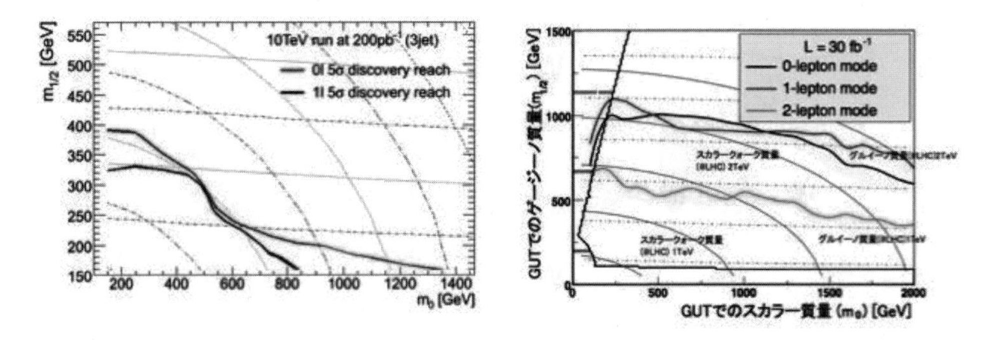

Fig. 11. 5σ-discovery potential in m_0-$m_{1/2}$ plane($\tan\beta$=10) for (a) $\sqrt{s}$=10TeV L=200 pb^{-1} and (b) $\sqrt{s}$=14TeV L=30 fb^{-1}.

5. EWSB scenario 2: Strong Coupling Gauge Theory

Alternative scenario of the EWSB is the Strong Coupling Gauge Theory (SCGT). The di-boson and dilepton resonance are typical event signals at LHC.

5.1. $\mathrm{W}^+\mathrm{W}^-$ *and* $\mathrm{W}Z$ *scatter*

$\mathrm{W}^+\mathrm{W}^-$ or $\mathrm{W}Z$ scatters will be affected in the Strong Coupling Gauge Theory. These scatter processes will be observed in VBF processes as shown in Fig.12(a). Chiral Lagrangian model is the effective theory and two parameters of the anomalous coupling (α_4 and α_5) are useful for VV scatters. Vector(ρ) and scalar(σ) resonance masses can be written:

$$\mathrm{M}_\rho^2 = \frac{\mathrm{v}^2}{4(\alpha_4 - 2\alpha_5)} \tag{1}$$

$$\mathrm{M}_\sigma^2 = \frac{3\mathrm{v}^2}{4(7\alpha_4 + 11\alpha_5)} \tag{2}$$

Lower mass state is scalar and higher mass is vector resonance. Non-resonance processes are also possible. The differential cross-section of the various resonances(non-resonance) are shown in Fig.12(b). The cross-section is expected to small even at $\sqrt{s}$=14 TeV.

The emitted W and Z bosons decay into quark pair or lepton pair ($\mathrm{W}^\pm \to \ell\nu$ $\mathrm{Z}^0 \to \ell^+\ell^-$). Tri-lepton topology ($\mathrm{W}Z \to \ell\nu\ell^+\ell^-$) is the promising channel as shown in Fig.13(a), since the background contribution is expected to be small. But the branching fraction of tri-lepton is also suppressed and the tri-lepton mode is promising only for the lighter case(M ~500 GeV). For the heavier case (M>800 GeV), dilepton+jets topology ($\mathrm{W}Z \to jj\ell^+\ell^-$) becomes promising as shown in Fig.13(b) and (c). The integrated luminosity of 100 fb^{-1} is necessary for M = 500–800 GeV resonance case (5σ discovery potential.) More luminosity of 200 fb^{-1} is necessary for M> 1 TeV resonance case and it is difficult at LHC for non-resonance case.

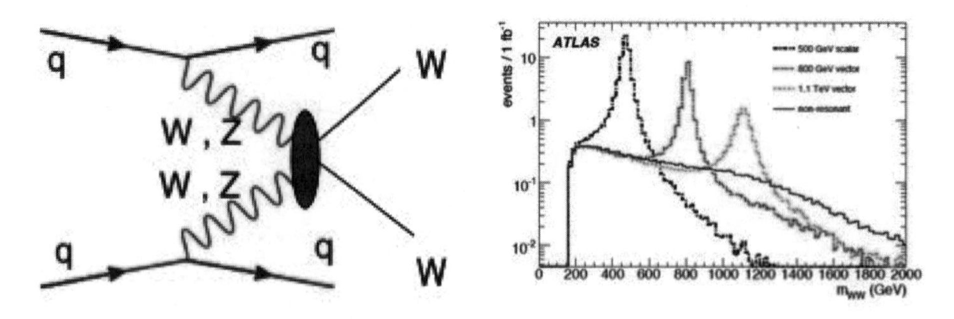

Fig. 12. (a) VBF process for VV scatter (b) di-Boson invariant mass distributions for various models. Scalar and vector resonances (Mass=500,800,1100 GeV) and non-resonance case.

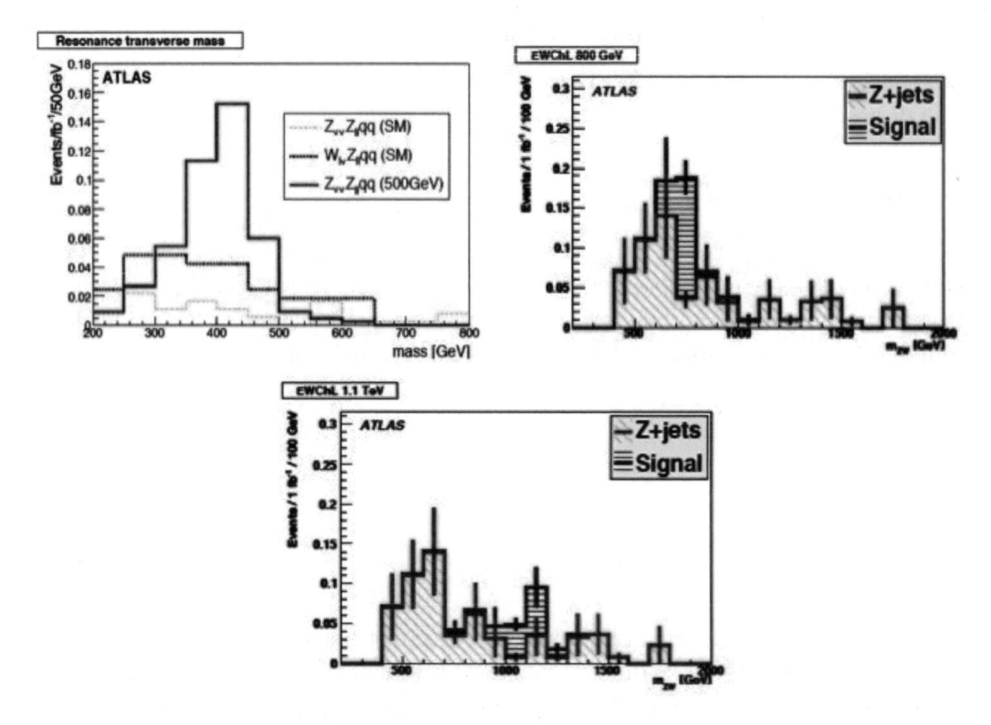

Fig. 13. The invariant mass distribution (a) MT mass for trilepton and missing. The red show the signal of 500GeV scalar. (b) and (c) The invariant mass of $\ell^+\ell^- qq$ for 800 GeV scalar and 1.1 TeV vector resonance. Blue shows the signal and the red shows the background contributions of $Z^0 + jets$

5.2. *Low mass technicolor model*

There might be many TC fermions (N_D) to make the walking model, otherwise TC models are strictly conflict with the EW precise measurements at LEP. TC scale $\Lambda_{TC} \sim 250GeV/\sqrt{N_D}$ becomes smaller and mass scale of Technicolor particles becomes light.

Nambu-Gladstone-Boson (π_{TC}) is mixed with Z_L and W_L. Technifermion bound state, ρ_{TC}, ω_{TC} and a_{TC} will couple with W,Z,γ. They decay into VV or fermion pair and make the narrow resonance. Mass of these bound state is around 400-800GeV and the decay modes and J^{CP} are listed in the next table. L=10 or 5 fb^{-1} is necessary for di-boson decay mode and di-lepton decay mode, respectively. These luminosity is expected around 2013 or 2014.

Table 3. TC boson and decay mode.

name	J^{PC} Spin	decay mode
$\rho^{\pm 0}_{TC}$	$(1^{--}, \mathrm{I}=1)$	$\rightarrow \mathrm{W}^+\mathrm{W}^-, \mathrm{WZ}, ZZ$
		$\rightarrow \ell^+\ell^-$
ω^0_{TC}	$(1^{--}, \mathrm{I}=0)$	$\rightarrow \gamma Z$
$a^{\pm 0}_{TC}$	$(1^{++}, \mathrm{I}=1)$	$\rightarrow \gamma W$
		$\rightarrow \ell^+\ell^-$

6. Conclusion

LHC is back now and we have real data at $\sqrt{s}$=900 GeV and 2.36 TeV. The ATLAS and CMS (also ALICE and LHCb) detectors work well and the good performance have been obtained. The LHC will be operated at $\sqrt{s}$=7 TeV for 2010 and 2011 due to the bad connection of splicing. The integrated luminosity is about 1 fb^{-1} in these two years.

There are two major scenarios for the ElectroWeak Symmetry Breaking:
(1) light Higgs boson and SUSY
(2) diboson and dilepton resonance in Strong Coupling Gauge Theory
They can be discovered at LHC before 2013–2015, and we can perform the crucial test.

Continuum Superpartners from Supersymmetric Unparticles

Hsin-Chia Cheng

Department of Physics, University of California
Davis, California 95616, USA
E-mail: cheng@physics.ucdavis.edu

In an exact conformal theory there is no particle. The excitations have continuum spectra and are called "unparticles" by Georgi. We consider supersymmetric extensions of the Standard Model with approximate conformal sectors. The conformal symmetry is softly broken in the infrared which generates a gap. However, the spectrum can still have a continuum above the gap if there is no confinement. Using the AdS/CFT correspondence this can be achieved with a soft wall in the warped extra dimension. When supersymmetry is broken the superpartners of the Standard Model particles may simply be a continuum above gap. The collider signals can be quite different from the standard supersymmetric scenarios and the experimental searches for the continuum superpartners can be very challenging.

Keywords: Supersymmetry, unparticle, LHC.

1. Introduction

The Large Hadron Collider (LHC) has started running. With its unprecedented center of mass energy, there is a high hope that it will discover new physics which revolutionizes high energy physics. However, LHC is a complicated machine with very high luminosities. In order to find new physics, we need to know what signals the new physics may give rise to, and how to search for them with the enormous Standard Model (SM) backgrounds.

A major motivation for new physics at the TeV scale is the hierarchy problem – the stability of the electroweak scale under radiative corrections. Many models for TeV-scale new physics are invented to address this problem, *e.g.*, supersymmetry (SUSY), technicolor models, large and warped extra dimensions, little Higgs models, etc. A lot of collider phenomenology studies have been devoted to these models. However, these models certainly do not cover all possible new physics that may appear at the LHC. There is no theorem that at the TeV scale we should only see the minimal models which just address the hierarchy problem.

From the experimental point of view, the collider searches should be signal-based. Even though there are experimental studies targeted for some specific models, generic searches for new particles such as new gauge bosons, new quarks and leptons apply to a wide range of models. As the LHC is starting, the recent model-building efforts have shifted from "solving problems of the Standard Model" to studying models which can give rise to "unexpected signals" (*e.g.*, hidden valleys,[1] quirks,[2] unparticles,[3] etc.) irrespective of whether they are related to any particular problem of the Standard Model. It is possible that the TeV-scale physics includes some of these extra sectors in additional to the standard scenario. The presence of the extra sector may obscure the experimental signals of the standard scenario if there is mixing between them, so it is imperative to study these possibilities and their experimental consequences. In this talk, consider such a case with a (super)conformal sector softly broken at the TeV scale and mixed with the supersymmetric Standard Model. As we will see, the mixings can make the superpartners of the SM particles have continuum spectra, and hence make their experimental consequences quite different from the standard SUSY scenario. This presentation is based on the work done in collaboration with Haiying Cai, Anibal Medina, and John Terning.[4]

2. Unparticles

We first give a brief introduction to unparticles. In Ref. [3] Georgi considered the possibility that a hidden conformal field theory (CFT) sector coupled to the SM through higher-dimensional operators,

$$\frac{C_U \Lambda_U^{d_{UV}-d_U}}{M_U^k} \mathcal{O}_{SM} \mathcal{O}_U, \tag{1}$$

where M_U is the scale where the interaction is induced, Λ_U is the scale where the hidden sector becomes conformal, $d_{UV} = k + 4 - d(\mathcal{O}_{SM})$, and d_U is the scaling dimension of the operator $\mathcal{O}_U$. There is no "particle" in a CFT. One can only talk about operators with some scaling dimensions. The spectral densities are continuous and hence Georgi called them "unparticles."

The phase space of an unparticle looks like a fractional number of particles. For a scalar unparticle,

$$\begin{aligned} \langle \mathcal{O}_U(x) \mathcal{O}_U(0) \rangle &= \int \frac{d^4 P}{(2\pi)^4} e^{-iPx} |\langle 0 | \mathcal{O}_U | P \rangle|^2 \rho(P^2), \\ |\langle 0 | \mathcal{O}_U | P \rangle|^2 \rho(P^2) &= A_{d_U} \theta(P^0) \theta(P^2) (P^2)^{d_U - 2}, \end{aligned} \tag{2}$$

where A_{d_U} is a normalization constant. In the limit $d_U \to 1$, it becomes the phase space of a single massless particle, $\delta(P^2)$. For $d_U \neq 1$, the spectral density is continuous.

Because the unparticle spectral density is continuous down to zero, many phenomenological consequences and constraints are similar to other scenarios with very light (or massless) degrees of freedom, *e.g.*, large extra dimensions proposed by Arkani-Hamed, Dimopoulos, and Dvali.[5] A list of things to be considered contains[a]

[a] There is a huge literature on phenomenology tests and constraints on unparticles. We only list a few sample references due to the space limit. Please check the citation list of the original Gerogi's paper for more complete references.

- invisible decays, *e.g.*, from Z, heavy mesons, quarkonia, neutrinos, ...,[6,7]
- missing energy in high energy collisions,[8]
- higher-dimensional operators induced by unparticles,[8–10]
- astrophysics: star (SN1987A) cooling,[11,12]
- cosmology: Big Bang nucleosynthesis (BBN),[11,13]
- long-range forces induced by unparticles.[14–16]

In particular, unparticles cannot carry SM charges if the spectral density continues down to zero without any gap. Otherwise they would have been copiously produced and observed.

However, as pointed by Fox, Rajaraman, and Shirman,[17] coupling to the Higgs sector will break the conformal invariance of the unparticles in the infrared (IR), which will induce a gap in general. Above the gap, there are two possibilities:

- If the theory confines, then it would produce QCD-like resonances.
- If the theory does not confine, one then expect that the spectral density is still continuous above the mass gap. A simple ansatz to describe it is to introduce an IR cutoff m,[17,18]

$$\begin{aligned}\Delta(p,m,d) &\equiv \int d^4x e^{ipx}\langle 0|T\mathcal{O}(x)\mathcal{O}^\dagger(0)|0\rangle \\ &= \frac{A_d}{2\pi}\int_{m^2}^{\infty}(M^2-m^2)^{d-2}\frac{i}{p^2-M^2+i\epsilon}dM^2. \end{aligned} \quad (3)$$

Such a spectral density arises when a massive particle couples to massless degrees of freedom. For example, quark jets can be considered as unparticles.[19]

With a mass gap, many low-energy phenomenological constraints which result from new light degrees of freedom no longer apply. In particular, with a mass gap, one can consider the possibility that unparticles carry SM charges[18] (if the gap is larger than $\mathcal{O}(100)$ GeV), which can be more interesting phenomenologically. This is the scenario that will be discussed in this talk.

3. Unparticles and AdS/CFT

A useful tool to study the (large N) conformal field theory is the AdS/CFT correspondence.[20] The metric of a 5D anti de Sitter (AdS$_5$) space in conformal coordinates is given by

$$ds^2 = \frac{R^2}{z^2}(dx_\mu^2 - dz^2), \quad (4)$$

where R is the curvature radius and z is the coordinate in the extra dimension. Using the AdS/CFT correspondence, Georgi's unparticle scenario can be described by the Randall-Sundrum II (RS2)[21] setup. The 5-dimensional (5D) bulk fields correspond to the 4-dimensional (4D) CFT operators, and the 5D bulk mass of a bulk field is related to the scaling dimension of the corresponding CFT operator. In RS2, the space is cut off by a UV brane (located at $z = z_{UV} \equiv \epsilon$) but there is no IR brane which means that the conformal invariance is good down to zero energy. The momentum in the extra dimension of a bulk field is not quantized. From the 4D point of view, the momentum in the extra dimension becomes the mass in 4D, so the 4D mass spectrum is continuous. SM fields are localized on the UV brane. They can interact with the bulk fields (CFT operators) through higher-dimensional interactions. Such a representation allows us to perform calculations involving unparticles using the ordinary (5D) field theory.[22–24]

In this talk we are interested in unparticles carrying SM charges with mass gaps. Some modifications of the RS2 setup are necessary. In particular, SM fields have to propagate in the extra dimension as well and a soft wall is needed to produce the gap in the IR. Before going to these discussions, we first describe how to formulate SUSY in the AdS bulk.

3.1. *SUSY in AdS bulk*

It is convenient to describe a 5D supersymmetric theory using the 4D $N = 1$ superspace formalism.[25,26] The 5D $N = 1$ SUSY corresponds to $N = 2$ SUSY in 4D. A 5D $N = 1$ hypermultiplet contains two chiral multiplets in 4D, $\Phi = \{\phi, \chi, F\}$, $\Phi_c = \{\phi_c, \psi, F_c\}$. The action of the 5D $N = 1$ hypermultiplet is given by[26]

$$S = \int d^4x\, dz \left\{ \int d^4\theta \left(\frac{R}{z}\right)^3 [\Phi^* \Phi + \Phi_c \Phi_c^*] + \right. \\ \left. + \int d^2\theta \left(\frac{R}{z}\right)^3 \left[\frac{1}{2}\, \Phi_c\, \partial_z \Phi - \frac{1}{2}\, \partial_z \Phi_c\, \Phi + m\frac{R}{z}\, \Phi_c\, \Phi\right] + h.c. \right\}, \tag{5}$$

It is convenient to define a dimensionless bulk mass $c \equiv mR$. It is related to the dimension of the corresponding operator in the 4D CFT picture.[27–29] For a left-handed CFT operator $\mathcal{O}_L$ (which corresponds to the 5D bulk field Φ), we have $d_s = 3/2 - c$, $d_f = 2 - c$ for $c \leq 1/2$, where d_s and d_f are the scaling dimensions of the scalar and fermion operators respectively. For $c < -1/2$ ($d_s > 2$), the correlator diverges as we take the UV brane to the AdS boundary ($\epsilon \to 0$). A counter term is needed on the UV brane to cancel the divergence, which implies UV sensitivity. On the other hand, for $c > 1/2$ the CFT becomes a free field theory with $d_s = 1$, $d_f = 3/2$. We will focus on the most interesting range $-1/2 \leq c \leq 1/2$ ($1 \leq d_s \leq 2$). For the right-handed CFT operator, we have the similar interpretations but with $c \to -c$.

In solving the equations of motion (EOMs), we include a z-dependent mass $m(z)$ which represents the soft wall. The EOMs for the fermions and the F-terms are

$$-i\bar{\sigma}^\mu \partial_\mu \chi - \partial_z \bar{\psi} + (m(z)R + 2)\frac{1}{z}\bar{\psi} = 0, \tag{6}$$

$$-i\sigma^\mu \partial_\mu \bar{\psi} + \partial_z \chi + (m(z)R - 2)\frac{1}{z}\chi = 0. \tag{7}$$

$$F_c^* = -\partial_z\phi + \left(\frac{3}{2} - m(z)R\right)\frac{1}{z}\phi, \tag{8}$$

$$F = \partial_z\phi_c^* - \left(\frac{3}{2} + m(z)R\right)\frac{1}{z}\phi_c^*. \tag{9}$$

Using the F-term equations we can find the second order EOM for the scalars:

$$\partial_\mu\partial^\mu\phi - \partial_z^2\phi + \frac{3}{z}\,\partial_z\phi + \left(m(z)^2R^2 + m(z)R - \frac{15}{4}\right)\frac{1}{z^2}\,\phi - (\partial_z m(z))\,\frac{R}{z}\phi = 0\,. \tag{10}$$

We can decompose the 5D field into a product of the 4D field and a profile in the extra dimensions

$$\chi(p,z) = \chi_4(p)\left(\frac{z}{z_{UV}}\right)^2 f_L(p,z), \quad \phi(p,z) = \phi_4(p)\left(\frac{z}{z_{UV}}\right)^{3/2} f_L(p,z), \tag{11}$$

$$\psi(p,z) = \psi_4(p)\left(\frac{z}{z_{UV}}\right)^2 f_R(p,z), \quad \phi_c(p,z) = \phi_{c4}(p)\left(\frac{z}{z_{UV}}\right)^{3/2} f_R(p,z), \tag{12}$$

where $p \equiv \sqrt{p^2}$ represents the 4D momentum. Then we find the profiles in the extra dimension f_L, f_R satisfy the Schrödinger-like equations with the potential determined by the mass term. For a constant mass, the potential scales as $1/z^2$ for large z, so there is a continuum of solutions starting from zero energy (4D mass). We can get different behaviors if $m \sim z^\alpha$ for large z:

- For $\alpha < 1$, the potential still goes to zero for large z, so we have continuum solutions without gap.
- For $\alpha > 1$, the potential grows without bounds at large z, so we get discrete solutions.[30]
- For $\alpha = 1$, the potential approaches a positive constant for large z, so we have a continuum with a gap.[22]

We are interested in the last possibility so we consider the case $m(z) = c + \mu z$. In the UV (small z), $m \approx c$, we have the usual CFT interpretation. At large z, $m(z) \propto z$, the conformal invariance in broken in the IR and a mass gap is developed.

We can easily obtain the zero mode solutions for $m(z) = c + \mu z$:

$$f_L(0,z) \sim e^{-\mu z} z^{-c}, \qquad f_R(0,z) \sim e^{\mu z} z^{c}. \tag{13}$$

One can see that only one of them has a normalizable zero mode. For definiteness we choose $\mu > 0$, then only the left-handed field has a zero mode, which is identified as the SM fermion.

The nonzero modes satisfy the Schrödinger-like equations,

$$\frac{\partial^2}{\partial z^2} f_R + \left(p^2 - \mu^2 - 2\frac{\mu c}{z} - \frac{c(c-1)}{z^2}\right) f_R = 0, \tag{14}$$

$$\frac{\partial^2}{\partial z^2} f_L + \left(p^2 - \mu^2 - 2\frac{\mu c}{z} - \frac{c(c+1)}{z^2}\right) f_L = 0. \tag{15}$$

They have the same form as the radial wave equation of the hydrogen atom,

$$\frac{\partial^2}{\partial r^2}u + \left(2ME + 2\frac{M\alpha}{r} - \frac{\ell(\ell+1)}{r^2}\right)u = 0, \tag{16}$$

except that c is in general a fractional number instead of an integer. We can immediately see the pattern of the solutions from our knowledge of the hydrogen atom. For $c\mu > 0$ which corresponds to a repulsive Coulomb potential, we have continuum solutions above the gap, $p^2 > \mu^2$. For $c\mu < 0$ on the other hand, we get discrete hydrogen-like spectrum below the gap in addition to the continuum above the gap.

3.2. *Holographic boundary action*

One can derive the unparticle propagator by probing the CFT with a boundary source field Φ_c^0. The holographic boundary action can be obtained by integrating out the bulk using the bulk EOMs with the boundary condition $\Phi_c(z_{UV}) = \Phi_c^0 = (\phi_c^0, \psi^0, F_c^0)$,[22]

$$S_{holo} = -\int d^4x[\phi_c^{0*}\Sigma_{\phi_c}\phi_c^0 + F_c^{0*}\Sigma_{F_c}F_{c0} + \psi_0^*\Sigma_\psi\psi_0] \tag{17}$$

where

$$\Sigma_{\phi_c} = \left(\frac{R}{z_{UV}}\right)^3 p\frac{f_L}{f_R}, \quad \Sigma_\psi = \left(\frac{R}{z_{UV}}\right)^4 \frac{p_\mu\sigma^\mu}{p}\frac{f_L}{f_R}, \quad \Sigma_{F_c} = \left(\frac{R}{z_{UV}}\right)^3 \frac{1}{p}\frac{f_L}{f_R}. \tag{18}$$

From the CFT point of view, the right-handed superfield Φ_c^0 is a source of the left-handed CFT operator which correspond to Φ. Since F_c^0 is the source for the scalar component, the propagator for the scalar CFT operator is

$$\Delta_s(p) \propto -\Sigma_{F_c}(p), \tag{19}$$

with

$$\Sigma_{F_c} = \frac{\epsilon(\mu + \sqrt{\mu^2 - p^2})}{p^2} \cdot \frac{W\left(-\frac{c\mu}{\sqrt{\mu^2-p^2}}, \frac{1}{2}+c, 2\sqrt{\mu^2-p^2}\epsilon\right)}{W\left(-\frac{c\mu}{\sqrt{\mu^2-p^2}}, \frac{1}{2}-c, 2\sqrt{\mu^2-p^2}\epsilon\right)}, \tag{20}$$

where $W(\kappa, m, \zeta)$ is the Whittaker function of the second kind. The fermionic and the F-component CFT correlators are simply $\Delta_f = p_\mu\sigma^\mu\Delta_s$ and $\Delta_F = p^2\Delta_s$ by SUSY relations.

We are interested in the conformal limit $\epsilon \to 0$, so we expand Σ_{F_c} for small ϵ and focus on the range $-1/2 < c < 1/2$. After properly rescaling the correlator by a power of ϵ to account for the correct dimension of the correlator, we find

$$\Delta_s^{-1} \approx \frac{p^2}{1-2c}\epsilon^{1-2c} - \frac{2^{-1+2c}p^2(\mu^2-p^2)^{-1/2+c}\Gamma(1-2c)\Gamma(c+\frac{c\mu}{\sqrt{\mu^2-p^2}})}{\Gamma(2c)\Gamma(1-c+\frac{c\mu}{\sqrt{\mu^2-p^2}})}. \tag{21}$$

The first term vanishes in the conformal limit for $-1/2 < c < 1/2$. We can see that there is a pole at $p^2 = 0$ which represents the zero mode, and a branch cut

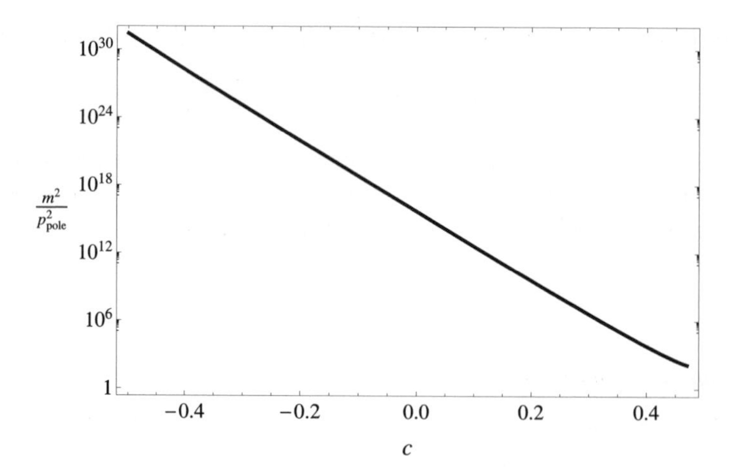

Fig. 1. Plot of m^2/p^2_{pole} vs c. Notice that as c gets closer to 1/2, for a given value of p^2_{pole}, the value of m^2 decreases.

for $p^2 > \mu^2$ which represents the continuum in the propagator. For $c < 0$, there are addition poles below the gap μ^2 which correspond to the hydrogen-like bound states discussed in the previous subsection.

3.3. *SUSY breaking*

Now we introduce SUSY breaking on the UV boundary by a scalar mass term,

$$\delta S = \frac{1}{2}\int d^4x \left(\frac{R}{z_{UV}}\right)^3 \int dz \left(m^2 z_{UV}\cdot \phi^*\phi + h.c.\right)\delta(z - z_{UV})\,. \tag{22}$$

The scalar propagator is modified due to the modified boundary conditions,

$$F_c(z_{UV}) = F_c^0 + m^2 z_{UV}\,\phi^*(z_{UV}), \qquad \psi(z_{UV}) = \psi^0, \qquad \phi_c(z_{UV}) = \phi_c^0 \tag{23}$$

One can repeat the analysis of the previous subsection and finds for $-1/2 < c < 1/2$ in the limit $\epsilon \to 0$,

$$\Delta_s^{-1}(p^2) \approx m^2\epsilon^{1-2c} - \frac{p^2}{2c-1}\epsilon^{1-2c} - \frac{2^{-1+2c}p^2(\mu^2-p^2)^{-1/2+c}\Gamma(1-2c)\Gamma(c+\frac{c\mu}{\sqrt{\mu^2-p^2}})}{\Gamma(2c)\Gamma(1-c+\frac{c\mu}{\sqrt{\mu^2-p^2}})}. \tag{24}$$

For small m^2, the pole which was at zero mass is shifted to

$$p^2_{pole} = \frac{2(2c-1)m^2(\mu\epsilon)^{1-2c}}{(-4^c + 2^{1+2c}c)\Gamma(1-2c)}. \tag{25}$$

The shift is smaller for smaller (more negative) c (Fig. 1), because the zero mode wave function is localized farther away from the UV brane ($f_{L,0} \propto e^{-\mu z} z^{-c}$).

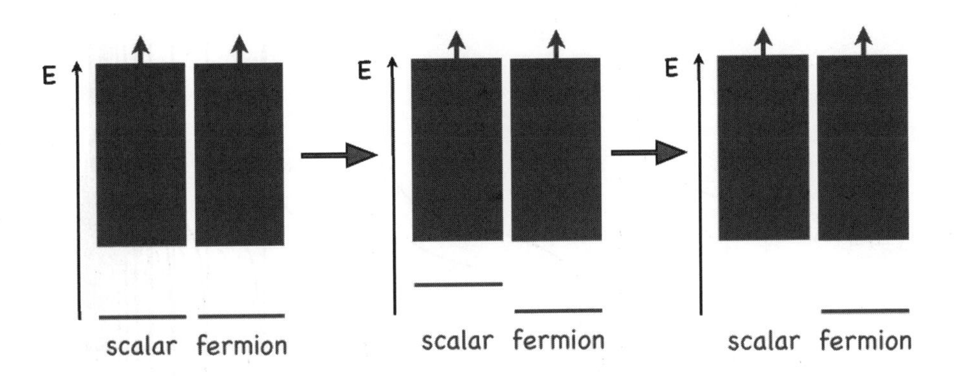

Fig. 2. The spectra of the scalar and the fermion when we increase the scalar SUSY-breaking mass term on the UV brane.

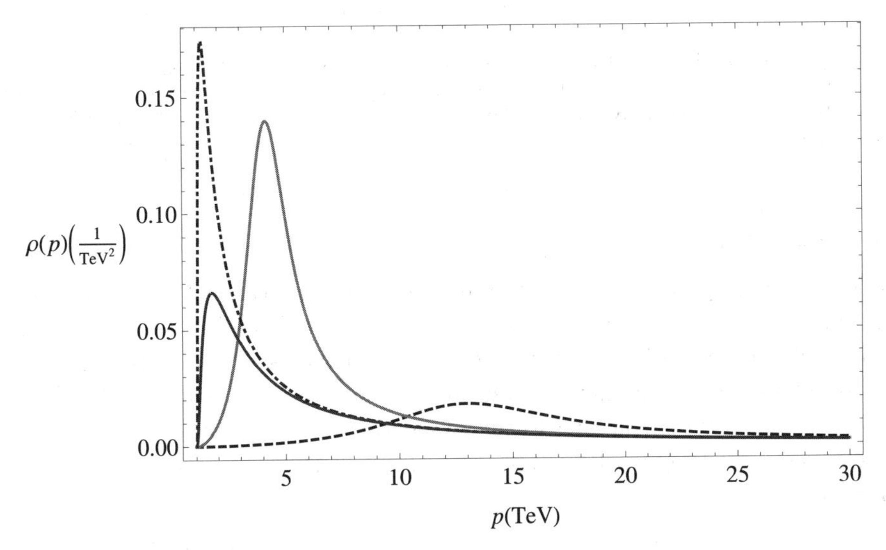

Fig. 3. Spectral function for four examples of boundary UV SUSY breaking masses $m = 2 \times 10^7$ GeV (blue curve, dashed), $m = 8 \times 10^6$ GeV (green curve, solid), $m = 2 \times 10^6$ GeV (purple curve, dot-dashed) and $m = 10^5$ GeV (red curve, dotted). The red and purple curves correspond to zero-mode poles localized at $p \approx 50$ GeV (red curve) and $p \approx 950$ GeV (purple curve), that haven't merged with the continuum. On the other hand, the green and blue curves correspond to the cases where the pole has merged into the continuum. In the examples, $\epsilon = 10^{-19}$ GeV^{-1}, $\mu = 1$ TeV and $c = 0.3$. We can see how the continuum peaks to higher momenta as the SUSY breaking mass m increases, specially as the pole merges into the continuum.

For $0 < c < 1/2$, as we increase m^2, the pole eventually merge into the continuum and only a continuum superpartner is left (Fig. 2). The continuum parts of the spectral functions for several choices of m^2 are shown in Fig. 3.

For $-1/2 < c < 0$, the discrete poles below the gap move towards the gap as m^2 increase (Fig. 4). However, they do not merge into the continuum for arbitrarily large m^2.

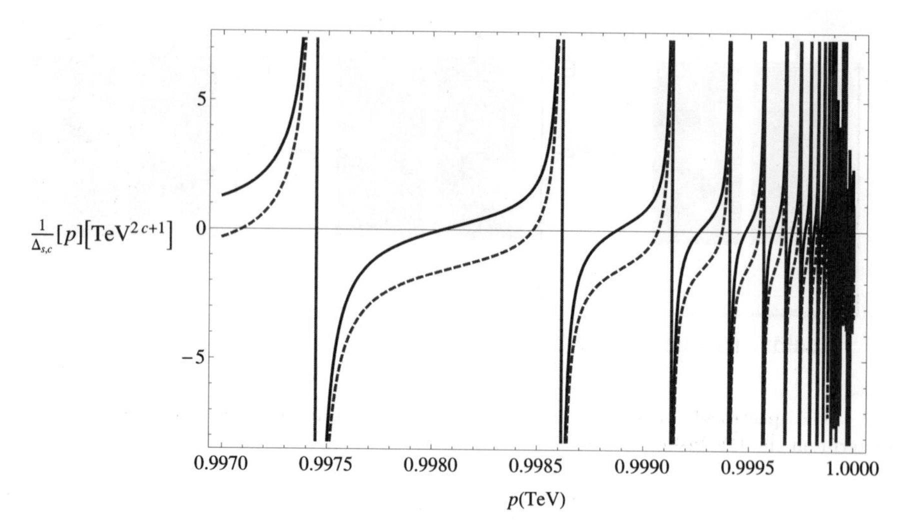

Fig. 4. Inverse correlator for two examples of boundary UV SUSY breaking masses: $m = 10^{13}$ GeV (blue curve) and $m = 2 \times 10^{14}$ GeV (red curve, dashed). In the examples, $\epsilon = 10^{-19}$ GeV^{-1}, $\mu = 1$ TeV and $c = -0.2$. We can see how the series of poles shift into the continuum with increasing values of m.

3.4. *Gauge fields*

A 5D $N = 1$ vector supermultiplet can be decomposed into a 4D $N = 1$ vector supermultiplet $V = (A_\mu, \lambda_1, D)$ and a chiral supermultiplet $\sigma = ((\Sigma + iA_5)/\sqrt{2}, \lambda_2, F_\sigma)$. We cannot proceed as before by introducing a bulk mass term for the gauge field because of the gauge invariance. The same effect can be obtained with a dilaton profile $\langle\Phi\rangle = e^{-2uz}/g_5^2$ coupling to the gauge kinetic term, which softly breaks the conformal symmetry in the IR.[23] The action for the 5D $N = 1$ vector supermultiplet can be written as

$$S_V = \int d^4x dz \cdot \frac{R}{z}\left\{\frac{1}{4}\int d^2\theta W_\alpha W^\alpha \Phi + h.c. + \frac{1}{2}\int d^4\theta \left(\partial_z V - \frac{R}{z}\frac{(\sigma + \sigma^\dagger)}{\sqrt{2}}\right)^2 (\Phi + \Phi^\dagger)\right\}. \tag{26}$$

We can obtain the bulk action in components after rescaling the fields, $A_5 \to \frac{z}{R}A_5$, $\lambda_1 \to (\frac{R}{z})^{3/2}\lambda_1$, and $\lambda_2 \to i(\frac{R}{z})^{1/2}\lambda_2$. We find a z-dependent bulk gaugino mass $1/2 + uz$ induced by the dilaton profile. It leads to a continuum with a mass gap as we found for the matter fields, and c is fixed to be equal to $1/2$ in this case because the 4D gauge field must have dimension one by gauge invariance.

Adding a SUSY-breaking gaugino mass term on the UV brane will lift the gaugino zero mode. For small Majorana gaugino mass m on the UV brane, the zero mode pole is shifted to

$$p_{\mathrm{pole}}^2 \approx \frac{m^2}{(\gamma_E + \ln(2u\epsilon))^2}. \tag{27}$$

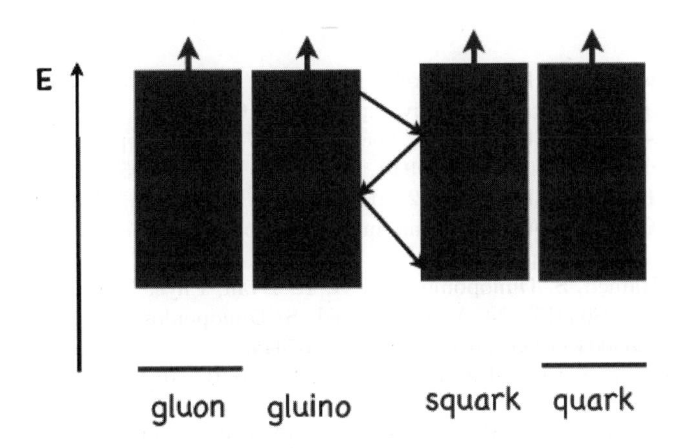

Fig. 5. Extened decay chains between the gluino and squark continua.

For large gaugino mass the zero mode will merge into the continuum just as what we saw for the scalar superpartners of matter fields with $c > 0$.

4. Phenomenology and Conclusions

We discussed a novel possibility for the supersymmetric extension of the Standard Model, where there are continuum excitations of the SM particles and their superpartners arising from conformal dynamics. The properties of the superpartners can be quite different in this type of models. The superpartner of a Standard Model particle can be either a discrete mode below a continuum, or the first of a series of discrete modes, or just a continuum. The continuum superpartners will be quite challenging experimentally. It would be difficult to reconstruct any peak or edge because of the additional smearing of the mass by the continuous spectrum.

At a high energy collider, if the superpartners are produced well above the threshold of the continuum, there is a possibility of extended decay chains as shown in Fig. 5. These events are expected to have large multiplicities and more spherical shapes as a reflection of the underlying conformal theory.[31–34]

The experimental searches and verifications of this scenario will be very challenging. It is a topic under current investigation. The LHC will be the only high energy collider in the foreseeable future. We need to be prepared for any surprises and challenges that new physics may present to us at the LHC.

Acknowledgments

I would like to thank the organizers for inviting me to give a talk at the SCGT 09 workshop. This work is supported in part by the Department of Energy Grant DE-FG02-91ER40674.

References

1. M. J. Strassler and K. M. Zurek, Phys. Lett. B **651**, 374 (2007) [arXiv:hep-ph/0604261]; M. J. Strassler, arXiv:hep-ph/0607160; T. Han, Z. Si, K. M. Zurek and M. J. Strassler, JHEP **0807**, 008 (2008) [arXiv:0712.2041 [hep-ph]].
2. J. Kang and M. A. Luty, JHEP **0911**, 065 (2009) [arXiv:0805.4642 [hep-ph]].
3. H. Georgi, Phys. Rev. Lett. **98**, 221601 (2007) [arXiv:hep-ph/0703260].
4. H. Cai, H. C. Cheng, A. D. Medina and J. Terning, Phys. Rev. D **80**, 115009 (2009) [arXiv:0910.3925 [hep-ph]].
5. N. Arkani-Hamed, S. Dimopoulos and G. R. Dvali, Phys. Lett. B **429**, 263 (1998) [arXiv:hep-ph/9803315]; N. Arkani-Hamed, S. Dimopoulos and G. R. Dvali, Phys. Rev. D **59**, 086004 (1999) [arXiv:hep-ph/9807344].
6. T. M. Aliev, A. S. Cornell and N. Gaur, JHEP **0707**, 072 (2007) [arXiv:0705.4542 [hep-ph]].
7. S. L. Chen, X. G. He and H. C. Tsai, JHEP **0711**, 010 (2007) [arXiv:0707.0187 [hep-ph]].
8. K. Cheung, W. Y. Keung and T. C. Yuan, Phys. Rev. Lett. **99**, 051803 (2007) [arXiv:0704.2588 [hep-ph]]; K. Cheung, W. Y. Keung and T. C. Yuan, Phys. Rev. D **76**, 055003 (2007) [arXiv:0706.3155 [hep-ph]].
9. H. Georgi, Phys. Lett. B **650**, 275 (2007) [arXiv:0704.2457 [hep-ph]].
10. M. Luo and G. Zhu, Phys. Lett. B **659**, 341 (2008) [arXiv:0704.3532 [hep-ph]].
11. H. Davoudiasl, Phys. Rev. Lett. **99**, 141301 (2007) [arXiv:0705.3636 [hep-ph]].
12. S. Hannestad, G. Raffelt and Y. Y. Y. Wong, Phys. Rev. D **76**, 121701 (2007) [arXiv:0708.1404 [hep-ph]].
13. J. McDonald, JCAP **0903**, 019 (2009) [arXiv:0709.2350 [hep-ph]].
14. Y. Liao and J. Y. Liu, Phys. Rev. Lett. **99**, 191804 (2007) [arXiv:0706.1284 [hep-ph]].
15. H. Goldberg and P. Nath, Phys. Rev. Lett. **100**, 031803 (2008) [arXiv:0706.3898 [hep-ph]].
16. N. G. Deshpande, S. D. H. Hsu and J. Jiang, Phys. Lett. B **659**, 888 (2008) [arXiv:0708.2735 [hep-ph]].
17. P. J. Fox, A. Rajaraman and Y. Shirman, Phys. Rev. D **76**, 075004 (2007) [arXiv:0705.3092 [hep-ph]].
18. G. Cacciapaglia, G. Marandella and J. Terning, JHEP **0801**, 070 (2008) [arXiv:0708.0005 [hep-ph]].
19. M. Neubert, Phys. Lett. B **660**, 592 (2008) [arXiv:0708.0036 [hep-ph]].
20. J. M. Maldacena, Adv. Theor. Math. Phys. **2**, 231 (1998) [Int. J. Theor. Phys. **38**, 1113 (1999)] [arXiv:hep-th/9711200]; S. S. Gubser, I. R. Klebanov and A. M. Polyakov, Phys. Lett. B **428**, 105 (1998) [arXiv:hep-th/9802109]; E. Witten, Adv. Theor. Math. Phys. **2**, 253 (1998) [arXiv:hep-th/9802150].
21. L. Randall and R. Sundrum, Phys. Rev. Lett. **83**, 4690 (1999) [arXiv:hep-th/9906064].
22. G. Cacciapaglia, G. Marandella and J. Terning, JHEP **0902**, 049 (2009) [arXiv:0804.0424 [hep-ph]].
23. A. Falkowski and M. Perez-Victoria, Phys. Rev. D **79**, 035005 (2009) [arXiv:0810.4940 [hep-ph]].
24. A. Friedland, M. Giannotti and M. Graesser, Phys. Lett. B **678**, 149 (2009) [arXiv:0902.3676 [hep-th]].
25. N. Arkani-Hamed, T. Gregoire and J. G. Wacker, JHEP **0203**, 055 (2002) [arXiv:hep-th/0101233].
26. D. Marti and A. Pomarol, Phys. Rev. D **64**, 105025 (2001) [arXiv:hep-th/0106256].
27. O. Aharony, S. S. Gubser, J. M. Maldacena, H. Ooguri and Y. Oz, Phys. Rept. **323**, 183 (2000) [arXiv:hep-th/9905111].

28. R. Contino and A. Pomarol, JHEP **0411**, 058 (2004) [arXiv:hep-th/0406257].
29. G. Cacciapaglia, G. Marandella and J. Terning, JHEP **0906**, 027 (2009) [arXiv:0802.2946 [hep-th]].
30. A. Karch, E. Katz, D. T. Son and M. A. Stephanov, Phys. Rev. D **74**, 015005 (2006) [arXiv:hep-ph/0602229].
31. J. Polchinski and M. J. Strassler, JHEP **0305**, 012 (2003) [arXiv:hep-th/0209211]; M. J. Strassler, arXiv:0801.0629 [hep-ph].
32. D. M. Hofman and J. Maldacena, JHEP **0805**, 012 (2008) [arXiv:0803.1467 [hep-th]].
33. Y. Hatta, E. Iancu and A. H. Mueller, JHEP **0805**, 037 (2008) [arXiv:0803.2481 [hep-th]].
34. C. Csaki, M. Reece and J. Terning, JHEP **0905**, 067 (2009) [arXiv:0811.3001 [hep-ph]].

Review of Minimal Flavor Constraints for Technicolor*

Hidenori S. Fukano† and Francesco Sannino
CP³-Origins, Campusvej 55, DK-5230 Odense M, Denmark

We analyze the constraints on the the vacuum polarization of the standard model gauge bosons from a minimal set of flavor observables valid for a general class of models of dynamical electroweak symmetry breaking. We will show that the constraints have a strong impact on the self-coupling and masses of the lightest spin-one resonances.

1. Introduction

Dynamical electroweak symmetry breaking constitutes one of the best motivated extensions of the standard model (SM) of particle interactions. Walking dynamics for breaking the electroweak symmetry was introduced in [2]. Studies of the dynamics of gauge theories featuring fermions transforming according to higher dimensional representations of the new gauge group has led to several phenomenological possibilities [3, 4] such as (Next) Minimal Walking Technicolor (MWT) [5]. The reader can find in [6] a comprehensive review of the current status of the phase diagram for chiral and nonchiral gauge theories needed to construct sensible extensions of the SM featuring a new strong dynamics. In [7] it was launched a coherent program to investigate different signals of minimal models of technicolor at the Large Hadron Collider experiment at CERN.

Whatever is the dynamical extension of the SM it will, in general, modify the vacuum polarizations of the SM gauge bosons. LEP I and II data provided direct constraints on these vacuum polarizations [8, 9]. In this talk, to be specific, we will assume that the vacuum polarizations are saturated by new spin-one states (techni-vector meson and techni-axial vector meson) and show that it is possible to provide strong constraints on their self-couplings and masses for a general class of models of dynamical electroweak symmetry breaking. Our results can be readily applied to any extension of the SM featuring new heavy spin-one states. In particular it will severely limit the possibility to have very light spin-one resonances to occur at the LHC even if the underlying gauge theory has vanishing S-parameter.

*This talk is based on [1] and given at 2009 Nagoya Global COE workshop " Strong Coupling Gauge Theories in LHC Era" (SCGT 09), December 8-11, 2009.

†Speaker, E-mail: hidenori@cp3.sdu.dk

2. Minimal $\Delta F = 2$ Flavor Corrections from Technicolor

Our goal is to compute the minimal contributions, i.e. coming just from the technicolor sector, for processes in which the flavor number F changes by two units, i.e. $\Delta F = 2$. Here we consider F to be either the strange or the bottom number.

Besides the intrinsic technicolor corrections to flavor processes one has also the corrections stemming out from extended technicolor models [10] which are directly responsible for providing mass to the SM fermions. In this talk, we are not attempting to provide a full theory of flavor but merely estimate the impact of a new dynamical sector, per se, on well known flavor observables. We will, however, assume that whatever is the correct mechanism behind the generation of the mass of the SM fermions it will lead to SM type Yukawa interactions [11]. This means that we will constrain models of technicolor with extended technicolor interactions entering in the general scheme of minimal flavor violation (MFV) theories [12].

We use the effective Lagrangian framework presented in [5] according to which the relevant interactions of the composite Higgs sector to the SM quarks up and down reads:

$$\mathcal{L}^{\text{quark}}_{\text{yukawa}} = \frac{\sqrt{2}\, m_{u_i}}{v} V_{ij} \cdot \bar{u}_{Ri} \pi^+ d_{Lj} - \frac{\sqrt{2}\, m_{d_i}}{v} V^*_{ji} \cdot \bar{d}_{Ri} \pi^- u_{Lj} + h.c.\,, \tag{1}$$

where $m_{ui}, (u_i = u, c, t)$ and $m_{di}, (d_i = d, s, b)$ are respectively the up and down quark masses of the ith generation. V_{ij} is the i, j element of the Cabibbo-Kobayashi-Maskawa (CKM) matrix. This is our starting point which will allow us to compute the $\Delta F = 2$ processes in Fig. 1 [a]. After the computation of all amplitudes in Fig. 1,

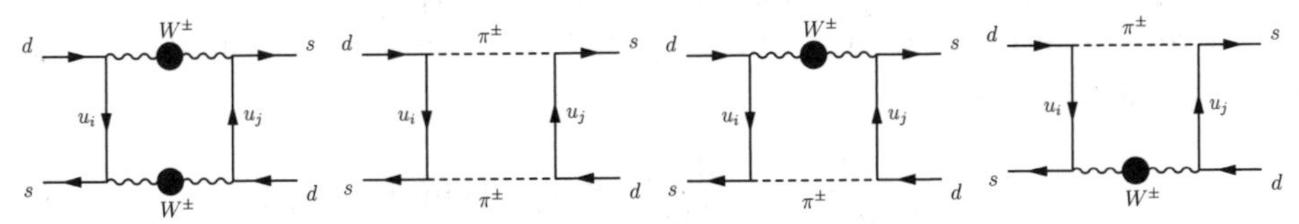

Fig. 1. Box diagrams for $\Delta S = 2$ *annihilation* processes. To obtain the $\Delta B = 2$ process, we should simply rename s with b and d with q $(q = d, s)$ in the various diagrams.

we obtain the effective Lagrangian describing these processes as

$$\mathcal{L}^{\Delta F=2}_{\text{eff}} = -\frac{G_F^2 M_W^2}{4\pi^2} \cdot A(a_V, a_A) \cdot Q_{\Delta F=2}\ , \quad A(a_V, a_A) \equiv \sum_{i,j=u,c,t} \left[\lambda_i \lambda_j \cdot E(a_i, a_j, a_V, a_A)\right]. \tag{2}$$

Here, we have expressed all the quantities by means of the following ratios $a_\alpha \equiv m_\alpha^2/M_W^2, (\alpha = i, j)$ and $a_v \equiv M_v^2/M_W^2, (v = V, A)$ and $m_i, (i = u, c, t)$ indicates the u_i mass while M_V, M_A are respectively the mass of the lightest techni-vector meson and techni-axial vector one. $Q_{\Delta F=2}$ and λ_i represent

[a]Note that the contribution of the *scattering* process to the invariant amplitude is equal to that of the *annihilation* one.

$\{Q_{\Delta F=2}, \lambda_i\} = \left\{(\bar{s}_L\gamma^\mu d_L)(\bar{s}_L\gamma_\mu d_L), V_{id}V_{is}^*\right\}$ for $K^0 - \bar{K}^0$ system and $\{Q_{\Delta F=2}, \lambda_i\} = \left\{(\bar{b}_L\gamma^\mu q_L)(\bar{b}_L\gamma_\mu q_L), V_{iq}V_{ib}^*\right\}$ for $B_q^0 - \bar{B}_q^0$ system. Moreover, $E(a_i, a_j, a_V, a_A)$ keeps track of the technicolor-modified gauge bosons propagators.

Indicating with $g_{\rm EW}$ the weak-coupling constant and $\tilde{g}$ the coupling constant governing the massive spin-one self interactions and by expanding up to the order in $O(g_{\rm EW}^4/\tilde{g}^4)$ one can rewrite $E(a_i, a_j, a_V, a_A)$ as:

$$E(a_i, a_j, a_V, a_A) = E_0(a_i, a_j) + \frac{g_{\rm EW}^2}{\tilde{g}^2}\Delta E(a_i, a_j, a_V, a_A). \tag{3}$$

The SM contribution is fully contained in E_0 which are consistent with the results in [13] and the technicolor one appear first in ΔE. The latter can be divided into a vector and an axial-vector contribution as follows: $\Delta E(a_i, a_j, a_V, a_A) = h(a_i, a_j, a_V) + (1-\chi)^2 \cdot h(a_i, a_j, a_A)$ where the quantity χ was introduced first in [14] and the axial-vector decay constant is directly proportional to the quantity $(1-\chi)^2$. The vector and axial decay constant are: $f_V^2 = M_V^2/\tilde{g}^2$, $f_A^2 = (1-\chi)^2 M_A^2/\tilde{g}^2$.

We also write:

$$A(a_V, a_A) = A_0 + \frac{g_{\rm EW}^2}{\tilde{g}^2}\cdot \Delta A(a_V, a_A). \tag{4}$$

Upon taking into account the unitarity of the CKM matrix and setting $a_u \to 0$ one has $A_0 = \eta_1 \cdot \lambda_c^2 \cdot \bar{E}_0(a_c) + \eta_2 \cdot \lambda_t^2 \cdot \bar{E}_0(a_t) + \eta_3 \cdot 2\lambda_c\lambda_t \cdot \bar{E}_0(a_c, a_t)$ and $\Delta A(a_V, a_A) = \eta_1 \cdot \lambda_c^2 \cdot \Delta\bar{E}(a_c, a_V, a_A) + \eta_2 \cdot \lambda_t^2 \cdot \Delta\bar{E}(a_t, a_V, a_A) + \eta_3 \cdot 2\lambda_c\lambda_t \cdot \Delta\bar{E}(a_c, a_t, a_V, a_A)$ where $\eta_{1,2,3}$ are the QCD corrections to $\bar{E}_0$ and $\Delta\bar{E}$. The explicit expressions for the functions E_0, ΔE, h, $\bar{E}$ and $\Delta\bar{E}$ various expressions can be found in [1].

We recall that the absolute value of the CP-violation parameter in the $K^0 - \bar{K}^0$ system is given by [15]:

$$(|\epsilon_K|)_{\rm full} = \frac{G_F^2 M_W^2}{12\sqrt{2}\pi^2} \times \left[\frac{M_K}{\Delta M_K}\right]_{\rm exp.} \times B_K f_K^2 \times [-{\rm Im}A(a_V, a_A)], \tag{5}$$

and the meson mass difference in the $Q^0 - \bar{Q}^0$, $Q = (K, B_d, B_s)$ system is given by

$$\left(\Delta M_Q\right)_{\rm full} \equiv 2 \cdot \left|\langle \bar{Q}^0| - \mathcal{L}_{\rm eff}^{\Delta F=2}|Q^0\rangle\right| = \frac{G_F^2 M_W^2}{6\pi^2} \cdot f_Q^2 \cdot M_Q \times B_Q \times |A(a_V, a_A)|, \tag{6}$$

where f_Q is the decay constant of the Q-meson and M_Q is its mass. B_Q is identified with the QCD bag parameter. This bag parameter is an intrinsic QCD contribution and we assume that the technicolor sector does not contribute to the bag parameter [b]. The experimental values of $G_F, M_W, f_Q, M_Q, \Delta M_Q$ and the bag parameter B_Q are [16] $G_F = 1.1664 \times 10^{-5}\,{\rm GeV}^{-2}$, $M_W = 80.398\,{\rm GeV}$, $f_K = 155.5\,{\rm MeV}$, $B_K = 0.72 \pm 0.040$, $f_{B_d}\sqrt{B_{B_d}} = 225 \pm 35\,{\rm MeV}$, $f_{B_s}\sqrt{B_{B_s}} = 270 \pm 45\,{\rm MeV}$, $M_K = 497.61 \pm 0.02\,{\rm MeV}$, $M_{B_d} = 5279.5 \pm 0.3\,{\rm MeV}$, $M_{B_s} = 5366.3 \pm 0.6\,{\rm MeV}$.

[b]This is a particularly good approximation when the technicolor sector does not have techiquarks charged under ordinary color. The best examples are Minimal Walking Technicolor models.

It is convenient to define the following quantities:

$$\delta_\epsilon \equiv \frac{g^2_{\rm EW}}{\tilde{g}^2} \cdot \frac{{\rm Im}\Delta A(a_V,a_A)}{{\rm Im}A_0} \quad , \quad \delta_{M_Q} \equiv \frac{g^2_{\rm EW}}{\tilde{g}^2} \cdot \frac{\Delta A(a_V,a_A)}{A_0} \,. \tag{7}$$

Using these expressions we write $(|\epsilon_K|)_{\rm full}$ and $(\Delta M_Q)_{\rm full}$ as $(|\epsilon_K|)_{\rm full} = (|\epsilon_K|)_{\rm SM} \times (1+\delta_\epsilon)$, $\left(\Delta M_Q\right)_{\rm full} = \left(\Delta M_Q\right)_{\rm SM} \times \left|1+\delta_{M_Q}\right|$ where $(|\epsilon_K|)_{\rm SM}$ and $\left(\Delta M_Q\right)_{\rm SM}$ are the SM expressions corresponding to the representations in Eqs.(5,6) with $A = A_0$ and their values are $(|\epsilon_K|)_{\rm SM} = (2.08^{+0.14}_{-0.13}) \times 10^{-3}$, $(\Delta M_K)_{\rm SM} = (3.55^{+1.09}_{-1.00})$ ns^{-1}, $(\Delta M_{B_d})_{\rm SM} = (0.56^{+0.19}_{-0.16})$ ps^{-1}, $(\Delta M_{B_s})_{\rm SM} = (17.67^{+6.38}_{-5.40})$ ps^{-1}. The experimental values are [16] $|\epsilon_K| = (2.229 \pm 0.012) \times 10^{-3}$, $\Delta M_K = 5.292 \pm 0.0009{\rm ns}^{-1}$, $\Delta M_{B_d} = 0.507 \pm 0.005{\rm ps}^{-1}$, $\Delta M_{B_s} = 17.77 \pm 0.10{\rm ps}^{-1}$.

We are now ready to compare the SM values with the experimental one. We can read off the constrain on δ_ϵ which is:

$$\delta_\epsilon = \left(7.05^{+7.93}_{-7.07}\right) \times 10^{-2} \quad (68\%\,{\rm C.L.})\,. \tag{8}$$

In order to compare the corrections associate to the kaon mass ΔM_K we formally separate the short distance contribution from the long distance one and write $\Delta M_K = (\Delta M_K)_{\rm SD} + (\Delta M_K)_{\rm LD}$. Here $(\Delta M_K)_{\rm SD}$ encodes the short distance contribution which must be confronted with the technicolor one Eq.(7) and the long distance contribution, $(\Delta M_K)_{\rm LD}$, corresponds to the exchange of the light pseudoscalar mesons. It is difficult to pin-point the $(\Delta M_K)_{\rm LD}$ contribution [15, 17] and hence we can only derive very weak constraints from δ_{M_K}. In fact we simply require that $(\Delta M_K)_{\rm SD} = (\Delta M_K)_{\rm SM}\ |1+\delta_{M_K}| \leq (\Delta M_K)_{\rm exp.}$. This means that:

$$|1+\delta_{M_K}| \leq 2.08 \quad (68\%\,{\rm C.L.})\,. \tag{9}$$

On the other hand the short distance contribution dominates the $B^0_q - \bar{B}^0_q$ mass difference [17] yielding the following constraints:

$$|1+\delta_{M_{Bd}}| = 0.91^{+0.38}_{-0.24} \quad , \quad |1+\delta_{M_{Bs}}| = 1.01^{+0.44}_{-0.27} \qquad (68\%\,{\rm C.L.})\,. \tag{10}$$

3. Constraining Models of Dynamical Electroweak Symmetry Breaking

We will now use the minimal flavor experimental information to reduce the parameter space of a general class of models of dynamical electroweak symmetry breaking.

If the underlying technicolor theory is QCD like we can impose the standard 1st Weinberg sum rule (WSR): $f_V^2 - f_A^2 = f_\pi^2 = \left(v_{\rm EW}/\sqrt{2}\right)^2$ and 2nd WSR: $f_V^2 M_V^2 - f_A^2 M_A^2 = 0$ with f_V and f_A the vector and axial decay constants as shown in [5, 14, 18]. Using the explicit expressions of the decay constants in terms of the coupling $\tilde{g}$ and vector masses provided in [5, 14, 18] and imposing the above sum rules we derive: $1/a_V = (g^2_{\rm EW} S)/(16\pi) - 1/a_A$ (≥ 0), with the S-parameter [8] reading [5, 14, 18]: $S \equiv 8\pi \left[f_V^2/M_V^2 - f_A^2/M_A^2\right] = \left(8\pi/\tilde{g}^2\right)\left[1-(1-\chi)^2\right]$. The condition above yields the following additional constraint for $\tilde{g}$ by simply noting that the quantity $(1-\chi)^2$ is positive: $\tilde{g} < \sqrt{8\pi/S}$. The constraints on $(M_A, \tilde{g})$ induced by WSRs and theoretical constraints are stronger, for a given S, than the ones deriving from flavor experiments and expressed in (8)-(10). This is not surprising given that in an ordinary technicolor theory the spin-one states are very heavy.

The situation changes when allowing for a walking behavior. Besides the flavor constraints one has also the ones due to the electroweak precision measurements [19] as well as the unitarity constraint of $W_L - W_L$ scattering [20]. We will consider all of them. As for the walking technicolor case we reduce the number of independent parameters at the effective Lagrangian level via the 1st WSR : $f_V^2 - f_A^2 = f_\pi^2 = \left(v_{\rm EW}/\sqrt{2}\right)^2$ and the 2nd modified [14] WSR : $f_V^2 M_V^2 - f_A^2 M_A^2 = a \cdot (16\pi^2 f_\pi^4)/d(\mathrm{R})$ where a is a number expected to be positive and $O(1)$ [14] and $d(\mathrm{R})$ is the dimension of the representation of the underlying technifermions as shown in [5]. We have now $M_A^2 < 8\pi f_\pi^2/S \cdot (2-\chi)/(1-\chi) = 8\pi f_\pi^2/S \cdot \left[1 + (1 - \tilde{g}^2 S/(8\pi))^{-1/2}\right]$.

In Fig. 2, we show the allowed region in the $(M_A, \tilde{g})$-plane after having imposed the minimal flavor constraints due to the experimental values of $|\epsilon_K|$ and ΔM_Q obtained using Eqs. (8)-(10) together with the theoretical constraints for $\tilde{g}, M_A^2$. To obtain Fig. 2, we used the expressions for A_0 and $\Delta A(a_V, a_A)$ in which $a_V = \left[1 - \tilde{g}^2 S/(8\pi)\right] \cdot a_A + 2\tilde{g}^2/g_{\rm EW}^2$. Given that the upper bound for δ_{M_K} is always larger than the theoretical estimate in the region $M_A > 200$ GeV we conclude that the ΔM_K constraint is not yet very severe and hence it is not displayed in Fig. 2. To make the plots we need also the value of the S parameters and hence we analyzed as explicit example minimal walking technicolor models. In the case of MWT, we use for S the naive MWT estimate, i.e. $S = 1/(2\pi)$ [3] while $\tilde{g}$ is constrained to be $\tilde{g} < 12.5$. In the case of Next to Minimal Walking Technicolor (NMWT), we use the the naive S is approximately $1/\pi$ [3] and the constraint on $\tilde{g}$ yields $\tilde{g} < 8.89$.

We have plotted the various constraints on the $(M_A, \tilde{g})$-plane for MWT(NMWT) in the upper and lower left (right) panel of Fig. 2. In the upper (lower) left figure we compare the 68% C.L. (95% C.L.) allowed regions coming from the minimal flavor constraints (the darker region above the dotted line) with the ones from LEP II data (region above the dashed line). It is clear that the flavor constraints are stronger for the 68% C.L. case but are weaker for the 95% C.L. one with respect to the constraints from LEP II data.

In the limit $M_A = M_V = M$ and $\chi = 0$ the effective theory acquires a new symmetry [21]. This new symmetry relates a vector and an axial field and can be shown to work as a custodial symmetry for the S parameter [21]. The only non-zero electroweak parameters are W, Y parameters. It was already noted in [19] that a custodial technicolor model cannot be easily achieved via an underlying walking dynamics and should be interpreted as an independent framework. This is so since custodial technicolor models do not respect the WSRs [c]. We directly compare in the Fig. 3 the constraints on the custodial technicolor parameter region $(M, \tilde{g})$ coming from LEP II and flavor constraints and find a similar trend as for the other cases.

[c] One can, of course, imagine more complicated vector spectrum leading to such a symmetry.

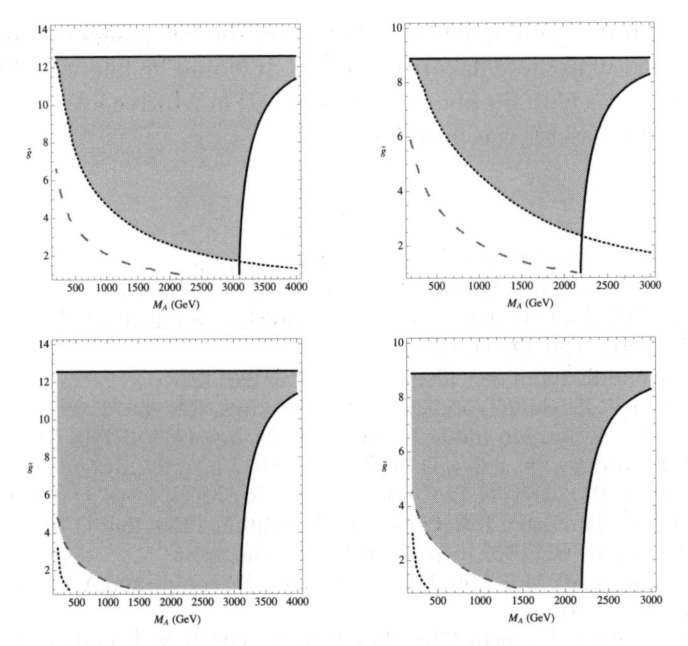

Fig. 2. The upper and lower left panels represents the allowed region in the $(M_A, \tilde{g})$-plane for MWT respectively for the 68% C.L. and 95% C.L.. A similar analysis is shown for NMWT in the right hand upper and lower panels. The region above the straight solid line is forbidden by the condition $\tilde{g} < 12.5$ for MWT and $\tilde{g} < 8.89$ for NMWT while the region below the solid curve (on the right corner) is forbidden by theoretical upper bound for M_A. In the two upper (lower) plots the dotted lines correspond to the 68% C.L. (95% C.L.) flavor constraints while the dashed lines are the 68% C.L. (95% C.L.) from LEP II data. The flavor constraints come only from ϵ_K since the ones from ΔM_{B_q} are not as strong.

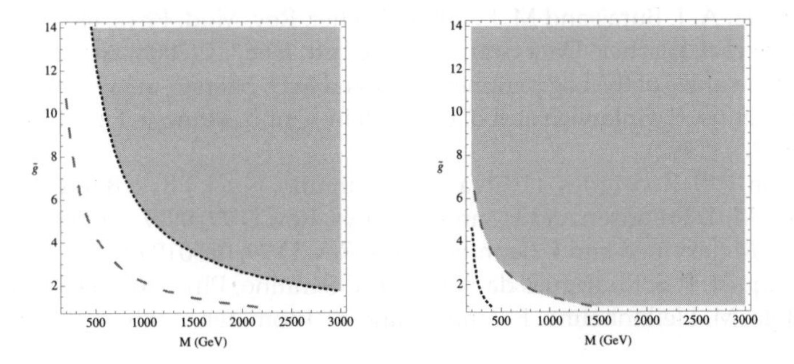

Fig. 3. The left (right) panel represents the allowed region in the $(M_A, \tilde{g})$-plane for CT respectively for the 68% C.L. (95% C.L.). In the two upper (lower) plots the dotted lines correspond to the 68% C.L. (95% C.L.) flavor constraints while the dashed lines are the 68% C.L. (95% C.L.) from LEP II data. The flavor constraints come only from ϵ_K since the ones from ΔM_{B_q} are not as strong.

4. Summary

Flavor constraints are relevant for models of dynamical electroweak symmetry breaking with light spin-one resonances, in fact, any model featuring spin-one resonances with the same quantum numbers of the SM gauge bosons will have to be confronted with these flavor constraints. It would be interesting to combine the present analysis with the one presented in [22] in which a low energy effective theory for the ETC sector was introduced.

References

[1] H. S. Fukano and F. Sannino, arXiv:0908.2424 [hep-ph].
[2] B. Holdom, Phys. Rev. D **24**, 1441 (1981). ; K. Yamawaki, M. Bando and K. i. Matumoto, Phys. Rev. Lett. **56**, 1335 (1986). ; T. W. Appelquist, D. Karabali and L. C. R. Wijewardhana, Phys. Rev. Lett. **57**, 957 (1986).
[3] F. Sannino and K. Tuominen, Phys. Rev. D **71**, 051901 (2005).
[4] D. D. Dietrich, F. Sannino and K. Tuominen, Phys. Rev. D **72**, 055001 (2005); *ibid* ; D. D. Dietrich, F. Sannino and K. Tuominen, Phys. Rev. D **73**, 037701 (2006); D. D. Dietrich and F. Sannino, Phys. Rev. D **75**, 085018 (2007) ; T. A. Ryttov and F. Sannino, Phys. Rev. D **76**, 105004 (2007) ; N. D. Christensen and R. Shrock, Phys. Lett. B **632**, 92 (2006).
[5] R. Foadi, M. T. Frandsen, T. A. Ryttov and F. Sannino, Phys. Rev. D **76**, 055005 (2007).
[6] F. Sannino, arXiv:0911.0931 [hep-ph]; 0804.0182 [hep-ph].
[7] A. Belyaev, R. Foadi, M. T. Frandsen, M. Jarvinen, F. Sannino and A. Pukhov, Phys. Rev. D **79**, 035006 (2009).
[8] M. E. Peskin and T. Takeuchi, 'Phys. Rev. D **46**, 381 (1992); M. E. Peskin and T. Takeuchi, Phys. Rev. Lett. **65**, 964 (1990).
[9] R. Barbieri, A. Pomarol, R. Rattazzi and A. Strumia, Nucl. Phys. B **703**, 127 (2004).
[10] S. Dimopoulos and L. Susskind, Nucl. Phys. B **155**, 237 (1979) ; E. Eichten and K. D. Lane, Phys. Lett. B **90**, 125 (1980).
[11] R. S. Chivukula and H. Georgi, Phys. Lett. B **188**, 99 (1987).
[12] G. D'Ambrosio, G. F. Giudice, G. Isidori and A. Strumia, Nucl. Phys. B **645**, 155 (2002).
[13] T. Inami and C. S. Lim, Prog. Theor. Phys. **65**, 297 (1981) [Erratum-ibid. **65**, 1772 (1981)].
[14] T. Appelquist and F. Sannino, Phys. Rev. D **59**, 067702 (1999).
[15] G. Buchalla, A. J. Buras and M. E. Lautenbacher, Rev. Mod. Phys. **68**, 1125 (1996).
[16] C. Amsler *et al.* [Particle Data Group], Phys. Lett. B **667**, 1 (2008) and for definitiveness, we use the values of the bag parameters quoted in U. Nierste, arXiv:0904.1869 [hep-ph].
[17] J. F. Donoghue, E. Golowich and B. R. Holstein, Camb. Monogr. Part. Phys. Nucl. Phys. Cosmol. **2**, 1 (1992).
[18] Z. y. Duan, P. S. Rodrigues da Silva and F. Sannino, Nucl. Phys. B **592**, 371 (2001).
[19] R. Foadi, M. T. Frandsen and F. Sannino, Phys. Rev. D **77**, 097702 (2008).
[20] R. Foadi, M. Jarvinen and F. Sannino, Phys. Rev. D **79**, 035010 (2009).
[21] T. Appelquist, P. S. Rodrigues da Silva and F. Sannino, Phys. Rev. D **60**, 116007 (1999).
[22] M. Antola, M. Heikinheimo, F. Sannino and K. Tuominen, arXiv:0910.3681 [hep-ph].

Standard Model and High Energy Lorentz Violation

Damiano Anselmi

Dipartimento di Fisica "Enrico Fermi", Università di Pisa
Largo Pontecorvo 3, I-56127 Pisa, Italy
and
INFN, Sezione di Pisa, Pisa, Italy
E-mail: damiano.anselmi@df.unipi.it

If Lorentz symmetry is violated at high energies, interactions that are usually non-renormalizable can become renormalizable by weighted power counting. The Standard Model admits a CPT invariant, Lorentz violating extension containing two scalar-two fermion interactions (which can explain neutrino masses) and four fermion interactions (which can explain proton decay). Suppressing the elementary scalar fields, we can use a dynamical symmetry breaking mechanism, in the Nambu–Jona-Lasinio spirit, to generate composite Higgs bosons and masses for fermions and gauge bosons. The low-energy effective action is uniquely determined and predicts relations among parameters of the Standard Model, which allows us to make indirect experimental tests of the high-energy Lorentz violation.

Keywords: Standard Model, Lorentz symmetry, Lorentz violation, renormalization.

Lorentz symmetry is a basic ingredient of the Standard Model of particle physics and one of the best tested and more precise symmetries of Nature.[1] However, the possibility that it might be violated at very high energies is still open and has inspired several investigations about the new physics that could emerge.

In quantum field theory, the violation of Lorentz symmetry at high energies allows us to renormalize otherwise non-renormalizable interactions,[2–4] such as two scalar-two fermion vertices and four fermion vertices. Terms with higher space derivatives modify the dispersion relations and generate propagators with improved ultraviolet behaviors. A "weighted" power counting, which assigns different weights to space and time, allows us to prove that the theory is renormalizable and consistent with (perturbative) unitarity, namely that no counterterms with higher time derivatives are generated. The theory remains also local, polynomial and causal.

Using these tools, we can formulate a Standard Model extension with the following properties:[5] it is CPT invariant, but Lorentz violating at high energies, it is unitary and renormalizable by weighted power counting; it contains the vertex $(LH)^2/\Lambda_L$, which gives Majorana masses to the neutrinos after symmetry breaking, but no right-handed neutrinos, nor other extra fields; it contains four fermion vertices, which can explain proton decay. The scale Λ_L is interpreted as the scale of

Lorentz violation. Inverse powers of Λ_L multiply operators of higher dimension. At energies much smaller than Λ_L Lorentz symmetry is recovered if certain parameters are appropriately fine-tuned. If neutrino masses have the origin we claim, then $\Lambda_L \sim 10^{14}$GeV. More generally, we expect that the scale of Lorentz violation is located below the Planck scale.

The model has two "weighted" dimensions, which means that at high energies its power counting resembles the one of a two-dimensional quantum field theory. In particular, only the four fermion vertices are strictly renormalizable, while the gauge and Higgs interactions are super-renormalizable. Then at energies $\gg \Lambda_L$ all gauge bosons and the Higgs field become free and decouple, and what remains is a (Lorentz violating) four fermion model in two weighted dimensions. It is then natural to inquire what physical effects are induced, at lower energies, by a dynamical symmetry breaking mechanism, in the Nambu–Jona-Lasinio spirit.[6] If we suppress the elementary scalar field, we obtain a model that is candidate to reproduce the observed low energy physics, predict relations among otherwise independent parameters, and possibly predict new physics detectable at LHC.[7]

We assume that invariance under rotations in preserved. We decompose coordinates x^μ as $(\hat{x}^\mu, \bar{x}^\mu)$, where $\hat{x}^\mu$, or simply $\hat{x}$, denotes the time component, and $\bar{x}^\mu$ denote the space components. Similarly, we decompose the space time index μ as $(\hat{\mu}, \bar{\mu})$, the partial derivative ∂_μ as $(\hat{\partial}_\mu, \bar{\partial}_\mu)$, and gauge vectors A_μ as $(\hat{A}_\mu, \bar{A}_\mu)$. The Lorentz violating theory is renormalizable by weighted power counting[2–4] in $đ = 1+3/n$ "weighted dimensions", where energy has weight 1 and the space components of momenta have weight $1/n$. Scalar propagators have weight -2 and fermion propagators have weight -1.

The model The "Standard-Extended Model" of ref. [5] has $n = 3$ and therefore weighted dimension 2. The lagrangian of its simplest version reads

$$\mathcal{L} = \mathcal{L}_Q + \mathcal{L}_{\text{kin}f} + \mathcal{L}_H + \mathcal{L}_Y - \frac{\bar{g}^2}{4\Lambda_L}(LH)^2 - \sum_{I=1}^{5} \frac{1}{\Lambda_L^2} g\bar{D}\bar{F}\,(\bar{\chi}_I \bar{\gamma} \chi_I) + \frac{Y_f}{\Lambda_L^2}\bar{\chi}\chi\bar{\chi}\chi - \frac{g}{\Lambda_L^2}\bar{F}^3, \tag{1}$$

where

$$\mathcal{L}_Q = \frac{1}{4}\sum_G \left(2F^G_{\hat{\mu}\bar{\nu}}\eta^G(\bar{\Upsilon})F^G_{\hat{\mu}\bar{\nu}} - F^G_{\bar{\mu}\bar{\nu}}\tau^G(\bar{\Upsilon})F^G_{\bar{\mu}\bar{\nu}}\right),$$

$$\mathcal{L}_{\text{kin}f} = \sum_{a,b=1}^{3}\sum_{I=1}^{5} \bar{\chi}^a_I\, i\left(\delta^{ab}\hat{\not{D}} - \frac{b_0^{Iab}}{\Lambda_L^2}\bar{\not{D}}^3 + b_1^{Iab}\bar{\not{D}}\right)\chi^b_I,$$

$$\mathcal{L}_H = |\hat{D}_{\hat{\mu}}H|^2 - \frac{a_0}{\Lambda_L^4}|\bar{D}^2\bar{D}_{\bar{\mu}}H|^2 - \frac{a_1}{\Lambda_L^2}|\bar{D}^2H|^2 - a_2|\bar{D}_{\bar{\mu}}H|^2 - \mu_H^2|H|^2 - \frac{\lambda_4\bar{g}^2}{4}|H|^4,$$

$$\mathcal{L}_Y = -\bar{g}\Omega_i H^i + \text{h.c.}, \qquad \Omega_i = \sum_{a,b=1}^{3} Y_1^{ab}\bar{L}^{ai}\ell^b_R + Y_2^{ab}\bar{u}^a_R Q^{bj}_L\varepsilon^{ji} + Y_3^{ab}\bar{Q}^{ai}_L d^b_R, \tag{2}$$

i, j are $SU(2)_L$ indices, $\chi_1^a = L^a = (\nu_L^a, \ell_L^a)$, $\chi_2^a = Q_L^a = (u_L^a, d_L^a)$, $\chi_3^a = \ell_R^a$, $\chi_4^a = u_R^a$ and $\chi_5^a = d_R^a$. Moreover, $\nu^a = (\nu_e, \nu_\mu, \nu_\tau)$, $\ell^a = (e, \mu, \tau)$, $u^a = (u, c, t)$ and $d^a = (d, s, b)$. The sum $\sum_G$ is over the gauge groups $SU(3)_c$, $SU(2)_L$ and $U(1)_Y$, and the last three terms of (1) are symbolic. Finally, $\bar{\Upsilon} \equiv -\bar{D}^2/\Lambda_L^2$, where Λ_L is the scale of Lorentz violation, and η^G, τ^G are polynomials of degree 2 and 4, respectively. Gauge anomalies cancel out exactly as in the Standard Model.[5] The "boundary conditions" such that Lorentz invariance is recovered at low energies are that b_1^{Iab} tend to δ^{ab} and a_2, η^G and τ^G tend to 1 (four such conditions can be trivially fulfilled normalizing the gauge fields and the space coordinates $\bar{x}$).

Since propagators contain higher powers of momenta the dispersion relations are modified. At $n = 3$ a typical modified dispersion relation reads

$$E = \sqrt{m^2 + a\bar{p}^2 + b\frac{\bar{p}^2}{\Lambda_L^2} + c\left(\frac{\bar{p}^2}{\Lambda_L^2}\right)^2},$$

where a, b and c are constants and of course $m = 0$ for gauge fields. The ultraviolet behaviors are improved enough to make the theory renormalizable. Since the weight of a scalar field vanishes in $ꟈ = 2$ a constant $\bar{g}$ of weight $1/2$ is attached to the scalar legs to ensure renormalizability. The gauge coupling g has weight 1. The weights of all other parameters are determined so that each lagrangian term has weight 2 ($= ꟈ$). We have neutrino masses $\sim v^2/\Lambda_L$, v being the Higgs vev, assuming that all other parameters involved in the vertex $(LH)^2/\Lambda_L$ are of order 1. Reasonable estimates of the neutrino masses (a fraction of eV) give $\Lambda_L \sim 10^{14}$GeV.

Alternative model An alternative model can be obtained rearranging the weight assignments to simplify the gauge sector. Specifically, we replace η^G with unity and τ^G with a polynomial of degree 2, which we denote by τ'^G. In a suitable "Feynman" gauge the gauge-field propagator becomes reasonably simple to be used in practical computations.[8] The price is a more complicated Higgs sector, because g and $\bar{g}$ get a lower weight (1/3). The simplest version of the alternative model has lagrangian

$$\begin{aligned} \mathcal{L}' = {} & \mathcal{L}'_Q + \mathcal{L}_{\mathrm{kin}f} + \mathcal{L}'_H + \mathcal{L}_Y - \frac{\bar{g}^2}{4\Lambda_L}(LH)^2 - \sum_{I=1}^{5} \frac{1}{\Lambda_L^2} g\bar{D}\bar{F}\,(\bar{\chi}_I \bar{\gamma} \chi_I) + \frac{Y_f}{\Lambda_L^2} \bar{\chi}\chi\bar{\chi}\chi \\ & - \frac{g}{\Lambda_L^2}\bar{F}^3 - \frac{1}{\Lambda_L^2} g\bar{g}\bar{\chi}\chi\bar{F}H - \frac{1}{\Lambda_L^2}\left(\bar{g}^3\bar{\chi}\chi H^3 + \bar{g}^2\bar{\chi}\bar{D}\chi H^2 + \bar{g}\bar{\chi}\bar{D}^2\chi H\right) \\ & - \frac{1}{\Lambda_L^4}\left(g\bar{D}^2\bar{F} + g^2\bar{F}^2\right) H^\dagger H, \end{aligned} \tag{3}$$

where

$$\mathcal{L}'_Q = \frac{1}{4}\sum_G \left(2F^G_{\hat{\mu}\bar{\nu}}F^G_{\hat{\mu}\bar{\nu}} - F^G_{\bar{\mu}\bar{\nu}}\tau'^G(\bar{\Upsilon})F^G_{\bar{\mu}\bar{\nu}}\right),$$

and

$$\mathcal{L}'_H = \mathcal{L}_H - \frac{\lambda_4^{(3)}\bar{g}^2}{4\Lambda_L^2}|H|^2|\bar{D}_{\bar{\mu}}H|^2 - \frac{\lambda_4^{(2)}\bar{g}^2}{4\Lambda_L^2}|H^\dagger\bar{D}_{\bar{\mu}}H|^2 - \frac{\bar{g}^2}{4\Lambda_L^2}\left[\lambda_4^{(1)}(H^\dagger\bar{D}_{\bar{\mu}}H)^2 + \text{h.c.}\right]$$
$$-\frac{\lambda_6\bar{g}^4}{36\Lambda_L^2}|H|^6.$$

The QED subsector of this model is considered in ref. [8], where the high- and low- energy renormalizations are calculated at one loop and compared. It is also shown that at low energies power-like divergences in Λ_L get multiplied by arbitrary (i.e. scheme-dependent) constants, therefore can be removed bypassing the hierarchy problem.

The most general lagrangian is (3) plus the extra terms

$$\begin{gathered}\bar{g}^6H^8,\quad \bar{g}^4\bar{D}^2H^6,\quad \bar{g}^2\bar{D}^4H^4,\quad g\bar{g}^2\bar{D}^2\bar{F}H^4,\quad g\bar{D}^4\bar{F}H^2,\quad g^2\bar{g}^2\bar{F}^2H^4,\quad g^2\bar{F}^4,\\ g^2\bar{D}^2\bar{F}^2H^2,\quad g^3\bar{F}^3H^2,\quad g\bar{D}^2\bar{F}^3,\quad g\bar{\varepsilon}\tilde{F}\bar{D}^2H^2,\quad g^2\bar{\varepsilon}\tilde{F}\bar{F}H^2,\quad \bar{\varepsilon}\tilde{F}\bar{D}^2\bar{F},\\ g\bar{\varepsilon}\bar{F}\tilde{F}\bar{F},\quad \bar{\varepsilon}\tilde{F}\hat{D}\bar{D}\bar{F},\quad g\bar{g}\bar{\chi}\chi\bar{F}H,\quad \bar{g}\bar{\chi}\chi\bar{D}^2H,\quad \bar{g}^2\bar{\chi}\chi\bar{D}H^2,\quad \bar{g}^3\bar{\chi}\chi H^3,\end{gathered}\tag{4}$$

and those obtained suppressing some fields and/or derivatives, where $\bar{\varepsilon}$ is the ε-tensor with three space indices. The extra terms (4) can be consistently dropped, because they are not generated back by renormalization.

Scalarless model The scalarless Standard-Extended Model[7] reads

$$\mathcal{L}_{\text{noH}} = \mathcal{L}'_Q + \mathcal{L}_{\text{kin}f} - \sum_{I=1}^{5}\frac{1}{\Lambda_L^2}g\bar{D}\bar{F}\,(\bar{\chi}_I\bar{\gamma}\chi_I) + \frac{Y_f}{\Lambda_L^2}\bar{\chi}\chi\bar{\chi}\chi - \frac{g}{\Lambda_L^2}\bar{F}^3,\tag{5}$$

and is obtained suppressing the Higgs field in (3). Obviously, the gauge anomalies of (5) still cancel. We see that the simplification is considerable.

If we suppress the Higgs field in (1), the only difference is that $\mathcal{L}_Q$ appears instead of $\mathcal{L}'_Q$ in (5). We keep the simpler model (5), but our arguments do not depend on this choice.

Very-high-energy model At very high energies ($\gg \Lambda_L$) gauge and Higgs fields become free and decouple, because their interactions are super-renormalizable, so all theories (1), (3) and (5) become a four fermion model in two weighted dimensions, with lagrangian

$$\mathcal{L}_{4\text{f}} = \sum_{a,b=1}^{3}\sum_{I=1}^{5}\bar{\chi}_I^a\, i\left(\delta^{ab}\hat{\partial\!\!\!/} - \frac{b_0^{Iab}}{\Lambda_L^2}\bar{\partial\!\!\!/}^3 + b_1^{Iab}\bar{\partial\!\!\!/}\right)\chi_I^b + \frac{Y_f}{\Lambda_L^2}\bar{\chi}\chi\bar{\chi}\chi.\tag{6}$$

We have kept also the terms multiplied by b_1^{Iab}, since they are necessary to recover Lorentz invariance at low energies. The high-energy renormalization of this model is studied in ref. [9].

Predictivity The virtue of our approach to the dynamical electroweak symmetry breaking is that the high-energy physics of our model is unambiguous, encoded in (6). Since (6), as well as (5), (1) and (3), are renormalizable by weighted power counting, we do not need to consider other sectors of unknown physics beyond them. Thus, our model is predictive, and actually provides a viable renormalizable environment for the Nambu–Jona-Lasinio mechanism. With respect to other approaches to composite Higgs bosons, such as technicolor, or the introduction of extra heavy gauge bosons to renormalize four fermion vertices, it has the advantage of being conceptually more economic.

We could worry that the dynamical symmetry breaking might reverberate the Lorentz violation down to low energies. Nevertheless, we have proved that the violation of Lorentz symmetry remains highly suppressed even when the dynamical symmetry breaking takes place. The minimum of the effective potential is Lorentz invariant and no Lorentz violating interactions are drawn down to low energies.

We assume exact CPT invariance. While (in local, Hermitian) theories a CPT violation implies also the violation of Lorentz symmetry, the converse is not true, in general, except for special subclasses of terms. Thus, we have to introduce two a priori different energy scales, a scale of Lorentz violation Λ_L, and a scale of CPT violation Λ_{CPT}, with

$$\Lambda_{\mathrm{CPT}} \geqslant \Lambda_L.$$

Recent bounds on Lorentz violation suggested by the analysis of γ-ray bursts,[10] which claim that the first correction

$$c(E) \sim c\left(1 - \frac{E}{\bar{M}}\right) \tag{7}$$

to the velocity of light involves an energy scale $\bar{M} \geqslant 1.3 \cdot 10^{18}\mathrm{GeV}$. In the realm of local perturbative quantum field theory, a dispersion relation giving (7) must contain odd powers of the energy, therefore it must also violate CPT. Thus, we are lead to interpret the results of ref. [10] as bounds on $\Lambda_{\mathrm{CPT}} = \bar{M}$ rather than Λ_L. It is conceivable that there exists an energy region $\Lambda_L \leqslant E \leqslant \Lambda_{\mathrm{CPT}}$ where Lorentz symmetry is violated but CPT is still conserved. Assuming $\Lambda_{\mathrm{CPT}} \geqslant M_{Pl}$, we still expect Λ_L to be located below the Planck mass and that the energy region $\Lambda_L < E < \Lambda_{\mathrm{CPT}}$ can span four or five orders of magnitude.

Composite Higgs bosons Adapting an old suggestion due to Nambu,[11] Miransky *et al.*[12] and Bardeen *et al.*[13] to our case, we can explore the following scenario. When gauge interactions are switched off, the dynamical symmetry mechanism produces fermion condensates $\langle \bar{q}q \rangle$. The effective potential can be calculated in the large N_c limit and has a Lorentz invariant (local) minimum, which gives masses to the fermions. Massive scalar bound states (composite Higgs bosons) emerge, together with Goldstone bosons. At a second stage, gauge interactions are switched back on, so the Goldstone bosons associated with the breaking of $SU(2)_L \times U(1)_Y$ to $U(1)_Q$ are "eaten" by the $W^{\pm}$ and Z bosons, which become massive.

The low-energy effective action is Lorentz invariant and uniquely determined. It looks like the Standard Model, but its parameters are not all independent. Said differently, our mechanism predicts relations among parameters of the Standard Model. The naivest predictions are obtained in the leading order of the large N_c expansion, with gauge interactions switched off, and considering just the top and bottom quarks. In this approach, which has a good 50% of error, the composite Higgs boson has mass $\sim 2m_t$. Taking into account of the error in our calculation, we anyway conclude that our model favors a very heavy Higgs, say with mass above 180GeV.

On the other hand, the relation between m_t and the Fermi constant turns out to be in "too-good" agreement with the experimental value. The formula reads

$$\frac{1}{G_F} = \frac{N_c m_t^2}{4\pi^2\sqrt{2}} \ln \frac{\Lambda_L^2}{m_t^2}. \tag{8}$$

Using our estimated value $\Lambda_L = 10^{14}$GeV, we find $m_t = 171.6$GeV. We do not understand why this prediction is so right.

Our mechanism can of course take place also in the Higgsed models (1) and (3), if the four fermion vertices are chosen appropriately. There, its effects sum to those of the elementary Higgs doublet.

References

1. V.A. Kostelecký, Ed., *Proceedings of the Fourth Meeting on CPT and Lorentz Symmetry*, World Scientific, Singapore, 2008;
   V.A. Kostelecký and N. Russell, *Data tables for Lorentz and CTP violation*, *ibid.* p. 308 and arXiv:0801.0287 [hep-ph].
2. D. Anselmi and M. Halat, Renormalization of Lorentz violating theories, Phys. Rev. D 76 (2007) 125011 and arXiv:0707.2480 [hep-th].
3. D. Anselmi, Weighted power counting and Lorentz violating gauge theories. I. General properties, Ann. Phys. 324 (2009) 874 and arXiv:0808.3470 [hep-th].
4. D. Anselmi, Weighted power counting and Lorentz violating gauge theories. II. Classification, Ann. Phys. 324 (2009) 1058 and arXiv:0808.3474 [hep-th].
5. D. Anselmi, Weighted power counting, neutrino masses and Lorentz violating extensions of the Standard Model, Phys. Rev. D 79 (2009) 025017 and arXiv:0808.3475 [hep-ph].
6. Y. Nambu and G. Jona-Lasinio, Dynamical model of elementary particles based on an analogy with superconductivity. I, Phys. Rev. 122 (1961) 345.
7. D. Anselmi, Standard Model without elementary scalars and high-energy Lorentz violation, EPJC 65 (2010) 523 and arXiv:0904.1849 [hep-ph].
8. D. Anselmi and M. Taiuti, Renormalization of high-energy Lorentz violating QED, arXiv:0912.0113 [hep-ph].
9. D. Anselmi and E. Ciuffoli, Renormalization of high-energy Lorentz violating four fermion models, arXiv:1002.2704 [hep-ph].
10. The Fermi LAT and Fermi GBM Collaborations, Fermi observations of High-energy gamma-ray emission from GRB 080916C, Science 323 (1009) 1688 and DOI: 10.1126/science.1169101.

11. Y. Nambu, in New Frontiers of Physics, proceedings of the XI International Symposium on Elementary Particle Physics, Kazimierz, Poland, 1988, Z. Ajduk, S. Pokorski and A. Trautman, Ed.s, World Scientific, Singapore, 1989.
12. V.A. Miransky, M. Tanabashi and K. Yamawaki, Is the t quark responsible for the mass of W and Z bosons?, Mod. Phys. Lett. A4 (1989) 1043;
13. W.A. Bardeen, C.T. Hill and M. Lindner, Minimal dynamical symmetry breaking of the standard model, Phys. Rev. D 41 (1990) 1647.

Dynamical Electroweak Symmetry Breaking and Fourth Family

Michio Hashimoto

Theory Center, IPNS, KEK
1-1 Oho, Tsukuba, Ibaraki 305-0801, Japan
E-mail: michioh@post.kek.jp

We propose a dynamical model with a $(2+1)$-structure of composite Higgs doublets: two nearly degenerate composites of the fourth family quarks t' and b', $\Phi_{t'} \sim \bar{t'}_R(t',b')_L$ and $\Phi_{b'} \sim \bar{b'}_R(t',b')_L$, and a heavier top-Higgs resonance $\Phi_t \sim \bar{t}_R(t,b)_L$. This model naturally describes both the top quark mass and the electroweak symmetry breaking. Also, a dynamical mechanism providing the quark mass hierarchy can be reflected in the model. The properties of these composites are analyzed in detail.

1. Introduction

Repetition of the generation structure of quarks and leptons is a great mystery in particle physics. Although three generation models have been widely accepted, the basic principle of the standard model (SM) allows the sequential fourth generation (family).[1] Also, the electroweak precision data does not exclude completely existence of the fourth family.[2,3] Noticeable is that the LHC has a potential for discovering the fourth family quarks at early stage.

If the fourth generation exists, we can naturally consider a scenario that the condensates of the fourth generation quarks t' and b' dynamically trigger the electroweak symmetry breaking (EWSB):[4] The Pagels-Stokar (PS) formula suggests that their contributions to the EWSB should not be small, because the masses of t' and b' should be heavy, $M_{t'} > 311$ GeV and $M_{b'} > 338$ GeV, respectively.[5]

On the other hand, the role of the top quark is rather subtle, i.e., although the contribution of the top is obviously much larger than the other three generation quarks, b, c and etc., it is estimated around 10-20% of the EWSB scale.

Recently, utilizing the dynamics considered in Ref. 6, we introduced a new class of models in which the top quark plays just such a role.[7] Its signature is the existence of an additional top-Higgs doublet Φ_t composed of the quarks and antiquarks of the third family, $\Phi_t \sim \bar{t}_R(t,b)_L$. In the dynamical EWSB scenario with the fourth family, the top-Higgs Φ_t is heavier than the fourth generation quark composites, $\Phi_{t'} \sim \bar{t}'_R(t',b')_L$ and $\Phi_{b'} \sim \bar{b}'_R(t',b')_L$. However, in general, Φ_t is not necessarily ultraheavy and decoupled from the TeV-scale physics. This leads to a model with three composite Higgs doublets.[8] We explore such a possibility, based on Refs. 7, 8.

As for the fourth family leptons, we assume that their masses are around 100 GeV,[9] and thus their contributions to the EWSB are smaller than that of the top quark. For the dynamics with very heavy fourth family leptons, and thereby with a lepton condensation, one needs to incorporate more Higgs doublets, say, a five composite Higgs model. Also, the Majorana condensation of the right-handed neutrinos should be reanalyzed. This possibility will be studied elsewhere.

2. Model

Based on the dynamical model in Ref. 7, we study a Nambu-Jona-Lasinio (NJL) type model with the third and fourth family quarks:[8]

$$\mathcal{L} = \mathcal{L}_f + \mathcal{L}_g + \mathcal{L}_{\rm NJL}, \tag{1}$$

where $\mathcal{L}_g$ represents the Lagrangian density for the SM gauge bosons, the fermion kinetic term is

$$\mathcal{L}_f \equiv \sum_{i=3,4} \bar{\psi}_L^{(i)} i\not{D}\psi_L^{(i)} + \sum_{i=3,4} \bar{u}_R^{(i)} i\not{D}u_R^{(i)} + \sum_{i=3,4} \bar{d}_R^{(i)} i\not{D}d_R^{(i)}, \tag{2}$$

and the NJL interactions are described by

$$\begin{aligned}\mathcal{L}_{\rm NJL} = {} & G_{t'}(\bar{\psi}_L^{(4)}t'_R)(\bar{t}'_R\psi_L^{(4)}) + G_{b'}(\bar{\psi}_L^{(4)}b'_R)(\bar{b}'_R\psi_L^{(4)}) + G_t(\bar{\psi}_L^{(3)}t_R)(\bar{t}_R\psi_L^{(3)}) \\ & + G_{t'b'}(\bar{\psi}_L^{(4)}t'_R)(\bar{b}'^{c}_R i\tau_2(\psi_L^{(4)})^c) + G_{t't}(\bar{\psi}_L^{(4)}t'_R)(\bar{t}_R\psi_L^{(3)}) \\ & + G_{b't}(\bar{\psi}_L^{(3)}t_R)(\bar{b}'^{c}_R i\tau_2(\psi_L^{(4)})^c) + (\text{h.c.}).\end{aligned} \tag{3}$$

Here $\psi_L^{(i)}$ denotes the weak doublet quarks of the i-th family, and $u_R^{(i)}$ and $d_R^{(i)}$ represent the right-handed up- and down-type quarks, respectively.

As was shown in Ref. 7, the diagonal parts of the NJL interactions, $G_{t'}$, $G_{b'}$ and G_t, can be generated from the topcolor interactions.[10] Following the dynamical model in Ref. 7, we assume that the coupling constants $G_{t'}$ and $G_{b'}$ are supercritical and mainly responsible for the EWSB, while the four-top coupling G_t is also strong, but subcritical.[7] The mixing term $G_{t't}$ can be generated by a flavor-changing-neutral (FCN) interaction between t' and t.[7] On the other hand, $G_{t'b'}$ may be connected with topcolor instantons.[10] In this case, in order to produce the four-fermion type operator, an appropriate dynamical model should be chosen. Although we keep the $G_{b't}$ term in a general discussion, it will be ignored in the numerical analysis. Since these four-fermion mixing terms provide the off-diagonal mass terms of the composite Higgs fields in low energy, at least two of them are required so as to evade (pseudo) Nambu-Goldstone (NG) bosons.

3. (2 + 1)-Higgs doublets

3.1. *Low energy effective model*

In low energy, the model introduced in the previous section yields an approximate $(2+1)$-structure in the sector of the Higgs quartic couplings. Indeed, in the bubble

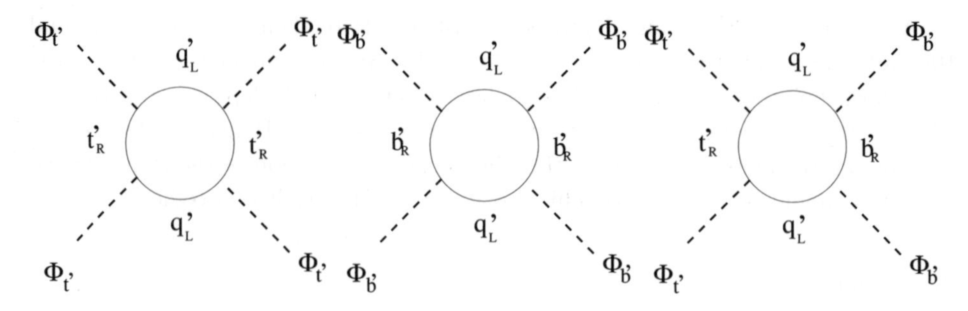

Fig. 1. Higgs quartic couplings for $\Phi_{t'}$ and $\Phi_{b'}$. We defined $q'_L \equiv (t', b')_L$.

approximation, the composite $\Phi_{t'(b')}$ couples only to $\psi_L^{(4)} \equiv q'_L = (t', b')_L$ and $t'_R(b'_R)$, while the top-Higgs Φ_t couples only to $\psi_L^{(3)} \equiv q_L = (t, b)_L$ and t_R. This leads to such a (2 + 1)-structure. (See Figs. 1 and 2.) Although the electroweak (EW) gauge interactions violate this structure, the breaking effects are suppressed, because the yukawa couplings are much larger than the EW gauge ones.

Let us study the low energy effective model.

It is convenient to introduce auxiliary fields at the NJL scale. In low energy these composite Higgs fields develop kinetic terms and hence acquire the dynamical degrees of freedom. The Lagrangian of the low energy model is then

$$\mathcal{L} = \mathcal{L}_f + \mathcal{L}_g + \mathcal{L}_s + \mathcal{L}_y, \tag{4}$$

with

$$\mathcal{L}_s = |D_\mu \Phi_{b'}|^2 + |D_\mu \Phi_{t'}|^2 + |D_\mu \Phi_t|^2 - V, \tag{5}$$

and

$$-\mathcal{L}_y = y_{b'} \bar{\psi}_L^{(4)} b'_R \tilde{\Phi}_{b'} + y_{t'} \bar{\psi}_L^{(4)} t'_R \Phi_{t'} + y_t \bar{\psi}_L^{(3)} t_R \Phi_t + (\text{h.c.}), \tag{6}$$

where V represents the Higgs potential and $\Phi_{t',b',t}$ are the renormalized composite Higgs fields ($\tilde{\Phi}_{b'} \equiv -i\tau_2 \Phi_{b'}^*$). Taking into account the renormalization group (RG) improved analysis, we study the following Higgs potential:[8]

$$V = V_2 + V_4, \tag{7}$$

with

$$\begin{aligned} V_2 = {} & M_{\Phi_{b'}}^2 (\Phi_{b'}^\dagger \Phi_{b'}) + M_{\Phi_{t'}}^2 (\Phi_{t'}^\dagger \Phi_{t'}) + M_{\Phi_t}^2 (\Phi_t^\dagger \Phi_t) \\ & + M_{\Phi_{t'}\Phi_{b'}}^2 (\Phi_{t'}^\dagger \Phi_{b'}) + M_{\Phi_{b'}\Phi_t}^2 (\Phi_{b'}^\dagger \Phi_t) + M_{\Phi_{t'}\Phi_t}^2 (\Phi_{t'}^\dagger \Phi_t) + (\text{h.c.}), \end{aligned} \tag{8}$$

$$\begin{aligned} V_4 = {} & \lambda_1 (\Phi_{b'}^\dagger \Phi_{b'})^2 + \lambda_2 (\Phi_{t'}^\dagger \Phi_{t'})^2 + \lambda_3 (\Phi_{b'}^\dagger \Phi_{b'})(\Phi_{t'}^\dagger \Phi_{t'}) + \lambda_4 |\Phi_{b'}^\dagger \Phi_{t'}|^2 \\ & + \frac{1}{2}\left[\lambda_5 (\Phi_{b'}^\dagger \Phi_{t'})^2 + (\text{h.c.}) \right] + \lambda_t (\Phi_t^\dagger \Phi_t)^2 . \end{aligned} \tag{9}$$

The Higgs mass terms are connected with the inverse of the four-fermion couplings. While $M_{\Phi_{b'}}^2$ and $M_{\Phi_{t'}}^2$ are negative, the mass square $M_{\Phi_t}^2$ is positive, which

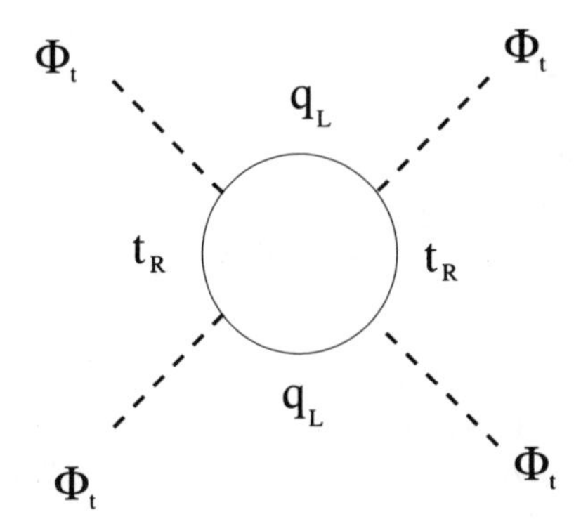

Fig. 2. Higgs quartic coupling for Φ_t. We defined $q_L \equiv (t, b)_L$.

reflects a subcritical dynamics of the t quark. Note that the top-Higgs Φ_t acquires a vacuum expectation value (VEV) due to its mixing with $\Phi_{t'}$. On the other hand, the quartic couplings $\lambda_{1-5,t}$ are induced in low energy as schematically shown in Figs. 1 and 2, and hence these values are dynamically determined.

The structure of the mass term part V_2 is general. On the other hand, the V_4 part is presented as the sum of the potential for the two Higgs doublets $\Phi_{t'}$ and $\Phi_{b'}$, and that for the one doublet Φ_t, i.e., it reflects the $(2+1)$-structure of the present model. In passing, when we consider general Higgs quartic couplings for the three Higgs, there appear 45 real parameters.[8]

3.2. *Mass spectra of the quarks and the Higgs bosons*

Let us analyze the mass spectra of the quarks and the Higgs bosons.

Note that the number of the physical Higgs bosons in our model are three for the CP even Higgs bosons (H_1, H_2 and H_3 with the masses $M_{H_1} \le M_{H_2} \le M_{H_3}$), two for the CP odd Higgs (A_1 and A_2 with the masses $M_{A_1} \le M_{A_2}$), and four for the charged ones ($H_1^\pm$ and $H_2^\pm$ with the masses $M_{H_1^\pm} \le M_{H_2^\pm}$). It turns out that the heavy Higgs bosons, $H_2^\pm, A_2$, and H_3, consist mainly of the components of the top-Higgs Φ_t.

The VEV's $v_{t',b',t}$ for the Higgs fields $\Phi_{t',b',t}$ are approximately determined by

$$\left[\lambda_2 + \frac{1}{2}(\lambda_3+\lambda_4+\lambda_5)\cot^2\beta_4\right] v_{t'}^2 \simeq -M_{\Phi_{t'}}^2 - M_{\Phi_{t'}\Phi_{b'}}^2 \cot\beta_4, \tag{10}$$

$$\left[\lambda_1 + \frac{1}{2}(\lambda_3+\lambda_4+\lambda_5)\tan^2\beta_4\right] v_{b'}^2 \simeq -M_{\Phi_{b'}}^2 - M_{\Phi_{t'}\Phi_{b'}}^2 \tan\beta_4, \tag{11}$$

$$v_t \simeq \frac{-M_{\Phi_{t'}\Phi_t}^2}{M_{\Phi_t}^2} v_{t'} + \frac{-M_{\Phi_{b'}\Phi_t}^2}{M_{\Phi_t}^2} v_{b'}, \tag{12}$$

where we defined the ratios of the VEV's by $\tan\beta_4 \equiv v_{t'}/v_{b'}$. Note that the relation $v^2 = v_{b'}^2 + v_{t'}^2 + v_t^2$ holds, where $v \simeq 246$ GeV.

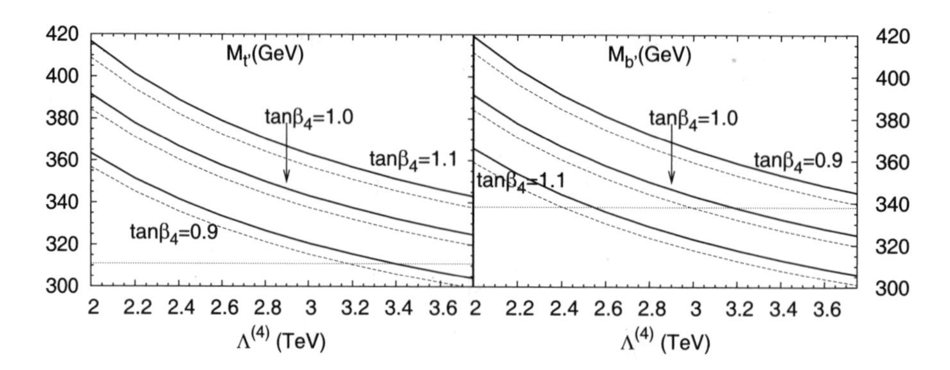

Fig. 3. $M_{t'}$ and $M_{b'}$. The bold and dashed curves are for $\Lambda^{(3)}/\Lambda^{(4)} = 1, 2$, respectively. The dotted lines correspond to the lower bounds for the masses of t' and b' at 95% C.L., $M_{t'} > 311$ GeV and $M_{b'} > 338$ GeV, respectively.

On the other hand, the masses of the CP odd and charged Higgs bosons are approximately given by

$$M_{A_1}^2 \simeq -2M_{\Phi_{t'}\Phi_{b'}}^2(1 - \tan^2\beta_{34}), \tag{13}$$

$$M_{A_2}^2 \simeq M_{\Phi_t}^2(1 + 2\tan^2\beta_{34}) + M_{A_1}^2 \tan^2\beta_{34}, \tag{14}$$

$$M_{H_1^\pm}^2 \approx M_{A_1}^2 + 2(m_{t'}^2 + m_{b'}^2)(1 - \tan^2\beta_{34}), \tag{15}$$

$$M_{H_2^\pm}^2 \approx M_{A_2}^2 + 2(m_{t'}^2 + m_{b'}^2)\tan^2\beta_{34}, \tag{16}$$

where we took $\tan\beta_4 = 1$ and defined $\tan\beta_{34} \equiv v_t/\sqrt{v_{t'}^2 + v_{b'}^2}$. The mass formulae for $H_{1,2,3}$ are quite complicated because of the 3×3 matrices.

In order to calculate the mass spectra more precisely, we employ the RGE's with the compositeness conditions:[11]

$$y_{t'}^2(\mu = \Lambda^{(4)}) = \infty, \quad y_{b'}^2(\mu = \Lambda^{(4)}) = \infty, \quad y_t^2(\mu = \Lambda^{(3)}) = \infty, \tag{17}$$

for the yukawa couplings, and

$$\left.\frac{\lambda_1}{y_{b'}^4}\right|_{\mu=\Lambda^{(4)}} = \left.\frac{\lambda_2}{y_{t'}^4}\right|_{\mu=\Lambda^{(4)}} = \left.\frac{\lambda_3}{y_{b'}^2 y_{t'}^2}\right|_{\mu=\Lambda^{(4)}} = \left.\frac{\lambda_4}{y_{b'}^2 y_{t'}^2}\right|_{\mu=\Lambda^{(4)}} = 0, \quad \left.\frac{\lambda_t}{y_t^4}\right|_{\mu=\Lambda^{(3)}} = 0, \tag{18}$$

for the Higgs quartic couplings, where $\Lambda^{(4)}$ and $\Lambda^{(3)}$ denote the composite scales for the fourth generation quarks and the top, respectively. The RGE's are given in Ref. 8. For consistency with the $(2+1)$-Higgs structure, we ignore the one-loop effects of the EW interactions. As for the Higgs loop effects, although they are of the $1/N$-subleading order, they are numerically relevant and hence incorporated.

Notice that the initial NJL model contains six four-fermion couplings: The EWSB scale and the pole ($\overline{\text{MS}}$) mass of the top quark are fixed to $v = 246$ GeV and $M_t = 171.2$ GeV ($m_t = 161.8$ GeV), respectively. We further convert the NJL

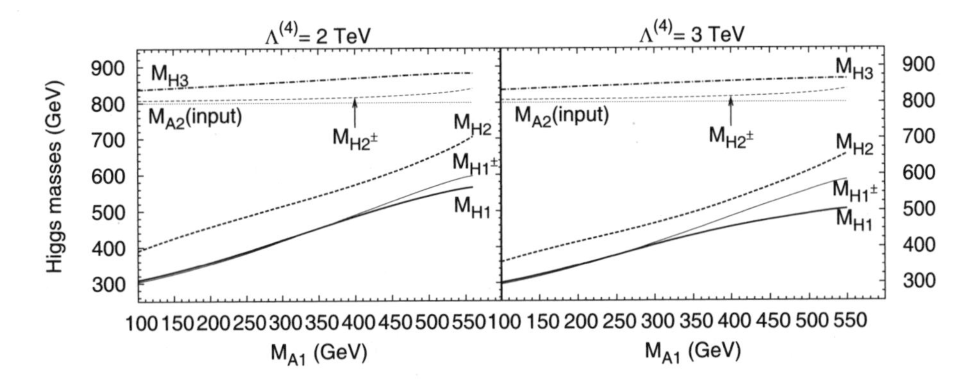

Fig. 4. Mass spectrum of the Higgs bosons for $\Lambda^{(4)} = 2, 3$ TeV . We took $\Lambda^{(3)}/\Lambda^{(4)} = 1.5$ and $\tan\beta_4 = 1$. $M_{A_2} = 800$ GeV is the input.

couplings into more physical quantities, M_{A_1}, M_{A_2} and $\tan\beta_4$. As for $M^2_{\Phi_{b'},\Phi_t}$, we fix $M^2_{\Phi_{b'},\Phi_t} = 0$. Numerically, it is consistent with $G_{b't} \approx 0$. For the numerical calculations, we use the QCD coupling constant $\alpha_3(M_Z) = 0.1176$.[9]

The results are illustrated in Figs. 3 and 4. The masses of t' and b' are essentially determined by the value of $\Lambda^{(4)}$, where we converted the $\overline{\text{MS}}$-masses $m_{t'}$ and $m_{b'}$ to the on-shell ones, $M_{t'(b')} = m_{t'(b')}[1 + 4\alpha_s/(3\pi)]$. As is seen in Fig. 3, their dependence on $\Lambda^{(3)}/\Lambda^{(4)}(= 1\text{–}2)$ is mild. When we vary $\tan\beta_4$ in the interval 0.9–1.1, the variations of $M_{t'}$ and $M_{b'}$ are up to 10% (see Fig. 3). The Higgs masses are relatively sensitive to the value of $\Lambda^{(4)}$ (see Fig. 4), while their sensitivity to $\Lambda^{(3)}/\Lambda^{(4)}(= 1\text{–}2)$ is low. The Higgs mass dependence on $\tan\beta_4$ is also mild, at most 5% for $\tan\beta_4 = 0.9\text{–}1.1$, and $\Lambda^{(4)} = 2\text{–}10$ TeV.

Since at the compositeness scale the yukawa couplings go to infinity, there could in principle be uncontrollable nonperturbative effects. By relaxing the compositeness conditions, we estimated such "nonperturbative" effects around 10 %. Since the loop effects of the EW interactions are expected to be much smaller, the uncertainties of the "nonperturbative" effects are dominant.

The 2σ-bound of R_b yields $M_{A_2} \geq 0.70, 0.58, 0.50$ TeV for $\Lambda^{(4)} = 2, 5, 10$ TeV. Following the (S,T) analysis a la LEP EWWG, we found that our model is within the 95% C.L. contour of the (S,T) constraint, when the fourth family lepton mass difference is $M_{\tau'} - M_{\nu'} \sim 150$ GeV.[3]

We can introduce the CKM structure in our model.[7,8] Since the mixing between the fourth family and the others is suppressed, $|V_{t'd}| \sim |V_{us}|m_c/m_{t'} \sim \mathcal{O}(10^{-3})$ and $|V_{t's}| \sim |V_{t'b}| \sim m_c/m_{t'} \sim \mathcal{O}(10^{-2})$, the contributions of the t'-loop to the B^0–$\bar{B}^0$ mixing, $b \to s\gamma$ and $Z \to \bar{b}b$ are negligible. Note also that the effects of the charged Higgs bosons are suppressed, because their masses are relatively heavy. As for the tree FCNC and FCCC, they are highly suppressed in the first and second families, because of the assumption that the top-Higgs is responsible for the top mass and does not couple to the other quarks, in the spirit of the $(2+1)$-Higgs structure.[8]

4. Summary

We have studied the (2+1) composite Higgs doublet model. It describes rather naturally both the top quark mass and the EWSB. We can incorporate the dynamical mechanism for the quark mass hierarchy and the CKM structure into the model.[7,8] It would be interesting to embed the present model into an extra dimensional one.[12]

The signature of the model is clear, i.e., as shown in Fig. 4, the masses of the four resonances are nearly degenerate and also the heavier top-Higgs bosons appear.

A noticeable feature is that due to the t' and b' contributions, the gluon fusion production of H_1 is considerably enhanced. For example, for $\Lambda^{(4)} = 3$ TeV, $\Lambda^{(3)}/\Lambda^{(4)} = 1.5$, $\tan\beta_4 = 1$, $M_{A_1} = 0.50$ TeV, and $M_{A_2} = 0.80$ TeV, we obtain $M_{t'} = M_{b'} = 0.33$ TeV and $M_{H_1} = 0.49$ TeV. In this case, $\sigma_{gg\to H_1}\mathrm{Br}(H_1 \to ZZ)$ is enhanced by 5.1, where the relative H_1ZZ and $H_1t\bar{t}$ couplings to the SM values are 0.86 and 2.0, respectively. Similarly, the CP odd Higgs production via the gluon fusion process should be enhanced. Also, the multiple Higgs bosons can be observed as $t\bar{t}$ resonances at the LHC. Detailed analysis will be performed elsewhere.

Acknowledgments

The research of M.H. was supported by the Grant-in-Aid for Science Research, Ministry of Education, Culture, Sports, Science and Technology, Japan, No. 16081211.

References

1. P. H. Frampton, P. Q. Hung and M. Sher, Phys. Rept. **330**, 263 (2000).
2. G.D.Kribs, T.Plehn, M.Spannowsky and T.M.P.Tait, Phys. Rev. D **76**, 075016 (2007).
3. M. Hashimoto, arXiv:1001.4335 [hep-ph].
4. B. Holdom, Phys. Rev. Lett. **57**, 2496 (1986) [Erratum-*ibid.* **58**, 177 (1987)]; Phys. Rev. D **54**, 721 (1996); JHEP **0608**, 076 (2006); C. T. Hill, M. A. Luty and E. A. Paschos, Phys. Rev. D **43**, 3011 (1991).
5. A. Lister [CDF Collaboration], arXiv:0810.3349 [hep-ex]; T. Aaltonen *et al.* [The CDF Collaboration], arXiv:0912.1057 [hep-ex].
6. R. R. Mendel and V. A. Miransky, Phys. Lett. B **268**, 384 (1991); V. A. Miransky, Phys. Rev. Lett. **69**, 1022 (1992).
7. M. Hashimoto and V. A. Miransky, Phys. Rev. D **80**, 013004 (2009).
8. M. Hashimoto and V. A. Miransky, arXiv:0912.4453 [hep-ph], to appear in PRD.
9. C. Amsler *et al.* [Particle Data Group], Phys. Lett. B **667**, 1 (2008).
10. C. T. Hill and E. H. Simmons, Phys. Rept. **381**, 235 (2003) [Err. *ibid.* **390**, 553 (2004)].
11. W. A. Bardeen, C. T. Hill and M. Lindner, Phys. Rev. D **41**, 1647 (1990).
12. M. Hashimoto, M. Tanabashi and K. Yamawaki, Phys. Rev. D **64**, 056003 (2001); *ibid.* **69**, 076004 (2004); V. Gusynin, M. Hashimoto, M. Tanabashi and K. Yamawaki, *ibid* **65**, 116008 (2002); M. Hashimoto and D. K. Hong, *ibid.* **71**, 056004 (2005).

Holmorphic Supersymmetric Nambu–Jona-Lasino Model and Dynamical Electroweak Symmetry Breaking*

Dong-Won Jung[1], Otto C. W. Kong[1] and Jae Sik Lee[2]

[1] *Department of Physics and Center for Mathematics and Theoretical Physics National Central University, Chung-li, 32054, Taiwan*

[2] *Physics Division, National Center for Theoretical Sciences, Hsinchu, 300, Taiwan*

We analyze the model of the dynamical electroweak symmetry breaking based on the supersymmetrized Nambu–Jona-Lasinio model with holomorphic dimension-five operators. The Minimal Supersymmetric Standard Model is derived as the low energy effective theory with both Higgs superfields as composites. A renormalization group analysis is performed, including the prediction of the Higgs mass range.

Keywords: Dynamical symmetry breaking, supersymmetry, Nambu–Jona-Lasinio model.

1. Introduction

Dynamical symmetry breaking has been a fascinating idea for symmetry breaking. In 1961, Nambu and Jona-Lasinio (NJL)[2] introduced the dimension-six four-fermion operators for the chiral symmetry breaking. They showed that bi-fermion condensate formed a scalar composite, and its non-trivial vacuum signified chiral symmetry breaking.

In the setup of Nambu and Jona-Lasinio, large coupling is required to make the fermions condensate, which is interpreted as the large Yukawa coupling. With this philosophy, top quark condensate models were proposed with the Higgs as an auxiliary field.[3–5] These models predicted the larger top quark mass than the recent observation. Some extenstions were also suggested, for instance, including bottom quark condensate[6] and one more fundamental scalar.[7]

Supersymmetrization of this idea started from the imbedment of the dimension-six four-fermion operator.[8–11] In the purely supersymmetric case, non-trivial vacuum is not possible because of the cancelations forced by supersymmetry,so the introduction of the soft supersymmetry breaking terms is inevitable. In the realistic analysis, infrared quasi fixed point nature of the renormalization group equation (RGE) was used, but the models failed to fit the recent observation of top quark and bottom quark masses.

---

*Based on the work by D.W. Jung et. al.[1]

So we propose a new scheme for the supersymmetric Nambu–Jona-Lasinio (SNJL) models.[1] Instead of dimension-six operators, we introduce the holomorphic dimension-five operators in the superpotential. In this case, scalar superpartners form condensates unlike the previous approaches. Moreover, it reproduces the minimal supersymmetric standard model (MSSM) as an effective low energy theory with both top and bottom quark Yukawa couplings. We provide the RGE analysis to show the phenomenological viabilities including Higgs mass spectrum.

2. Nambu–Jona-Lasino (NJL) Model

In this section, we provide the brief review of the NJL model follwoing the Ref. [9]. We start from the Lagrangian with the dimension-six four-fermion operator,

$$\mathcal{L} = i\partial_m\psi_+\sigma^m\overline{\psi}_+ + i\partial_m\psi_-\sigma^m\overline{\psi}_- + g^2\psi_+\psi_-\overline{\psi}_+\overline{\psi}_-. \tag{1}$$

With the auxiliary field ϕ, this can be rewritten

$$\begin{aligned}\mathcal{L} &\to \mathcal{L} - (\phi - g\overline{\psi}_+\overline{\psi}_-)(\phi - g\psi_+\psi_-) \\ &= i\partial_m\psi_+\sigma^m\overline{\psi}_+ + i\partial_m\psi_-\sigma^m\overline{\psi}_- + \phi^*\phi + g\phi\psi_+\psi_- + g\phi\overline{\psi}_+\overline{\psi}_-,\end{aligned} \tag{2}$$

where $\phi = g\overline{\psi}_+\overline{\psi}_-$ from the equartion of motion. From the one-loop effective potential for ϕ and $v = 16\pi^2\frac{V}{\Lambda^4}$, the symmetry breaking condition is

$$\frac{\partial v}{\partial\phi^*} = \phi\frac{2g^2}{\Lambda^2}\left[\frac{1}{\alpha} - 1 + \eta\ln\left(\frac{1}{\eta}+1\right)\right] = 0, \tag{3}$$

where $\eta = \frac{g^2}{\Lambda^2}\phi^*\phi$, $\alpha = \frac{g^2\Lambda^2}{8\pi^2}$. In order for the above equation to have the non-zero solution for η,

$$\frac{g^2\Lambda^2}{8\pi^2} > 1, \tag{4}$$

should be satisfied.

By expanding around the vacuum expectation value ϕ_0 and calculating the two-point function, we can obtain the effective action with $\phi(x) = \sqrt{\frac{1}{2}}\,(\sigma(x) + i\pi(x))$,

$$\Gamma = Z\int d^4x\left[\frac{1}{2}\sigma(x)\left(\Box - 4m^2\right)\sigma(x) + \frac{1}{2}\pi(x)\Box\pi(x)\right], \tag{5}$$

where $Z = \frac{g^2}{16\pi^2}\left(\ln\frac{\Lambda^2}{m^2} + \mathcal{O}(1)\right)$. So if the coupling constant is large enough, the chiral symmetry is broken by the condensation of fermions and the system can be described as effective Yukawa-like theory in the low energy. Scalar particles with masses $2m$ and zero appear, which can be interptreted as Higgs particle and Goldstone boson each.

3. Supersymmetric NJL (SNJL) Model

The supersymmetrization of NJL model was started in the 1980s. Let's first look into the simplest toy setup briefly. The basic model Lagrangian with chiral superfields is

$$\mathcal{L} = \int d^4\theta \left[\bar{\Phi}_+\Phi_+ + \bar{\Phi}_-\Phi_-\right] + \int d^4\theta \left[g^2\,\bar{\Phi}_+\bar{\Phi}_-\Phi_+\Phi_-\right]. \tag{6}$$

In componet field, one can check that it contains dimension-six operator,

$$\mathcal{L}_\psi = i\bar{\psi}_+\sigma^\mu\partial_\mu\psi_+ + i\bar{\psi}_-\sigma^\mu\partial_\mu\psi_- + g^2\bar{\psi}_+\bar{\psi}_-\psi_+\psi_-. \tag{7}$$

To apply auxilary field technique as in the previous section, two chiral superfields are necessary. The resulting Lagrangian is

$$\begin{aligned}\mathcal{L} = & \int d^4\theta \left[(\bar{\Phi}_+\Phi_+ + \bar{\Phi}_-\Phi_-)(1 - m^2\theta^2\bar{\theta}^2) + \bar{\Phi}_1\Phi_1\right] \\ & + \int d^2\theta \left[\mu\Phi_2(\Phi_1 + g\Phi_+\Phi_-)\right] + h.c.\end{aligned} \tag{8}$$

The equation of motion for Φ_2 gives Φ_1 as a composite, $\Phi_1 = -g\,\Phi_+\Phi_-$, which yields $\bar{\Phi}_1\Phi_1 = g^2\bar{\Phi}_+\bar{\Phi}_-\Phi_+\Phi_-$ Note that soft supersymmetry breaking terms are important, since the potential is zero if the supersymmetry is exact, which means no non-trivial vacuum.

In 1991, Carena et. al.[11] provided the most realistic model based on this idea. They used the action

$$\begin{aligned}\Gamma_\Lambda = \; & \Gamma_{YM} \\ & + \int dV \left[\bar{Q}e^{2V_T}Q + T^c e^{-2V_T}\bar{T}^c + \bar{B}^c e^{-2V_B}B^c\right]\left(1 - \Delta^2\theta\bar{\theta}^2\right) \\ & + \int dV \bar{H}_1 e^{2V_{H_1}} H_1 \left(1 - M_H^2\theta\bar{\theta}^2\right) \\ & - \int dS \left(m_0 H_1 H_2\left(1 + B_0\theta^2\right) - g_T H_2 Q T^c\left(1 + A_0\theta^2\right)\right) + h.c.\end{aligned} \tag{9}$$

Even though there is no kinetic term for H_2 now, it is generated from the one-loop,

$$Z_{H_2}\int dV \bar{H}_2 e^{2V_{H_2}} H_2\left(1 + A_0\theta^2 + A_0\bar{\theta}^2 + \left(2\Delta^2 + A_0^2\right)\theta^2\bar{\theta}^2\right), \tag{10}$$

where $Z_{H_2} = \frac{g_T^2 N_c}{16\pi^2}\ln\frac{\Lambda^2}{\mu^2}$. So after rescaling $H_2 \to H_2\left(1 - A_0\theta^2\right)/\sqrt{Z_{H_2}}$, we obtain the action which looks almost like the MSSM,

$$\begin{aligned}\Gamma_\Lambda = \; & \Gamma_{YM} \\ & + \int dV \left[\bar{Q}e^{2V_T}QT^c e^{-2V_T}\bar{T}^c + B^c e^{-2V_B}\right]\left(1 - \Delta^2\theta\bar{\theta}^2\right) \\ & + \int dV \bar{H}_1 e^{2V_{H_1}} H_1\left(1 - M_H^2\theta\bar{\theta}^2\right) \\ & - \int dS\left(mH_1H_2\left(1 + B_0\theta^2\right) - h_T H_2 Q T^c\left(1 + A_0\theta^2\right)\right) + h.c. \\ & + \int dV \bar{H}_2 e^{2V_{H_2}} H_2\left(1 + 2\Delta^2\theta^2\bar{\theta}^2\right),\end{aligned} \tag{11}$$

where $m = m_0/\sqrt{Z_{H_2}}, h_T = g_T/\sqrt{Z_{H_2}}$.

Starting from the above Lagrangian they performed the RGE analysis. In the RGE language, symmetry breaking or condensation condition translate into the boudary conditions at the compositenss scale Λ;

$$m, h_T \to \infty \ \text{ while keeping } \ \frac{h_T}{m} \ \text{fixed.} \tag{12}$$

Though the Lagrangian looks almost like the MSSM, it lacks the bottom quark Yukawa interaction in reality. In addtion, only very small $\tan\beta$ is allowed generally, which is almost excluded by the LEP II expriment.

4. Holomorphic SNJL (HSNJL) Model

So we consier an alternative for the SNJL, which utilizes dimension-six four-fermion operators. The holmorphic dimension-five operators in the superpotential is examined:

$$\mathcal{L} = \int d^4\theta \left[\bar{\Phi}_+\Phi_+ + \bar{\Phi}_-\Phi_-\right] - \int d^2\theta \left[\frac{G}{2}\Phi_+\Phi_-\Phi_+\Phi_-\right] + h.c. \tag{13}$$

With the auxiliary field Φ_0,

$$\begin{aligned}
\mathcal{L} = \quad & \int d^4\theta \left[(\bar{\Phi}_+\Phi_+ + \bar{\Phi}_-\Phi_-)(1 - m^2\theta^2\bar{\theta}^2)\right] \\
& + \int d^2\theta \Big[\frac{1}{2}(\sqrt{\mu}\Phi_0 + \sqrt{G}\Phi_+\Phi_-)(\sqrt{\mu}\Phi_0 + \sqrt{G}\Phi_+\Phi_-) - \frac{G}{2}\Phi_+\Phi_-\Phi_+\Phi_-\Big] + h.c. \\
= \quad & \int d^4\theta \left[(\bar{\Phi}_+\Phi_+ + \bar{\Phi}_-\Phi_-)(1 - m^2\theta^2\bar{\theta}^2)\right] \\
& + \int d^2\theta \left[\frac{\mu}{2}\Phi_0^2 + \sqrt{\mu G}\Phi_0\Phi_+\Phi_-\right] + h.c..
\end{aligned} \tag{14}$$

So the superpotential reduces to

$$W = -\frac{\mu}{2}\Phi_0^2 - y\Phi_0\Phi_+\Phi_-, \tag{15}$$

with $y = \sqrt{\mu G}$. Equation of motion for Φ_0 yields $\Phi_0 = -\sqrt{G/\mu}\,\Phi_+\Phi_-$ *i.e.* Φ_0 as composite of two chiral superfields as the case of Φ_1 in the dimension-six model.

Coleman-Weinberg effective potential can be calculated with the inclusion of the soft supersymmetry breaking terms, $W \to W(1+\zeta)$, $\zeta = B\theta^2$,

$$V_{eff} = V_{tree} + V_{1-loop}, \tag{16}$$

with $M_f = |y\phi_0|^2, M_s = |y\phi_0|^2 + m^2 \pm |By\phi_0|$. Tree-level potential is

$$V_{tree} = -\frac{B\mu}{2}\phi_0^2 + c.c.. \tag{17}$$

One-loop effective potential can be computed easily, but we drop its concrete form here because of its complexity. Due to the supersymmetry, the potential is zero when the soft supersymmetry breaking terms are absent and one-loop effective potential goes to zero as ϕ_0 becomes large. Unfortunately, this potential is not bounded from below, so the symmetry breaking cannot be achieved. In the case of gauged model, there are always effective D-terms which produce the quartic terms of the Higgs

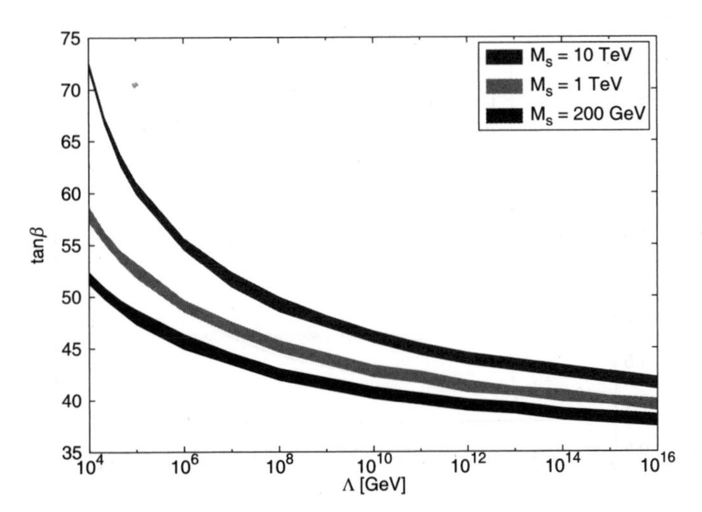

Fig. 1. We have included a SUSY threshold correction ϵ_b of value -0.01 in the running of y_b with $[(1+\epsilon_b \tan\beta) = \sqrt{2}m_b/(y_b\, v\, \cos\beta)]$.

fields. Those can make the potential bounded. In this way, we can construct the realistic model.

Let's construct the realistic model. Just for the third generation,

$$W = G\,\varepsilon_{\alpha\beta} Q_3^{\alpha a} U_3^{c\,a} Q_3^{\beta b} D_3^{c\,b}(1 + A\theta^2). \tag{18}$$

Two auxilaiary Higgs superfields are introduced, then

$$\begin{aligned} W &= -\mu(H_d - \lambda_t Q_3 U_3^c)(H_u - \lambda_b Q_3 D_3^c)(1 + A\theta^2) \\ &= (-\mu H_d H_u + y_t Q_3 H_u U_3^c + y_b H_d Q_3 D_3^c)(1 + A\theta^2), \end{aligned} \tag{19}$$

where $\mu\lambda_t = y_t$, $\mu\lambda_b = y_b$, $\mu\lambda_t\lambda_b = G$. Equation of motion for H_u yields $H_d = \lambda_t Q_3 U_3^c$ while that for H_d yields $H_u = \lambda_b Q_3 D_3^c$. It is exaltly same superpotential with the MSSM , with the boundary conditions for Yukawa couplings and μ at the compositeness scale Λ,

$$y_t, y_b, \mu \to \infty, \text{while keeping } \frac{y_t y_b}{\mu} = G. \tag{20}$$

Since the effective Lagragian at low energy is exactly the MSSM, we can analyse it with the MSSM RGEs.

5. Renormalization Group Equation Analysis

The RGE anaysis is done with the boundary conditions of Eq. (29). The phenomenological constraints are

$$m_t = 171.3 \pm 1.6\text{GeV}, \tag{21}$$

$$m_b = 4.20^{+0.17}_{-0.07}\,\text{GeV}. \tag{22}$$

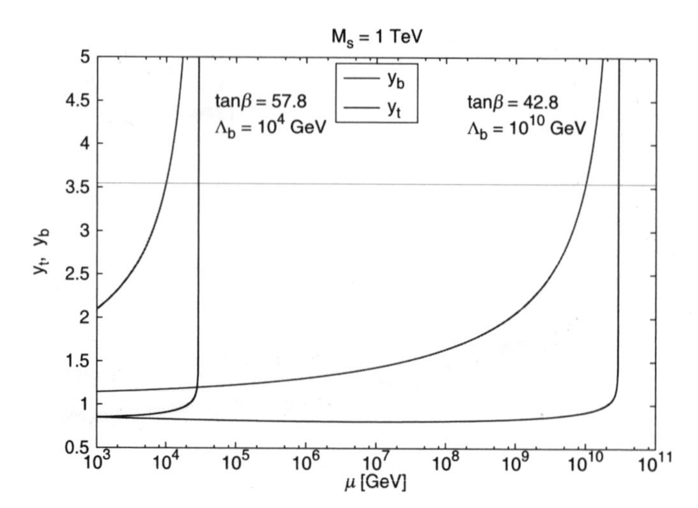

Fig. 2. Illustration of y_b and y_t runnings for a couple of cases.

For the numerical analysis we introduce the Λ_t and Λ_b for the compositen scale where $yt, b^2/4\pi = 1$, respectively.

We look for admissible cases of Λ_t and Λ_b values with the now precisely determined top and bottom masses implemented. Note that for large value of tanβ, the bottom Yukawa y_b is big. We find that there is always a window of tanβ value giving admissible solution, for any Λ_b we take (with supersymmetry scale M_S from 200 GeV to 10 TeV), as given in Fig. 1.

In Fig. 2, we show the y_t and y_b runnings for a couple of typical cases. Note that y_b plays the more important role reaching the perturbative limit before y_t. We have $\Lambda_t \sim 3\Lambda_b$, so we can say that compositeness scale is roughly arund $\Lambda_{b,t}$. It is also important to note that we have included a supersymmetry threshold correction ϵ_b [*cf.* $(1 + \epsilon_b \tan\beta) = \sqrt{2}m_b/(y_b\, v\, \cos\beta)$] of value -0.01 in the running of y_b. The negative value is important. For $\epsilon_b > 0$, solution is possble only with uncomfortably large tanβ. Even though its exact value is dependent on the details of the particle spectrums, we use $|\epsilon_b| = \alpha_s/3\pi \sim 0.01$[13] here as a reasonable estimate.

We provide the Higgs mass prediction in our model in Fig. 3. We determine the lightest Higgs mass as a function of M_S for $M_A > 100$ GeV. The value loses sensitivity to M_A as the latter get bigger. The result shows little sensitivity to Λ_b, which reflects the infrared quasi-fixed point nature of the MSSM RGEs.[12] We can conclude that the Higgs mass predicted in our model is admissible.[14] Further analysis can be done with the particle spectrums specified.

6. Summary

In this literauture we study the alternative for the conventional SNJL model. Instead of usual dimension-six four-fermion operartors, in our HSNJL model with dimesion-five operators, bi-scalar vacuum condensates play the role of Higgs particles which

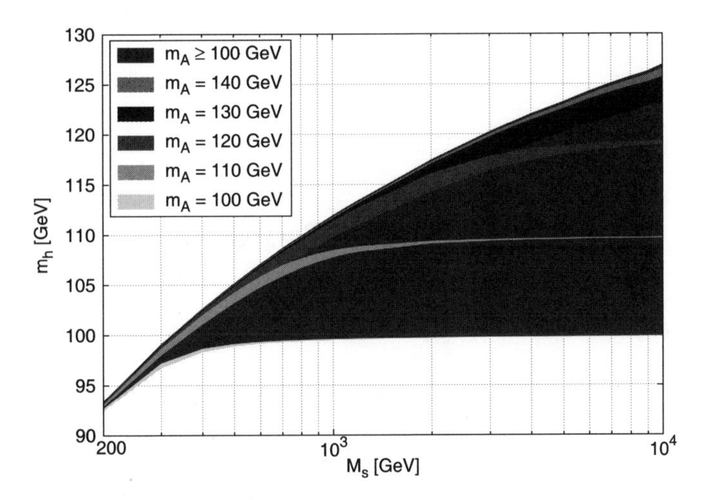

Fig. 3. Prediction for the lightest Higgs mass.

achieve the electroweak symmetry breaking. It provides the complete MSSM as low energy effective theory, which was not in the conventional SNJL approaches. The analysis with the RGEs shows that both top and bottom quark masses can be fit together. In this case, bottom Yukawa coupling drives the blow-up of top Yukawa coupling. Threshold correction is important, including sign. For $M_S < 1.5$ TeV, the Higgs mass is on the low side, but compatible with the MSSM Higgs search limits.

References

1. D. W. Jung, O. C. W. Kong and J. S. Lee, Phys. Rev. D **81**, 031701(R) (2010) arXiv:0906.3580 [hep-ph].
2. Y. Nambu and G. Jona-Lasinio, Phys. Rev. **122** (1961) 345.
3. W. A. Bardeen, C. T. Hill and M. Lindner, Phys. Rev. D **41** (1990) 1647.
4. V. A. Miransky, M. Tanabashi and K. Yamawaki, Phys. Lett. B **221** (1989) 177.
5. W. J. Marciano, Phys. Rev. Lett. **62** (1989) 2793.
6. M. A. Luty, Phys. Rev. D **41** (1990) 2893.
7. B. Chung, K. Y. Lee, D. W. Jung and P. Ko, JHEP **0605** (2006) 010 [arXiv:hep-ph/0510075].
8. W. Buchmuller and S. T. Love, Nucl. Phys. B **204** (1982) 213.
9. W. Buchmuller and U. Ellwanger, Nucl. Phys. B **245** (1984) 237.
10. T. E. Clark, S. T. Love and W. A. Bardeen, Phys. Lett. B **237** (1990) 235.
11. M. S. Carena, T. E. Clark, C. E. M. Wagner, W. A. Bardeen and K. Sasaki, Nucl. Phys. B **369** (1992) 33.
12. C.D. Froggatt *et.al.*, Phys. Lett. B **298**, 356 (1993).
13. L. Hall *et.al.*, Phys. Rev. D **50**, 7048 (1994).
14. S. Schael *et.al.*, Eur. Phys. J. C **47**, 547 (2006); see also A. Djouadi, Eur. Phys. J. C **59**, 389 (2009).

Ratchet Model of Baryogenesis

Tatsu Takeuchi*
Physics Department, Virginia Tech, Blacksburg, VA 24061, USA

Azusa Minamizaki and Akio Sugamoto
Physics Department, Ochanomizu University, 2-1-1 Ōtsuka, Bunkyo-ku, Tokyo 112-8610, Japan

We propose a toy model of baryogenesis which applies the 'ratchet mechanism,' used frequently in the theory of biological molecular motors, to a model proposed by Dimopoulos and Susskind.

Keywords: Baryogenesis, ratchet mechanism, Dimopoulos-Susskind model.

1. Introduction

The ratio of baryon-number to photon-number densities in our universe has been established via Big-Bang Nucleosynthesis (BBN) [1–3] and WMAP [4, 5] to be

$$\eta = \frac{n_B}{n_\gamma} \approx 6 \times 10^{-10} . \tag{1}$$

The more precise numbers are

$$\begin{aligned} \eta_{10}(\mathrm{BBN:D/H}) &= 5.8 \pm 0.3 \ , \\ \eta_{10}(\mathrm{WMAP:\ 7yr}) &= 6.18 \pm 0.15 \ , \end{aligned} \tag{2}$$

where $\eta_{10} = 10^{10}\,\eta$, and the BBN value is determined from the deutron abundance reported in Ref. [6, 7]. As we can see, the agreement is very good.

The objective of baryogenesis is to explain how the above number can come about from a universe initially with zero net baryon number. Since the pioneering work of Sakharov [8], very many proposals have been made as to what this baryogenesis mechanism could be.[a] Among the early ones was a model by Dimopoulos and Susskind [15] in which baryon number is generated via the coherent semi-classical time-evolution of a complex scalar field. Similar mechanisms have been employed by Affleck and Dine [16], Cohen and Kaplan [17], and Dolgov and Freese [18], of which the Affleck-Dine mechanism has been popular and intensely studied due to its natural implementability in SUSY models. Two of us have also considered the

---

*Presenting author.
[a]For recent reviews, see Refs. [9–14].

application of the Dimopoulos-Susskind model to the cosmological constant problem [19].

In this talk, I will discuss the Dimopoulos-Susskind model, how it satisfies Sakharov's three conditions for baryogenesis, in particular, how it uses the expansion of the universe to satisfy the third, and then propose the 'ratchet mechanism' [20–22] as an alternative for driving the model away from thermal equilibrium.

2. The Dimopoulos-Susskind Model

Consider the action of a complex scalar field given by

$$S = \int d^4x \sqrt{-g} \left[g^{\mu\nu} \partial_\mu \phi^\dagger \partial_\nu \phi - V(\phi, \phi^\dagger) \right] . \tag{3}$$

If the potential $V(\phi, \phi^\dagger)$ is invariant under the global change of phase

$$\phi \to e^{i\xi} \phi , \qquad \phi^\dagger \to e^{-i\xi} \phi^\dagger , \tag{4}$$

then the corresponding conserved current is

$$B_\mu = \sqrt{-g} \left(i \phi \overleftrightarrow{\partial_\mu} \phi^\dagger \right) . \tag{5}$$

If we identify B_0 with the baryon number density, then adding to the action a potential which is not invariant under the above phase change, such as

$$V_0(\phi, \phi^\dagger) = \lambda \left(\phi + \phi^\dagger \right) \left(\alpha \phi^3 + \alpha^* \phi^{\dagger 3} \right) , \qquad |\alpha| = 1 , \tag{6}$$

would lead to baryon number violation. Furthermore, unless $\alpha = \pm 1$, this potential also violates C and CP since ϕ transforms as

$$\begin{aligned} \phi(t, \vec{x}) &\xrightarrow{C} \phi^\dagger(t, \vec{x}) , \\ \phi(t, \vec{x}) &\xrightarrow{CP} \pm \phi^\dagger(t, -\vec{x}) , \end{aligned} \tag{7}$$

where the sign under CP depends on the parity of ϕ. (P is not violated.)

In Ref. [15], Dimopoulos and Susskind subject ϕ to the potential

$$V_n(\phi, \phi^\dagger) = \lambda \left(\phi \phi^\dagger \right)^n \left(\phi + \phi^\dagger \right) \left(\alpha \phi^3 + \alpha^* \phi^{\dagger 3} \right) . \tag{8}$$

The purpose of the factor $(\phi \phi^\dagger)^n$ is simply to give the coupling constant λ a negative mass dimension. Setting $\phi = \phi_r \, e^{i\theta}/\sqrt{2}$, the baryon number density becomes

$$n_B = B_0 = \sqrt{-g} \, \phi_r^2 \, \dot{\theta} , \tag{9}$$

which shows that to generate a non-zero baryon number n_B, one must generate a non-zero $\dot{\theta}$. The potential in the polar representation of ϕ is

$$V_n(\phi_r, \theta) = \lambda \left(\frac{\phi_r^2}{2} \right)^n \phi_r^4 \cos\theta \, \cos(3\theta + \beta) , \tag{10}$$

where we have set $\alpha = e^{i\beta}$. The θ-dependence of this potential for fixed ϕ_r is shown in Fig. 1 for the case $\beta = \pi/2$. Note that under B, C, and CP, the phase θ

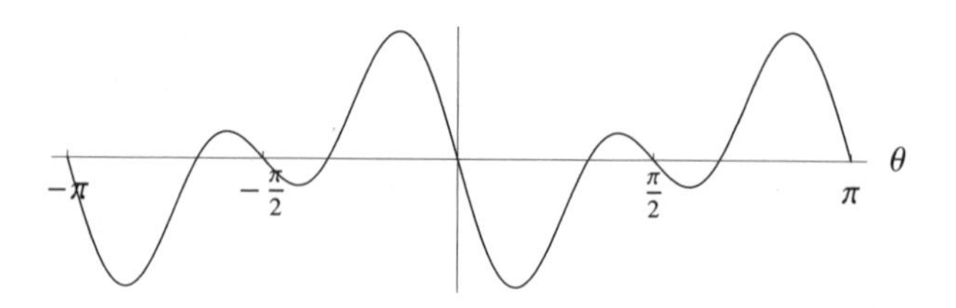

Fig. 1. θ-dependence of the B, C, and CP violating potential, Eq. (10), for the case $\beta = \pi/2$.

transforms as

$$\begin{aligned}\theta(t,\vec{x}) &\xrightarrow{B} \theta(t,\vec{x}) + \xi\ ,\\ \theta(t,\vec{x}) &\xrightarrow{C} -\theta(t,\vec{x})\ ,\\ \theta(t,\vec{x}) &\xrightarrow{CP} -\theta(t,-\vec{x})\ .\end{aligned} \tag{11}$$

If the parity of ϕ is negative, then θ will also be shifted by π under CP. So in terms of θ, the violation of B is due to the loss of translational invariance, and the violation of C and CP are due to the loss of left-right reflection invariance which happens when $\beta \neq 0, \pi$. The question is, can the asymmetric force provided by this potential make θ flow in one preferred direction thereby generate a non-zero $\dot{\theta}$? For that, one must move away from thermal equilibrium.

In the original Dimopoulos-Susskind paper [15], this shift away from thermal equilibrium is accomplished by the expansion of the universe. Consider a flat expanding universe with the Friedman-Robertson-Walker metric:

$$ds^2 = dt^2 - \left(\frac{a(t)}{a_0}\right)^2 d\vec{x}^2\ . \tag{12}$$

During a radiation dominated epoch, the scale factor evolves as

$$\frac{a(t)}{a_0} \sim \sqrt{2t}\ . \tag{13}$$

Introducing the conformal variable $\tau = \sqrt{2t}$, the line-element becomes

$$ds^2 = \tau^2\left(d\tau^2 - d\vec{x}^2\right)\ , \tag{14}$$

while the action simplifies to

$$S = \int d^3\vec{x}\, d\tau \left[\partial_\mu\hat{\phi}^\dagger\partial^\mu\hat{\phi} - \frac{1}{\tau^{2n}}V_n(\hat{\phi},\hat{\phi}^\dagger) + \cdots\right]\ . \tag{15}$$

Here, the scalar field has been rescaled to $\hat{\phi} \equiv \tau\phi$, and the ellipses represent total divergences and terms that depend only on $|\hat{\phi}|$.

At this point, a simplifying assumption is made that the dynamics of $|\hat{\phi}|$ is such that it is essentially constant and does not evolve with τ, leaving only the phase of $\hat{\phi}$ as the dynamic variable.[b] Setting $\hat{\phi} = e^{i\theta}/\sqrt{2}$, the action within a domain of spatially constant θ becomes

$$S = \int d^3\vec{x}\, d\tau \left[\frac{1}{2}\left(\frac{d\theta}{d\tau}\right)^2 - \frac{1}{\tau^{2n}} V_n(1,\theta) \right] . \tag{16}$$

The equation of motion for θ within that domain is then

$$\frac{d^2\theta}{d\tau^2} + \frac{1}{\tau^{2n}} \frac{\partial V_n}{\partial \theta} = 0 . \tag{17}$$

To this, a friction term, which is assumed to come from the self-interaction of $\hat{\phi}$, is added by hand as

$$\frac{d^2\theta}{d\tau^2} + \frac{1}{\tau^{2n}} \frac{\partial V_n}{\partial \theta} + \frac{\lambda^2}{\tau^{4n}} \frac{d\theta}{d\tau} = 0 , \tag{18}$$

where the coefficient of $d\theta/d\tau$ has been fixed simply by dimensional analysis. If $n > 0$, both force and friction terms vanish in the limit $\tau \to \infty$, and it is possible to show that a non-zero $n_B \sim d\theta/d\tau$ survives asymptotically, its final value depending on the initial value of θ. This initial value is expected to vary randomly from domain to domain, resulting in different asymptotic baryon numbers in each, and when summed results in an overall net baryon number. On the other hand, if $n = 0$, which would make the self-interactions of ϕ renormalizable, the friction term will eventually bring all motion to a full stop.

3. The Ratchet Mechanism

A striking feature of the Dimopoulos-Susskind model is its similarity with the problem of biased random walk one encounters in the modeling of biological motors [20–22]. An example of a biological motor is the myosin molecule which walks along actin filaments. This molecule is modeled as moving along a periodic sawtooth-shaped potential, similar to that shown in Fig. 1. Thermal equilibrium inside a living organism is broken by the presence of ATP (adenosine triphosphate) whose hydrolysis into ADP (adenosine diphosphate) and P (phosphate) provides the energy required to fuel the motion:

$$\text{ATP} \;\to\; \text{ADP} + \text{P} + \text{energy} . \tag{19}$$

This is often modeled as a randomly fluctuating temperature of the thermal bath: the molecule is excited out of a potential well during periods of high-temperature, allowing it to diffuse into the neighboring ones, and then drops back into a well during periods of low-temperature. Due to the asymmetry of the potential, this sequence can lead to biased motion depending on the depth and width of the repeating potential wells, and the height and frequency of the temperature fluctuations.

[b] This assumption that $|\hat{\phi}|$ is constant would require the magnitude of the unscaled field $|\phi|$ to evolve as $1/\tau = 1/\sqrt{2t}$.

Analogy with such 'temperature ratchet' models suggests a possible way to drive the evolution of θ in the Dimopoulos-Susskind model without relying on the non-renormalizability of the self-interaction of ϕ, or the expansion of the universe directly. Let us assume the existence of ATP- and ADP-like particles A and B which interact with ϕ via the reaction

$$A+\phi \leftrightarrow B+\phi+Q\,, \tag{20}$$

where Q is the energy released in the reaction. A and B are assumed to be stable (or highly meta-stable) states that have fallen out of thermal equilibrium at an earlier time in the evolution of the universe. Though they interact with ϕ, giving or taking energy away from it, their masses are such that the decay

$$A \rightarrow B+\phi+\bar{\phi} \tag{21}$$

is kinematically forbidden.

In order to isolate the effect of the presence of a bath of these particles, we neglect the expansion of the universe and subject $\phi=\phi_r\, e^{i\theta}/\sqrt{2}$ to the $n=0$ renormalizable Dimopoulos-Susskind potential $V_0(\phi_r,\theta)$. We again adopt the simplifying assumption that the evolution of ϕ_r is suppressed. Though the interactions between ϕ and the A and B particles occur randomly, we model their effect by a periodically fluctuating kinetic energy of θ [20] :

$$K(t) \;=\; K_0\Big[1+A\sin(\omega t)\Big]^2\,. \tag{22}$$

This function oscillates between $K_{\rm min} = K_0(1-A)^2$ and $K_{\rm max} = K_0(1+A)^2 = K_{\rm min}+Q$. Therefore,

$$Q \;=\; 4K_0 A\,. \tag{23}$$

Then, the equation of motion of θ in our model will be given by the Langevin equation

$$\phi_r^2\,\ddot{\theta} \;=\; -\frac{\partial V_0}{\partial\theta} - \eta\,\dot{\theta} + \sqrt{4\eta K(t)}\,\xi(t)\,, \tag{24}$$

where η is the coefficient of friction, and $\xi(t)$ is Gaussian white noise:

$$\langle\xi(t)\rangle \;=\; 0\,,\qquad \langle\xi(t)\xi(s)\rangle \;=\; \delta(t-s)\,. \tag{25}$$

The above Langevin equation is equivalent to the following Fokker-Planck equation governing the evolution of the probability density $p(\theta,t)$ and the probability current $j(\theta,t)$:

$$\begin{aligned} 0 &= \frac{\partial p(\theta,t)}{\partial t} + \frac{\partial j(\theta,t)}{\partial\theta}\,,\\ j(\theta,t) &= -\frac{1}{\eta}\left[\frac{\partial V_0}{\partial\theta}\,p(\theta,t) + 2K(t)\frac{\partial p(x,t)}{\partial x}\right]\,. \end{aligned} \tag{26}$$

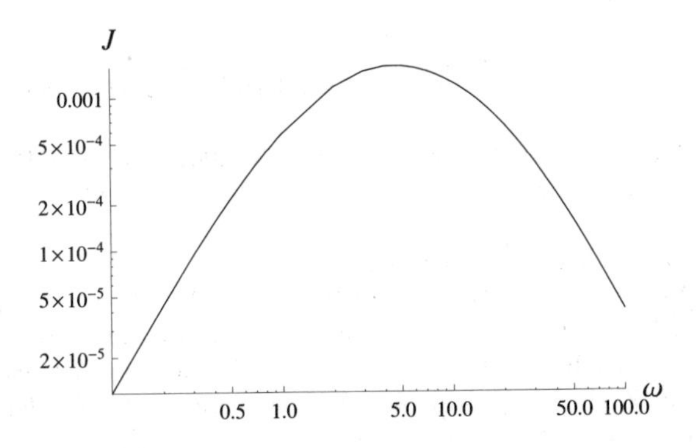

Fig. 2. ω-dependence of J for the case $\beta = \pi/2$, $K_{\min} = 0.5$, $Q = \eta = \phi_r = \lambda = 1$.

The quantity of interest for baryon number generation is the period-averaged probability current

$$J = \frac{1}{\mathcal{T}} \int_0^{\mathcal{T}} j(x,t)\, dt\ , \tag{27}$$

which is asymptotically independent of x and approaches a constant, a non-zero value signifying a non-zero baryon number. For the sake of simplicity, we set $\beta = \pi/2$, and ϕ_r, λ, and η all equal to one. We then solved these equations numerically for various values of $K_{\min}$, ω, and Q, and have found that non-zero J can be generated for a very wide range of parameter choices. As an example, we show the ω-dependence of J for the case $K_{\min} = 0.5$ and $Q = 1$ in Fig. 2. Further details of our analysis can be found in Ref. [23].

4. What is the ATP-like Particle?

Whether the ratchet mechanism we are proposing here can be embedded into a realistic scenario remains to be seen. Of particular difficulty may be maintaining a sufficiently large population of the ATP-like particles to drive the ratchet. But what can these ATP-like particles be? Several possibilities come to mind: First, it could be the inflaton at reheating, transferring energy to the ϕ field via parametric resonance. Second, they could be heavy Kaluza-Klein (KK) modes in some extra-dimension model. And third, perhaps they could be technibaryons transferring energy to technimeson ϕ's. Finally, regardless of what their actual identities are, if the ATP-like particles are highly stable and still around, they may constitute dark matter, thereby connecting baryogenesis with the dark matter problem. These, and other possibilities will be discussed elsewhere [24].

Acknowledgments

We would like to thank Philip Argyres and Daniel Chung for helpful suggestions. T.T. is supported by the U.S. Department of Energy, grant DE–FG05–92ER40709, Task A.

References

[1] S. Weinberg, "The First Three Minutes. A Modern View of the Origin of the Universe," (1977).
[2] G. Steigman, Ann. Rev. Nucl. Part. Sci. **57**, 463 (2007) [arXiv:0712.1100 [astro-ph]].
[3] B. D. Fields and S. Sarkar, in the "Review of Particle Physics," J. Phys. G **37**, 075021 (2010).
[4] E. Komatsu *et al.*, arXiv:1001.4538 [astro-ph.CO].
[5] D. J. Fixsen, Astrophys. J. **707**, 916 (2009) [arXiv:0911.1955 [astro-ph.CO]].
[6] J. M. O'Meara, S. Burles, J. X. Prochaska, G. E. Prochter, R. A. Bernstein and K. M. Burgess, Astrophys. J. **649**, L61 (2006) [arXiv:astro-ph/0608302].
[7] M. Pettini, B. J. Zych, M. T. Murphy, A. Lewis and C. C. Steidel, Mon. Not. Roy. Astron. Soc. **391**, 1499 (2008) [arXiv:0805.0594 [astro-ph]].
[8] A. D. Sakharov, JETP Letters, **5**, 24 (1967)
[9] A. Riotto and M. Trodden, Ann. Rev. Nucl. Part. Sci. **49**, 35 (1999) [arXiv:hep-ph/9901362].
[10] M. Dine and A. Kusenko, Rev. Mod. Phys. **76**, 1 (2004) [arXiv:hep-ph/0303065].
[11] J. M. Cline, arXiv:hep-ph/0609145.
[12] W. Buchmüller, arXiv:0710.5857 [hep-ph].
[13] M. Shaposhnikov, J. Phys. Conf. Ser. **171**, 012005 (2009).
[14] S. Weinberg, "Cosmology," Oxford University Press (2008).
[15] S. Dimopoulos and L. Susskind, Phys. Rev. D **18**, 4500 (1978).
[16] I. Affleck and M. Dine, Nucl. Phys. B **249**, 361 (1985).
[17] A. G. Cohen and D. B. Kaplan, Phys. Lett. B **199** (1987) 251.
[18] A. Dolgov and K. Freese, Phys. Rev. D **51**, 2693 (1995) [arXiv:hep-ph/9410346].
[19] A. Minamizaki and A. Sugamoto, Phys. Lett. B **659**, 656 (2008) [arXiv:0705.3682 [hep-ph]].
[20] P. Reimann, R. Bartussek, R. Häußler, P. Hänggi, Phys. Lett. A **215** (1996) 26.
[21] F. Julicher, A. Ajdari and J. Prost, Rev. Mod. Phys. **69**, 1269 (1997).
[22] P. Reimann, Phys. Rept. **361**, 57 (2002) [arXiv:cond-mat/0010237].
[23] A. Minamizaki, Ph.D. Thesis, Ochanomizu University, 2010 (in Japanese).
[24] A. Minamizaki, A. Sugamoto, and T. Takeuchi, in preparation.

Classical Solutions of Field Equations in Einstein Gauss-Bonnet Gravity

P. Suranyi, C. Vaz and L. C. R. Wijewardhana

Department of Physics, University of Cincinnati, Cincinnati, Ohio, USA

In this lecture we review recent analysis of the thermodynamic properties of classical black hole and brane solutions in Einstein Gauss-Bonnet (EGB) gravity. Spherically symmetric black hole solutions of EGB gravity in five dimensions have a lower bound for its mass below which the solution fails to exist. As the mass approaches the minimum value the Hawking temperature approaches zero making the configuration semiclassically stable. Uniform black brane solutions defined in a 4+d dimensional space with d compactified torroidal dimensions also have a minimum mass. As the brane approaches the minimum mass the temperature approaches a non zero value making these configurations semiclassically unstable. Unlike these black branes, black holes caged in a compact space would still have a minimum mass and leave stable remnants under evaporation. Such remnants could constitute one of the components of dark matter.

1. Gauss Bonnet Black Holes and Branes

Quantum theories of gravity like string theory give rise to additional higher curvature correction terms to the Einstein action.[1] Some higher order curvature terms when taken in isolation and quantized can potentially lead to problematic high energy behavior, such as the presence of negative norm states (ghosts). The simplest non trivial higher curvature term avoiding these problems is the Gauss-Bonnet (GB) term which makes a non trivial contribution to dynamics in space-time dimensions greater than or equal to five. When expanded about a flat background the equation of motion of Einstein Gauss-Bonnet (EGB) gravity contains at most second derivative terms avoiding the problem with negative norm states. In this lecture we report an analysis of the critical behavior of a class of classical solutions to EGB gravity. Such a term arises in the low energy gravity action in string theory when corrections to zero slope limit are calculated.[2]

The equation of motion of EGB gravity is given by

$$G_{ab} \equiv R_{ab} - \frac{1}{2} g_{ab} R + \alpha L_{ab} = 0,$$

where α is the coupling, the Lanczos tensor is

$$L_{ab} = -\frac{1}{2} g_{ab} \left(R^2 - 4R^{ab} R_{ab} + K \right) + 2RR_{ab} - 4R_{ac} R^c_b - 2R_{acde} R_b^{cde} + 4R_{acdb} R^{cd}, \tag{1}$$

and K is the Kretschmann scalar,

$$K = R_{abcd} R^{abcd}.$$

Spherically symmetric black hole solutions in EGB gravity in dimensions greater than or equal to 5 were found by Boulware and Deser.[3] Thermodynamics of such black holes were formulated by Myers and Simon and by Wiltshire.[4] It was seen that for $\alpha > 0$, in five dimensions , spherically symmetric black hole solutions do not exist below the mass

$$M_{\rm min} = \frac{3\pi\alpha}{4G_5}$$

and the Hawking temperature goes to zero as this mass is approached,

$$T \simeq \frac{r_H}{8\pi\alpha} = \frac{1}{8\pi\alpha}\sqrt{\frac{8G_5M}{3\pi} - 2\alpha}$$

Thus at the end of Hawking evaporation such black holes would leave behind stable remnants.

If our world is contained in a higher dimensional bulk objects that appear as black holes in 4 dimensions would be black branes in the higher dimensional spacetime. Do such objects leave stable remnants when they evaporate if higher curvature correction terms of the Gauss Bonnet type are included in the gravitational action? We have investigated this question in a couple of papers and here we summarize our conclusions.

In conventional Einstein gravity a $D = 4$ Schwarzschild black hole can be trivially extended to a black string ($D = 5$) or black brane solutions ($D > 5$). Solutions become more complicated if higher derivative terms are added to the Einstein action, such as the GB term. Though asymptotically Minkowski black hole solutions have been found in $D \geq 5$ theories,[3] no exact black string (brane) solutions are known in Einstein-Gauss-Bonnet (EGB) gravity. Yet the investigations of black strings (branes) is more important because with compact extra dimensions all but the lightest static objects must be black strings (branes).

Two alternative techniques have been used to investigate black brane solutions. Kobayashi and Tanaka[5] solved the 5 dimensional Einstein equations modified by Lanczos tensor (1) (which is the variation of the GB term), numerically. Using an expansion around the event horizon they also found an exact lower bound for the black string mass, $M_c \sim \sqrt{\alpha}$, where α is the coupling constant of the GB term. For $M < M_c$ the solutions do not have a horizon, they represent naked singularities.

In a subsequent work,[6] following,[5] we used a horizon expansion to investigate black strings in $D > 5$ EGB theory. We found that in every dimension, $D \geq 5$, a lower bound, similar to that in $D = 5$, exists for the mass of the black string. We also investigated the thermodynamics of the black string but the fifth order expansion in α, employed in our paper, was insufficient to get a definitive answer for the thermal behavior of the system when the mass of the black string approached the lower limit.

In the paper [7] we used improved numerical and expansion methods to investigate black string solutions. We used the following ansatz for the metric of a uniform string, made dimensionless by factoring out r_h^2, as

$$ds^2 = -f(\rho)dt^2 + \frac{g(\rho)}{f(\rho)}d\rho^2 + (\rho+1)^2 d\Omega^2 + h(\rho)dw^2,$$

ρ defined by

$$\rho = \frac{r - r_h}{r_h},$$

to replace the radial coordinate, r where r_h is the radius of the horizon. Here $d\Omega^2$ is the metric on the three sphere and w is the coordinate of the fifth dimension, which assumed to be periodic with a period L. Our discussion could easily be generalized to more than one extra dimensions compactified on a torus.

In our discussion we used a dimensionless GB coupling defined as

$$\beta = 1 - 8\frac{\alpha}{r_h^2}. \tag{2}$$

We find two critical points, $\beta_{\text{crit}} = 0$ and 3, where the condition $\beta = 0$ determines the critical radius. The critical radius determines a nonzero critical mass. We paid particular attention to the singularities at the endpoints of range, R, $0 \le \beta < 3$, where β is defined in (2). No black string solutions exists outside R. In particular, we found a critical solution at $\beta = 0$ which, in contrast to solutions at $\beta > 0$, has a horizon expansion that includes half-integer powers of the radial horizon variable ρ. Metric components and all scalars, including the Ricci scalar and the Kretschmann scalar are finite, but have $\sqrt{\beta}$ type singularities at $\beta = 0$.

At $\beta > 0$ the metric components are singular inside the horizon, at $\rho = -3\beta/7$. Unless $\beta = 0$ it is always possible to continue the metric to a region inside the horizon.

To further study the $\beta = 0$ limit of solutions we introduced Eddington-Finkelstein coordinates

$$u = t - \int d\rho \frac{\sqrt{g(\rho)}}{f(\rho)}$$

Then the metric takes the form

$$ds^2 = -f(\rho)du^2 + 2\sqrt{g(\rho)}du\, d\rho + (\rho+1)^2 d\Omega^2 + h(\rho)dw^2,$$

regular everywhere including a neighborhood of the horizon [a]. The singularity at $-3\beta/7$ moves to the horizon when $\beta \to 0$. We have not been able to construct a metric inside this singular point. We have also been unable to continue the critical metric, with $\beta = 0$, inside the horizon. Since Kruskal coordinates usually represent the maximal non-singular extension of the space-time, we define the Kruskal coordinates for the critical string such that the coefficient of $dUdV$ is -1. They can be

[a]Note that $g(\rho) > 0$ on the interval $-3\beta/7 < \rho < \infty$ for the whole range, R, of admissible values of β.

obtained in a asymptotic expansion from the Eddington-Finkelstein coordinates u and

$$v = t + \int d\rho \frac{\sqrt{g(\rho)}}{f(\rho)}.$$

We find

$$ds^2 = -dU\,dV + (\rho+1)^2 d\Omega^2 + h[\rho]dw^2,$$

where ρ has the following expression by the Kruskal coordinates U and V

$$\rho = -U\,V(1 + c\sqrt{U\,V} + \cdots),$$

where c is a non-vanishing constant. It is obvious that we cannot cross the light like surfaces $U = 0$ or $V = 0$, which represent the horizon. However, we have shown that a geodesic of a massive probe falling toward the horizon reaches it in finite proper time (at $\tau = 0$), but it cannot pass the horizon because its radial coordinate turns complex at $\tau > 0$ even though the curvature invariants are finite at that point. The meaning of this is unclear to us at this time. Either there is a different choice of coordinates in which the space-time may be extended through the horizon, but they are not the Kruskal coordinates and we have not been able to find them, or the critical solution is pathological and should be excluded by a censorship principle.

We used numerical techniques to calculate the solutions for the whole range R. We calculated the ADM mass, tension, Hawking temperature, and entropy. While the Hawking-temperature tends to a finite value at $\beta = 0$, it diverges at $\beta = 3$. The ADM mass and tension are modified only by a finite amount compared to pure Einstein gravity and bounded from above and below over the whole range, R. The entropy is a monotonically decreasing function of β, vanishing in the limit $\beta = 3$. This is not surprising in view of the divergence of the Hawking temperature. We also performed a stability analysis by considering linear perturbations about the black brane solutions. We found that for $\alpha > 0$ and small enough compactification size the solutions reach the critical mass before encountering a Gregory-Laflamme[8] type instability.

There are two unresolved problems concerning the fate of black strings. If $\alpha > 0$ they are driven by Hawking radiation toward the critical state at $\beta = 0$. That state, however is very problematic. Though g_{tt} has a linear zero and all scalars are finite at that zero, test particles that reach the $\rho = 0$ point in finite time have nowhere to go because $\rho(\tau)$ becomes complex as a function of proper time.

As finding solutions for black holes in compactified spaces is even more difficult than finding black string solutions we have no way of comparing the entropy and relative stability of these solutions. This is one of the reasons why we are not able to answer what happens with black strings that reach the end of their lives at the $\beta = 0$ line. Since the limit of their Hawking temperature is not zero, as a study based on numerical approximations and $1/\rho$ expansions lead us to believe, so they are certainly not frozen at that point. Their ADM mass is

$$M_c \simeq 1.23\frac{\sqrt{8\alpha}L}{2G}.$$

They cannot go into a black string state with a smaller ADM, mass, however, because such black string states do not exist. It seems that the only possibility is that some other, yet undiscovered, type of higher entropy static or possibly dynamic state exists, into which the black string can morph. Another possible solution, at least for $D \geq 6$, is that the inclusion of higher order Lovelock terms would cure this anomaly. The fate of black stings at and beyond $\beta = 0$ is one of the most important questions we would like to investigate in the future. Other questions include a possible numerical study of black holes in compactified spaces.

This work used a Kaluza-Klein type compactification. It would be more difficult, but perhaps more interesting to repeat this work in a Randall-Sundrum[9] type anti-de Sitter space. This work is funded in part by the U.S. Department of Energy under the grant DE-FG-02-84ER40153.

References

1. C. Callen, E. Martinec, M. Perry and D. Friedan, Nucl. Phys. **B262**, 593 (1985).
2. Rafael I. Nepomechie, Phys.Rev. D **32**, 3201 (1985); B. Zwiebach, Phys. Lett. **156B**, 315 (1985); B. Zumino, Phys. Rep. **137**, 109 (1986).
3. D. Boulware and S. Deser, Phys. Rev. Lett. **55**, 2656 (1985).
4. R. Myers and J. Simon, Phys. Rev. D **38**, 2434 (1988); D.L. Wiltshire, Phys. Rev. D **38**, 2445 (1988).
5. T. Kobayashi, T.Tanaka, Phys. Rev. D **71**, 084005 (2005).
6. C. Sahabandu, P. Suranyi, C. Vaz, and L.C.R. Wijewardhana, Phys. Rev. D **73**, 044009 (2006).
7. P.Suranyi, C.Vaz and L.C.R.Wijewardhana, Phys.Rev.D79:124046,2009. e-Print: arXiv:0810.0525 [hep-th]
8. R. Gregory and R. Laflamme, Phys. Rev. Lett. **70**, 2837, 1993.
9. L. Randall and R. Sundrum, Phys. Rev. Lett. 83:3370-3373, L. Randall and R. Sundrum, Phys. Rev. Lett. 83:4690-4693, 1999.

Black Holes Constitute All Dark Matter

Paul H. Frampton

Department of Physics and Astronomy, University of North Carolina, Chapel Hill, USA
and
Institute for the Mathematics and Physics of the Universe, University of Tokyo, Japan
E-mail: frampton@physics.unc.edu, paul.frampton@ipmu.jp

The dimensionless entropy , $\mathcal{S} \equiv S/k$, of the visible universe, taken as a sphere of radius 50 billion light years with the Earth at its "center", is discussed. An upper limit (10^{112}), and a lower limit (10^{102}), for $\mathcal{S}$ are introduced. It is suggested that intermediate-mass black holes (IMBHs) constitute all dark matter, and that they dominate $\mathcal{S}$.

Keywords: Dark matter, black hole, entropy, halo.

1. Introduction

Two references useful for further information about the material of this talk are:

(1) P.H.F. and T.W. Kephart. *Upper and Lower Bounds on Gravitational Entropy.* JCAP 06:008 (2008) and
(2) P.H.F. *Identification of All Dark Matter as Black Holes.* `arXiv:0905.3632 [hep-th]`. JCAP 0910:016 (2009).

2. The Entropy of the Universe

As interest grows in pursuing alternatives to the Big Bang, including cyclic cosmologies, it becomes more pertinent to address the difficult question of what is the present entropy of the universe?

Entropy is particularly relevant to cyclicity because it does not naturally cycle but has the propensity only to increase monotonically. In one recent proposal, the entropy is jettisoned at turnaround. In any case, for cyclicity to be possible there must be a gigantic reduction in entropy (presumably without violation of the second law of thermodynamics) of the visible universe at some time during each cycle.

Standard treatises on cosmology address the question of the entropy of the universe and arrive at a generic formula for a thermalized gas of the form

$$S = \frac{2\pi^2}{45} g_* V_U T^3 \tag{1}$$

where g_* is the number of degrees of freedom, T is the Kelvin temperature and V_U is the volume of the visible universe. From Eq.(1) with $T_\gamma = 2.7^0$K and $T_\nu = T_\gamma(4/11)^{1/3} = 1.9^0$K we find the entropy in CMB photons and neutrinos are roughly equal today

$$S_\gamma(t_0) \sim S_\nu(t_0) \sim 10^{88}. \tag{2}$$

Our topic here is the gravitational entropy, $S_{grav}(t_0)$. Following the same path as in Eqs. (1,2) we obtain for a thermal gas of gravitons $T_{grav} = 0.91^0$K and then

$$S_{grav}^{(thermal)}(t_0) \sim 10^{86} \tag{3}$$

This graviton gas entropy is a couple of orders of magnitude below that for photons and neutrinos. This graviton gas entropy is a couple of orders of magnitude below that for photons and neutrinos. But there are larger contributions to gravitational entropy from elsewhere!!!

3. Upper Limit on the Gravitational Entropy

We shall assume that dark energy has zero entropy and we therefore concentrate on the gravitational entropy associated with dark matter. The dark matter is clumped into halos with typical mass $M(halo) \simeq 10^{11} M_\odot$ where $M_\odot \simeq 10^{57} GeV \simeq 10^{30} kg$ is the solar mass and radius $R(halo) = 10^5 pc \simeq 3 \times 10^{18} km \simeq 10^{18} r_S(M_\odot)$. There are, say, 10^{12} halos in the visible universe whose total mass is $\simeq 10^{23} M_\odot$ and -corresponding Schwarzschild radius is $r_S(10^{23} M_\odot) \simeq 3 \times 10^{23} km \simeq 10 Gpc$. This happens to be the radius of the visible universe corresponding to the critical density. This has led to an upper limit for the gravitational entropy is for one black hole with mass $M_U = 10^{23} M_\odot$.

Using $S_{BH}(\eta M_\odot) \simeq 10^{77} \eta^2$ corresponds to the holographic principle for the upper limit on the gravitational entropy of the visible universe:

$$S_{grav}(t_0) \leq S_{grav}^{(HOLO)}(t_0) \simeq 10^{123} \tag{4}$$

which is 37 orders of magnitude greater than for the thermalized graviton gas in Eq.(3) and leads us to suspect (correctly) that Eq.(3) is a gross underestimate. Nevertheless, Eq.(4) does provide a credible upper limit, an overestimate yet to be refined downwards below, on the quantity of interest, $S_{grav}(t_0)$.

The reason why a thermalized gas of gravitons grossly underestimates the gravitational entropy is because of the 'clumping' effect on entropy. Because gravity is universally attractive its entropy is increased by clumping. This is somewhat counter-intuitive since the opposite is true for the familiar 'ideal gas'. It is best illustrated by the fact that a black hole always has 'maximal' entropy by virtue of the holographic principle.

4. Lower Limit on Gravitational Entropy

It is widely believed that most, if not all, galaxies contain at their core a supermassive black hole with mass in the range $10^5 M_\odot$ to $10^9 M_\odot$ with an average mass about $10^7 m_\odot$. Each of these carries an entropy S_{BH}(supermassive) $\simeq 10^{91}$. Since there are 10^{12} halos this provides the lower limit on the gravitational entropy of

$$S_{grav}(t_0) \geq 10^{103} \tag{5}$$

which, by now, provides an eight order of magnitude window for $S_{grav}(t_0)$.

The lower limit in Eq.(5) from the galactic supermassive black holes may be largest contributor to the entropy of the present universe but this seems to us highly unlikely because they are so very small. Each supermassive black hole is about the size of our solar system or smaller and it is intuitively unlikely that essentially all of the entropy is so concentrated.

Gravitational entropy is associated with the clumping of matter because of the long range unscreened nature of the gravitational force. This is why we propose that the majority of the entropy is associated with the largest clumps of matter: the dark matter halos associated with galaxies and cluster.

5. Intermediate Mass Black Holes

If we consider normal baryonic matter, other than black holes, contributions to the entropy are far smaller. The background radiation and relic neutrinos each provide $\sim 10^{88}$. We have learned in the last decade about the dark side of the universe. WMAP suggests that the pie slices for the overall energy are 4% baryonic matter, 24% dark matter and 72% dark energy. Dark energy has no known microstructure, and especially if it is characterized only by a cosmological constant, may be assumed to have zero entropy. As already mentioned, the baryonic matter other than the SMBHs contributes far less than $(S_U)^{min}$.

This leaves the dark matter which is concentrated in halos of galaxies and clusters.

It is counter to the second law of thermodynamics when higher entropy states are available that essentially all the entropy of the universe is concentrated in SMBHs. The Schwarzschild radius for a $10^7 M_\odot$ SMBH is $\sim 3 \times 10^7$ km and so 10^{12} of them occupy only $\sim 10^{-36}$ of the volume of the visible universe.

Several years ago important work by Xu and Ostriker showed by numerical simulations that IMBHs with masses above $10^6 M_\odot$ would have the property of disrupting the dynamics of a galactic halo leading to runaway spiral into the center. This provides an upper limit $(M_{IMBH})^{max} \sim 10^6 M_\odot$.

Gravitational lensing observations are amongst the most useful for determining the mass distributions of dark matter. Weak lensing by, for example, the HST shows the strong distortion of radiation from more distant galaxies by the mass of the dark matter and leads to astonishing three-dimensional maps of the dark matter trapped within clusters. At the scales we consider $\sim 3 \times 10^7$ km, however, weak lensing has no realistic possibility of detecting IMBHs in the forseeable future.

Gravitational microlensing presents a much more optimistic possibility. This technique which exploits the amplification of a distant source was first emphasized in modern times (Einstein considered it in 1912 unpublished work) by Paczynski. Subsequent observations found many examples of MACHOs, yet insufficient to account for all of the halo by an order of magnitude. These MACHO searches looked for masses in the range $10^{-6}M_{\odot} \leq M \leq 10^{2}M_{\odot}$.

The time t_0 of a microlensing event is given by

$$t_0 \equiv \frac{r_E}{v} \tag{6}$$

where r_E is the Einstein radius and v is the lens velocity usually taken as $v = 200$ km/s. The radius r_E is proportional to the square root of the lens mass and numerically one finds

$$t_0 \simeq 0.2y \left(\frac{M}{M_{\odot}}\right)^{1/2} \tag{7}$$

so that, for the MACHO masses considered, $2h \leq t_0 \leq 2y$.

6. Cosmological Entropy Considerations

The cosmological entropy range

$$102 \leq \log_{10} S_U \leq 112 \tag{8}$$

is the first of two interesting windows which are the subject. Conventional wisdom is $S_U \sim (S_U)^{min} = 10^{102}$.

7. Intermediate Mass Black Holes and Microlensing Longevity

$\log_{10} n_{max}$	$\log_{10} \eta$	$\log_{10} S_{halo}$	$\log_{10} S_U$	t_0 (years)
8	2	88	100	2
7	3	89	101	6
6	4	90	102	20
5	5	91	103	60
4	6	92	104	200

(Assumes $\rho_{IMBH} \sim 1\% \rho_{DM}$)

8. Observation of IMBHs

Since microlensing observations already impinge on the lower end of the range in the Table, it is likely that observations which look at longer time periods, have higher statistics or sensitivity to the period of maximum amplification can detect heavier mass IMBHs in the halo. If this can be achieved, and it seems a worthwhile enterprise, then the known entropy of the universe could be increased by more than two orders of magnitude.

There exists interesting other analyses pertinent to existence of massive halo objects:

J. Yoo, J, Chanamé and A. Gould, Astrophys. J. **601,** 311 (2004). astro-ph/0307437.

It is this entropy argument based on holography and the second law of thermodynamics which is the most compelling supportive argument for IMBHs. If each galaxy halo asymptotes to a black hole the final entropy of the universe will be $\sim 10^{112}$ as in Eq.(8) and the universe will contain just $\sim 10^{12}$ supergigantic black holes. Conventional wisdom is that the present entropy due entirely to SMBHs is only $\sim 10^{-10}$ of this asymptopic value. IMBHs can increase the fraction up to $\sim 10^{-8}$, closer to asymptopia and therefore more probable according to the second law of thermodynamics.

There are several previous arguments about the existence of IMBHs and they have put upper limits on their fraction of the halo mass. The entropy arguments are new and provide additional motivation to tighten these upper bounds or discover the halo black holes. One observational method is high longevity microlensing events. It is up to the ingenuity of observers to identify other, possibly more fruitful, methods some of which have already been explored in a preliminary way.

9. Post-conference update

Since the SCGT09 conference took place, the paper:

Paul H. Frampton, Masihiro Kawasaki, Fuminobu Takahashi and Tsutomu Yanagida
IPMU-09-0157 (December 2009).
Primordial Black Holes as All Dark Matter
arXiv:1001.2308 [hep-ph].

shows that it is possible to form black holes in the early universe with mass $10^5 M_\odot$ and with sufficient abundance to provide all of the dark matter.

Acknowledgments

This work was supported in part by the World Premier International Research Center Initiative (WPI initiative), MEXT, Japan and by U.S. Department of Energy Grant No. DE-FG02-05ER41418.

Electroweak Precision Test and $Z \to b\bar{b}$ in the Three Site Higgsless Model*

Tomohiro Abe

Department of Physics, Nagoya University
Nagoya, 464-8602, Japan
E-mail: abetomo@eken.phys.nagoya-u.ac.jp

The three site Higgsless model is a highly-deconstructed Higgsless model with only three sites. In this model, we show that the KK gauge boson mass, KK fermion mass, and the KK gauge boson couplings with light quarks and leptons are severely constrained by the electroweak precision data. Especially we find that perfect fermiphobity of KK gauge boson is ruled out by the precision data. We also compute the flavor dependent chiral logarithmic corrections to the decay $Z \to b\bar{b}$. We find that the phenomenological constraints on this model arising from measurements of $Z \to b\bar{b}$ are relatively mild, requiring only that the heavy Dirac fermion be heavier than 1 TeV or so, and are satisfied automatically in the range of parameters allowed by the electroweak precision data.

1. Introduction

The standard model (SM) describes the phenomenology among the elementary particles very well. But the Higgs boson has not observed yet even though it is needed for the electroweak symmetry breaking (EWSB) in the SM. So there is a possibility the EWSB can be occurred without any scalar bosons. To pursue this possibility we need to take into account for the perturbative unitarity of the longitudinal gauge bosons' scattering. To keep a theory perturbative, some new particles have to appear below 1TeV and contribute to those scattering. One of the possibilities is a Higgsless model.[1] This is based on a gauge theory in five-dimension which means an infinite number of the gauge bosons, KK gauge bosons, exist in four-dimension. These KK gauge bosons keep the perturbative unitarity.

An infinite number of the gauge bosons imply an infinite number of gauge symmetries in four-dimension.[2–5] In fact a gauge theory in five-dimension can be described as a gauge theory in four-dimension by a method called deconstruction.[6,7]

For the low energy phenomenology, a finite number of gauge bosons are sufficient. So the roughly deconstructed Higgsless model is suitable for the phenomenological analysis. The three site Higgsless model[8] is a highly deconstructed Higgsless model. We study the constraints to it from the electroweak precision test (EWPT).[9,10]

*This report is based on Ref. 9 and Ref. 10.

2. Three site Higgsless model

The three site Higgsless Model has $SU(2)_0 \times SU(2)_1 \times U(1)_2$ gauge symmetry. The particles in this model are SM particles, except the Higgs boson, and their KK partners. In the gauge sector, there are $W'^{\pm}$ and Z' in addition to $W^{\pm}$ and Z. To push up the perturbative unitarity bound, $W'^{\pm}$ and Z' have to be lighter than 1TeV and be coupled with $W^{\pm}$ and Z. We can get a lower bound of the $W'^{\pm}$ and Z' mass, $M_{W'}$, from the constraint to WWZ coupling constant $M_{W'} > 380$ GeV. As a result we can expect $W'^{\pm}$ and Z' might be found at the LHC experiment.

3. Electroweak precision test

Using the S and T parameters[11,12] we can get the lower bound of KK fermion mass, M_F, and the constraint for the coupling among $W'^{\pm}$ and the standard model fermions, $g_{W'ff}$.

In general, Higgsless models have a large correction to the S parameter at the tree level. If the SM fermions are not coupled with the KK gauge bosons, however, the correction to the S parameter becomes small. In this model, therefore, the $g_{W'ff}$ is proportional to the S parameter at the tree level and should be negligible small. This estimate is qualitative. For quantitative estimation, we need one-loop calculations. The results are following.

$$\begin{aligned}\alpha S &= -4s^2 \frac{M_W}{M_{W'}} \frac{g_{W'ff}}{g_{Wff}} \\ &\quad - \frac{\alpha}{24\pi} \frac{M_{W'}}{M_W} \frac{g_{W'ff}}{g_{Wff}} \ln \frac{M_{W'}^2}{M_F^2} - \frac{\alpha}{24\pi} \ln \frac{M_{W'}^2}{M_F^2} + \frac{\alpha}{12\pi} \ln \frac{\Lambda^2}{M_{Href}^2} \qquad (1) \\ \alpha T &= -\frac{\sqrt{2} G_F}{64\pi^2} \frac{M_t^4}{M_T^2} \frac{M_{W'}^4}{M_W^4} \frac{1}{\left[1 - \frac{M_{W'}}{M_W} \frac{g_{W'ff}}{g_{Wff}}\right]^2} - \frac{3\alpha}{32\pi c^2} \left(\ln \frac{M_{W'}^2}{M_{Href}^2} + \ln \frac{\Lambda^2}{M_{Href}^2} \right) \qquad (2)\end{aligned}$$

where Λ is a cutoff scale of this model. By the naive dimensional analysis, we can estimate $\Lambda < 4.3$ TeV. Comparing them with experimental constraint,[13] we can find $g_{W'ff}/g_{Wff} \sim \mathcal{O}(10^{-2})$ (Fig.1). This means we need the high luminosity for the Drell-Yan production of $W'^{\pm}$ and its decay into the SM fermions at LHC experiment. The lower bound of KK fermion mass can be found at 1.8 TeV (Fig.2). This implies that KK fermions are rarely produced at the LHC experiment.

4. $Zb\bar{b}$ coupling

$Zb\bar{b}$ coupling also tells us a lower bound of KK fermion mass. We use the constraint to R_b. So the flavor dependent correction to the $Zb\bar{b}$ coupling is sufficient. We parametrize $\delta g_L^{b\bar{b}}$ as follows,

$$g_{Zb\bar{b}} = g_Z \left(-\frac{1}{2} + \delta g_L^{b\bar{b}} + \frac{1}{3} \sin^2 \theta_W \right), \qquad (3)$$

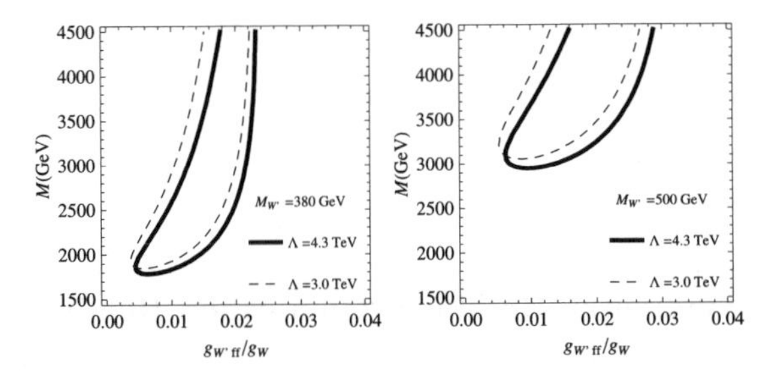

Fig. 1. Allowed region of $g_{W'ff}/g_{Wff}$ and KK fermion mass. Inner side of contours are experimentally allowed. The left panel is $M_{W'} = 380$ GeV case and the right panel is $M_{W'} = 500$ GeV case. In both case we can find $g_{W'ff}/g_{Wff} \sim \mathcal{O}(10^{-2})$.

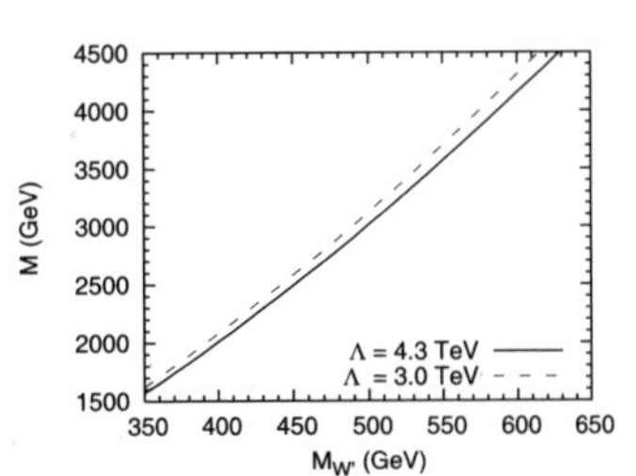

Fig. 2. Allowed region of $M_{W'}$ and KK fermion mass. The upper region is the allowed region. The region below the line is excluded.

and we find this correction in this model at the one-loop level is

$$\delta g_L^{b\bar{b}} = \frac{m_t^2}{(4\pi)^2 v^2}\left[1 + \frac{f_1^2 f_2^2}{2(f_1^2+f_2^2)^2}\log\left(\frac{\Lambda^2}{M_F^2}\right)\right], \tag{4}$$

where $1/v^2 = \sqrt{2}G_F$. We take $f_1 \sim f_2 \sim \sqrt{2}v$ for simplicity. From the experimental constraint to R_b, finally we find that $M_F > 1$ TeV. So the constraint from $Zb\bar{b}$ coupling is relatively mild and automatically satisfied with EWPT.

References

1. C. Csaki, C. Grojean, H. Murayama, L. Pilo and J. Terning, Phys. Rev. D **69**, 055006 (2004) [arXiv:hep-ph/0305237].
2. M. Bando, T. Kugo, S. Uehara, K. Yamawaki and T. Yanagida, Phys. Rev. Lett. **54**, 1215 (1985).
3. M. Bando, T. Kugo and K. Yamawaki, Prog. Theor. Phys. **73**, 1541 (1985).
4. M. Bando, T. Kugo and K. Yamawaki, Nucl. Phys. B **259**, 493 (1985).
5. M. Bando, T. Kugo and K. Yamawaki, Phys. Rept. **164**, 217 (1988).
6. N. Arkani-Hamed, A. G. Cohen and H. Georgi, Phys. Rev. Lett. **86**, 4757 (2001) [arXiv:hep-th/0104005].
7. C. T. Hill, S. Pokorski and J. Wang, Phys. Rev. D **64**, 105005 (2001) [arXiv:hep-th/0104035].
8. R. S. Chivukula, B. Coleppa, S. Di Chiara, E. H. Simmons, H. J. He, M. Kurachi and M. Tanabashi, Phys. Rev. D **74**, 075011 (2006) [arXiv:hep-ph/0607124].
9. T. Abe, S. Matsuzaki and M. Tanabashi, Phys. Rev. D **78**, 055020 (2008) [arXiv:0807.2298 [hep-ph]].
10. T. Abe, R. S. Chivukula, N. D. Christensen, K. Hsieh, S. Matsuzaki, E. H. Simmons and M. Tanabashi, Phys. Rev. D **79**, 075016 (2009) [arXiv:0902.3910 [hep-ph]].
11. M. E. Peskin and T. Takeuchi, Phys. Rev. Lett. **65**, 964 (1990).
12. M. E. Peskin and T. Takeuchi, Phys. Rev. D **46**, 381 (1992).
13. C. Amsler *et al.* [Particle Data Group], Phys. Lett. B **667**, 1 (2008).

Chiral Symmetry and BRST Symmetry Breaking, Quaternion Reality and the Lattice Simulation

Sadataka Furui
School of Science and Engineering, Teikyo University
Utsunomiya, 320-8551, Japan
E-mail: furui@umb.teikyo-u.ac.jp

We discuss that the deviation of the Kugo-Ojima color confinement parameter $u(0)$ from -1 in the case of quenched lattice simulation and the consistency with -1 in the case of full QCD simulation could be attributed to the boundary condition defined by fermions inside the region of $r < 1$fm. By using the domain wall fermion propagator in lattice simulation, we show that the chiral symmetry breaking in the infrared can become manifest when one assumes that the left-handed fermion on the left wall and the right-handed fermion on the right wall are correlated by a self-dual gauge field. The relation between the infrared fixed point of the running coupling measured in lattice simulations, the prediction of the BLM renormalization theory, the conformal field theory with use of the t'Hooft anomaly matching condition in non-SUSY supersymmetric theory and the quaternion real condition are discussed.

Keywords: Infrared fixed point, quaternion, triality, critical flavor number.

1. Introduction

The infrared (IR) QCD is characterized by the color confinement and the chiral symmetry breaking. We performed lattice simulations[1,2] using the gauge configuration produced with domain wall fermion[3] which preserves the chiral symmetry in the zero - mass limit and compared with results of staggered fermion. In these lattice simulations and in comparison with other works, we observed qualitative differences between quenched and unquenched simulations.

1.1. *BRST symmetry*

The Kugo-Ojima color confinement attracted renewed interest by a conjecture of possible BRST (Becchi-Rouet-Stora-Tyutin) symmetry breaking in the infrared.[4,5] These authors pay attention to the restriction of the gauge configuration to the fundamental modular region defined by Zwanziger to solve the Gribov problem.

Kondo[5] parametrized the horizon function

$$\langle (gf^{abc}A^b_\mu c^c)(gf^{def}A^e_\nu \bar{c}^f)\rangle_k = -\delta^{ad}\left(\delta^T_{\mu\nu}u(q^2) + F(q^2)\frac{q_\mu q_\nu}{q^2}(u(q^2)+w(q^2))\right) \quad (1)$$

and assuming $w(0) = 0$, obtained $u(0) = -2/3$. There is an argument against this result[8] showing that w and u are not multicatively renormalizable. On the lattice, however the first term of the r.h.s of eq.(1.1) i.e. $\delta^{ab}\delta^T_{\mu\nu}u(q^2)$ is $\delta^{ab}\delta^T_{\mu\nu}S(U_\mu)u(q^2)$ where $S(U_\mu)$ is $S(e^{A_\mu}) = \frac{A_\mu/2}{\tanh A_\mu/2}$ in the log$-U$ definition and $S(U_\mu)_{ab} = tr\left(\lambda^\dagger_a \frac{1}{2}\left\{\frac{U_\mu+U^\dagger_\mu}{2}, \lambda_b\right\}|_{traceless\, p}\right)$ in the $U-$linear definition.

The expectation value of S is proportional to e and e/d is about 0.95 in 56^4 lattice quenched SU(3) log$-U$ definition but about 0.88 in the $U-$linear definition.[6] When $S(0) \sim 0.85$ instead of 1 as in DSE, a solution of $w(0) = 0$ and $u(0) = -\frac{2}{3}$ is possible.

1.2. *QCD running coupling*

Recently Appelquist[9] claimed by performing lattice simulation of running coupling in the Schrödinger functional method that there is a critical number of flavors N_f^c below which both chiral symmetry breaking and confinement set in. He assigned $8 \leq N_f^c \leq 12$ and in the case of $N_f = 8$ the running coupling monotonically increase as β decreases and in the case of 16^4 lattice and $\beta \sim 4.65$, $\bar{g}^2(L)/4\pi \sim 1.7$, and that there is no sign of infrared fixed point at $N_f < 8$. However, applicability of the Schrödinger functional method in the region of q below and around $\Lambda \sim 213\pm40$MeV is not clear. Brodsky et al,[10] argue that in the DSE $q < \Lambda$ or $r > 1$fm, confinement is essential and the quark-gluon should not be treated as free.

In our lattice simulation of $N_f = 2+1$ DWF, $\alpha_s(q) \sim 1.7$ at $q \sim 0.6$GeV and the deviation of the running coupling from 2-loop perturbative calculation is significant below $q \sim 3$GeV due to A^2 condensates. The running coupling data of JLab[7] suggest presence of the infrared fixed point and that $N_f = 3$ is close to the N_f^c.

1.3. *Quaternion real condition and proton form factor*

The infrared fluctuation of the gluon propagator could be attributed to the chiral anaomly. t'Hooft discussed that the anomaly cancellation on the color triplet and color sextet components in the 3 quark baryon sector,[11] and formulations in non-SUSY supersymmetric theory were discussed by Sannino.[12] In standard DWF analysis, the limit of the distance between the left domain wall and the right domain wall approaching infinity corresponds to the continuum limit. I take a specific gauge that the fermion on the left domain wall and the right domain wall are correlated by a self-dual gauge field (instanton) that satisfy quaternion real condition.[13] Then I calculate the ratio of the three point function and the two point function of fermions using the SU(6) proton wave function. As shown in Fig.2, at zero momenrtum the form factor comes exclusively from the left-handed fermion.

A product of quaternions makes an octonion, whose automorphism is the exceptional lie group G_2 which has 3+6+5 dimensional stable manifolds and posesses the triality symmetry. The spontaneous supersymmetry breaking of massless fermions

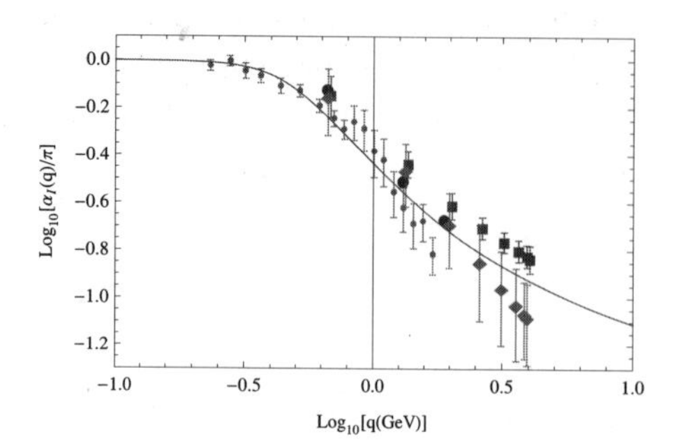

Fig. 1. The running coupling of the domain wall fermion. Coulomb gauge gluon-ghost coupling of $m_u = 0.01/a$(square), $0.02/a$(diamond), and quark-gluon coupling of $m_u = 0.01/a$(large disks). Small disks are the α_{s,g_1} derived from the spin structure function of the JLab group[7]and the solid curve is their fit.

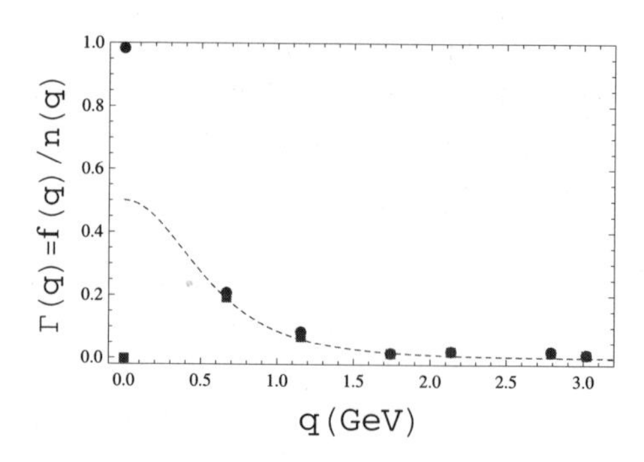

Fig. 2. The form factor of a proton using the DWF $m_u = 0.01/a$ gauge fixed such that the fermion on the left wall and the right wall are correlated by self-dual gauge field, and the dipole fit with $M^2 = 0.71\text{GeV}^2$. At zero momentum left-handed(LH) fermion dominates the form factor, and at finite momentum, LH and RH contribute almost equally.

on 5 dim manifolds and ghost, gluon on 3 dim and 6 dim color space, and the problem of the large critical N_f for the conformality by about a factor of 3, would be explained by the triality symmetry of the G_2 group of the octonion.[14] Whether the large critical N_f is an artefact, and how sensitive the results on the boundary conditions are further to be investigated.

Acknowledgments

I thank Dr. F. Sannino, Prof. S. Brodsky for helpful information and discussion , and Dr. A. Deur for sending me the JLab experimental data. Numerical calculation was done at KEK, YITP Kyoto Univ. and RCNP Osaka Univ.

References

1. S. Furui, arXiV:0912.5796 and references therein.
2. S. Furui, Few-Body Syst. **45**, 51(2009), Erratum:DOI 10.1007/s00601-009-0053-4.
3. C. Allton et al., Phys. Rev. D**76**,014504 (2007); arXiv:hep-lat/0701013.
4. D. Dudal,et al, Phys. Rev. D**79**,121701(R)(2009).
5. K-I. Kondo, Phys. Lett. B**678**,322(2009).
6. H. Nakajima and S. Furui, Nucl. Phys. **B**(Proc.supl.) 141, 34(2005), arXiv:hep-lat/0408001.
7. A. Deur, V. Burkert, J.P. Chen and W. Korsch, Phys. Lett. B**665**,349 (2008).
8. Ph. Boucaud et al., arXiv:hep-lat/0909.2615.
9. T. Appelquist, Prog. Theor. Phys.(Kyoto)**180**,72(2009).
10. S.J. Brodsky and R. Shrock, Phys. Lett. B**666**,95(2008),arXiv:0806.1535[hep-th].
11. G.'t Hooft, in "*Recent Developments in Gauge Theories*", p.135 Plenum Press (1980).
12. F. Sannino, Phys. Rev. D**80**,065011(2009); arXiv:0911.0931[hep-th].
13. E.Corrigan and P. Goddard, Comm. Math. Phys.**80**, 575(1981).
14. É. Cartan, "*The theory of Spinors*", Dover Pub. (1966)

Holographic Techni-Dilaton, or Conformal Higgs*

Kazumoto Haba[a], Shinya Matsuzaki[b] and Koichi Yamawaki[a]
[a] *Department of Physics, Nagoya University, Nagoya, 464-8602, Japan*
[b] *Department of Physics, Pusan National University, Busan 609-735, Korea*

We study a holographic model dual to walking/conformal technicolor (W/C TC) deforming a hard-wall type of bottom-up setup by including effects from techni-gluon condensation. We calculate masses of (techni-) ρ meson, a_1 meson, and flavor/chiral-singlet scalar meson identified with techni-dilaton (TD)/conformal Higgs boson, as well as the S parameter. It is shown that gluon contributions and large anomalous dimension tend to decrease specifically mass of the TD. In the typical model with $S \simeq 0.1$, we find $m_{\rm TD} \simeq 600$ GeV, while $m_\rho, m_{a_1} \simeq 4$TeV.

1. A holographic technicolor model with techni-gluon condensation

The origin of the mass is the most urgent issue of the particle physics today and is to be resolved at the LHC experiments. In order to account for the origin of mass dynamically without introducing the ad-hoc Higgs boson, here we consider the Walking/Conformal Technicolor (W/C TC)[1,2] having large anomalous dimension $\gamma_m \simeq 1$ due to the strong coupling which stays almost non-running ("walking") over wide energy range near the "conformal fixed point". In contrast to a folklore that TC is a "higgsless model" in analogy with the QCD, there actually exists "techni-dilaton' (TD)' [1] as a pseudo Nambu-Goldstone boson of the approximate conformal symmetry, which is relatively light compared with other bound states like techni-ρ and techni-a_1, etc. Here we shall estimate the mass of TD in the bottom-up holographic W/C TC by newly introducing a flavor-singlet bulk field corresponding to the techni-gluon condensate. Details are given in the forthcoming paper.[3]

Following a bottom-up approach of holographic-dual of QCD[4] and that of walking/conformal (W/C) TC,[5,6] we consider a five-dimensional gauge model possessing the $SU(N_f)_L \times SU(N_f)_R$ gauge symmetry. The model is defined on the five-dimensional anti-de-Sitter space with the curvature radius L, which is described by the metric $ds^2 = g_{MN}dx^M dx^N = (L/z)^2 \left(\eta_{\mu\nu}dx^\mu dx^\nu - dz^2\right)$ with $\eta_{\mu\nu} = \mathrm{diag}[1,-1,-1,-1]$. The fifth direction z is compactified on an interval extended from ultraviolet (UV) and infrared (IR) branes, $\epsilon \le z \le z_m$. In addition to the bulk left- (L_M) and right- (R_M) gauge fields, we introduce a bulk scalar field Φ which transforms as bifundamental representation under the $SU(N_f)_L \times SU(N_f)_R$

*Talk presented by K. Haba.

gauge symmetry so as to deduce the information concerning the chiral condensation-operator $\bar{T}T$. According to the holographic dictionary, the mass-parameter m_Φ is related to γ_m, the anomalous dimension of $\bar{T}T$, as $m_\Phi^2 = -(3-\gamma_m)(1+\gamma_m)/L^2$, where $\gamma_m \simeq 0$ corresponds to QCD and QCD-like TC and $\gamma_m \simeq 1$ to the W/C TC. Here we newly introduce an additional chiral-singlet and massless bulk scalar field Φ_X dual to techni-gluon condensate $\langle \alpha G_{\mu\nu}^2 \rangle$, where α is related to the TC gauge couping $g_{\rm TC}$ by $\alpha = g_{\rm TC}^2/(4\pi)$. We adopt a "dilaton-like" coupling for interaction terms involving Φ_X in such a way that all the fields couple to Φ_X in the exponential form like $e^{\Phi_X(z)}$. (Φ_X is *not* identified with techni-dilaton in this talk.)

Thus the five-dimensional action takes the form:

$$S_5 = \int d^4x \int_\epsilon^{z_m} dz\, \sqrt{-\det g_{MN}}\, \frac{1}{g_5^2} e^{c g_5^2 \Phi_X(z)} \Big(-\frac{1}{4}{\rm Tr}\left[L_{MN}L^{MN} + R_{MN}R^{MN} \right] + {\rm Tr}\left[D_M\Phi^\dagger D^M \Phi - m_\Phi^2 \Phi^\dagger \Phi \right] + \frac{1}{2}\partial_M \Phi_X \partial^M \Phi_X \Big), \quad (1)$$

where $D_M\Phi = \partial_M\Phi + iL_M\Phi - i\Phi R_M$, g_5 denotes the gauge coupling in five-dimension, and c is the dimensionless coupling constant. It turns out that requiring the present model to reproduce high-energy behaviors in the underlying theory leads to $(L/g_5^2) = N_{\rm TC}/(12\pi^2)$ and $c = -N_{\rm TC}/(192\pi^3)$: Our model exactly reproduces the high-energy behaviors for vector and axial-vector current correlators up till the terms of $1/Q^8$ expected from the OPE. For details, see Ref.3.

We ignore Kaluza-Klein (KK) modes of Φ_X (including the lowest mode) which are identified with massive glueballs with mass of order $\mathcal{O}(\Lambda_{\rm TC})$ which is much larger than the electroweak scale, $\Lambda_{\rm TC} \gg F_\pi$, in the case of W/C TC with $\gamma_m \simeq 1$. The techni-dilaton, a flavor-singlet scalar bound state of techni-fermion and anti-techni-fermion, will be identified with the lowest KK mode in the KK decomposition of Φ, $\Phi(x,z) = v(z) + \sigma^{(1)}(x)\sigma_1(z) + \cdots$, but not of Φ_X.

Following the holographic recipe, we obtain formulas for $\langle \alpha G_{\mu\nu}^2 \rangle$, $\langle \bar{T}T \rangle$, M_ρ, M_{a_1}, $M_{\sigma_1} (\equiv M_{\rm TD})$, F_π, and the S parameter, where M_ρ and M_{a_1} are the masses of the lowest KK modes for the vector and axial-vector mesons identified with the techni-ρ and-a_1 mesons. Once γ_m is specified, in the continuum limit $\epsilon \to 0$ those quantities involve only three undetermined parameters: z_m, and dimensionless quantities $\xi = \sqrt{2} L \langle \Phi \rangle |_{z=z_m}$ and $G = \langle e^{c g_5^2 \Phi_X(z)} \rangle |_{z=z_m} - 1$ which are related to $\langle \bar{T}T \rangle$ and $\langle \alpha G_{\mu\nu}^2 \rangle$, respectively. Some dimensionless quantities such as $\hat{S} \equiv S/(N_f/2)$ (S parameter per each techni-fermion doublet) and $M_{\rho,a_1,{\rm TD}}/F_\pi$ (ratios of meson masses to F_π) are expressed as a function of only ξ and G when γ_m fixed.

2. Mass of techni-dilaton

We first present a generic analysis of the effects of the gluon condensate and the anomalous dimension on $M_{\rho,a_1,{\rm TD}}/F_\pi$. The gluon contribution may be evaluated through a quantity $\Gamma = \Gamma(\gamma_m, \xi, G) \equiv \left(\left(\frac{1}{\pi}\langle \alpha G_{\mu\nu}^2 \rangle / F_\pi^4\right) / \left(\frac{1}{\pi}\langle \alpha G_{\mu\nu}^2 \rangle / f_\pi^4\right)_{\rm QCD} \right)^{1/4}$. Fig. 1 shows plots of $M_{\rho,a_1,{\rm TD}}/F_\pi$ as a function of γ_m and Γ and indicates that $M_{\rm TD}/F_\pi$ rapidly decreases as γ_m and/or Γ become larger in contrast to $M_{\rho,a_1}/F_\pi$.

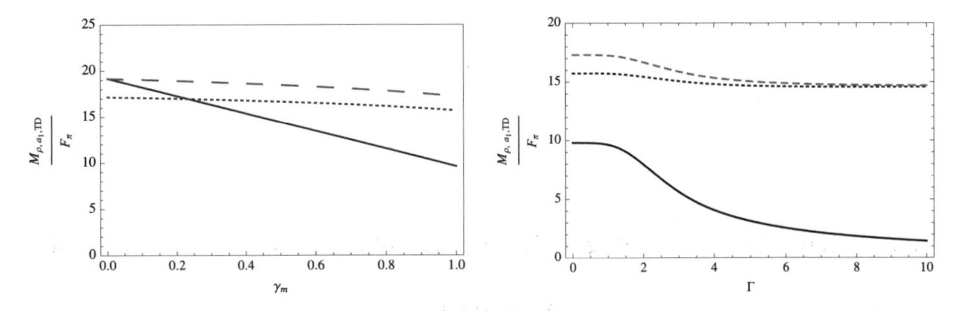

Fig. 1. Plots of $(M_{\rho,a_1,\mathrm{TD}}/F_\pi)$ in the case of $\hat{S} = 0.1$ with $N_{\mathrm{TC}} = 3$ fixed. The left panel is γ_m dependence with $\Gamma = 1$ and the right panel is Γ dependence with $\gamma_m = 1$. The dotted, dashed, solid lines respectively denote (M_ρ/F_π), (M_{a_1}/F_π), and (M_{TD}/F_π) in both panels.

We now consider a typical model of W/C TC, based on the Caswell-Banks-Zaks infrared fixed point in the large N_f QCD where we use the estimate $N_f \simeq 4N_{\mathrm{TC}} = 12(N_{\mathrm{TC}} = 3)$ from the two-loop beta function and ladder Schwinger-Dyson equation. In the W/C TC dynamics the techni-gluon condensate is not on order of the intrinsic scale Λ_{TC} but of the mass of techni-fermions $m(\ll \Lambda_{\mathrm{TC}})$ through the conformal anomaly associated with the mass generation, $\partial^\mu D_\mu = \theta^\mu_\mu = 4\theta^0_0 = \beta(\alpha)/(4\alpha^2) \cdot \langle \alpha G^2_{\mu\nu} \rangle$ and $\langle \theta^0_0 \rangle = -\frac{N_f N_{\mathrm{TC}}}{\pi^4} m^4$, where the nonperturbative beta function behaves as $\beta(\alpha) \to 0$ when $\Lambda_{\mathrm{TC}}/m \to \infty$ (conformal fixed point). This implies that the gluon condensate $\langle \alpha G^2_{\mu\nu} \rangle$ (or Γ) becomes large in that limit and hence $m_{\mathrm{TD}} \to 0$ as is seen in Fig. 1. Indeed Λ_{TC} is identified with the Extended TC (ETC) scale Λ_{ETC}: $\Lambda_{\mathrm{TC}} = \Lambda_{\mathrm{ETC}} \sim (10^4 - 10^5)F_\pi$ where $F_\pi (= \mathcal{O}(m))$ is given by $F_\pi = 246/\sqrt{N_f/2}$ GeV. Using the phenomenological input $S = 0.1$, we have $\Gamma \simeq 7$. Then we find $M_{\mathrm{TD}} \simeq 550$–$680$ GeV in contrast to $M_\rho \simeq M_{a_1} \simeq 3.8$–$3.9$ TeV.

This work was supported in part by the Global COE Program "Quest for Fundamental Principles in the Universe" and the Daiko Foundation. S.M. was supported by the Korean Research Foundation Grant (KRF-2008-341-C00008).

References

1. K. Yamawaki, M. Bando and K. Matumoto, Phys. Rev. Lett. **56**, 1335 (1986); M. Bando, K. Matumoto and K. Yamawaki, Phys. Lett. B **178**, 308 (1986).
2. T. Akiba and T. Yanagida, Phys. Lett. B **169**, 432 (1986); T. W. Appelquist, D. Karabali and L. C. R. Wijewardhana, Phys. Rev. Lett. **57**, 957 (1986). See also B. Holdom, Phys. Lett. B **150**, 301 (1985).
3. K. Haba, S. Matsuzaki, and K. Yamawaki, in preparation.
4. J. Erlich, E. Katz, D. T. Son and M. A. Stephanov, Phys. Rev. Lett. **95**, 261602 (2005); L. Da Rold and A. Pomarol, Nucl. Phys. B **721**, 79 (2005).
5. D. K. Hong and H. U. Yee, Phys. Rev. D **74**, 015011 (2006); M. Piai, arXiv:0608241[hep-ph].
6. K. Haba, S. Matsuzaki and K. Yamawaki, Prog. Theor. Phys. **120**, 691 (2008).

Phase Structure of Topologically Massive Gauge Theory with Fermion

Yuichi Hoshino

Kushiro National College of Technolofy, Otanoshike Nishi 2 -32-1, Kushiro City Hokkaido 084-0916, Japan
E-mail: hoshino@ippan.kushiro-ct.ac.jp

Using Bloch-Nordsieck approximation fermion propagator in 3-dimensional gauge theory with topological mass is studied. Infrared divergence of Chern-Simon term is soft,which modifies anomalous dimension. In unquenched QCD with 2-component spinor anomalous dimension has fractional value,where order parmeter is divergent.

Keywords: Topologically massive gauge theory, parity.

1. Introduction

The Lagrangeans of Topologically massive gauge theory with fermion are[1]

$$L = \frac{1}{4}F_{\mu\nu}F^{\mu\nu} + \frac{1}{4}\theta\epsilon^{\mu\nu\rho}F_{\mu\nu}A_\rho + \overline{\psi}(i\gamma\cdot(\partial - ieA) - m)\psi + \frac{1}{2d}(\partial\cdot A)^2, \quad (1)$$

$$\begin{aligned} L = &\frac{1}{4g^2}tr(F_{\mu\nu}F^{\mu\nu}) - \frac{\theta}{4g^2}\epsilon^{\mu\nu\rho}tr(F_{\mu\nu}A_\rho - \frac{2}{3}A_\mu A_\nu A\rho) \\ &+ \overline{\psi}(i\gamma\cdot(\partial - ieA) - m)\psi + \frac{1}{2d}(\partial\cdot A)^2\,, \end{aligned} \quad (2)$$

where $\theta = (g^2/4\pi)n, (n = 0, \pm 1, \pm 2...)$ with γ matrix for 4-comonent fermion.[2] In comparison with massless QED_3, QCD_3 topological mass term seems to soften the infrared divergence of massive fermion near its on-shell. In Minkowski metric we have 5 γ matrices $\{\gamma_\mu, \gamma_\nu\} = 2g_{\mu\nu}, (\mu = 0, 1, 2)$

$$\begin{aligned} &\gamma_0 = \begin{pmatrix} \sigma_3 & 0 \\ 0 & -\sigma_3 \end{pmatrix}, \gamma_{1,2} = -i\begin{pmatrix} \sigma_{1,2} & 0 \\ 0 & -\sigma_{1,2} \end{pmatrix}, \gamma_4 = \begin{pmatrix} 0 & I \\ I & 0 \end{pmatrix}, \gamma_5 = \begin{pmatrix} 0 & -iI \\ iI & 0 \end{pmatrix}, \\ &\tau \equiv \frac{-i}{2}[\gamma_4, \gamma_5] = diag\,(I, -I)\,. \end{aligned} \quad (3)$$

There are two redundant matrices γ_4 and γ_5 which anticommutes with other three γ matrices. There exists two kind of chiral transformation $\psi \to \exp(i\alpha\gamma_4)\psi, \psi \to \exp(\alpha\gamma_5)\psi$. The matrices $\{\gamma_4, \gamma_5, I_4, \tau\}$ generate a $U(2)$ chiral symmetry containing massless spinor fields. This $U(2)$ symmetry is broken down to $U(1) \times U(1)$ by a spinor mass term $m_e\overline{\psi}\psi$ and parity violating mass $m_o\overline{\psi}\tau\psi$, where $\overline{\psi}\tau\psi$ is a spin density. Here after we take 4-component spinors to study chiral symmetry breaking by $m^e\overline{\psi}\psi$ in pure QED$_3$, where $\overline{\psi}\tau\psi$ is invariant under chiral transformation.After that the effects of Chern-Simon term will be studied by 2-component spinors.

2. Fermion spectral function

The spectral function of 4-component fermion and photon propagator[3] are defined as

$$S_F(x) = S_F^0(x)\exp(F(x)), S_F^0(x) = -(i\gamma\cdot\partial + m)\frac{\exp(-m\sqrt{-x^2})}{4\pi\sqrt{x^2}}. \tag{4}$$

$$D_F^0(k) = -i(\frac{g_{\mu\nu} - k_\mu k_\nu/k^2 - i\theta\epsilon_{\mu\nu\rho}k^\rho/k^2}{k^2-\theta^2+i\epsilon}) + id\frac{k_\mu k_\nu}{k^4}, \tag{5}$$

where F is an $O(e^2)$ matrix element $|T_1|^2$ for the process electron $(p+k) \rightarrow$ electron (p) + photon (k) as

$$T_1 = -ie\frac{\epsilon_\mu(k,\lambda)}{\gamma\cdot(p+k)-m}\gamma^\mu\exp(i(p+k)\cdot x)U(p,s), \tag{6}$$

$$\sum_{\lambda,S} T_1\overline{T_1} = -\frac{\gamma\cdot p+m}{2m}e^2[\frac{m^2}{(p\cdot k)^2} + \frac{1}{p\cdot k} + \frac{(d-1)}{k^2}] - \frac{\gamma\cdot p}{m}\frac{e^2}{4\theta}\frac{m}{p\cdot k}, \tag{7}$$

$$F = \int\frac{d^3k}{(2\pi)^2}\exp(ikx)\theta(k_0)\delta(k^2-\theta^2)\sum_{\lambda,S}T_1\overline{T_1}. \tag{8}$$

Here we use the retarded propagator to derive the function F

$$D_+(x) = \int\frac{d^3k}{i(2\pi)^2}\exp(ik\cdot x)\theta(k_0)\delta(k^2-\theta^2) = \frac{\exp(-\theta\sqrt{-x^2})}{8\pi i\sqrt{-x^2}} \tag{9}$$

The function F is evaluated by α integration for pure QED$_3$.[3,4]

$$\begin{aligned} F &= ie^2m^2\int_0^\infty \alpha d\alpha D_F(x+\alpha p) - e^2\int_0^\infty d\alpha D_F(x+\alpha p) - i(d-1)e^2\frac{\partial}{\partial\theta^2}D_F(x,\theta^2) \\ &= \frac{e^2}{8\pi}[\frac{\exp(-\theta|x|) - \theta|x|E_1(\theta|x|)}{\theta} - \frac{E_1(\theta|x|)}{m} + \frac{(d-1)\exp(-\theta\,|x|))}{2\theta}]. \end{aligned} \tag{10}$$

It is well known that function F has linear and logarithmic infrared divergence with respect to θ where θ is a bare photon mass. Here we notice the followings. **(1)** $\exp(F)$ includes all infrared divergences. **(2)** quenched propagator has linear and logarithmic infrared divergences. Linear divegence is absent in a special gauge. In unquenched case θ dependence of $\exp(F)$ is modified by dressed boson spectral function to $\int ds\rho_\gamma(s)\exp(F(x,\sqrt{s})$, where $\pi\rho_\gamma(s) = -\Im(s-\Pi(s))^{-1}$.

3. Phase strucutures

For short and long distance we have the approximate form of the function F

$$F_S \sim A - \theta|x| + (D+C|x|)\ln(\theta\,|x|)) - \frac{(d+1-2\gamma)e^2|x|}{16\pi}, (\theta|x| \ll 1), F_L \sim 0, (1 \ll \theta|x|). \tag{11}$$

From the above formulae we have

$$\exp(F) = A(\theta|x|)^{D+C|x|}(\theta|x| \ll 1), A = \exp(\frac{e^2(1+d)}{16\pi\theta} + \frac{e^2\gamma}{8\pi m}),$$
$$C = \frac{e^2}{8\pi}, D = \frac{e^2}{8\pi m}, \tag{12}$$

where m is the physical mass and γ is an Euler's constant.Here we see D acts to change the power of $|x|$. For $D = 1, S_F(0)$ is finite and we have $\langle\overline{\psi}\psi\rangle \neq 0$.[4,6] Thus if we require $D = 1$, we obtain $m = e^2/8\pi$. In the same way we add the conrtribution of Chern-Simon term. For simplicity we consider the 2-component spinors in (7). In the condensed phase we have the modified anomalous dimension for $\theta > 0$

$$D = \frac{e^2}{8\pi m} + \frac{e^2}{32\pi\theta} = 1, m^{'} = \frac{e^2}{8\pi}/(1 - \frac{e^2}{32\pi\theta}), \Delta m = \frac{e^2}{8\pi}\frac{e^2}{32\pi\theta}/(1 - \frac{e^2}{32\pi\theta}). \tag{13}$$

In unquenched case there are parity even and odd spectral function of gauge boson by vacuum polarization.[1] In this case we separate D into parity even and odd contribution $D^e = e^2/8\pi m, D^O = e^2/32\pi\theta$. For Topologically Massive QCD, θ is quantized with $n(0, \pm1, \pm2, ..)$. Thus we have $D = e^2/(8\pi m) + 1/8n = 1$ for quenched case, and $D^e = e^2/8\pi m = 1, D^O = 1/(8n)$ for unquenched case, which leads to $\langle\overline{\psi}\psi\rangle = \infty$ for any $n \neq 0$. In the 4-component spinor free fermion propagator is decomposed into chiral representation

$$S_F(p) = \frac{1}{m_e I + m_O\tau - \gamma\cdot p} = \frac{(\gamma\cdot p + m_+)\tau_+}{p^2 - m_+^2 + i\epsilon} + \frac{(\gamma\cdot p + m_-)\tau_-}{p^2 - m_-^2 + i\epsilon}. \tag{14}$$

The difference in two spectral functions is a opposite sign of each Chern-Simon contribution for $\tau_\pm$. In Toplogically massive gauge theory dynamical mass is parity even and Chern-Simon term shifts mass of different chirality with opposite sign. However shifted mass may not strongly depend on θ but $\langle\overline{\psi}\psi\rangle$ is proportional to $\theta(12)$.[5] Our approximation is convenient for unquenched case by the use of gauge boson spectral function.[6]

References

1. S. Deser, R. Jackiw & Templeton, Annals of Physics **281,** 409-449 (2000).
2. C. J. Burden, Nuclear Physics **B**387 (1992) 419-446, Kei-ichi Kondo, Int. J. Mod. Phys. **A11**; 777-822, 1996.
3. R. Jackiw, L. Soloviev, Phys. Rev. **173**.5 (1968) 1485.
4. Yuichi Hoshino, in CONTINUOUS ADVANCES IN QCD 2008; 361-372.
5. Toyoki Matsuyama, Hideko Nagahiro, Mod. Phys. Lett. A15 (2000) 2373-2386.
6. C. S. Fisher, Reinhart Alkofer, T. Darm, P. Maris, Phys. Rev. **D70**: 073007, 2004.

New Regularization in Extra Dimensional Model and Renormalization Group Flow of the Cosmological Constant

Shoichi Ichinose

Laboratory of Physics, School of Food and Nutritional Sciences
University of Shizuoka, Yada 52-1, Shizuoka 422-8526, Japan
E-mail: ichinose@u-shizuoka-ken.ac.jp

Casimir energy is calculated for 5D scalar theory in the *warped* geometry. A new regularization, called *sphere lattice regularization*, is taken. The regularized configuration is *closed-string like*. We numerically evaluate Λ(4D UV-cutoff), ω(5D bulk curvature, warp parameter) and T(extra space IR parameter) dependence of Casimir energy. 5D Casimir energy is *finitely* obtained after the *proper renormalization procedure*. The *warp parameter* ω suffers from the *renormalization effect*. We examine the cosmological constant problem.

1. Introduction In the quest for the unified theory, the higher dimensional (HD) approach is a fascinating one from the geometrical point. Historically the initial successful one is the Kaluza-Klein model, which unifies the photon, graviton and dilaton from the 5D space-time approach. The HD theories , however, generally have the serious defect as the quantum field theory(QFT) : un-renormalizability. The HD quantum field theories, at present, are not defined within the QFT.

In 1983, the Casimir energy in the Kaluza-Klein theory was calculated by Appelquist and Chodos.[1] They took the cut-off (Λ) regularization and found the quintic (Λ^5) divergence and the finite term. The divergent term shows the *unrenormalizability* of the 5D theory, but the finite term looks meaningful[2] and, in fact, is widely regarded as the right vacuum energy which shows *contraction* of the extra axis.

In the development of the string and D-brane theories, a new approach to the renormalization group was found. It is called *holographic renormalization.* We regard the renormalization flow as a curve in the bulk. The flow goes along the extra axis. The curve is derived as a dynamical equation such as Hamilton-Jacobi equation. It originated from the AdS/CFT correspondence. Spiritually the present basic idea overlaps with this approach.

2. Casimir Energy of 5D Scalar Theory In the warped geometry, $ds^2 = \frac{1}{\omega^2 z^2}(\eta_{\mu\nu}dx^\mu dx^\nu + dz^2)$, we consider the 5D massive *scalar* theory with $m^2 = -4\omega^2$. $\mathcal{L} = \sqrt{-G}(-\frac{1}{2}\nabla^A\Phi\nabla_A\Phi - \frac{1}{2}m^2\Phi^2)$. The Casimir energy E_{Cas} is given by

$$e^{-T^{-4}E_{Cas}} = \int \mathcal{D}\Phi \exp\{i\int d^5X\mathcal{L}\}\Big|_{\text{Euclid}} = \exp\sum_{n,p}\{-\frac{1}{2}\ln(p_E^2 + M_n^2)\}\ , \quad (1)$$

where M_n is the eigenvalues of the following operator.

$$\{s(z)^{-1}\hat{L}_z + {M_n}^2\}\psi_n(z) = 0 \ , \quad \hat{L}_z \equiv \frac{d}{dz}\frac{1}{(\omega z)^3}\frac{d}{dz} - \frac{m^2}{(\omega z)^5} \ , \tag{2}$$

where $s(z) = \frac{1}{(\omega z)^3}$. Z_2 parity is imposed as: $\psi_n(z) = -\psi_n(-z)$ for $P = -$; $\psi_n(z) = \psi_n(-z)$ for $P = +$. The expression (1) is the familiar one of the Casimir energy. It is re-expressed in a *closed* form using the heat-kernel method and the propagator. First we can express it, using the heat equation solution, as follows ($\omega/T = \mathrm{e}^{\omega l}$).

$$\mathrm{e}^{-T^{-4}E_{Cas}} = (\mathrm{const}) \times \exp\left[T^{-4}\int \frac{d^4p}{(2\pi)^4} 2\int_0^\infty \frac{1}{2}\frac{dt}{t}\mathrm{Tr}\ H_p(z,z';t)\right] ,$$

$$\mathrm{Tr}\ H_p(z,z';t) = \int_{1/\omega}^{1/T} s(z)H_p(z,z;t)dz \ , \ \{\frac{\partial}{\partial t} - (s^{-1}\hat{L}_z - p^2)\}H_p(z,z';t) = 0 \ . \tag{3}$$

The heat kernel $H_p(z,z';t)$ is formally solved, using the Dirac's bra and ket vectors $(z|, |z)$, as $H_p(z,z';t) = (z|\mathrm{e}^{-(-s^{-1}\hat{L}_z + p^2)t}|z')$. We here introduce the position/momentum propagators $G_p^{\mp}$: $G_p^{\mp}(z,z') \equiv \int_0^\infty dt\ H_p(z,z';t)$. They satisfy the following differential equations of *propagators*.

$$(\hat{L}_z - p^2 s(z))G_p^{\mp}(z,z') = \begin{cases} \epsilon(z)\epsilon(z')\hat{\delta}(|z| - |z'|) & \mathrm{P}= -1 \\ \hat{\delta}(|z| - |z'|) & \mathrm{P}=1 \end{cases} \tag{4}$$

$G_p^{\mp}$ can be expressed in a *closed* form. Taking the *Dirichlet* condition at all fixed points, the expression for the fundamental region ($1/\omega \le z \le z' \le 1/T$) is given by

$$G_p^{\mp}(z,z') = \mp\frac{\omega^3}{2}z^2 z'^2 \frac{\{\mathbf{I}_0(\frac{\tilde{p}}{\omega})\mathbf{K}_0(\tilde{p}z) \mp \mathbf{K}_0(\frac{\tilde{p}}{\omega})\mathbf{I}_0(\tilde{p}z)\}\{\mathbf{I}_0(\frac{\tilde{p}}{T})\mathbf{K}_0(\tilde{p}z') \mp \mathbf{K}_0(\frac{\tilde{p}}{T})\mathbf{I}_0(\tilde{p}z')\}}{\mathbf{I}_0(\frac{\tilde{p}}{T})\mathbf{K}_0(\frac{\tilde{p}}{\omega}) - \mathbf{K}_0(\frac{\tilde{p}}{T})\mathbf{I}_0(\frac{\tilde{p}}{\omega})} , \tag{5}$$

where $\tilde{p} \equiv \sqrt{p^2}, p^2 \ge 0$. We can express Casimir energy as,

$$-E_{Cas}^{\Lambda,\mp}(\omega,T) = \int \frac{d^4p_E}{(2\pi)^4}\bigg|_{\tilde{p}\le\Lambda} \int_{1/\omega}^{1/T} dz\ F^{\mp}(\tilde{p},z), F^{\mp}(\tilde{p},z) = \frac{2}{(\omega z)^3}\int_{\tilde{p}}^{\Lambda} \tilde{k}\ G_{\tilde{k}}^{\mp}(z,z)d\tilde{k} , \tag{6}$$

where $\tilde{p} = \sqrt{p_E^2}$. The momentum symbol p_E indicates Euclideanization. Here we introduce the UV cut-off parameter Λ for the 4D momentum space.

3. UV and IR Regularization and Evaluation of Casimir Energy The integral region of the above equation (6) is displayed in Fig.1. In the figure, we introduce the regularization cut-offs for the 4D-momentum integral, $\mu \le \tilde{p} \le \Lambda$. For simplicity, we take the following IR cutoff of 4D momentum : $\mu = \Lambda \cdot \frac{T}{\omega} = \Lambda \mathrm{e}^{-\omega l}$. Importantly, (6) shows the *scaling* behavior for large values of Λ and $1/T$. From a *close* numerical analysis, we have confirmed : (4A) $E_{Cas}^{\Lambda,-}(\omega,T) = \frac{2\pi^2}{(2\pi)^4} \times \left[-0.0250\frac{\Lambda^5}{T}\right]$. The Λ^5-divergence, (4A), shows the notorious problem of the higher dimensional theories. We have proposed an approach to solve this problem and given

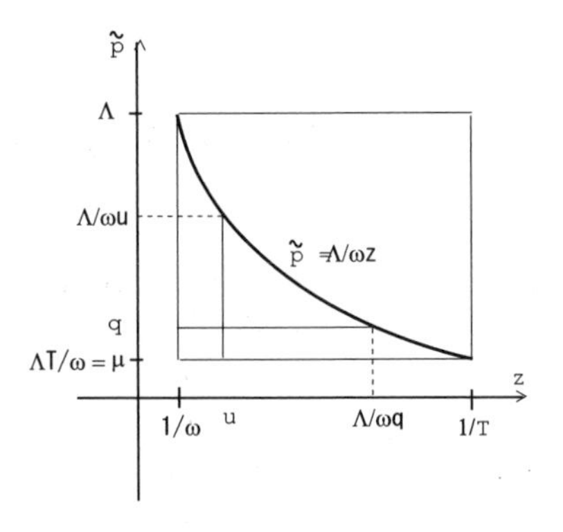

Fig. 1. Space of $(z,\tilde{p})$ for the integration. The hyperbolic curve was proposed.[3]

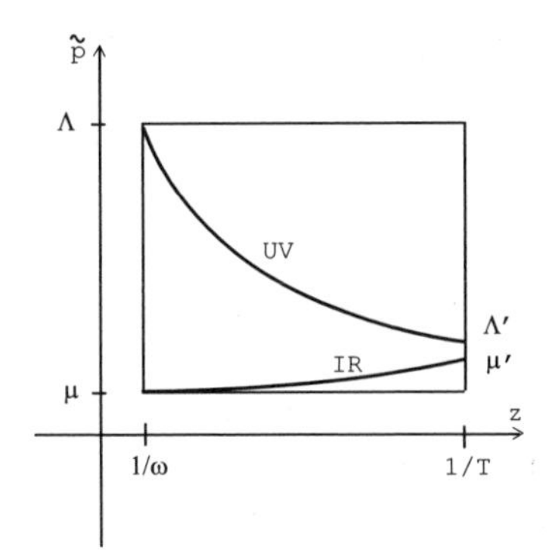

Fig. 2. Space of $(\tilde{p}$,z) for the integration (present proposal).

a legitimate explanation within the 5D QFT.[4,5] See Fig.2. The IR and UV cutoffs change along the etra axis. Their S^3-radii are given by $r_{IR}(z)$ and $r_{UV}(z)$. The 5D volume region bounded by B_{UV} and B_{IR} is the integral region of the Casimir energy E_{Cas}. The forms of $r_{UV}(z)$ and $r_{IR}(z)$ can be determined by the *minimal area principle*: $3+\frac{4}{z}r'r-\frac{r''r}{r'^2+1}=0, r'\equiv\frac{dr}{dz}, r''\equiv\frac{d^2r}{dz^2}, 1/\omega\leq z\leq 1/T$. We have confirmed, by numerically solving the above differential eqation (Runge-Kutta), those curves that show the flow of renormalization really appear. The results imply the *boundary conditions* determine the property of the renormalization flow.

4. Weight Function and the Meaning We consider another approach which respects the *minimal area principle.* Let us introduce, instead of restricting the integral region, a *weight function* $W(\tilde{p},z)$ in the $(\tilde{p},z)$-space for the purpose of suppressing UV and IR divergences of the Casimir Energy.

$$-E^{\mp\,W}_{Cas}(\omega,T)\equiv\int\frac{d^4p_E}{(2\pi)^4}\int_{1/\omega}^{1/T}dz\ W(\tilde{p},z)F^{\mp}(\tilde{p},z)\ ,\quad \tilde{p}=\sqrt{p_4^2+p_1^2+p_2^2+p_3^2}\ ,$$

$$\begin{cases}(N_1)^{-1}\mathrm{e}^{-(1/2)\tilde{p}^2/\omega^2-(1/2)z^2T^2}\equiv W_1(\tilde{p},z),\ N_1=1.711/8\pi^2 & \text{elliptic suppr.}\\ (N_2)^{-1}\mathrm{e}^{-\tilde{p}zT/\omega}\equiv W_2(\tilde{p},z),\ N_2=2\frac{\omega^3}{T^3}/8\pi^2 & \text{hyperbolic suppr.1}\\ (N_8)^{-1}\mathrm{e}^{-1/2(\tilde{p}^2/\omega^2+1/z^2T^2)}\equiv W_8(\tilde{p},z),\ N_8=0.4177/8\pi^2 & \text{reciprocal suppr.1}\end{cases}\tag{7}$$

where $F^{\mp}(\tilde{p},z)$ are defined in (6). They (except W_2) give, after normalizing the factor Λ/T, *only* the *log-divergence.*

$$E^W_{Cas}/\Lambda T^{-1}=-\alpha\omega^4\left(1-4c\ln(\Lambda/\omega)-4c'\ln(\Lambda/T)\right)\ ,\tag{8}$$

where the numerical values of α, c and c' are obtained depending on the choice of the weight function.[6] This means the 5D Casimir energy is *finitely* obtained by the ordinary renormalization of the warp factor ω.

In the previous work,[5] we have presented the following idea to define the weight function $W(\tilde{p},z)$. In the evaluation (7), the $(\tilde{p},z)$-integral is over the rectangle region shown in Fig.1 (with $\Lambda\rightarrow\infty$ and $\mu\rightarrow 0$). Following Feynman,[7] we can replace the integral by the summation over all possible paths $\tilde{p}(z)$.

$$-E^W_{Cas}(\omega,T) = \int \mathcal{D}\tilde{p}(z) \int_{1/\omega}^{1/T} dz\, S[\tilde{p}(z),z], S[\tilde{p}(z),z] = \frac{2\pi^2}{(2\pi)^4}\tilde{p}(z)^3 W(\tilde{p}(z),z) F^{\mp}(\tilde{p}(z),z). \tag{9}$$

There exists the *dominant path* $\tilde{p}_W(z)$ which is determined by the minimal principle: $\delta S = 0$. Dominant Path $\tilde{p}_W(z)$: $\frac{d\tilde{p}}{dz} = -\frac{\partial \ln(WF)}{\partial z} / \left(\frac{3}{\tilde{p}} + \frac{\partial \ln(WF)}{\partial \tilde{p}}\right)$. Hence it is fixed by $W(\tilde{p},z)$. On the other hand, there exists another independent path: the minimal surface curve $r_g(z)$. Minimal Surface Curve $r_g(z)$: $3 + \frac{4}{z}r'r - \frac{r''r}{r'^2+1} = 0$, $\frac{1}{\omega} \leq z \leq \frac{1}{T}$. It is obtained by the *minimal area principle*: $\delta A = 0$ where

$$ds^2 = (\delta_{ab} + \frac{x^a x^b}{(rr')^2})\frac{dx^a dx^b}{\omega^2 z^2} \equiv g_{ab}(x)dx^a dx^b, A = \int \sqrt{g}\, d^4x = \int_{1/\omega}^{1/T} \frac{\sqrt{r'^2+1}\, r^3}{\omega^4 z^4} dz. \tag{10}$$

Hence $r_g(z)$ is fixed by the *induced geometry* $g_{ab}(x)$. Here we put the *requirement:*[5] (4A) $\tilde{p}_W(z) = \tilde{p}_g(z)$, where $\tilde{p}_g \equiv 1/r_g$. This means the following things. We *require* the dominant path coincides with the minimal surface line $\tilde{p}_g(z) = 1/r_g(z)$ which is defined independently of $W(\tilde{p},z)$. $W(\tilde{p},z)$ is defined here by the induced geometry $g_{ab}(x)$. In this way, we can connect the integral-measure over the 5D-space with the geometry. We have confirmed the coincidence by the numerical method.

In order to most naturally accomplish the above requirement, we can go to a *new step.* Namely, we *propose* to *replace* the 5D space integral with the weight W, (7), by the following *path-integral.* We *newly define* the Casimir energy in the higher-dimensional theory as follows.

$$-\mathcal{E}_{Cas}(\omega,T,\Lambda) = \int_{1/\Lambda}^{1/\mu} d\rho \int_{\substack{r(\omega^{-1}) \\ = r(T^{-1}) \\ = \rho}} \prod_{a,z} \mathcal{D}x^a(z) F(\frac{1}{r},z) \exp\left[-\int_{1/\omega}^{1/T} \frac{\sqrt{r'^2+1}\, r^3}{2\alpha'\omega^4 z^4} dz\right], \tag{11}$$

where $\mu = \Lambda T/\omega$ and the limit $\Lambda T^{-1} \to \infty$ is taken. The string (surface) tension parameter $1/2\alpha'$ is introduced. (Note: Dimension of α' is [Length]^4.) $F(\tilde{p},z)$ is defined in (6) and shows the *field-quantization* of the bulk scalar (EM) fields.

5. Discussion and Conclusion When c and c' are sufficiently small we find the renormalization group function for the warp factor ω as

$$\omega_r = \omega(1 - c\ln(\Lambda/\omega) - c'\ln(\Lambda/T)) \ , \ \beta \equiv \frac{\partial}{\partial(\ln\Lambda)} \ln\frac{\omega_r}{\omega} = -c - c' \ . \tag{12}$$

No local counterterms are necessary.

Through the Casimir energy calculation, in the higher dimension, we find a way to quantize the higher dimensional theories within the QFT framework. The quantization *with respect to the fields* (except the gravitational fields $G_{AB}(X)$) is done in the standard way. After this step, the expression has the summation *over the 5D space(-time) coordinates or momenta* $\int dz \prod_a dp^a$. We have proposed that this summation should be replaced by the *path-integral* $\int \prod_{a,z} \mathcal{D}p^a(z)$ with the *area* action

(Hamiltonian) $A = \int \sqrt{\det g_{ab}} d^4x$ where g_{ab} is the *induced* metric on the 4D surface. This procedure says the 4D momenta p^a (or coordinates x^a) are *quantum statistical* operators and the extra-coordinate z is the inverse temperature (Euclidean time). We recall the similar situation occurs in the standard string approach. The space-time coordinates obey some uncertainty principle.[8]

Recently the dark energy (as well as the dark matter) in the universe is a hot subject. It is well-known that the dominant candidate is the cosmological term. The cosmological constant λ appears as: (5A) $R_{\mu\nu} - \frac{1}{2}g_{\mu\nu}R - \lambda g_{\mu\nu} = T_{\mu\nu}^{matter}, S = \int d^4x\sqrt{-g}\{\frac{1}{G_N}(R+\lambda)\} + \int d^4x\sqrt{-g}\{\mathcal{L}_{matter}\}, g = \det\, g_{\mu\nu}$. We consider here the 3+1 dim Lorentzian space-time ($\mu, \nu = 0,1,2,3$). The constant λ observationally takes the value : (5B) $\frac{1}{G_N}\lambda_{obs} \sim \frac{1}{G_N {R_{cos}}^2} \sim m_\nu^4 \sim (10^{-3}eV)^4, \lambda_{obs} \sim \frac{1}{R_{cos}^2} \sim 4\times 10^{-66}(eV)^2$, where $R_{cos} \sim 5\times 10^{32}\mathrm{eV}^{-1}$ is the cosmological size (Hubble length), m_ν is the neutrino mass. On the other hand, we have theoretically so far : (5C) $\frac{1}{G_N}\lambda_{th} \sim \frac{1}{{G_N}^2} = {M_{pl}}^4 \sim (10^{28}eV)^4$. We have the famous huge discrepancy factor : (5D) $\frac{\lambda_{th}}{\lambda_{obs}} \sim N_{DL}^2, N_{DL} \equiv M_{pl}R_{cos} \sim 6\times 10^{60}$, where N_{DL} is the Dirac's large number. If we use the present result, we can obtain a natural choice of T, ω and Λ as follows. By identifying $T^{-4}E_{Cas} = -\alpha_1 \Lambda T^{-1}\omega^4/T^4$ with $\int d^4x\sqrt{-g}(1/G_N)\lambda_{ob} = R_{cos}^2(1/G_N)$, we obtain the following relation: (5E) $N_{DL}^2 = R_{cos}^2\frac{1}{G_N} = -\alpha_1\frac{\omega^4\Lambda}{T^5}$. The warped ($\mathrm{AdS}_5$) model predicts the cosmological constant *negative*, hence we have interest only in its absolute value. We take the following choice for Λ and ω : (5F) $\Lambda = M_{pl} \sim 10^{19}GeV, \omega \sim \frac{1}{\sqrt[4]{G_N {R_{cos}}^2}} = \sqrt{\frac{M_{pl}}{R_{cos}}} \sim m_\nu \sim 10^{-3}\mathrm{eV}$.

As shown above, we have the standpoint that the cosmological constant is mainly made from the Casimir energy. We do not yet succeed in obtaining the value α_1 negatively, but succeed in obtaining the finiteness of the cosmological constant and its gross absolute value. The smallness of the value is naturally explained by the renormalization flow. Because we already know the warp parameter ω *flows* (12), the $\lambda_{obs} \sim 1/R_{cos}^2 \propto \omega^4$, says that the *smallness of the cosmological constant comes from the renormalization flow* for the non asymptotic-free case ($c + c' < 0$ in (12)).

The IR parameter T, the normalization factor Λ/T in (8) and the IR cutoff $\mu = \Lambda\frac{T}{\omega}$ are given by : (5G) $T = R_{cos}^{-1}(N_{DL})^{1/5} \sim 10^{-20}eV, \frac{\Lambda}{T} = (N_{DL})^{4/5} \sim 10^{50}, \mu = M_{pl}N_{DL}^{-3/10} \sim 1GeV \sim m_N$, where m_N is the nucleon mass. The degree of freedom of the universe (space-time) is given by : (5H) $\frac{\Lambda^4}{\mu^4} = \frac{\omega^4}{T^4} = N_{DL}^{6/5} \sim 10^{74} \sim (\frac{M_{pl}}{m_N})^4$.

References

1. T. Appelquist and A. Chodos, Phys. Rev. Lett. **50** (1983) 141
2. S. Ichinose, Phys. Lett. **152B** (1985), 56
3. L. Randall and M.D. Schwartz, *JHEP* **0111** (2001) 003, hep-th/0108114
4. S. Ichinose and A. Murayama, Phys. Rev. **D76** (2007) 065008, hep-th/0703228
5. S. Ichinose, Prog. Theor. Phys. **121** (2009) 727, ArXiv:0801.3064v8[hep-th]
6. S. Ichinose, ArXiv:0812.1263[hep-th]
7. R.P. Feynman, *Statistical Mechanics*, W.A. Benjamin, Inc., Massachusetts, 1972
8. T. Yoneya, Prog. Theor. Phys. **103** (2000) 1081

Spectral Analysis of Dense Two-Color QCD

T. Kanazawa*, T. Wettig† and N. Yamamoto*

*Department of Physics, The University of Tokyo
Tokyo 113-0033, Japan
†Department of Physics, University of Regensburg
93040 Regensburg, Germany
E-mails: tkanazawa@nt.phys.s.u-tokyo.ac.jp,
tilo.wettig@physik.uni-regensburg.de,
yamamoto@nt.phys.s.u-tokyo.ac.jp

We study spectral properties of two-color QCD with an even number of flavors at high baryon density. We construct the low-energy effective Lagrangian for the Nambu-Goldstone bosons, derive Leutwyler-Smilga-type spectral sum rules and construct a suitable random matrix theory. Our results can in principle be tested in lattice QCD simulations.

Keywords: Chiral symmetry breaking, diquark condensate, nonzero chemical potential, chiral random matrix theory.

1. Introduction

Understanding the phase structure of Quantum Chromodynamics (QCD) at low temperature (T) and nonzero quark chemical potential (μ) is an important subject relevant to many areas of physics,[1] but the severe fermion sign problem in first-principle lattice simulations is hindering progress in this field considerably.

In this report we analyse dense two-color QCD with an even number of flavors. Our principal motivation comes from the fact that two-color QCD can be simulated on the lattice even at $\mu \neq 0$. At large μ, two-color matter exists in the form of a BCS superfluid, which is the genuine counterpart of the color superconducting phase for three-color QCD. Below, we will construct the low-energy effective Lagrangian for the Nambu-Goldstone (NG) bosons, define a new ε-regime, obtain exact spectral sum rules for the Dirac eigenvalues, and introduce a chiral random matrix theory (ChRMT) for the new ε-regime.[2,3] Testing of our results on the lattice would provide the first signature of BCS pairing and give the magnitude of the fermion gap Δ. For the corresponding work in dense three-color QCD, see Ref. 4.

2. Low-energy effective theory

At $\mu \gg \Lambda_{\rm QCD}$, perturbative one-gluon exchange indicates that the color antisymmetric channel is attractive, implying instability of the Fermi surface. A gap Δ

appears in the spectrum of quasiquarks near the Fermi surface, while the diquark condensate forms in the color- and flavor-antisymmetric channel, breaking chiral symmetry spontaneously according to

$$\mathrm{SU}(N_f)_L \times \mathrm{SU}(N_f)_R \times \mathrm{U}(1)_B \times \mathrm{U}(1)_A \to \mathrm{Sp}(N_f)_L \times \mathrm{Sp}(N_f)_R\,, \tag{1}$$

where we neglected the $\mathrm{U}(1)_A$-anomaly (since μ is large) and assumed that N_f is even and ≥ 4. (See Ref. 2 for $N_f = 2$.) The NG fields originating from (1),

$$\Sigma_L \in \mathrm{SU}(N_f)_L/\mathrm{Sp}(N_f)_L, \quad \Sigma_R \in \mathrm{SU}(N_f)_R/\mathrm{Sp}(N_f)_R, \quad V \in \mathrm{U}(1)_B, \quad A \in \mathrm{U}(1)_A,$$

are gapless in the chiral limit and govern the low-energy physics near the Fermi surface. From symmetry principles and weak-coupling calculations, the chiral Lagrangian valid at energy scales below Δ is given by[2]

$$\begin{aligned}\mathcal{L} = {} & \frac{f_H^2}{2}\left\{|\partial_0 V|^2 - v_H^2|\partial_i V|^2\right\} + \frac{N_f f_{\eta'}^2}{2}\left\{|\partial_0 A|^2 - v_{\eta'}^2|\partial_i A|^2\right\} \\ & + \frac{f_\pi^2}{2}\,\mathrm{Tr}\left\{|\partial_0 \Sigma_L|^2 - v_\pi^2|\partial_i \Sigma_L|^2 + (L \leftrightarrow R)\right\} - \frac{3\Delta^2}{4\pi^2}\left\{A^2\,\mathrm{Tr}(M\Sigma_R M^T \Sigma_L^\dagger) + \text{c.c.}\right\},\end{aligned} \tag{3}$$

where M is a generalized $N_f \times N_f$ quark mass matrix. The absence of an $O(M)$ term in the chiral Lagrangian is a consequence of the $\mathbb{Z}(2)_L \times \mathbb{Z}(2)_R$ symmetry of the diquark pairing. As V and the gluon fields decouple from the other NG modes, they will be neglected in the following.

3. Partition function at finite volume

Next, we consider two-color QCD in a Euclidean box of size L^4 ($\equiv V_4$) and let m_{NG} denote the masses of NG fields. The dynamics simplifies drastically in the regime

$$\frac{1}{\Delta} \ll L \ll \frac{1}{m_{\mathrm{NG}}} \tag{4}$$

since the zero modes of the NG fields dominate the physics.[2,5] Thus the partition function in the new ε-regime (4) is simply given by

$$Z(M) = \int_{\mathrm{U}(1)_A} dA \int_{\mathrm{SU}(N_f)_L/\mathrm{Sp}(N_f)_L} d\Sigma_L \int_{\mathrm{SU}(N_f)_R/\mathrm{Sp}(N_f)_R} d\Sigma_R \exp\left[\frac{3}{2\pi^2} V_4 \Delta^2\, \mathrm{Re}[A^2\,\mathrm{Tr}(M\Sigma_R M^T \Sigma_L^\dagger)]\right]. \tag{5}$$

4. Spectral sum rules for the Dirac operator

Let us denote the complex Dirac eigenvalues by $i\lambda_n$. Starting from the microscopic Lagrangian of two-color QCD, we find the normalized partition function to be

$$Z(M) = \left\langle \prod_n{}' \det\left(1 + \frac{M^\dagger M}{\lambda_n^2}\right)\right\rangle, \tag{6}$$

where $\langle \cdots \rangle$ represents expectation values with respect to the measure in the chiral limit. $\prod_n'$ (and later $\sum_n'$) denotes the product (sum) over all eigenvalues with

$\mathrm{Re}\,\lambda_n > 0$. Equating the above two expressions for $Z(M)$, we can derive a number of novel spectral sum rules, e.g.,

$$\left\langle \sum_n{}' \frac{1}{\lambda_n^2} \right\rangle = \left\langle \sum_{m<n}{}' \frac{1}{\lambda_m^2 \lambda_n^2} \right\rangle = \left\langle \sum_n{}' \frac{1}{\lambda_n^6} \right\rangle = 0, \quad \left\langle \sum_n{}' \frac{1}{\lambda_n^4} \right\rangle = \frac{9(V_4\Delta^2)^2}{4\pi^4(N_f-1)^2}. \tag{7}$$

The vanishing of many spectral sums is a salient feature of the high-density limit.

We define the microscopic spectral density $\rho_s(z)$ by

$$\rho_s(z) \equiv \lim_{V_4\to\infty} \frac{\pi^2}{3V_4\Delta^2}\,\rho\!\left(\frac{\pi z}{\sqrt{3V_4\Delta^2}}\right) \quad \text{with} \quad \rho(\lambda) \equiv \left\langle \sum_n \delta^2(\lambda - \lambda_n) \right\rangle, \tag{8}$$

Using $\rho_s(z)$ the last spectral sum rule in (7) can be cast into a form that involves the dimensionless variable z only. The existence of the limit (8) leads to the nontrivial observation that the smallest Dirac eigenvalue has a magnitude of order $1/\sqrt{V_4\Delta^2}$ on average. Since it is well known at $\mu = 0$ that the microscopic spectral density is a universal function fixed solely by global symmetries and their breaking,[6] we expect that $\rho_s(z)$ defined above is also universal and can be obtained from the corresponding ChRMT, which is much simpler than QCD. We propose[3]

$$Z(\{\hat{m}_f\}) = \int dA\,dB\; \mathrm{e}^{-N\alpha^2\,\mathrm{Tr}(AA^T+BB^T)} \prod_{f=1}^{N_f} \det\begin{pmatrix} \hat{m}_f\mathbf{1} & A \\ B & \hat{m}_f\mathbf{1} \end{pmatrix} \tag{9}$$

as the corresponding ChRMT, where A and B are real $N \times N$ matrices. Indeed it embodies the symmetry breaking pattern (1) correctly and reduces for $N \gg 1$ to a form that is identical to (5) if we identify

$$N\alpha^2\hat{\mathbf{m}}^2 \iff \frac{3}{4\pi^2}V_4\Delta^2 M^2 \tag{10}$$

with $\hat{\mathbf{m}} \equiv \mathrm{diag}(\hat{m}_1, \ldots, \hat{m}_{N_f})$. Since the mass dependence of the partition function is identical, the microscopic spectral density obtained from (9) must satisfy the spectral sum rules derived in the ε-regime of dense two-color QCD. The determination of the concrete form of $\rho_s(z)$ based on (9) is left for future work.

Acknowledgments

TK and NY are supported by the Japan Society for the Promotion of Science for Young Scientists. TW acknowledges support by DFG.

References

1. M. G. Alford, A. Schmitt, K. Rajagopal and T. Schäfer, *Rev. Mod. Phys.* **80**, 1455 (2008).
2. T. Kanazawa, T. Wettig and N. Yamamoto, *JHEP* **08**, p. 003 (2009).
3. T. Kanazawa, T. Wettig and N. Yamamoto, arXiv:0912.4999 [hep-ph].
4. N. Yamamoto and T. Kanazawa, *Phys. Rev. Lett.* **103**, p. 032001 (2009).
5. J. Gasser and H. Leutwyler, *Phys. Lett.* **B188**, p. 477 (1987).
6. J. J. M. Verbaarschot and T. Wettig, *Ann. Rev. Nucl. Part. Sci.* **50**, 343 (2000).

NJL Model with Dimensional Regularization at Finite Temperature

T. Fujihara[a], T. Inagaki[a], D. Kimura[a,b], H. Kohyama[c,d] and A. Kvinikhidze[e]

[a] *Department of Physics, Hiroshima University*
Higashi-Hiroshima, 739-8526, Japan
[b] *Learning Support Center, Hiroshima Shudo University*
Hiroshima, 731-3195, Japan
[c] *Institute of Physics, Academia Sinica, Taipei 115, Taiwan*
[d] *Physics Division, National Center for Theoretical Sciences, Hsinchu 300, Taiwan*
[e] *A. Razmadze Mathematical Institute of Georgian Academy of Sciences*
M. Alexidze Str. 1, 380093 Tbilisi, Georgia

We consider two and three flavor Nambu–Jona-Lasinio (NJL) model at finite temperature by using the dimensional regularization. It is known that physical results produced by the NJL model depend on the method of regularization (Fujihara *et al.* Prog. Theor. Phys. Suppl. 174, 72 (2008), Phys. Rev. D 79, 096008 (2009)). In the dimensional regularization scheme we regard the space-time dimensions as one of the parameters in the effective theory. We obtain the η meson mass and the topological susceptibility in such regularization and compare the results with ones obtained in the cut-off regularization.

1. Introduction

Nambu–Jona-Lasinio (NJL) model[1] is a low energy effective theory of QCD.[2–4] This model has the same chiral symmetry as QCD, and the symmetry is broken down dynamically. NJL model contains a four-fermion interaction corresponding to a dimension six operator in the Lagrangian. The model is nonrenormalizable in four space-time dimensions. Thus the regularization parameter can not be removed away from some physical results. It is necessary to discuss the regularization dependence.

In the low energy region below 1 GeV QCD gives better description if three-flavor quarks are involved as compared with only two-flavor quarks. The chiral symmetry $U_L(3) \otimes U_R(3)$ is broken down to $SU_L(3) \otimes SU_R(3) \otimes U_V(1)$ due to the axial anomaly $U_A(1)$. The $U_A(1)$ anomaly plays essential role to describe the $\eta - \eta'$ mixing property and the heavy mass of η' meson, $m_{\eta'} \simeq 958$ MeV.

Most analysis have been using the momentum cut-off regularization. Unfortunately, the cut-off scale often breaks some of the symmetries of the model. Moreover, the critical chemical potential where the color superconductivity takes place and the η' meson mass are of the order of the cut-off scale. In such situation it is expected that the sharp cut-off artifact may have non-negligible effect on the analysis. That is why our analysis is based on the dimensional regularization in the NJL model (see Refs. 5–7 and also 8, 9).

2. Two-flavor NJL model

Two-flavor NJL model is defined by the Lagrangian,

$$\mathcal{L}_{2f} = \bar{\psi}(i\partial\!\!\!/ - m_2)\psi + G\{(\bar{\psi}\psi)^2 + (\bar{\psi}i\gamma_5\vec{\tau}\psi)^2\}, \tag{1}$$

where m is the quark mass matrix, $m_2 = \mathrm{diag}(m_u, m_d)$, G is the coupling constant and $\vec{\tau}$ represents the isospin Pauli matrices.

To examine the phase structure of the chiral symmetry, we evaluate the effective potential at finite temperature, T, by using the cut-off and the dimensional regularizations. To reproduce the observed values, $m_u = m_d = 4.5$ MeV, $m_\pi = 138$ MeV and $f_\pi = 92.4$ MeV at $T = 0$, we fix the parameters of the model, $\Lambda = 720$ MeV and $G = 3.80 \times 10^{-6}$ MeV^{-2} for the cut-off regularization. In the chiral limit, the critical temperature, T_c, becomes 170 MeV with these parameters.

Dimensional regularization is applied to the loop integration momenta (corresponding to quark internal lines) by making the transition from the 4- to D-dimensional integration, $d^4k \to M_{\mathrm{ren}}^{4-D} d\Omega_{D-1} dk_0 k^{D-2} dk$. These two parameters, the space-time dimension D and a renormalization scale M_{ren}, appear to regularize the theory. Two parameters and the effective coupling can be fixed by the above physical observables and T_c. We obtain $D = 2.28$, $M_{\mathrm{ren}} = 127$ MeV, and $G = -0.0178$ MeV^{2-D}.[a] In Table 1. we show the parameter sets of the model, $\langle \bar{u}u \rangle$ and T_c for $m_u = m_d = 3.3,\ 4.5,\ 5.5$ MeV in the dimensional regularization scheme.

Table 1. The parameter sets of the 2-flavor NJL model, $\langle \bar{u}u \rangle$ and T_c for $m_u = m_d = 3.3,\ 4.5,\ 5.5$ MeV.

$m_{u,d}$ (MeV)	D	M_{ren} (MeV)	G (MeV^{2-D})	$-\langle \bar{u}u \rangle$ (MeV^3)	T_c (MeV)
3.3	2.19	130	−0.0194	291^3	150
4.5	2.28	127	−0.0178	262^3	170
5.5	2.40	117	−0.0124	245^3	212

3. Three-flavor NJL model

The Lagrangian of the three-flavor NJL model is given by

$$\mathcal{L}_{3f} = \bar{\psi}(i\partial\!\!\!/ - m_3)\psi + G\{(\bar{\psi}\lambda^a\psi)^2 + (\bar{\psi}i\gamma_5\lambda^a\psi)^2\} - K[\det_{i,j} \bar{\psi}_i(1-\gamma_5)\psi_j + \mathrm{h.c.}], \tag{2}$$

where λ^a $(a = 0, 1, 2, \cdots, 8)$ are the Gell-Mann matrices with $\lambda^0 = \sqrt{2/3}\, I_3$. Third term in Eq. (2) breaks the $U_A(1)$ symmetry.[10–12] Comparing the QCD axial current with the NJL axial current, one can find the topological charge density,[4] $Q(x)$, and the topological susceptibility,[13] $\chi = \int d^4x \langle 0|TQ(x)Q(0)|0\rangle$. To obtain m_η and χ in the dimensional regularization, we first study the case $D = 3.4$, $m_u = m_d = 5.5$ MeV as a test case. Following the procedure presented in Refs. 3 and 4, we fit the remaining parameters G, K, m_s and M_{ren} by using the physical quantities $m_\pi = 138$ MeV, $f_\pi = 93$ MeV, $m_K = 495$ MeV and $m_{\eta'} = 957.5$ MeV as inputs.

[a] After renormalizing the coupling constant, we adjust it to a positive value.

Then we calculate m_η and χ, and compare the results with the ones in the cut-off method and experimental/empirical values,[4,13–15] which are shown in Table 2.

Table 2. Comparison of dimensional regularization with cut-off regularization results and experimental/empirical values in the 3-flavor model.

	m_s (MeV)	m_η (MeV)	$\chi^{1/4}$ (MeV)	$-\langle\bar{u}u\rangle$ (MeV^3)
dimensional reg.	141	525	191	247^3
cut-off reg.	136	487	166	245^3
exp./emp. values	70-130	548	170±7	$(225\pm25)^3$

4. Summary

We have discussed the characteristic features of the extended NJL model with the dimensional and the cut-off regularizations. The dimensional regularization is applied to the loop integrals with respect to the momenta corresponding to internal fermion lines. We should notice that only the radiative corrections are evaluated in the space-time dimensions less than four to keep the four-dimensional properties in the real world. We fixed the parameters to reproduce the up and down quark mass and light meson properties and calculate T_c, $\langle\bar{u}u\rangle$, m_s, m_η and χ using two regularizations. These results based on the dimensional regularization method are still tentative ($D = 3.4$) for the three-flavor NJL model, but they almost reproduce the experimental/empirical values as well as ones obtained in the cut-off regularization method. The input parameters, D and $M_{\rm ren}$, should be fixed phenomenologically. The work is being continued in hope of better choice of parameters to explain the experimental data.

References

1. Y. Nambu and G. Jona-Lasinio, Phys. Rev. **122**, 345 (1961), *ibid.* **124**, 246 (1961).
2. U. Vogl and W. Weise, Prog. Part. Nucl. Phys. **27** 195 (1991).
3. S.P. Klevansky, Rev. Mod. Phys. **64** 649 (1992).
4. T. Hatsuda and T. Kunihiro, Phys. Rept. **247** 221 (1994).
5. T. Inagaki, T. Kouno and T. Muta, Int. J. Mod. Phys. A **10**, 2241 (1995).
6. T. Inagaki, D. Kimura and A. Kvinikhidze, Phys. Rev. D **77**, 116004 (2008).
7. T. Fujihara, T. Inagaki, D. Kimura and A. Kvinikhidze, Prog. Theor. Phys. Suppl. **174**, 72 (2008), Phys. Rev. D **79** 096008 (2009).
8. S. Krewald and K. Nakayama, Annals Phys. **216**, 201 (1992).
9. R.G. Jafarov, and V.E. Rochev, Russ. Phys. J. **49**, 364 (2006),
10. M. Kobayashi and T. Maskawa, Prog. Theor. Phys. **44**, 1422 (1970), M. Kobayashi, H. Kondo and T. Maskawa, Prog. Theor. Phys. **45**, 1955 (1971).
11. G. 't Hooft, Phys. Rev. D **14**, 3432 (1976), Phys. Rept. **142**, 357 (1986).
12. M.A. Shifman, A.I. Vainshtein and Valentin I. Zakharov, Nucl. Phys. B **163**, 46 (1980).
13. K. Fukushima, K. Ohnishi and K. Ohta, Phys. Rev. C **63**, 045203 (2001).
14. B. Alles, M. D'Elia and A. Di Giacomo, Nucl. Phys. B **494**, 281 (1997), Erratum-*ibid.* B **679**, 397 (2004).
15. C. Amsler *et al.* (Particle Data Group), Phys. Lett. B **667**, 1 (2008).

A New Method of Evaluating the Dynamical Chiral Symmetry Breaking Scale and the Chiral Restoration Temperature in General Gauge Theories by Using the Non-Perturbative Renormalization Group Analyses with General 4-Fermi Effective Interaction Space

Ken-Ichi Aoki* and Daisuke Sato†

Institute for Theoretical Physics, Kanazawa University, Kanazawa, 920-1192, Japan
*E-mails: *aoki@hep.s.kanazawa-u.ac.jp, †satodai@hep.s.kanazawa-u.ac.jp*

Kazuhiro Miyashita

Department of Library and Information Science, Aichi Shukutoku University
Nagakute, Aichi, 480-1197, Japan
E-mail: miya@asu.aasa.ac.jp

We propose a non-perturbative renormalization group method to investigate the dynamical chiral symmetry breaking in general gauge theories. We setup the effective interaction space to contain all possible 4-fermi interactions allowed by the exact chiral symmetry with N_f flavours. The renormalization group beta functions are calculated without assuming the large-N_f or large-N_c dominance. We solve the renormalization group equation with a small bare mass operator and directly evaluate the induced chiral condensate and the effective mass. We study N_f- and N_c-dependence of the chiral breaking scale including the case having the infrared fixed point for the gauge coupling constant. Also we extend our method to the finite temperature case and obtain the critical temperature of the chiral symmetry restoration.

Keywords: Chiral symmetry, renormalization group, gauge theory.

1. Non-perturbative renormalization group

We work for $\mathrm{SU}(N_\mathrm{c})$ gauge theory with N_f flavours, which has the chiral symmetry $\mathrm{SU}(N_\mathrm{c}) \times \mathrm{SU}(N_\mathrm{f})_\mathrm{L} \times \mathrm{SU}(N_\mathrm{f})_\mathrm{R} \times \mathrm{U}(1)_\mathrm{V} \times \mathrm{Parity}$. Non-perturbative renormalization group is defined for the following Wilsonian effective action,

$$S_\mathrm{eff}[\bar{\psi}, \psi, A; t] = \int d^4x \Big\{ \bar{\psi}\,(\partial\!\!\!/ - g_\mathrm{s} A\!\!\!/)\,\psi + \frac{1}{2N_\mathrm{f}N_\mathrm{c}} \sum_i G_i(t)\mathcal{O}_i(\bar{\psi}, \psi; t) + \frac{1}{4}\left(F^a_{\mu\nu}\right)^2 + \frac{1}{2\alpha}(\partial_\mu A_\mu)^2 \Big\}\,, \quad (1)$$

where t is a parameter representing the renormalization scale by $\Lambda(t) = \Lambda_0 e^{-t}$. The effective 4-fermi operators $\mathcal{O}_i$ are given by

$$\begin{aligned} \mathcal{O}_1 &= (\bar{\psi}\gamma_\mu\psi)^2 - (\bar{\psi}\gamma_5\gamma_\mu\psi)^2, & \mathcal{O}_2 &= (\bar{\psi}\gamma_\mu\psi)^2 + (\bar{\psi}\gamma_5\gamma_\mu\psi)^2, \\ \mathcal{O}_{\mathrm{c}1} &= (\bar{\psi}\gamma_\mu T^a\psi)^2 - (\bar{\psi}\gamma_5\gamma_\mu T^a\psi)^2, & \mathcal{O}_{\mathrm{c}2} &= (\bar{\psi}\gamma_\mu T^a\psi)^2 + (\bar{\psi}\gamma_5\gamma_\mu T^a\psi)^2. \end{aligned} \quad (2)$$

These operators are complete due to the chiral symmetry. Note that we omit operators which are not generated by gauge interactions. The β functions of renormalization group equation are calculated through the diagrams in Fig. 1.

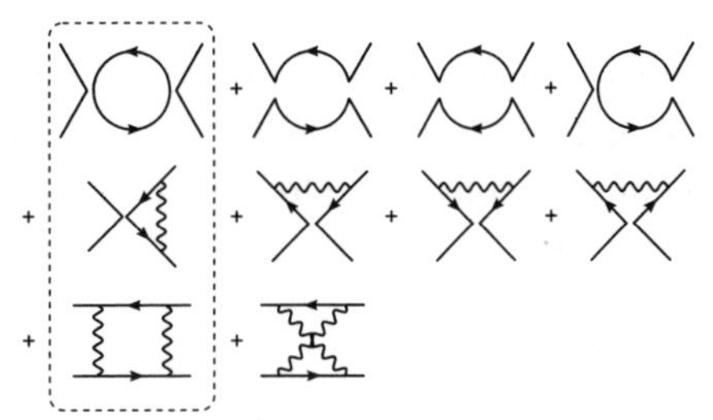

Fig. 1. Diagrams of beta functions for four-fermi operators.

2. Evaluation of the effective mass and the chiral condensates

We introduce a mass term in the effective action, $S_{\text{eff}}[\text{massless}] + m(m_0;t)\bar{\psi}\psi$, where $m(m_0;0) = m_0$ is the bare mass. We take the zero bare mass limit after all calculation. The effective mass is given by a limit : $m_{\text{phys}} \equiv \lim_{m_0 \to +0} \lim_{t \to \infty} m(m_0;t)$. We define a scale-dependent quantity $\langle\bar{\psi}\psi\rangle(m_0;t) \equiv (\partial S_{\text{eff}}(m_0;t)/\partial m_0)/L^4$, whose evolution equation is derived as follows:

$$\frac{d}{dt}\langle\bar{\psi}\psi\rangle(m_0;t) = 3\langle\bar{\psi}\psi\rangle(m_0;t) - \frac{m}{2\pi^2(1+m^2)} e^{-t} \frac{\partial m}{\partial m_0} \, . \tag{3}$$

The chiral condensates are given by a limit : $\langle\bar{\psi}\psi\rangle_{\text{phys}} = \lim_{m_0 \to +0} \lim_{t \to \infty} \langle\bar{\psi}\psi\rangle(m_0;t)$. We plot the running of the mass in Fig. 2, which shows the spontaneous chiral symmetry breaking below some renormalization scale.

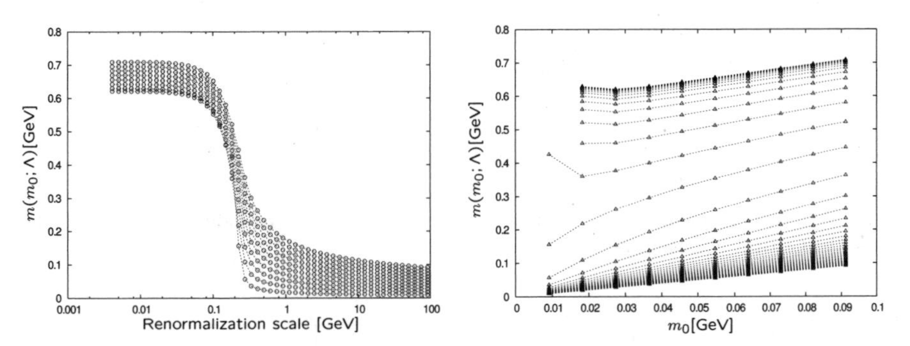

Fig. 2. Spontaneous mass generation,as a function of the renormalization scale, and as a function of the bare mass. ($N_c = N_f = 3$)

3. Chiral symmetry restoration at finite temperature

Using the finite temperature propagators, we observe the symmetry restoration at $T = T_c$ (Fig. 3). We see the critical scaling with $\beta_m = 0.36$, which is largely deviated from the mean field value 0.5. This comes from the fact that our analyses take account of all possible contribution including non-ladder type diagrams.

4. N_c, N_f dependences

We analyse the N_f, N_c dependences of the results. Fig. 4 (a) shows the effective mass of $N_f = 3$ case barely depends on N_c and is almost equal to $N_c = \infty$ results. Using

2-loop gauge running, we find the critical N_f (Fig. 5 (b)). As for T_c, we have almost N_c-independent results for $N_f = 3$. For all N_c, N_f cases, the ratio of $T_c/m(T = 0)$ is almost constant. On the other hand, the critical exponent β_m depends N_c, N_f, except for $N_f = 3$ case.

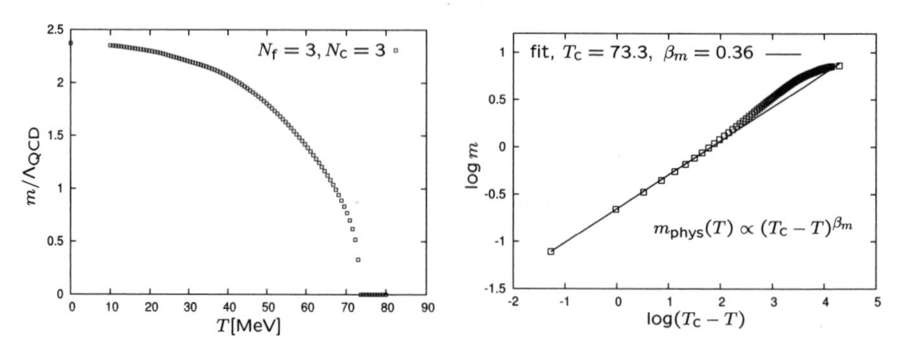

Fig. 3. Behavior of the critical symmetry restoration. $T_c = 73$ MeV, $\beta_m = 0.36$. ($N_c = N_f = 3$)

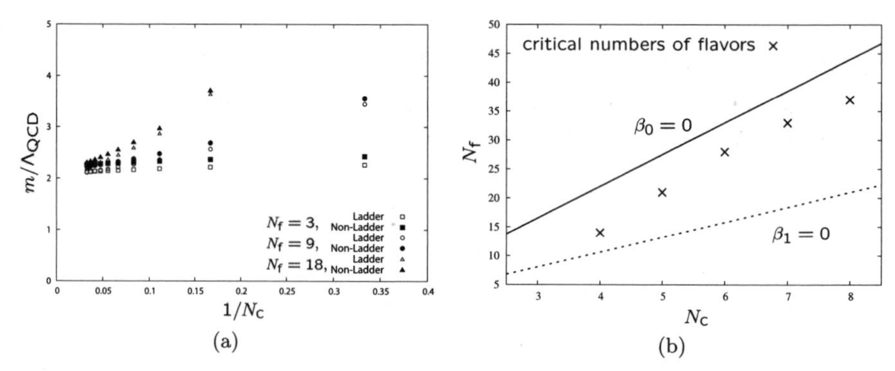

Fig. 4. N_c, N_f dependences of $T = 0$ results. (a) Comparison of the ladder with the non-ladder results. (b) Critical N_f, using 2-loop gauge running.

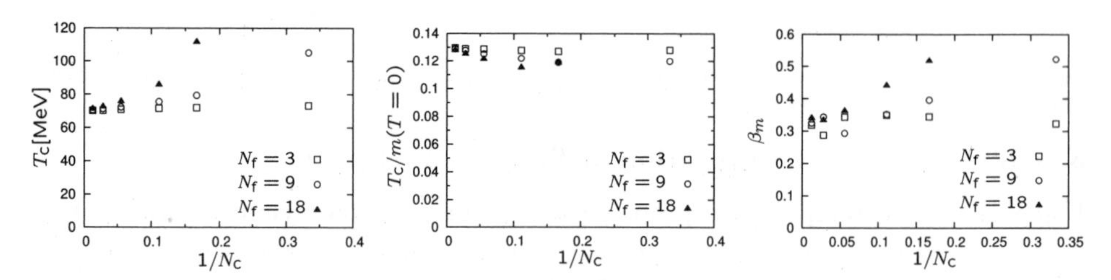

Fig. 5. N_c, N_f dependences of the finite temperature properties.

References

1. K-I. Aoki and K. Miyashita, *Prog. Theor. Phys.* **121**, 875 (2009), and references therein.

The Effective Chiral Lagrangian with Vector Mesons and Hadronic τ Decays

D. Kimura

Hiroshima Shudo-University, Hiroshima, 731-3195, Japan

Kang Young Lee

Division of Quantum Phases and Devices, School of Physics
Konkuk University, Seoul, 143-701, Korea

T. Morozumi* and K. Nakagawa**

Graduate School of Science, Hiroshima University
Higashi-Hiroshima, 739-8526, Japan
** E-mail: morozumi@hiroshima-u.ac.jp*
*** Present affiliation is Mitsubishi-Material Co.*

We study the chiral Lagrangian including the vector mesons such as K^* and ρ. The one loop corrections to vector form factors including the chiral breaking is studied. Once we obtain the form factors, one can study the hadronic tau decay $\tau \to K\pi\nu$ and compute the hadronic invariant distribution.

1. Hadronic τ decays and vector mesons

In two body hadronic τ decays, such as $\tau^- \to \pi^-\pi^0\nu$ and $\tau^- \to K^-\pi^0\nu$, the vector mesons (ρ^-, K^{-*}) contribute as intermediate states in the tree and loop diagrams. The hadronic invariant mass distributions are measured by Belle and Babar and the distributions show the dominant contribution $M_{had} < 1$(GeV) coming from the decay chains; $\tau^- \to \rho^-\nu \to \pi^-\pi^0\nu$ and $\tau^- \to K^{*-}\nu \to K^-\pi^0\nu$. Predicting the decay distributions is a challenging task. Here, we study a chiral Lagrangian with vector mesons including radiative corrections and apply the model to the τ hadronic decays. Beyond the standard models such as two Higgs doublet model, CP violation of the hadronic τ decays are also expected. To predict the direct CP violation, the understanding the strong phase shifts and the precise predictions of them are very important issues. In the center of mass frame of two hadrons in final states, the spin unpolarized τ decay distribution can be parameterized by θ (outgoing kaon direction with respect to incoming τ direction defined in hadronic CM frame) and by s (the hadronic invariant mass squared). The double differential rate for $\tau \to K\pi\nu$ with respect to the angular variable θ and the hadronic invariant mass squared s is given by the formulae,

$$\frac{d^2\Gamma}{d\sqrt{s}d\cos\theta} = \frac{G_F^2|V_{us}|^2}{2^5\pi^3}\frac{(m_\tau^2-s)^2}{m_\tau^3}p_K(s)\left((\frac{m_\tau^2}{s}\cos^2\theta+\sin^2\theta)p_K(s)^2|F(s)|^2\right.$$
$$\left.+\frac{m_\tau^2}{4}|F_s(s)|^2 - \frac{m_\tau^2}{\sqrt{s}}p_K(s)\cos\theta\mathrm{Re}(FF_s^*)\right), \tag{1}$$

where $p_K(s)$ is the three momentum of kaon in the hadronic cm frame. For $\tau \to \pi\pi\nu$, one can replace the CKM element V_{us} with V_{ud} in Eq.(1). F and F_S are the vector and the scalar form factors defined below,

$$\langle K^-(p_K)\pi(p_\pi)|\bar{s}\gamma_\mu u|0\rangle = F(Q^2)q^\mu + \left(F_S(Q^2) - \frac{\Delta_{K\pi}}{Q^2}F(Q^2)\right)Q^\mu, \tag{2}$$

with $Q^\mu = (p_K+p_\pi)^\mu$, $q^\mu = (p_K-p_\pi)^\mu$, and $\Delta_{K\pi} = m_K^2 - m_\pi^2$. In the isospin limit, i.e., $m_u = m_d$, for $\tau^- \to \pi^-\pi^0\nu$, only the vector form factor contributes to the decay.

2. The chiral Lagrangian with vector mesons and the form factors

In what follows, we show how one can derive the form factors. We start with the following Lagrangian.

$$\mathcal{L} = \frac{f^2}{4}\mathrm{Tr}DUDU^\dagger + B\mathrm{Tr}M(U+U^\dagger) - \frac{1}{2}\mathrm{Tr}F_{\mu\nu}F^{\mu\nu} + M_V^2\mathrm{Tr}(V_\mu - \frac{\alpha_\mu}{g})^2, \tag{3}$$

where $U = \exp(2i\frac{\pi}{f}) = \xi^2$ is the SU(3) unitary matrix and π corresponds to the SU(3) octet Nambu Goldstone bosons,(π, K, η_8) while V_μ corresponds to SU(3) octet of the vector mesons (ρ, K^*). α_μ are given as, $\alpha_\mu = \frac{\xi^\dagger D_{L\mu}\xi + \xi\partial_\mu\xi^\dagger}{2i} = \alpha_\mu^0 + \frac{\xi^\dagger A_{L\mu}\xi}{2}$. $F_{\mu\nu}$ corresponds to the field strength of vector mesons and defined as $F_{\mu\nu} = \partial_\mu V_\nu - \partial_\mu V_\nu + ig[V_\mu, V_\nu]$. The digrams within one loop level is shown in Fig.(1). The Feynman diagrams in the first two lines of Fig.(1) have the corresponding diagrams in chiral perturbation theory. However, to reproduce the result of chiral perturbation theory, one must add the diagrams with a vector meson propagator by replacing the massive vector meson propagator as,

$$D_{\mu\nu} = -i\frac{g_{\mu\nu} - \frac{k_\mu k_\nu}{M_V^2}}{k^2 - M_V^2} \to i\frac{g_{\mu\nu}}{M_V^2}. \tag{4}$$

Then by adding the contribution to the diagrams in the first two lines in Fig.(1), one obtains the same result as the one loop computation of the chiral perturbation theory. To go beyond the chiral perturbation result, one must include the remaining contribution of the vector meson propagator and also include the Feynman diagrams with more than one vector meson propagator. The former contribution is included by using the dynamical part of the propagator of the vector mesons.

$$D_{\mu\nu} - i\frac{g_{\mu\nu}}{M_V^2} = -i\frac{k^2 g_{\mu\nu} - k_\mu k_\nu}{M_V^2(k^2 - M_V^2)}. \tag{5}$$

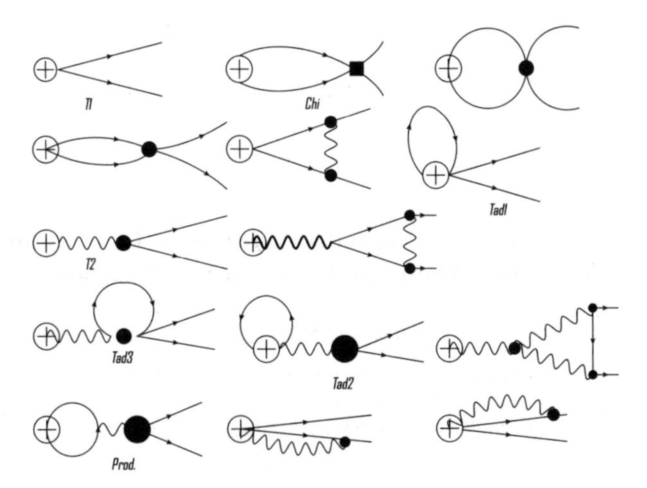

Fig. 1. The diagrams contributing to the vector form factors. The cross denotes the insertion of the vector current $\bar{s}\gamma_\mu u$. The wavy lines denote the vector meson propagators. The solid lines correspond to the pseudo-scalar meson propagators.

3. Vector meson self-energy

In addition to the pseudo Nambu Goldstone loop corrections, we include the vector mesons loop corrections. To count the degree of the divergence, it is convenient to split the massive propagator of the vector mesons into the two parts,

$$-i\frac{g_{\mu\nu} - \frac{k_\mu k_\nu}{M_V^2}}{k^2 - M_V^2} = -i\frac{g_{\mu\nu} - \frac{k_\mu k_\nu}{k^2}}{k^2 - M_V^2} + i\frac{k_\mu k_\nu}{M_V^2}\frac{1}{k^2}. \tag{6}$$

The first term is the same form as that of massive renormalizable propagators while the second term corresponds to the would be Nambu Goldstone boson. Since the second term does not fall in the limit $k \to \infty$, this gives the higher superficial degree of divergence than that arisen from the first term. Then as a result, the counter terms of the self-energy corrections for the vector mesons must contain the following higher derivative terms in addition to the wave function renormalization of field strength,

$$\mathcal{L}_c = -\frac{1}{2}((Z_V - 1)\mathrm{Tr}F_{\mu\nu}F^{\mu\nu} - Z_V^{(2)}\mathrm{Tr}F_{\mu\nu}\frac{\Box}{M_V^2}F^{\mu\nu} + Z_V^{(4)}\mathrm{Tr}\frac{\Box}{M_V^2}F^{\mu\nu}\frac{\Box}{M_V^2}F_{\mu\nu}). \tag{7}$$

Using the counter terms and including all the Feynman diagrams shown above, we will present the analytical and numerical results of the form factors elsewhere in the future publication.

Acknowledgment

We would like to thank organizers of SCGT2009. This work of T.M. is supported by KAKENHI, Grant-in-Aid for Scientific Research on Priority Areas, New development for flavor physics (No.20039008), MEXT, Japan.

Spontaneous SUSY Breaking with Anomalous $U(1)$ Symmetry in Metastable Vacua and Moduli Stabilization*

Hiroyuki Nishino

Department of Physics, Nagoya University
Nagoya 464-8602, Japan
E-mail: hiroyuki@eken.phys.nagoya-u.ac.jp

We consider a SUSY breaking model with anomalous $U(1)$ symmetry. We discard R-symmetry and allow non-renormalizable terms for the model. It will be shown that certain class of models, where the number of positively charged fields is larger than that of negatively charged fields, can have meta-stable SUSY breaking vacuum. And we consider moduli stabilization.

Keywords: Supersymmetry, supersymmetric model.

1. Introduction

Minimal supersymmetric standard model (MSSM) is one of the candidate as new physics beyond the SM and that can solve the fine-tuning promlem of Higgs mass in the SM In addition, the MSSM has attractive features: for example, it give a candidate of dark matter and it realize the unification of the SM gauge couplings.[1] The MSSM suggests grand unified theory (GUT) at unifiaction scale. This GUT (SUSY GUT) is an attractive model because that can unify gauge couplings and matter fields and solve some problems of the SM. As one of such model, there is a simple SUSY $SU(5)$ GUT model.[2,3] But simple SUSY $SU(5)$ GUT model has a serious problem which is called the doublet-triplet splitting problem of the Higgs mass. It is difficult to solve this problem in simple SUSY $SU(5)$ GUT model, naturally.[4]
However, in recent years, it has been found that $SO(10)$ (or E_6) GUT with anomalous $U(1)$ gauge symmetry can solve the doublet-triplet splitting problem, origins of Yukawa hierarchy and mixing angles in the SM, where anomalous $U(1)$ gauge symmetry plays an important role in solving above problems.[5–7] We will review some feature of anomalous $U(1)$ gauge symmetry later.
On the other hand, it is known that SUSY can be spontaneously broken in a model with anomalous $U(1)$ symmetry, where the potential in this model contains only finite number of generic interactin terms because R symmetry is imposed.[8] One of the main obstacle for the unification is R symmetry. In breaking SUSY, R symmetry plays an important role but in most of phenomenologically viable model, R

*This talk is based on work in collaboration with S.-G. Kim, N. Maekawa and K. Sakurai.

symmetry is not imposed. We will propose a spontaneous SUSY breaking model with anomalous $U(1)$ symmetry, but R symmetry is not imposed.

2. Review of anomalous $U(1)$ symmetry

Anomalous $U(1)$ gauge symmetry has some features and that can solve some problems in the standard model and grand unified theories.[5–7] We review $U(1)$ gauge symmetry as follow.

(1): This theory has Fayet-Iliopoulos D term,

$$\xi^2 D = \xi^2 \int d^4\theta V, \tag{1}$$

where V is gauge superfield.

(2): The magnitude of VEV is decided by corresponding anomalous $U(1)$ charge.

$$\langle S \rangle = 0, \quad \langle Z \rangle = \xi^{-z} \equiv \lambda^{-z} << 1, \quad (\lambda << 1) \tag{2}$$

where each small letter represents anomalous $U(1)$ charge of each field ($s > 0, z < 0$) and we take cut off $\Lambda = 1$. This feature is important to explain the origin of Yukawa hierarchy and to solve GUT problems. Therefore, we will consider SUSY breaking model keeping this features.

3. Spontaneous SUSY breaking with anomalous $U(1)$ symmetry

In this section, we consider a SUSY breaking model with anomalous $U(1)$ symmetry and without R symmetry. As a result, we will show that this model has meta-stable SUSY breaking vacuum. For simplicity, we consider a model that has two fields S and Θ which is anomalous $U(1)$ charge $s > 0$ and -1. This model has the following superpotential,

$$W = \sum_n (\Theta^s S)^n, \tag{3}$$

where we take cutoff $\Lambda = 1$. Note that, here, since we don't impose R symmetry, we allow power of $(S\Theta^s)$.

In this model, there are SUSY vacua at $\langle S \rangle \sim \langle \Theta \rangle \sim O(1)$. the superpotential in this model approximately become

$$W \sim \Theta^s S \tag{4}$$

at $S, \Theta << 1$. (4) is approximately the same superpotential as in the model with R symmetry, $R(S) = 2, R(\Theta) = 0$. F-terms of field S and D-term of (4) are as follows.

$$F_S^* = -\frac{\partial W}{\partial S} = -\Theta^s, D_A = g\left(\xi^2 - |\Theta|^2 + s|S|^2\right). \tag{5}$$

We find that $\langle \Theta \rangle$ should be $\xi \sim \lambda$ in order to make the potential minimized. Then, $\langle F_S \rangle \neq 0$. As above, $\langle F_S \rangle = 0$ and $\langle D_A \rangle = 0$ can not be simultaneously zero. Therefore, SUSY is broken at $S, \Theta << 1$. This SUSY breaking vacuum is meta-stable vacuum. The VEVs of each field become $\langle S \rangle \sim 0$ and $\langle \Theta \rangle \sim \lambda$. These values of VEVs are satisfied with the important VEV relations (2). We checked that the lifetime of this meta-stable vacuum is longer than the universe age by using reference.[9]

4. Moduli stabilization in a model with anomalous $U(1)$ symmetry

In our model, we can stabilize moduli field. Path integral in anomalous $U(1)$ GUT is invariant under anomalous $U(1)$ symmetry, $V_A \to V_A + \frac{i}{2}(\Lambda - \Lambda^\dagger)$, $D \to D + \frac{i}{2}\delta_{GS}\Lambda$, where V_A is anomalous $U(1)$ gauge field,D is moduli field, Λ is a gauge parameter field. And δ_{GS} is parameter. Therefore, we find that $D + D^\dagger - \delta_{GS}V_A$ is invariant. We consider that Kähler potential of the S,

$$K_S = S^\dagger S f(D + D^\dagger - \delta_{GS}V_A) \tag{6}$$

for moduli stabilization. $f(x)$ is a function of x.If the function $f(x)$ is given by $f(x) = c(x - x_0)^2 + \epsilon$, where $c \sim O(1), 0 < \epsilon << 1$, then the $\frac{\partial^2 K_S}{\partial S \partial S^\dagger}$ become much smaller than 1 at $x = x_0$. The potential can be written by using the Kähler potential(6) as,

$$V \sim \left(\frac{\partial^2 K_S}{\partial S \partial S^\dagger}\right)^{-1} \left|\frac{\partial W}{\partial S}\right|^2 \sim \frac{\xi^{2s}}{c(x - x_0)^2 + \epsilon}, \tag{7}$$

where $x = D + D^\dagger$ and $\xi^2(D)$ is reduction function of D.Therefore, moduli is stabilized at $x \sim x_0$. This vacuum is meta-stable vacuum. We checked that the lifetime of this meta-stable vacuum is longer than the universe age.

5. Summary and conclusion

We considered a SUSY breaking model with anomalous $U(1)$ symmetry without R symmetry. As a result, we found out that this model has a meta-stable vacuum at $\langle S \rangle \sim 0$ and $\langle \Theta \rangle \sim \lambda$. Till now, we could explain Yukawa hierarchy and solve GUT problem by using VEV relation $\langle S_i \rangle \sim 0$ and $\langle Z_j \rangle \sim \lambda^{-z_j}$. But, we found that SUSY can be easily broken by adding to one positive field in supersymmetric model and the magnitude of each VEV is satisfied with the important VEV relations. And moduli is also stabilized by using Kähler potential of the S in our model.

References

1. For a review, see Stephen P . Martin, hep-ph/9709356
2. H. Georgi, S.L. Glashow, Phys. Rev. Lett. 32 438 (1974)
3. S. Dimopoulos, H. Georgi Nuc. Phys. B193, 150 (1981)
4. For a review, see L. Randall, C. Csa'ki, hep-ph/9508208
5. Nobuhiro Maekawa and Toshifumi Yamashita, Prog. Theor. Phys. 107 1201, (2002)
6. Nobuhiro Maekawa and Toshifumi Yamashita, Prog. Theor. Phys. 110 93, (2003)
7. Nobuhiro Maekawa, Prog. Theor. Phys. 107 597, (2003)
8. P. Fayet and J. Iliopoulos, Phys. Lett. **B51** 461 (1974)
9. M. J. Duncan and Lars Gerhard Jensen, Phys. Lett. **B 291** 109 (1992).

A New Description of the Lattice Yang-Mills Theory and Non-Abelian Magnetic Monopole Dominance in the String Tension*

Akihiro Shibata

Computing Research Center, High Energy Accelerator Research Organization (KEK)
Oho 1-1, Tsukuba, Ibaraki, 305-0801 Japan

We propose a new description of the $SU(N)$ Yang-Mills theory on a lattice, which enables one to explain quark confinement based on the dual superconductivity picture in a gauge independent way. We apply this fomulation to the $SU(3)$ Yang-Mills theory on a lattice. We give preliminary numerical results showing the dominance of the non-Abelian magnetic monopole in the string tension obtained from the Wilson loop in the fundamental representation.

Keywords: Dual superdonductivity, quark confinment, non-Abelian magnetic monopole.

1. New variables on a lattice

We construct a lattice formulation in which an ordinary link variable $U_{x,\mu} \in G = SU(N)$ is decomposed in a gauge-independent manner into two variables $X_{x,\mu}$, $V_{x,\mu} \in G$, i.e., $U_{x,\mu} = X_{x,\mu} V_{x,\mu}$, so that only the variable $V_{x,\mu}$ carries the dominant contribution for quark confinement in agreement with the dual superconductivity picture. $V_{x,\mu}$ is defined as a link variable and transforms just like the original Yang-Mills (YM) link variable $U_{x,\mu}$, while $X_{x,\mu}$ is defined like a site variable representing a matter field and transforms according to the adjoint representation.

Here, we summarize the $SU(N)$ $(N = 3)$ case[12] as an extension of Cho-Faddev-Niemi-Shavanov (CFNS) decomposition in $SU(2)$ case[45].[6] For $SU(3)$ case, we have two options of decomposition corresponding to its stability groups $\tilde{H}$; the maximal option (corresponding to the conventional Abelian projection in the maximal Abelian gauge) $\tilde{H} = U(1) \times U(1)$ and the minimal option $\tilde{H} = U(2)$ which gives the "Abelian" part corresponding to the Wilson loop for the fundamental representation.[9]

By introducing a *single* type of the color field $\mathbf{h}_x := \Theta_x \frac{1}{2} \lambda^8 \Theta_x^\dagger \in G/\tilde{H} = SU(3)/U(2)$, the defining equations for the decomposition are given by

$$D_\mu^\epsilon[V]\mathbf{h}_x := \frac{1}{\epsilon}\left(V_{x,\mu}\mathbf{h}_{x+\mu} - \mathbf{h}_x V_{x,\mu}\right) = 0, \tag{1}$$

$$\mathrm{tr}(X_{x,\mu}\mathbf{h}_x) = 0, \tag{2}$$

*This is based on the work in collaboration with K.-I. Kondo, S. Kato, T. Shinohara.

where eq.(1) represents that $\mathbf{h}(x)$ is covariant constant in the background $V_{x,\mu}$, and eq.(2) means that $X_{x,\mu}$ has the vanishing $\tilde{H}$-commutative part[a]. The decomposition is determined by solving the defining equation for given $U_{x,\mu}$ and $\mathbf{h}_x$ by way of a newly defined variable $\tilde{V}_{x,\mu}$ which does not belong to $SU(3)$ (which is given by the exact solution[3]):

$$\tilde{V}_{x,\mu} := U_{x,\mu} + \frac{2\sqrt{3}}{5}\left(\mathbf{h}_x U_{x,\mu} + U_{x,\mu}\mathbf{h}_{x+\mu}\right) + \frac{24}{5}\mathbf{h}_x U_{x,\mu}\mathbf{h}_{x+\mu}, \tag{3}$$

$$\underline{V_{x,\mu}} := \left(\tilde{V}_{x,\mu}\tilde{V}^{\dagger}_{x,\mu}\right)^{-1/2}\tilde{V}_{x,\mu}, \quad V_{x,\mu} = \underline{V_{x,\mu}}\left(\det \underline{V_{x,\mu}}\right)^{-1/3}, \quad X_{x,u} = U_{x,\mu}V^{\dagger}_{x,\mu}. \tag{4}$$

In order to obtain the equipollent theory (written in terms of the new variables) with the original Yang-Mills theory, we impose the reduction condition which plays the role of reducing the extended gauge symmetry $SU(3)_\Omega \times [SU(3)/U(2)]_\Theta$ to the original gauge symmetry $SU(3)_{\Omega=\Theta}$ so that the new variables $V_{x,\mu}, X_{x,\mu}$ and $\mathbf{h}_x$ transform under the same group $\Omega = \Theta \in SU(3)$.[8]

2. "Abelian" dominance in the Wilson loop

We apply this formulation for the lattice data. The configuration of YM field can be generated by using the standard algorithm for the Wilson action, since color field $\mathbf{h}_x$ can is determined so as to fulfill the reduction condition, and the decomposition $U_{x,\mu} = V_{x,\mu}X_{x,\mu}$ in a gauge-invariant manner from eqs.(3-4).

First, we measure correlation function of new variable. The vacuum expectation value of the color field $\mathbf{h}_x$ and its correlation functions shows the preserving color symmetry; $\langle h^A \rangle = 0$, $\langle h^A(x)h^B(0)\rangle = \delta^{A,B}D(x)$. The propagator (correlation functions) of new variables ($\mathbf{V}_{x,\mu}, \mathbf{X}_{x,\mu}$) and original gauge YM fields ($\mathbf{A}_{x,\mu}$) in Landau gauge. We obtain that the propagator of $\mathbf{A}_{x,\mu}$ and $\mathbf{V}_{x,\mu}$ dump slowly, while $\mathbf{X}_{x,\mu}$ dumps quickly for long, i.e., V-dominance ("Abelian"-dominance) in the infrared region.[8]

Then we move to main part of this paper, Abelian dominance in Wilson loop (or static inter-quark potential): $\langle W_C[U]\rangle \cong \langle W_C[V]\rangle$. From non-Abelian Stokes theorem, the Wilson loop operator in the fundamental representation is rewritten into the Wilson loop by new variables of minimal option.[9] The right panel of Fig.1 shows the (preliminary) measurement of Wilson loop of the new variable, $V_{x,\mu}$: the distance (R) v.s. $\ln(\langle W_C[V](R,T)\rangle)/T$ is plotted, where Wilson loop C is given by the $R \times T$ rectangle. The Wilson loop is fitted by the potential $\langle W_C[V](R,T)\rangle = \exp(-V(R;T))$. The static inter-quark potential, $V(R) := \sigma R + b + c/R$, is obtained by the extrapolation $\lim_{T\to\infty} V(R;T)/T$ after fitting. By comparing the data by Edward et.el.,[11] the string tension by YM fields σ_{full} is reproduced by the string tension by $V_{x,\mu}$ fields, σ_V; $\sigma_V/\sigma_{full} \approx 0.95$. The right panel of Fig.1 shows the

[a]By using a non-Abelian Stokes theorem that a set of defining equations is obtained as a necessary and sufficient condition for the Wilson loop operator to be dominated by the decomposed variable $V_{x,\mu}$ in the sense $W_C[U_{x,\mu}] \cong (\text{const.})W_C[V_{x,\mu}]$:[10]

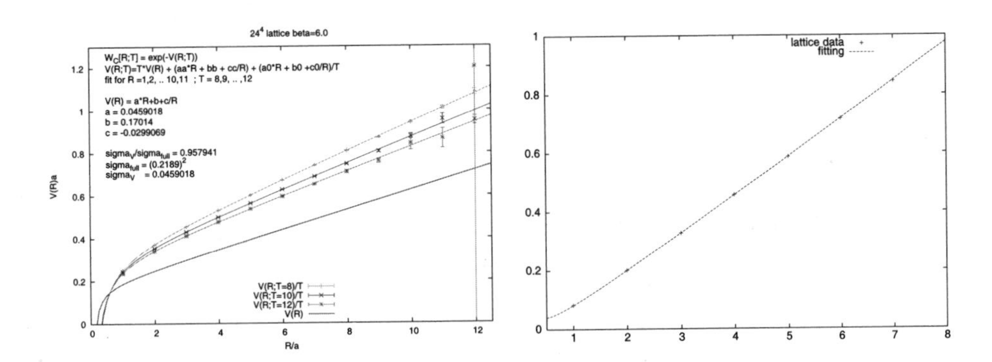

Fig. 1. (Left panel) The average of Wilson loop of by V field. The inter quark potential is fitted by $V(R,T)$.(Right panel) The inter-quark statc potential calvurated by non-Abelian magnetic monopole.(preliminaly)

inter-quark static potential calculated by non-Abelian magnetic monopole which is defined by the defined from the field strength of the non-Abelian gauge field, i.e.,V-field;

$$k_\mu = \frac{1}{2}\epsilon^{\mu\lambda\alpha\beta}\partial_\lambda F_{\alpha\beta}[V], \qquad \exp(-ig\epsilon F_{\alpha\beta}[V]\mathbf{h}_x) = V_{x,\mu}V_{x+\mu,\nu}V^\dagger_{x+\nu,\mu}V^\dagger_{x,\nu}.$$

It shows that string tension is reproduced by the Wilson loop by magnetic monopole.

Acknowledgment

This work is financially supported by Grant-in-Aid for Scientific Research (C) 21540256 from Japan Society for the Promotion of Science (JSPS) and the Large Scale Simulation Program No. 09/10-19 (FY2009/2010) of High Energy Accelerator Research Organization (KEK).

References

1. K.-I. Kondo, T. Shinohara, T. Murakami, Prog.Theor.Phys.120 1-50 (2008)
2. A.Shibata, S.Kato, K.-I.Kondo, T.Shinohara, S.Ito, Phys.Lett.B669 107-118 (2008).
3. A.Shibata, K.-I. Kondo, T.Shinohara, KEK-PREPRINT-2009-32, CHIBA-EP-181, arXiv:0911.5294 [hep-lat]
4. S.Ito, S.Kato, K.-I. Kondo, A. Shibata, T.Shinohara, Phys.Lett. B645 67-74 (2007)
5. A. Shibata, S. Kato, K.-I. Kondo, T. Murakami, T. Shinohara, S. Ito, Phys.Lett. B653 101-108 (2007)
6. S. Kato, K.-I. Kondo, A. Shibata, T. Shinohara, S. Ito, PoS(LAT2009) 228
7. A. Shibata, S. Kato, K.-I. Kondo, T. Shinohara and S. Ito, POS(LATTICE-2007) 331, arXiv:0710.3221 [hep-lat]
8. A. Shibata, S. Kato,K. Kondo, T. Shinohara and S. Ito, POS(LATTICE-2008) 331, arXiv:0810.0956 [hep-lat]
9. K.-I. Kondo, Phys.Rev.D77 085029 (2008) (arXiv:0801.1274 [hep-th])
10. K.-I.Kondo and A.Shibata, arXiv:0801.4203 [hep-th]
11. R.G.Edward, U.M. Heller and T.R.Klassen, Nucl.Phys. N617 (1988) 377-392 (arXir-hep-lat/9711003v2)

Thermodynamics with Unbroken Center Symmetry in Two-Flavor QCD*

S. Takemoto* and M. Harada

Department of Physics, Nagoya University
Nagoya, 464-8602, Japan
**E-mail: takemoto@hken.phys.nagoya-u.ac.jp*

C. Sasaki

Physik Department, Technische Universität München
D-85747 Gerching, Germany
and
Frankfurt Institute for Advanced Studies, J.W. Goethe University
D-60438 Frankfurt am Main, Germany
E-mail: sasaki@fias.uni-frankfurt.de

We study general features of thermodynamic quantities and hadron mass spectra in a possible phase where the chiral $SU(2)_L \times SU(2)_R$ symmetry is spontaneously broken while its center Z_2 symmetry remains unbroken.

Keywords: QCD phase diagram, chiral symmetry breaking.

1. Introduction

Chiral symmetry breaking plays a very important role in acquiring hadron masses. The bilinear quark condensate $\langle \bar{q}q \rangle$ breaks the chiral symmetry $SU(N_f)_L \times SU(N_f)_R$ down to the diagonal subgroup $SU(N_f)_V$. On the other hand, the quartic quark condensates

$$\langle O_1 \rangle = \langle \bar{q}_L \lambda_a \gamma_\mu q_L \cdot \bar{q}_R \lambda_a \gamma^\mu q_R \rangle \,, \quad \langle O_2 \rangle = \langle \bar{q}_R \lambda_a q_L \cdot \bar{q}_L \lambda_a q_R \rangle \,, \tag{1}$$
$$(\lambda_a \, (a = 1, 2, \cdots, N_f^2 - 1) \text{ are generators of } SU(N_f) \,)$$

break the chiral symmetry down to $SU(N_f)_V \times (Z_{N_f})_A$. This unorthodox pattern of chiral symmetry breaking is, however, strictly ruled out by QCD inequalities, which is valid both at zero and finite temperatures but at zero density.[2] QCD with finite baryon number density cannot exclude this unorthodox pattern, so the phase with broken chiral symmetry and unbroken center $(Z_{N_f})_A$ symmetry can appear in a high density region of the QCD phase diagram. In this report, we will show our results on the phase diagram, thermodynamic quantities and hadron mass spectra

*This report is based on Ref. [1].

using a chiral model. It has the same chiral symmetry $SU(2)_L \times SU(2)_R$ as in two-flavor QCD with massless quarks and allows unorthodox ($\to SU(2)_V \times (Z_2)_A$) as well as orthodox ($\to SU(2)_V$) pattern of chiral symmetry breaking.

2. Phase Diagram

Based on the discussion in Section 1, we introduce a 2-quark state M_{ij} $(i,j=1,2)$ and a 4-quark state Σ_{ab} $(a,b=1,2,3)$, which transform under $(Z_2)_A$ [a] as

$$M_{ij} \to -M_{ij}\,, \quad \Sigma_{ab} \to \Sigma_{ab}\,. \tag{2}$$

Constructing a potential up to the fourth order in these fields, and taking a mean field approximation as

$$M_{ij} = \frac{1}{\sqrt{2}}\sigma\delta_{ij}\,, \quad \Sigma_{ab} = \frac{1}{\sqrt{3}}\chi\delta_{ab}\,, \tag{3}$$

one obtains a Ginzburg-Landau potential

$$V(\sigma,\chi) = \sigma^4 + A\sigma^2 + \chi^4 + B\chi^2 - \sigma^2\chi + D\chi^3 + F\sigma^2\chi^2\,. \tag{4}$$

Phase diagrams of this potential with $F=0$ are shown in Fig. 1. In addition to the chiral and the center symmetry broken phase (phase I) and the chiral restored phase (phase III), the phase where the chiral symmetry is broken by a quartic quark condensate but the center $(Z_2)_A$ symmetry is restored by a vanishing bilinear quark condensate appears. We call this phase (phase II) the Z_2 symmetric phase. It is clearly shown that the number of critical points and the order of phase transitions between phase I and II change significantly with the value of D. On the other hand, it is apparent from the potential (4) that the topology of the phase diagram is almost independent of the value of F. A new tricritical point (TCP) associated with the restoration of the center symmetry appears for $-1 < D \leq 0$. The appearance of a critical point CP_1 for negative D instead of a tricritical point TCP_1 without any explicit breaking is also an interesting feature of this model.

3. Thermodynamic Quantities and Hadron Mass Spectra in the Z_2 Symmetric Phase

Let us now see the general features of thermodynamic quantities and hadron mass spectra in the phases appearing in this model.

We have calculated the quark number susceptibility along a path running from phase I to phase III via phase II. It is more enhanced at the restoration point of the center symmetry rather than that of the chiral symmetry. Thus, we expect a rapid change in the quark number density when the center symmetry is restored. This situation is similar to what we see in the phase transition from the hadronic phase to the quarkyonic phase[3] in the large N_c limit. In our model, this feature

[a]This transformation corresponds to $q_L \to -q_L$, $q_R \to q_R$ in quark fields.

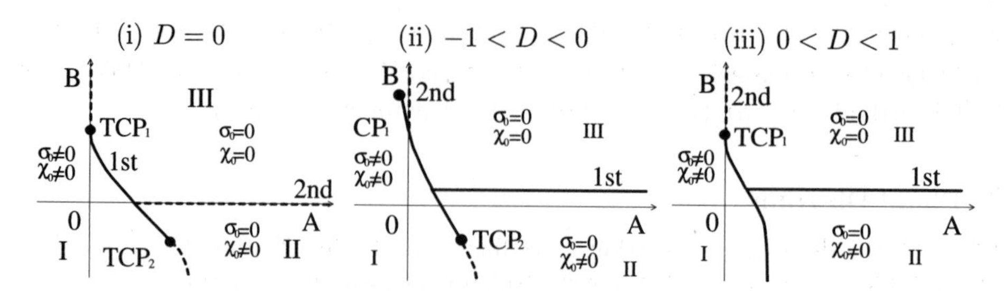

Fig. 1. Phase diagrams with $F = 0$ for various values of D. The unbroken symmetry of phase I is $SU(2)_V$, that of phase II is $SU(2)_V \times (Z_2)_A$ and that of phase III is $SU(2)_L \times SU(2)_R$. Solid and dashed lines represent first and second order phase boundaries, respectively. TCP_2 in (ii) as well as TCP_1 in (iii) coincides with a triple point for $|D| \geq 1$.

originates from the fact that unlike σ, the Yukawa coupling of χ to baryons ($\chi \bar{B} B$) is not allowed by the Z_2 invariance.

We then have investigated hadron mass spectra in phase I and II. The Nambu-Goldstone boson in phase I is a mixture of 2-quark and 4-quark states, while that in phase II is a pure 4-quark state. It is noticeable that the pion decay constant remains finite and scalar and pseudo-scalar mesons differ in mass in phase II despite $\sigma_0 = 0$. Moreover, nucleons in phase II cannot gain masses by chiral symmetry breaking owning to the unbroken center $(Z_2)_A$ symmetry. These are the features which distinguish the Z_2 symmetric phase from other phases.

4. Conclusions

We have discussed a new phase in dense QCD where the chiral symmetry is spontaneously broken while its center symmetry is restored. We have shown that the quark number susceptibility exhibits a strong enhancement at the restoration point of the center symmetry rather than that of the chiral symmetry. Then we have compared hadron mass spectra in two phases with broken and unbroken center symmetry. It is of particular interest to explore if such a phase appears in the phase diagram of the microscopic models like Nambu–Jona-Lasino models or Schwinger-Dyson equations, which will be reported in future papers.

Acknowledgments

The work of S.T. has been supported in part by Global COE Program "Quest for Fundamental Principles in the Universe" of Nagoya University (G07).

References

1. M. Harada, C. Sasaki and S. Takemoto, Phys. Rev. D **81**, 016009 (2010).
2. I. I. Kogan, A. Kovner and M. A. Shifman, Phys. Rev. D **59**, 016001 (1999).
3. L. McLerran and R. D. Pisarski, Nucl. Phys. A **796**, 83 (2007), Y. Hidaka, L. D. McLerran and R. D. Pisarski, Nucl. Phys. A **808**, 117 (2008).

Masses of Vector Bosons in Two-Color QCD Based on the Hidden Local Symmetry*

T. Yamaoka*, M. Harada and C. Nonaka

Department of Physics, Nagoya University, Nagoya, 464-8602, Japan
** E-mail: yamaoka@hken.phys.nagoya-u.ac.jp*

We study the dependence of the "vector" boson masses (mesons with $J^P = 1^-$ and diquark baryons with $J^P = 1^+$) on the baryon number density μ_B for two-color QCD. We show the μ_B-dependence signals the phase transition of $U(1)_B$ breaking and it gives information about mixing among "vector" bosons. We also discuss the comparison with lattice data.

Keywords: Chiral symmetry, two-color QCD, finite density.

1. Introduction

Quantum chromodynamics (QCD) shows various phases under extreme conditions. However, it is difficult to study the phase structure especially near the critical temperature and/or density directly from QCD, because of the strong coupling. Lattice QCD simulation is one of powerful theoretical tools, but it is not applicable in the finite density region due to the sign problem. The problem in simulations of real-life QCD at finite baryon density produces interest in the two-color QCD which is free from the sign problem. Two-color QCD has some interesting features in the following respect: Color-singlet baryons appear together with ordinary mesons as pseudo Namubu-Goldstone (NG) bosons associated with the spontaneous breaking of the chiral symmetry and their interactions are determined uniquely by the low-energy theorem. This allows us to construct low-energy effective theories including the baryons as light degrees of freedom naturally, and to investigate properties of hadrons even at finite baryon number density μ_B using them.

In Ref. [1], we construct a low energy effective Lagrangian for the two-color QCD including the "vector" bosons (mesons with $J^P = 1^-$ and diquark baryons with $J^P = 1^+$) in addition to the pseudo-NG bosons (mesons with $J^P = 0^-$ and baryons with $J^P = 0^+$) having the degenerate mass M_π. The effective Lagrangian is composed based on the chiral symmetry breaking pattern of $SU(2N_f) \to Sp(2N_f)$ in the framework of the hidden local symmetry (HLS),[2] and the "vector" bosons are introduced as the gauge bosons of the $Sp(2N_f)$ HLS. In this report, we show

*This report is based on Ref. [1].

our result on the vacuum structure of our model and the μ_B-dependence of "vector" boson masses in the case of $N_f = 2$.

2. "Vector" boson masses at leading order

In the case of $N_f = 2$, 10 "vector" bosons are introduced as the gauge bosons of the $Sp(4)$ HLS. They are classified as follows; isovector "ρ"-meson, isoscalar "ω"-meson, isovector baryon (V_{B_+}), and isovector anti-baryon (V_{B_-}). We assume that the spatial rotational symmetry is not broken and study the vacuum structure of HLS Lagrangian at the leading order. In Ref. [1], we find that there is a condensation of baryonic-NG boson with $J^P = 0^+$ and $U(1)_B$ symmetry is broken spontaneously for $\mu_B > M_\pi$ similarly to the case with only the pseudo-NG bosons included.[3] We also find that the time component of ω meson has a VEV for any μ_B.

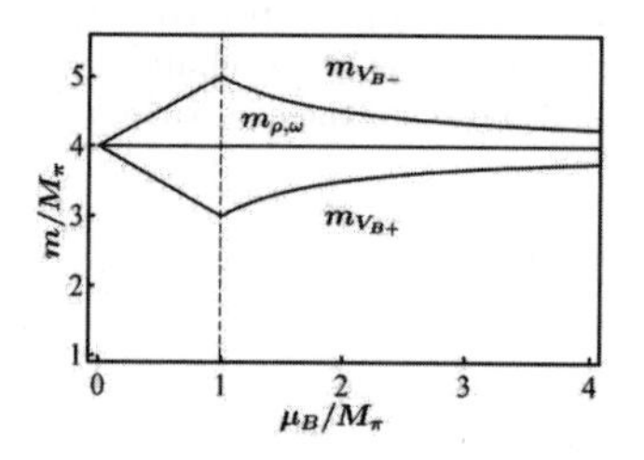

Fig. 1. The masses of the "vector" bosons in unit of M_π as a function of μ_B/M_π.

We show an example of the μ_B dependence of "vector" bosons in Fig. 1. Vector mesons (ρ and ω) with $J^P = 1^-$ do not change their masses at all. On the other hand, we find that the mass of the baryon (V_{B_+}) with $J^P = 1^+$ decreases for $\mu_B < M_\pi$ and turns to increase for $\mu_B > M_\pi$, and the mass of anti-baryon (V_{B_-}) with $J^P = 1^+$ shows the opposite behavior.

We stress that the baryon V_{B_+} does not mix with the anti-baryon V_{B_-} even though the baryon number $U(1)_B$ symmetry is spontaneously broken for $\mu_B > M_\pi$ at the leading order.

3. Effect of higher order terms

In this section, we discuss the effects of higher order terms in the HLS. When we assume that the vacuum structure is not changed, the corrections of the higher order terms appear with two free parameters C_1 and C_2 for $\mu_B > M_\pi$. These parameters are not determined by the symmetry structure. The choice of $C_1 = 0$ and $C_2 > 0$ ($C_2 < 0$) implies that the higher order terms provide a negative (positive) contribution to all the "vector" boson masses equally.

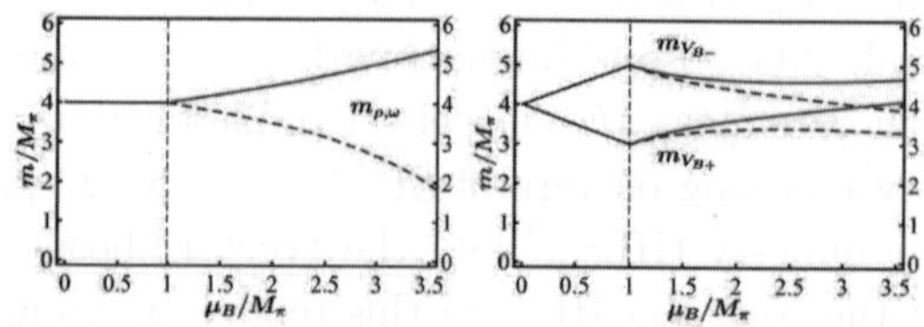

Fig. 2. The masses of the "vector" bosons including the effect of higher order terms. Dashed (Solid) curves stand for $C_2 = 4$ ($C_2 = -4$).

As an example, we show the μ_B-dependence of "vector" boson masses in Fig. 2. The curves on the left figure are for the degenerate ρ and ω mesons and those on the right figure are for baryonic and anti-baryonic "vector" bosons.

On the other hand, the effect of nonzero C_1 produces the mass difference between ρ and ω, and this difference is linked to the mixing strength between V_{B_+} and V_{B_-} as

$$m_\rho^2 - m_\omega^2 = \frac{C_1 g^2(\mu_B^4 - M_\pi^4)}{2\mu_B^2}, \tag{1}$$

where, m_ρ and m_ω are mass eigenvalues, and g is the HLS gauge coupling constant. This relation is obtained from the symmetry breaking pattern and the assumption that all the bosons other than π and V are heavy enough to be neglected in the Lagrangian. Thus, a violation of this relation may signal a new phase transition.

Furthermore, in the case of relatively heavy πs, we compare our result with lattice data.[4] We find that our model cannot simultaneously reproduce the μ_B-dependence of "vector" bosons in the lattice data: It is difficult to make $m_{V_{B+}}$ stable against μ_B for $\mu_B > M_\pi$ without introducing the difference between m_ρ and m_ω. This strongly suggests that there is a large mixing among "vector" baryons ($V_{B\pm}$) with $J^P = 1^+$ and axial-vector mesons with $J^P = 1^+$ which are not included in our model. Also, this indicates that the mass of isovector axial-vector meson with $J^P = 1^+$ shown in Fig. 2 of Ref. [3] is nothing but $m_{V_{B+}}$ observed through the large mixing.

4. Summary

We investigated the μ_B-dependences of the "vector" boson masses and found that the mass of anti-baryon with $J^P = 1^+$ (V_{B_-}) increases for $\mu_B < M_\pi$ and turns to decrease for $\mu_B > M_\pi$. The mass of baryon with $J^P = 1^+$ (V_{B_+}) shows the opposite behavior. These behaviors of the baryons signal the phase transition of $U(1)_B$ breaking in two-color QCD at finite density. The present analysis is valid when the "axial-vector" bosons (mesons with $J^P = 1^+$ and baryons with $J^P = 1^-$) are heavy. When the "axial-vector" bosons are light, we need to include these states using the generalized HLS.[5] Using this formalism, we may investigate the phase structure in the range of μ_B wider than that studied in the present analysis.

The work of T.Y. has been supported in part by Global COE Program "Quest for Fundamental Principles in the Universe" of Nagoya University (G07).

References

1. M. Harada, C. Nonaka and T. Yamaoka, arXiv:1002.4705 [hep-ph]
2. M. Bando, T. Kugo and K. Yamawaki, Phys. Rept. **164**, 217 (1988). M. Harada and K. Yamawaki, Phys. Rept. **381**, 1 (2003).
3. J. B. Kogut, M. A. Stephanov, D. Toublan, J. J. M. Verbaarschot and A. Zhitnitsky, Nucl. Phys. B **582**, 477 (2000).
4. S. Hands, P. Sitch and J. I. Skullerud, Phys. Lett. B **662**, 405 (2008).
5. M. Bando, T. Kugo and K. Yamawaki, Nucl. Phys. B **259**, 493 (1985); M. Bando, T. Fujiwara and K. Yamawaki, Prog. Theor. Phys. **79**, 1140 (1988); N. Kaiser and U. G. Meissner, Nucl. Phys. A **519**, 671 (1990).

Walking Dynamics from String Duals

Maurizio Piai
Swansea University

A large class of solutions of the background equations for a specific system of D5-branes shows many of the properties of a walking theory. The gauge coupling is almost constant over an intermediate range of energies, the coupling is strong, the theory confines, a condensate forms. We study the Wilson loops and the spectrum of scalar glueballs in these backgrounds, finding quite surprising results. This model, and possibly its extensions, provide a suitable laboratory in which to study the (strongly coupled) properties of walking dynamics, which lies at the core of walking technicolor.

The Quark Mass Dependence of the Nucleon Mass in AdS/QCD

Hyo Chul Ahn
Pusan National University

We study the quark mass dependence of the baryon mass using Bottom- up approach of holographic QCD. We find that nucleon masses are linear to pion mass square and the slope of ground state, Roper state and N(1535) are 0.74 GeV^1, 0.47 GeV^1 and 0.35 GeV^1 respectively. Then we compare our result with Lattice QCD and results of Top-down approach.

Structure of Thermal Quasi-Fermion in QED/QCD from the Dyson-Schwinger Equation

Hisao Nakkagawa

Nara University

We non-perturbatively study the structure of quasi-fermion in thermal QED/QCD by solving the Dyson-Schwinger equation in the wide range of temperature as well as of coupling constant. The three-peak structure of the quasi-fermion spectrum, suggested by Harada and Nemoto in a chiral invariant linear sigme model, is clearly seen in QED/QCD. Behavior of the decay width as a function of temperature as well as coupling constant shows the clear deviation from the HTL results. Change of the structure across the boundary of chiral phase transition is also studied in detail.

Critical Behaviors of Sigma-Mode and Pion in Holographic Superconductors

Cheonsoo Park

Pusan National University

We study the critical behaviors of holographic superconductors in 2+1 dimensions. We find the temperature dependences of pion decay constants and mass of sigma-mode, and compute their critical exponents.

LIST OF PARTICIPANTS

Abe Tomohiro	Nagoya University
Ahn Hyo Chul	Pusan National University
Anselmi Damiano	Pisa U. & INFN, Pisa
Appelquist Thomas	Yale University
Asai Shoji	University of Tokyo
Borzumati Francesca	National Taiwan University
Brodsky Stanley J.	SLAC, Stanford University
Cheng Hsin-Chia	University of California, Davis
Chivukula R. Sekhar	Michigan State University
Daisuke Yamamoto	The University of Tokyo
Enomoto Seishi	Nagoya University
Fleming George Tamminga	Yale University
Frampton Paul H.	IPMU and UNC
Fukano Hidenori Sakuma	CP3-Origins, Univ of Southern Denmark
Fukaya Hidenori	Nagoya University
Furui Sadataka	Teikyo University
Haba Kazumoto	Nagoya
Harada Masayasu	Nagoya University
Hashi Manami	University of Tokyo, Komaba
Hashimoto Koji	RIKEN
Hashimoto Michio	KEK
Hayakawa Masashi	Nagoya University
Holland Kieran Michael	University of the Pacific
Hong Deog Ki	Pusan National University
Hosek Jiri	Nuclear Physics Institute
Hoshino Yuichi	Kushiro National College of Technology
Hosotani Yutaka	Osaka University
Ichinose Shoichi	University of Shizuoka
Iijima Toru	Nagoya University
Inagaki Tomohiro	Hiroshima University
Ito Motoharu	Nagoya University
Jung Dong-Won	National Central University, Taiwan
Kanaya Kazuyuki	University of Tsukuba
Kanazawa Takuya	University of Tokyo
Kawase Hidetoshi	Nagoya University

Kikukawa Yoshio	Institute of Physics, University of Tokyo
Kimura Daiji	Hiroshima Shudo University
Kitano Ryuichiro	Tohoku University
Kondo Kei-Ichi	Chiba University
Konishi Kenichi	University of Pisa /INFN, Pisa
Konishi Yasufumi	Kyoto Sangyo University
Kozo Koizumi	Kyoto Sangyo University
Kugo Taichiro	YITP, Kyoto University
Kurachi Masafumi	Los Alamos National Laboratory
Kurata Yasuhiro	Nagoya University
Kusafuka Yuki	Nara Women's University
Kuti Julius	University of California, San Diego
Kuwakino Shogo	Nagoya University
Lombardo Maria Paola	LNF INFN
Maekawa Nobuhiro	Nagoya Univ.
Maru Nobuhito	Chuo University
Maskawa Toshihide	Nagoya Univ, Kyoto Sangyo Univ.
Matsuzaki Shinya	Pusan National University
Miransky Vladimir A	The University of Western Ontario
Miyashita Kazuhiro	Aichi Shukutoku University
Morozumi Takuya	Hiroshima University
Nakkagawa Hisao	Nara University
Nishino Hiroyuki	Nagoya University
Nojiri Shin'ichi	Nagoya University
Nonaka Chiho	Nagoya University
Onogi Tetsuya	Osaka University
Park Cheonsoo	Pusan National University
Piai Maurizio	Swansea University
Rho Mannque	IPhT, CEA Saclay, Hanyang University
Sakai Tadakatsu	Nagoya University
Sannino Francesco	CP3-Origins, Univ of Southern Denmark
Sato Daisuke	Kanazawa Univ, Institute for Theoretical Physics
Semenoff Gordon Walter	University of British Columbia
Shibata Akihiro	KEK
Shintani Eigo	Osaka University
Simmons Elizabeth H.	Michigan State University
Sinclair Donald Keith	Argonne National Laboratory
Sogami Ikuo S.	Kyoto Sangyo University
Suganuma Hideo	Kyoto University
Takei Kazuaki	Nagoya University
Takemoto Shinpei	Nagoya University
Takeuchi Tatsu	Virginia Tech
Tanabashi Masaharu	Nagoya University

Teraguchi Shunsuke	Nagoya University
Terao Haruhiko	Nara Women's University
Tobe Kazuhiro	Nagoya University
Tuominen Kimmo	CP3-Origins
Uno Shumpei	Nagoya University
Wijewardhana L.C. Rohana	University of Cincinnati
Yamada Norikazu	KEK
Yamamoto Kazuhiro	Osaka City University
Yamaoka Tetsuro	Nagoya University
Yamashita Toshifumi	Nagoya University
Yamawaki Koichi	Nagoya University
Yoda Hiroshi	Nagoya University
Yoshida Koji	Nara University
Yoshikawa Tadashi	Nagoya University